Handbook of
Pharmaceutical Analysis

DRUGS AND THE PHARMACEUTICAL SCIENCES

Executive Editor

James Swarbrick

AAI, Inc.
Wilmington, North Carolina

Advisory Board

DRUGS AND THE PHARMACEUTICAL SCIENCES

A Series of Textbooks and Monographs

1. Pharmacokinetics, *Milo Gibaldi and Donald Perrier*
2. Good Manufacturing Practices for Pharmaceuticals: A Plan for Total Quality Control, *Sidney H. Willig, Murray M. Tuckerman, and William S. Hitchings IV*
3. Microencapsulation, *edited by J. R. Nixon*
4. Drug Metabolism: Chemical and Biochemical Aspects, *Bernard Testa and Peter Jenner*
5. New Drugs: Discovery and Development, *edited by Alan A. Rubin*
6. Sustained and Controlled Release Drug Delivery Systems, *edited by Joseph R. Robinson*
7. Modern Pharmaceutics, *edited by Gilbert S. Banker and Christopher T. Rhodes*
8. Prescription Drugs in Short Supply: Case Histories, *Michael A. Schwartz*
9. Activated Charcoal: Antidotal and Other Medical Uses, *David O. Cooney*
10. Concepts in Drug Metabolism (in two parts), *edited by Peter Jenner and Bernard Testa*
11. Pharmaceutical Analysis: Modern Methods (in two parts), *edited by James W. Munson*
12. Techniques of Solubilization of Drugs, *edited by Samuel H. Yalkowsky*
13. Orphan Drugs, *edited by Fred E. Karch*
14. Novel Drug Delivery Systems: Fundamentals, Developmental Concepts, Biomedical Assessments, *Yie W. Chien*
15. Pharmacokinetics: Second Edition, Revised and Expanded, *Milo Gibaldi and Donald Perrier*
16. Good Manufacturing Practices for Pharmaceuticals: A Plan for Total Quality Control, Second Edition, Revised and Expanded, *Sidney H. Willig, Murray M. Tuckerman, and William S. Hitchings IV*
17. Formulation of Veterinary Dosage Forms, *edited by Jack Blodinger*
18. Dermatological Formulations: Percutaneous Absorption, *Brian W. Barry*
19. The Clinical Research Process in the Pharmaceutical Industry, *edited by Gary M. Matoren*
20. Microencapsulation and Related Drug Processes, *Patrick B. Deasy*
21. Drugs and Nutrients: The Interactive Effects, *edited by Daphne A. Roe and T. Colin Campbell*
22. Biotechnology of Industrial Antibiotics, *Erick J. Vandamme*

23. Pharmaceutical Process Validation, *edited by Bernard T. Loftus and Robert A. Nash*
24. Anticancer and Interferon Agents: Synthesis and Properties, *edited by Raphael M. Ottenbrite and George B. Butler*
25. Pharmaceutical Statistics: Practical and Clinical Applications, *Sanford Bolton*
26. Drug Dynamics for Analytical, Clinical, and Biological Chemists, *Benjamin J. Gudzinowicz, Burrows T. Younkin, Jr., and Michael J. Gudzinowicz*
27. Modern Analysis of Antibiotics, *edited by Adjoran Aszalos*
28. Solubility and Related Properties, *Kenneth C. James*
29. Controlled Drug Delivery: Fundamentals and Applications, Second Edition, Revised and Expanded, *edited by Joseph R. Robinson and Vincent H. Lee*
30. New Drug Approval Process: Clinical and Regulatory Management, *edited by Richard A. Guarino*
31. Transdermal Controlled Systemic Medications, *edited by Yie W. Chien*
32. Drug Delivery Devices: Fundamentals and Applications, *edited by Praveen Tyle*
33. Pharmacokinetics: Regulatory • Industrial • Academic Perspectives, *edited by Peter G. Welling and Francis L. S. Tse*
34. Clinical Drug Trials and Tribulations, *edited by Allen E. Cato*
35. Transdermal Drug Delivery: Developmental Issues and Research Initiatives, *edited by Jonathan Hadgraft and Richard H. Guy*
36. Aqueous Polymeric Coatings for Pharmaceutical Dosage Forms, *edited by James W. McGinity*
37. Pharmaceutical Pelletization Technology, *edited by Isaac Ghebre-Sellassie*
38. Good Laboratory Practice Regulations, *edited by Allen F. Hirsch*
39. Nasal Systemic Drug Delivery, *Yie W. Chien, Kenneth S. E. Su, and Shyi-Feu Chang*
40. Modern Pharmaceutics: Second Edition, Revised and Expanded, *edited by Gilbert S. Banker and Christopher T. Rhodes*
41. Specialized Drug Delivery Systems: Manufacturing and Production Technology, *edited by Praveen Tyle*
42. Topical Drug Delivery Formulations, *edited by David W. Osborne and Anton H. Amann*
43. Drug Stability: Principles and Practices, *Jens T. Carstensen*
44. Pharmaceutical Statistics: Practical and Clinical Applications, Second Edition, Revised and Expanded, *Sanford Bolton*
45. Biodegradable Polymers as Drug Delivery Systems, *edited by Mark Chasin and Robert Langer*
46. Preclinical Drug Disposition: A Laboratory Handbook, *Francis L. S. Tse and James J. Jaffe*
47. HPLC in the Pharmaceutical Industry, *edited by Godwin W. Fong and Stanley K. Lam*
48. Pharmaceutical Bioequivalence, *edited by Peter G. Welling, Francis L. S. Tse, and Shrikant V. Dinghe*

49. Pharmaceutical Dissolution Testing, *Umesh V. Banakar*
50. Novel Drug Delivery Systems: Second Edition, Revised and Expanded, *Yie W. Chien*
51. Managing the Clinical Drug Development Process, *David M. Cocchetto and Ronald V. Nardi*
52. Good Manufacturing Practices for Pharmaceuticals: A Plan for Total Quality Control, Third Edition, *edited by Sidney H. Willig and James R. Stoker*
53. Prodrugs: Topical and Ocular Drug Delivery, *edited by Kenneth B. Sloan*
54. Pharmaceutical Inhalation Aerosol Technology, *edited by Anthony J. Hickey*
55. Radiopharmaceuticals: Chemistry and Pharmacology, *edited by Adrian D. Nunn*
56. New Drug Approval Process: Second Edition, Revised and Expanded, *edited by Richard A. Guarino*
57. Pharmaceutical Process Validation: Second Edition, Revised and Expanded, *edited by Ira R. Berry and Robert A. Nash*
58. Ophthalmic Drug Delivery Systems, *edited by Ashim K. Mitra*
59. Pharmaceutical Skin Penetration Enhancement, *edited by Kenneth A. Walters and Jonathan Hadgraft*
60. Colonic Drug Absorption and Metabolism, *edited by Peter R. Bieck*
61. Pharmaceutical Particulate Carriers: Therapeutic Applications, *edited by Alain Rolland*
62. Drug Permeation Enhancement: Theory and Applications, *edited by Dean S. Hsieh*
63. Glycopeptide Antibiotics, *edited by Ramakrishnan Nagarajan*
64. Achieving Sterility in Medical and Pharmaceutical Products, *Nigel A. Halls*
65. Multiparticulate Oral Drug Delivery, *edited by Isaac Ghebre-Sellassie*
66. Colloidal Drug Delivery Systems, *edited by Jörg Kreuter*
67. Pharmacokinetics: Regulatory • Industrial • Academic Perspectives, Second Edition, *edited by Peter G. Welling and Francis L. S. Tse*
68. Drug Stability: Principles and Practices, Second Edition, Revised and Expanded, *Jens T. Carstensen*
69. Good Laboratory Practice Regulations: Second Edition, Revised and Expanded, *edited by Sandy Weinberg*
70. Physical Characterization of Pharmaceutical Solids, *edited by Harry G. Brittain*
71. Pharmaceutical Powder Compaction Technology, *edited by Göran Alderborn and Christer Nyström*
72. Modern Pharmaceutics: Third Edition, Revised and Expanded, *edited by Gilbert S. Banker and Christopher T. Rhodes*
73. Microencapsulation: Methods and Industrial Applications, *edited by Simon Benita*
74. Oral Mucosal Drug Delivery, *edited by Michael J. Rathbone*
75. Clinical Research in Pharmaceutical Development, *edited by Barry Bleidt and Michael Montagne*

76. The Drug Development Process: Increasing Efficiency and Cost Effectiveness, *edited by Peter G. Welling, Louis Lasagna, and Umesh V. Banakar*

77. Microparticulate Systems for the Delivery of Proteins and Vaccines, *edited by Smadar Cohen and Howard Bernstein*

78. Good Manufacturing Practices for Pharmaceuticals: A Plan for Total Quality Control, Fourth Edition, Revised and Expanded, *Sidney H. Willig and James R. Stoker*

79. Aqueous Polymeric Coatings for Pharmaceutical Dosage Forms: Second Edition, Revised and Expanded, *edited by James W. McGinity*

80. Pharmaceutical Statistics: Practical and Clinical Applications, Third Edition, *Sanford Bolton*

81. Handbook of Pharmaceutical Granulation Technology, *edited by Dilip M. Parikh*

82. Biotechnology of Antibiotics: Second Edition, Revised and Expanded, *edited by William R. Strohl*

83. Mechanisms of Transdermal Drug Delivery, *edited by Russell O. Potts and Richard H. Guy*

84. Pharmaceutical Enzymes, *edited by Albert Lauwers and Simon Scharpé*

85. Development of Biopharmaceutical Parenteral Dosage Forms, *edited by John A. Bontempo*

86. Pharmaceutical Project Management, *edited by Tony Kennedy*

87. Drug Products for Clinical Trials: An International Guide to Formulation • Production • Quality Control, *edited by Donald C. Monkhouse and Christopher T. Rhodes*

88. Development and Formulation of Veterinary Dosage Forms: Second Edition, Revised and Expanded, *edited by Gregory E. Hardee and J. Desmond Baggot*

89. Receptor-Based Drug Design, *edited by Paul Leff*

90. Automation and Validation of Information in Pharmaceutical Processing, *edited by Joseph F. deSpautz*

91. Dermal Absorption and Toxicity Assessment, *edited by Michael S. Roberts and Kenneth A. Walters*

92. Pharmaceutical Experimental Design, *Gareth A. Lewis, Didier Mathieu, and Roger Phan-Tan-Luu*

93. Preparing for FDA Pre-Approval Inspections, *edited by Martin D. Hynes III*

94. Pharmaceutical Excipients: Characterization by IR, Raman, and NMR Spectroscopy, *David E. Bugay and W. Paul Findlay*

95. Polymorphism in Pharmaceutical Solids, *edited by Harry G. Brittain*

96. Freeze-Drying/Lyophilization of Pharmaceutical and Biological Products, *edited by Louis Rey and Joan C. May*

97. Percutaneous Absorption: Drugs–Cosmetics–Mechanisms–Methodology, Third Edition, Revised and Expanded, *edited by Robert L. Bronaugh and Howard I. Maibach*

98. Bioadhesive Drug Delivery Systems: Fundamentals, Novel Approaches, and Development, *edited by Edith Mathiowitz, Donald E. Chickering III, and Claus-Michael Lehr*
99. Protein Formulation and Delivery, *edited by Eugene J. McNally*
100. New Drug Approval Process: Third Edition, The Global Challenge, *edited by Richard A. Guarino*
101. Peptide and Protein Drug Analysis, *edited by Ronald E. Reid*
102. Transport Processes in Pharmaceutical Systems, *edited by Gordon L. Amidon, Ping I. Lee, and Elizabeth M. Topp*
103. Excipient Toxicity and Safety, *edited by Myra L. Weiner and Lois A. Kotkoskie*
104. The Clinical Audit in Pharmaceutical Development, *edited by Michael R. Hamrell*
105. Pharmaceutical Emulsions and Suspensions, *edited by Francoise Nielloud and Gilberte Marti-Mestres*
106. Oral Drug Absorption: Prediction and Assessment, *edited by Jennifer B. Dressman and Hans Lennernäs*
107. Drug Stability: Principles and Practices, Third Edition, Revised and Expanded, *edited by Jens T. Carstensen and C. T. Rhodes*
108. Containment in the Pharmaceutical Industry, *edited by James P. Wood*
109. Good Manufacturing Practices for Pharmaceuticals: A Plan for Total Quality Control from Manufacturer to Consumer, Fifth Edition, Revised and Expanded, *Sidney H. Willig*
110. Advanced Pharmaceutical Solids, *Jens T. Carstensen*
111. Endotoxins: Pyrogens, LAL Testing, and Depyrogenation, Second Edition, Revised and Expanded, *Kevin L. Williams*
112. Pharmaceutical Process Engineering, *Anthony J. Hickey and David Ganderton*
113. Pharmacogenomics, *edited by Werner Kalow, Urs A. Meyer, and Rachel F. Tyndale*
114. Handbook of Drug Screening, *edited by Ramakrishna Seethala and Prabhavathi B. Fernandes*
115. Drug Targeting Technology: Physical • Chemical • Biological Methods, *edited by Hans Schreier*
116. Drug–Drug Interactions, *edited by A. David Rodrigues*
117. Handbook of Pharmaceutical Analysis, *edited by Lena Ohannesian and Anthony J. Streeter*

ADDITIONAL VOLUMES IN PREPARATION

Pharmaceutical Process Scale-Up, *edited by Michael Levin*

Modern Pharmaceutics: Fourth Edition, Revised and Expanded, *edited by Gilbert S. Banker and Christopher T. Rhodes*

Dermatological and Transdermal Formulations, *edited by Kenneth A. Walters*

Clinical Drug Trials and Tribulations, Second Edition, Revised and Expanded, *edited by Allen Cato, Lynda Sutton, and Allen Cato III*

Handbook of
Pharmaceutical Analysis

edited by

Lena Ohannesian
McNeil Consumer Healthcare Company
Fort Washington, Pennsylvania

Anthony J. Streeter
The R. W. Johnson Pharmaceutical Research Institute
Spring House, Pennsylvania

MARCEL DEKKER, INC.   NEW YORK · BASEL

Second Indian Reprint 2008

ISBN-10: 0-8247-0462-2
ISBN-13: 978-0-8247-0462-9

Headquarters
Marcel Dekker, Inc.
270 Madison Avenue, New York, NY 10016
tel: 212-696-9000; fax: 212-685-4540

World Wide Web
http://www.dekker.com

Printed and bound in India by Replika Press Pvt. Ltd.

FOR SALE IN INDIAN SUBCONTINENT ONLY.

To our daughters, Taleen, Sareen, and Aleen, for their patience and understanding for all the time, on top of our very demanding careers, that we took up working on assembling this book. May their lives be filled with the same curiosity and joy that we find in science, in whatever experiences the future brings to them.

Preface

Twenty years have passed since the publication of the first volume of *Pharmaceutical Analysis*, which has since evolved into the *Handbook of Pharmaceutical Analysis*. The original two volumes succeeded in concisely filling the gap between undergraduate text and detailed monograph on pharmaceutical analysis for practitioners of the pharmaceutical sciences. As with other branches of science, the technologies employed in pharmaceutical analysis today have advanced tremendously over the last two decades, with methodologies becoming routine that were purely experimental a few years ago.

Our goals in preparing this revised version were to bring the text up to date with the most important developments in the field of pharmaceutical analysis while still preserving the scope and level of coverage of our previous texts. We have tried to maintain the intermediate level of coverage throughout the book, with each chapter containing detailed descriptions of theory, instrumentation, and applications, as well as pertinent references. We decided to combine the chapters into a single volume for convenience and thus it is being published as a single, authoritative handbook. This book is not intended to be a comprehensive resource in itself, but to give the reader background information on the most widely used techniques and, with almost 2000 references, to direct the reader to more detailed sources in the scientific literature.

We have included many of the chapters from *Pharmaceutical Analysis*, in updated form, since they continue to provide the backbone of pharmaceutical analysis for the book. However, because of their limited applications in modern pharmaceutical analysis, we have elected not to include some chapters such as those on thin-layer chromatography, pyrolysis-gas chromatography, gas chromatography, and functional group analysis.

Before you can perform any of the analytical techniques described in this book, it is first necessary to obtain your drug substance. It is becoming clearer that many drugs can exist in different polymorphic forms, which can behave very differently from one another during formulation and after administration. Chapter 1 addresses this topic. Chapter 2 deals with the techniques that are often necessary for separating the drug from the other components of a pharmaceutical formulation or biological sample and/or preparing the drug sample for one of the analytical methods that are covered in later chapters. This chapter also introduces the reader to the concept of validation for assays of drugs in plasma, which, since the results of these measurements are critical in determining the safety margins for exposure in humans versus animals in toxicity studies, is subject to close scrutiny by regulatory authorities during the approval of new drugs. The sound scientific principles that underlie the FDA guidances in this area have, by a process of osmosis, influenced government and academia as well.

Following Chapter 3 on high-performance liquid chromatography, we have included a chapter on the many new applications of mass spectrometry. This technique, usually in combination with high-performance liquid chromatography, has become the mainstay of metabolite or decomposition product identification, and its great potential for selectivity and sensitivity has proven invaluable for many drug and metabolite assays. The various techniques applied to high-performance liquid chromatography for removing the chromatographic solvents and introducing the solutes directly into the mass spectrometer, without the need for extraction, derivatization or heating, make high-performance liquid chromatography coupled with mass spectrometry especially valuable for peptides and heat labile molecules.

Chapter 5 covers ultraviolet–visible spectroscopy and Chapter 6, on immunoassay techniques, emphasizes the wide array of new methodologies that do not use radioisotopes. Chapter 7 discusses one of the most novel techniques for chromatographic separation of molecules—capillary electrophoresis—and its widespread applications to pharmaceuticals. Chapter 8, ''Atomic Spectroscopy,'' and Chapter 9, ''Luminescence Spectroscopy,'' contain current information on these important technologies.

Chapter 10, on nuclear magnetic resonance spectroscopy was included because of the unique role that nuclear magnetic resonance spectroscopy has in the study of polymorphisms as well as in verification of compound identity and the structural confirmation of metabolic or decomposition products of drugs. Chapter 11, ''Vibrational Spectroscopy,'' reflects the invaluable contributions that those techniques such as infrared spectroscopy have made to the analysis of formulations, among other studies.

Finally, we could not ignore another area in which the regulatory authorities have had a major impact on pharmaceutical analysis in the pharmaceutical industry. Our concluding chapter deals with process validation as it applies to pharmaceutical analysis.

Lena Ohannesian
Anthony J. Streeter

Contents

Preface   *v*

Contributors   *xi*

1. Form Selection of Pharmaceutical Compounds   1
   Ann W. Newman and G. Patrick Stahly

2. Preparation of Drug Samples for Analysis   59
   David E. Nadig

3. High-Performance Liquid Chromatography   87
   Thomas H. Stout and John G. Dorsey

4. Mass Spectroscopy in Pharmaceutical Analysis   151
   Frank J. Belas and Ian A. Blair

5. Ultraviolet–Visible Spectroscopy   187
   John H. Miyawa and Stephen G. Schulman

6. Immunoassay Techniques   225
   Jean W. Lee and Wayne A. Colburn

7. Applications of Capillary Electrophoresis Technology in the
    Pharmaceutical Industry   313
    Charles J. Shaw and Norberto A. Guzman

8. Atomic Spectroscopy   387
    Helen E. Taylor and Stephen G. Schulman

9. Luminescence Spectroscopy   427
    John H. Miyawa and Stephen G. Schulman

10. Solid-State Nuclear Magnetic Resonance Spectroscopy   467
    David E. Bugay

11. Vibrational Spectroscopy   501
    David E. Bugay and W. Paul Findlay

12. Statistical Considerations in Pharmaceutical Process
    Development and Validation   537
    Gerald J. Mergen

Drug Index   *571*
Subject Index   *579*

Contributors

Frank J. Belas*   Center for Cancer Pharmacology, University of Pennsylvania, Philadelphia, Pennsylvania

Ian A. Blair   Center for Cancer Pharmacology, University of Pennsylvania, Philadelphia, Pennsylvania

David E. Bugay   SSCI, Inc., West Lafayette, Indiana

Wayne A. Colburn   MDS Pharma Services, Phoenix, Arizona

John G. Dorsey   Department of Chemistry, Florida State University, Tallahassee, Florida

W. Paul Findlay   Department of Industrial and Physical Pharmacy, Purdue University, West Lafayette, Indiana

Norberto A. Guzman   The R. W. Johnson Pharmaceutical Research Institute, Raritan, New Jersey

Jean W. Lee   MDS Pharma Services, Lincoln, Nebraska

---

* Current affiliation: Lilly Research Laboratories, Lilly Corporate Center, Indianapolis, Indiana.

Gerald J. Mergen   McNeil Consumer Healthcare Company, Fort Washington, Pennsylvania

John H. Miyawa   College of Pharmacy, University of Florida, Gainesville, Florida

David E. Nadig   The R. W. Johnson Pharmaceutical Research Institute, Raritan, New Jersey

Ann W. Newman   SSCI, Inc., West Lafayette, Indiana

Stephen G. Schulman   College of Pharmacy, University of Florida, Gainesville, Florida

Charles J. Shaw   The R. W. Johnson Pharmaceutical Research Institute, Raritan, New Jersey

G. Patrick Stahly   SSCI, Inc., West Lafayette, Indiana

Thomas H. Stout*   Department of Chemistry, University of Cincinnati, Cincinnati, Ohio

Helen E. Taylor   College of Pharmacy, University of Florida, Gainesville, Florida

---

* *Current affiliation*: Eurand America, Inc., Vandalia, Ohio.

Handbook of
Pharmaceutical Analysis

1

Form Selection of Pharmaceutical Compounds

Ann W. Newman and G. Patrick Stahly
SSCI, Inc., West Lafayette, Indiana

I. INTRODUCTION

The drug development process involves a number of activities which are carried out simultaneously, as shown by the oversimplified depiction in Fig. 1. Once a molecule is discovered that has desirable biological activity, the process of creating a pharmaceutical drug product from this molecule begins. As toxicology and efficacy studies are undertaken, methods for manufacture of the active molecule and for its delivery in therapeutic doses are sought. Critical to the latter effort is finding a form of the active molecule which exhibits appropriate physical properties. The form ultimately selected, called the active pharmaceutical ingredient (API), or drug substance, must be stable and bioavailable enough to be formulated into a drug product, such as a tablet or suspension. This formulation must be effective at delivering the active molecule to the targeted biosystem.

This chapter describes methodology useful in selection of the appropriate solid form of a drug substance for inclusion in a drug product. Form selection is commonly considered among the primary goals of a preformulation study. However, the investigative techniques discussed herein also have application in early drug substance and drug product development activities (shown by the circled area in Fig. 1).

Solid form selection involves the preparation and property evaluation of many derivatives of an active molecule. Drug substance properties of importance in the drug development process may be categorized as shown in Table 1. These properties depend on the nature of the drug substance and the final formulation. Many bioactive organic molecules contain ionizable groups such as carboxylic

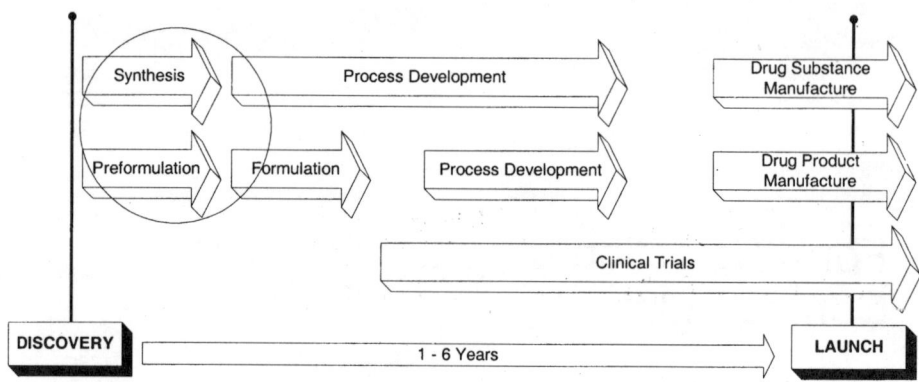

Fig. 1   The drug development process.

acid or amino groups. Reaction of these compounds with acids or bases produce salts, which have much different physical properties than the neutral parents. A single molecular entity, be it a salt or a neutral molecule, often exists in multiple solid forms, each of which exhibits unique physical properties. The properties of many such forms need to be evaluated relative to the intended formulation. A lyophilized product that will be dissolved and injected needs to be chemically stable in the dry state and adequately soluble in the carrier. On the other hand, the drug substance in a tablet formulation needs to be processable, chemically stable, and physically stable in the dry state, as well as having adequate solubility for delivery.

Form selection activities should be started as early in the development process as material availability allows. Salt selection, including preparation and eval-

Table 1   Some Important Properties of Drug Substances

Bioavailability	Chemical and physical stability	Processibility
Dissolution rate	Excipient compatibility	Color
Solubility	Hygroscopicity	Compactibility
Toxicity	Oxidative stability	Density
	Photostability	Ease of drying
	Thermodynamic stability	Filterability
	Crystal form	Flowability
		Hardness
		Melting point
		Particle size

uation of samples, and polymorph screening can be carried out with as little as half a gram of active compound. Results of form selection include information that can be used in planning the final step of the manufacturing process (often crystallization) as well as information that is critical to formulation development.

The nature and extent of work to be performed during development can be modeled after the draft International Committee on Harmonization (ICH) Q6A document on specifications, which can be found on the Food and Drug Administration (FDA) website (www.fda.cder.gov). This document outlines the specifications needed for a New Drug Application and contains several decision trees to guide the selection of specifications. The Q6A decision tree 4 (Fig. 2) describes

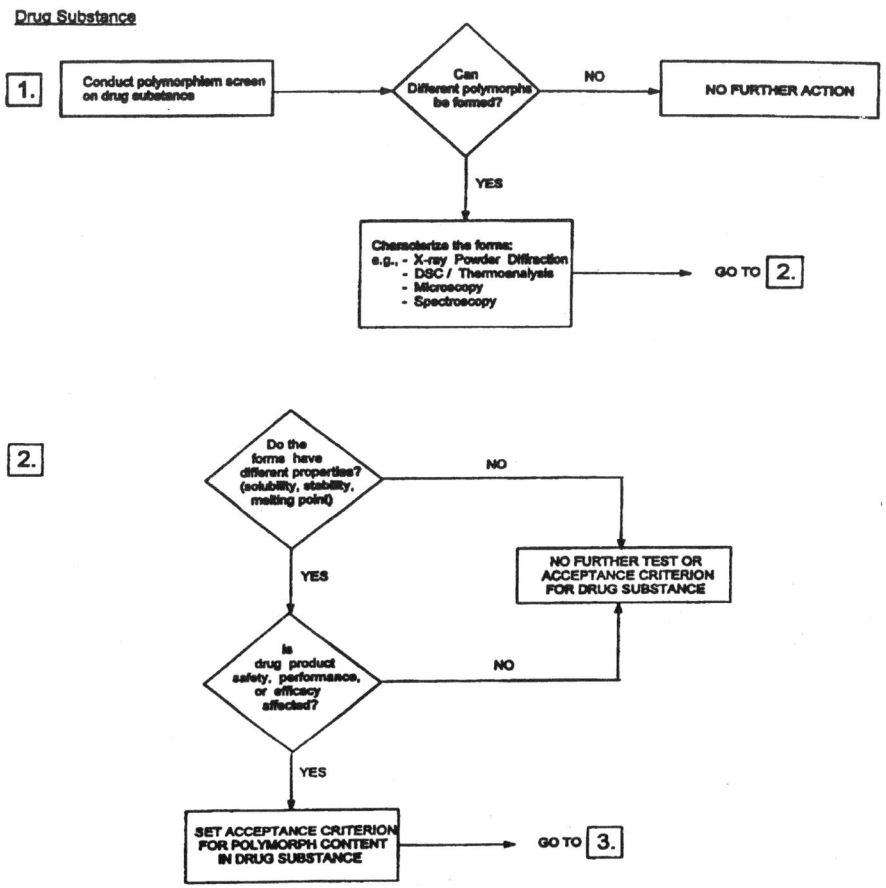

Fig. 2 Flow chart 4 from the ICH Q6A document (www.fda.cder.gov).

3.

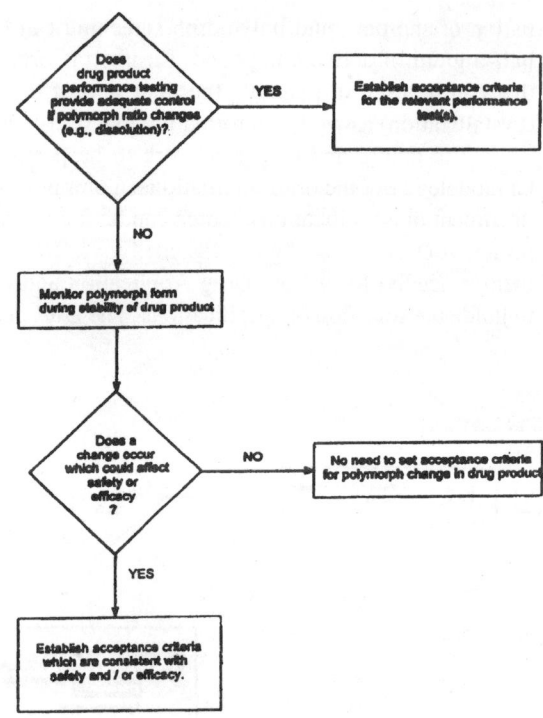

Fig. 2 Continued

methods for the study of solids for a polymorph screen as well as characterization of the drug substance in the drug product. Other decision trees have also been reported in the literature (1).

In this chapter we describe the form selection process. A short review of the analytical techniques commonly employed is followed by sections covering salt and solid form selection. Form selection should be approached in a planned, rational manner, but it is important to realize that not all compounds will allow adherence to a single experimental plan. The exercise is a scientific one, and it will yield the best results only if carried out with judgment and flexibility.

II. ANALYTICAL TECHNIQUES

A number of analytical techniques are commonly used in form selection studies. Various publications (2–4) and books (5,6) describe physical characterization of solid-state pharmaceuticals. A brief description of common methods will be presented in this section.

A. X-Ray Diffraction

Crystalline organic solids are made up of molecules which are packed or ordered in a specific arrangement. These molecules are held together by relatively weak forces, such as hydrogen bonding and van der Waals interactions. The arrangement of the molecules is defined by a unit cell, which is the smallest repeating unit of a crystal. The unit cell can be divided into planes, as shown in Fig. 3.

X-ray diffraction techniques used for characterizing pharmaceutical solids

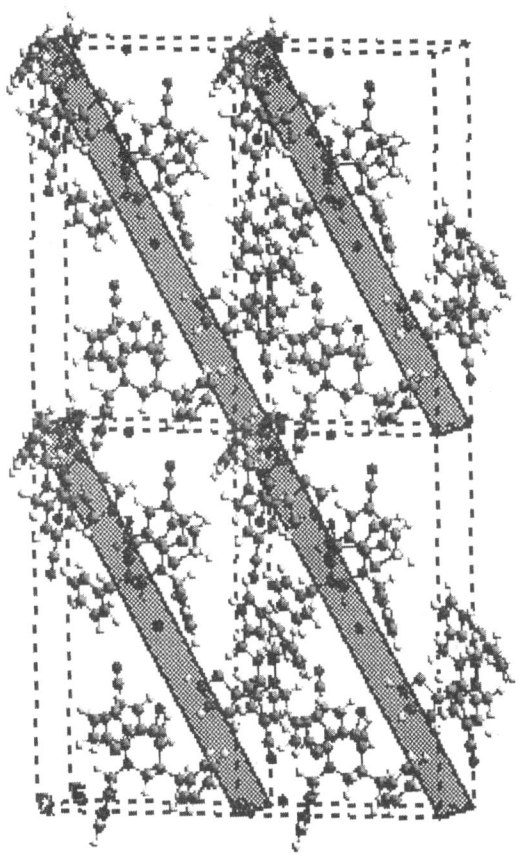

Fig. 3   A packing diagram of unit cells divided into planes.

include the analysis of single crystals and powders. The electrons surrounding the atoms diffract X-rays in a manner described by the Bragg equation:

$$n\lambda = 2d \sin \theta \quad (n = 1, 2, 3, \ldots) \tag{1}$$

where

λ = X-ray wavelength
d = spacing between the diffracting planes
θ = diffraction angle

A schematic of the diffraction phenomenon is given in Fig. 4. X-rays will be diffracted at an angle defined as θ. Knowing the diffraction angle and the X-ray wavelength, the spacing between the planes can be calculated. Conditions of the Bragg equation must be satisfied to achieve constructive interference of the diffracted X-rays and produce a beam that can be measured by the detector. If the conditions of the Bragg equation are not satisfied, diffracted waves interfere destructively, with a net diffracted intensity of zero.

For single-crystal diffraction, a good-quality single crystal of the sample of interest is required. From the angles and intensities of diffracted radiation, the structure of the crystal can be elucidated and the positions of the molecules in the unit cell can be determined. The result is often displayed graphically as the asymmetric unit, which is the smallest part of a crystal structure from which the complete structure can be obtained using space-group symmetry operations.

X-rays   Detector

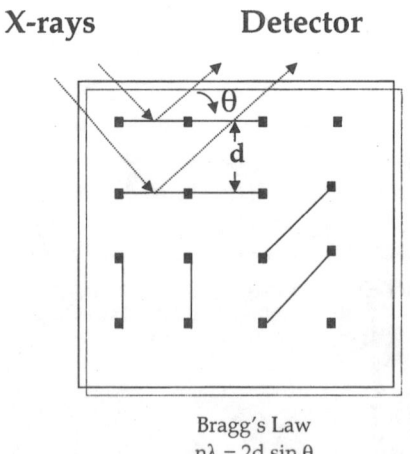

Bragg's Law
$n\lambda = 2d \sin \theta$

Fig. 4   A schematic representation of X-ray diffraction.

The unit cell parameters, the lengths (a, b, c) as well as the angles (α, β, γ) of the unit cell are also determined from the crystal structure. There are seven classes of unit cells: triclinic, monoclinic, orthorhomic, tetragonal, hexagonal, rhombohedral, and cubic. For pharmaceutics, only triclinic ($a \neq b \neq c$, $\alpha \neq \beta \neq \gamma \neq 90°$), monoclinic ($a \neq b \neq c$, $\alpha \neq \gamma \neq 90°$, $\beta = 90°$), and orthorhombic ($a \neq b \neq c$, $\alpha = \beta = \gamma = 90°$) unit cells are commonly observed.

The unit cells can be "packed" into a three-dimensional display of the crystal lattice. The orientation of the molecules is responsible for various properties of the crystalline substance. For example, hydrogen bonding networks may provide high stability, and spaces in the structure may allow easy access of small molecules to provide hydrated or solvated forms.

Crystal structures provide important and useful information about solid-state pharmaceutical materials. Unfortunately, it is not always possible to grow suitable single crystals of a drug substance. In these cases, X-ray diffraction of powder samples can be used for comparison of samples.

X-ray powder diffraction (XRPD) is the analysis of a powder sample. The typical output is a plot of intensity versus the diffraction angle (2θ). Such a plot can be considered a fingerprint of the crystal structure, and is useful for determination of crystallographic sameness of samples by pattern comparison. A crystalline material will exhibit peaks indicative of reflections from specific atomic planes. The patterns are representative of the structure, but do not give positional information about the atoms in the molecule. One peak will be exhibited for all repeating planes with the same spacing. An amorphous sample, on the other hand, will exhibit a broad hump in the pattern called an amorphous halo, as shown in Fig. 5.

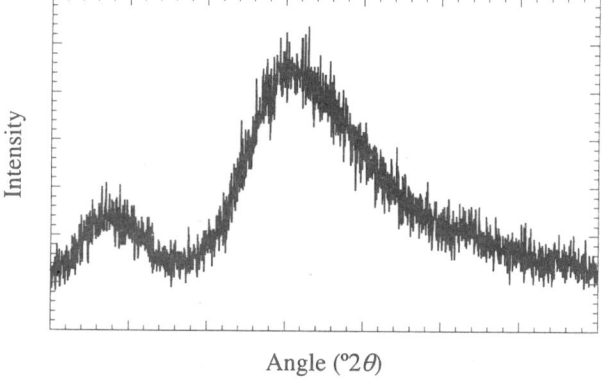

Angle (°2θ)

Fig. 5 The XRPD pattern exhibited by an amorphous material.

XRPD is dependent on a random orientation of the particles during analysis to obtain a representative powder pattern. The sample, as well as sample preparation, can greatly effect the resulting pattern. Large particles or certain particle morphologies, such as needles or plates, can result in preferred orientation. Preferred orientation is the tendency of crystals to pack against each other with some degree of order and it can affect relative peak intensities, but not peak positions, in XRPD patterns. If a powder is packed into an XRPD sample holder and the surface is smoothed with a microscope slide or similar device, crystals at the surface can become aligned so that a nonstatistical arrangement of crystal faces is presented to the X-ray beam. The result is that some reflections are artificially intensified and others are artificially weakened. One way to determine if preferred orientation is causing relative peak intensity changes is to grind and reanalyze samples. Grinding reduces particle size and disrupts the crystal habit, both of which tend to minimize preferred orientation effects. However, grinding can cause crystal form changes, so care must be taken to interpret the patterns with this in mind. The effects of preferred orientation can be profound, as illustrated by the XRPD patterns shown in Fig. 6.

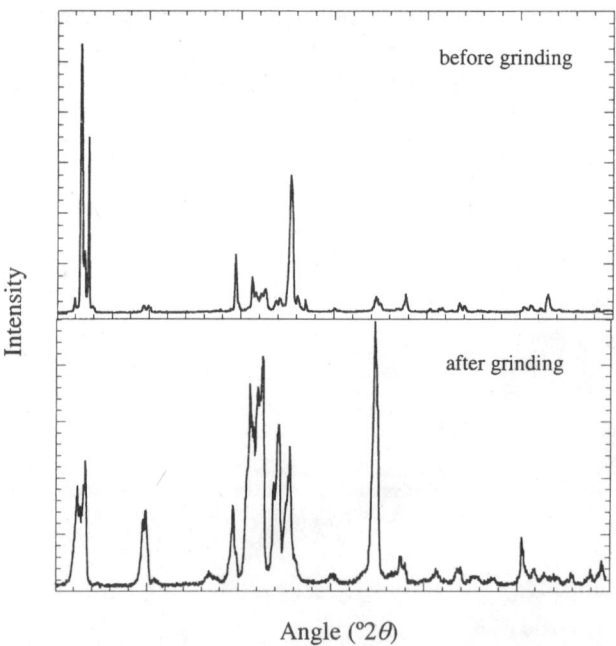

Fig. 6   XRPD patterns of the same sample before (top) and after (bottom) grinding. The polymorphic form of the sample was not changed by grinding.

Qualitative analysis of powder patterns can be used to determine if multiple samples are the same crystal form or if multiple crystal forms have been produced. Mixtures of samples can also be evaluated. When mixtures are obtained, XRPD can also be used in a quantitative mode to calculate the amount of each phase present.

B. Thermal Methods

Thermal methods of analysis discussed in this section are differential scanning calorimetry (DSC), thermogravimetry (TG), and hot-stage microscopy (HSM). All three methods provide information upon heating the sample. Heating can be static or dynamic in nature, depending on the information required.

Differential scanning calorimetry monitors the energy required to maintain the sample and a reference at the same temperature as they are heated. A plot of heat flow (W/g or J/g) versus temperature is obtained. A thermal transition which absorbs heat (melting, volatilization) is called endothermic. If heat is released during a thermal transition (crystallization, degradation), it is called exothermic. The area under a DSC peak is directly proportional to the heat absorbed or released and integration of the peak results in the heat of transition.

Samples are loaded into pans for DSC analysis. Pan configuration (open, crimped, hermetically sealed, hermetically sealed with a.pinhole, etc.) and scan rate can result in variations in position and intensity of the thermal events. These variations can be used to gain further information about the sample as well as other crystal forms.

The observance of thermal transitions by DSC is insufficient to fully characterize the behavior of a substance on heating. It is not known if an endothermic transition observed in the DSC is a volatilization or a melt without corroborating information, such as TG or HSM data. It is important to understand the origin of the DSC transitions to fully characterize the system and understand the relationship between various solid forms.

Thermogravimetry measures the weight change of a sample as a function of temperature. A total volatile content of the sample is obtained, but no information on the identity of the evolved gas is provided. The evolved gas must be identified by other methods, such as gas chromatography, Karl Fisher titration (specifically to measure water), TG–mass·spectroscopy, or TG–infrared spectroscopy. The temperature of the volatilization and the presence of steps in the TG curve can provide information on how tightly water or solvent is held in the lattice. If the temperature of the TG volatilization is similar to an endothermic peak in the DSC, the DSC peak is likely due or partially due to volatilization. It is usually necessary to utilize multiple techniques to determine if more than one thermal event is responsible for a given DSC peak.

Hot-stage microscopy is a technique that supplements DSC and TG. Events observed by DSC and/or TG can be readily characterized by HSM. Melting, gas evolution, and solid–solid transformations can be visualized, providing the most straightforward means of identifying thermal events. Many polymorphic systems have been investigated using only these thermal methods, as illustrated by the publications of Kuhnert-Brandstätter (7). Details of the methodologies used in hot-stage microscopy have also been reviewed (8).

Thermal analysis can be used to determine the melting points, recrystallizations, solid-state transformations, decompositions, and volatile contents of pharmaceutical materials. DSC can also be used to analyze mixtures quantitatively.

C. Vibrational Spectroscopy

Common methods used to characterize drugs and excipients are infrared (IR) and Raman spectroscopy. These techniques are sensitive to the structure, conformation, and environment of organic compounds. Because of this sensitivity, they are useful characterization tools for pharmaceutical crystal forms. Qualitative as well as quantitative analysis can be performed with both techniques.

Infrared spectroscopy is based on the conversion of IR radiation into molecular vibrations. For a vibration to be IR-active, it must involve a changing molecular dipole (asymmetric mode). For example, vibration of a dipolar carbonyl group is detectable by IR spectroscopy. Whereas IR has been traditionally used as an aid in structure elucidation, vibrational changes also serve as probes of intermolecular interactions in solid materials.

Sampling techniques for IR include pellets, mulls, and diffuse reflectance. Diffuse reflectance is the best choice for crystal form determination, due to the minimal sample manipulation required. Mulls can also be used for form identification, but peaks due to the suspension medium may interfere with the peaks of interest.

Raman spectroscopy is based on the inelastic scattering of laser radiation with loss of vibrational energy by a sample. A vibrational mode is Raman-active when there is a change in the polarizability during the vibration. Symmetric modes tend to be Raman-active. For example, vibrations about bonds between the same atom, such as in alkynes, can be observed by Raman spectroscopy.

Small amounts of samples can be analyzed by Raman spectroscopy and a variety of sample holders are available, ranging from stainless steel holders to glass NMR tubes. The samples are analyzed neat, eliminating the need for sample preparation procedures that may induce solid form changes. Since a laser is used, only a small portion of the sample is in the beam during analysis.

D. Nuclear Magnetic Resonance (NMR) Spectroscopy

NMR spectroscopy probes atomic environments based on the different resonance frequencies exhibited by nuclei in a strong magnetic field. Many different nuclei are observable by the NMR technique, but those of hydrogen and carbon atoms are most frequently studied. NMR spectroscopy of solutions is commonly used for structure elucidation. However, solid-state NMR measurements are extremely useful for characterizing the crystal forms of pharmaceutical solids.

Nuclei that are typically analyzed with this technique include those of ^{13}C, ^{31}P, ^{15}N, ^{25}Mg, and ^{23}Na. Different crystal structures of a compound can result in perturbation of the chemical environment of each nucleus, resulting in a unique spectrum for each form. Once resonances have been assigned to specific atoms of the molecule, information on the nature of the polymorphic variations can be obtained. This can be useful early in drug development, when the single-crystal structure may not be available. Long data acquisition times are common with solid-state NMR, so it is often not considered for routine analysis of samples. However, it is usually a very sensitive technique, and sample preparation is minimal. NMR spectroscopy can be used either qualitatively or quantitatively, and can provide structural data, such as the identity of solvents bound in a crystal.

E. Moisture Sorption/Desorption and Hygroscopicity

Hygroscopicity and the formation of hydrated crystal forms can be investigated by means of moisture sorption/desorption methods. Sample analysis may be carried out using automated equipment or by periodic weighing of samples kept over saturated salt solutions providing various relative humidities (RHs). In either case, water taken in or released by a sample is detected as a change in sample weight. If a material readily loses water of hydration at low relative humidity, the stability of the hydrate may need to be investigated further. If a material readily gains moisture at ambient or high relative humidity, hygroscopicity studies will be needed to determine if a change in crystal form is associated with the water uptake. This is done by characterizing material equilibrated under various relative humidity conditions using techniques suitable for detection of crystal form, such as XRPD, TG, DSC, and IR spectroscopy. Changes in water content and crystal form may lead to definition of specific handling conditions under which a change in form will not occur.

F. Summary

Only a brief description of selected techniques for solid-state characterization has been given above. Many other techniques are available. It is imperative that a multidisciplinary approach be applied to the characterization solids; no single analytical method can provide all the information necessary to understand the nature and properties of solid pharmaceutical compounds.

III. SALT SELECTION

A. Factors Guiding Salt Selection

Salt selection is a critical part of the drug development process because selection of an appropriate salt can significantly reduce time to market. Changing salts in the middle of a development program may require repeating most of the biological, toxicological, formulation, and stability studies performed initially. However, continuing the development of a nonoptimal salt may lead to increased developmental and production costs, even product failure. Selection of the correct salt early in the development process will avoid these problems and facilitate downstream development activities. In addition, salts that exhibit advantageous properties are usually patentable as new chemical entities.

Salts are used to alter the physical, chemical, biological, and economic properties of a drug substance. The change in crystal structure accomplished by forming a salt can lead to greatly improved properties. The advantages of using salt forms in pharmaceutical formulations have been extensively reviewed (9). A variety of factors can guide the salt selection process and a partial list of considerations is given in Table 2.

A change in the solubility of a drug substance is often a major reason for choosing a salt. In many cases, substances containing free acid or base groups have poor aqueous solubility which saltification of these groups can improve, leading ultimately to greater bioavailability. Increasing the solubility of a weak acid–base drug substance by forming a variety of salts has been reported for

Table 2 Factors Guiding the Salt Selection
Process

Bioavailability	pH of salt solutions
Chemical stability	Physical stability
Crystallinity	Processing properties
Dissolution rate	Purity
Cost	Solubility
Handling properties	Taste
Hygroscopicity	Toxicity
Melting point	Wettability
Intended formulation	Yield

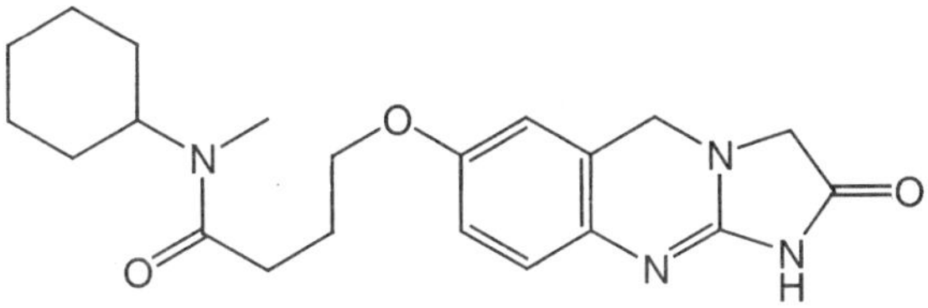

Fig. 7   The structure of RS-82856.

RS-82856 (Fig. 7) (10). Five salts (chloride, hydrogen sulfate, phosphate, sodium, and potassium) exhibited significantly higher solubility and dissolution rates than the parent drug. Based on a variety of physical parameters (solubility, dissolution rate, melting point, hygroscopicity, and chemical stability), the hydrogen sulfate form was recommended for development. A bioavailability study in dogs comparing the parent drug and the hydrogen sulfate salt resulted in the salt being absorbed approximately two to three times more efficiently than the parent drug. The solubility and dissolution data were good indicators of the bioavailability of this material.

For some drugs, preparation of stable salts may not be feasible, or free acid or free base forms may be preferred. A reported example compares the free base and hydrochloride salt of the poorly water-soluble drug, α-pentyl-3-(2-quinolinylmethoxy)benzenemethanol, known as REV 5901 (Fig. 8) (11). For this drug substance, lower solubility of the chloride salt, along with equivalent dissolution rates, resulted in the free base being chosen for development.

It should be noted that a salt usually exhibits a higher melting point than the free acid or base, which can result in greater stability and easier processing. However, there is often a relationship between melting point and aqueous solubility. Gould, in his study of the salts of basic drugs, concluded that "ideal solubility of a drug in all solvents decreases by an order of magnitude with an increase of 100°C in its melting point" (12). An example of this phenomenon is the antimalarial drug α-(2-piperidyl)-3,6-bis(trifluoromethyl)-9-phenanthrenemethanol hy-

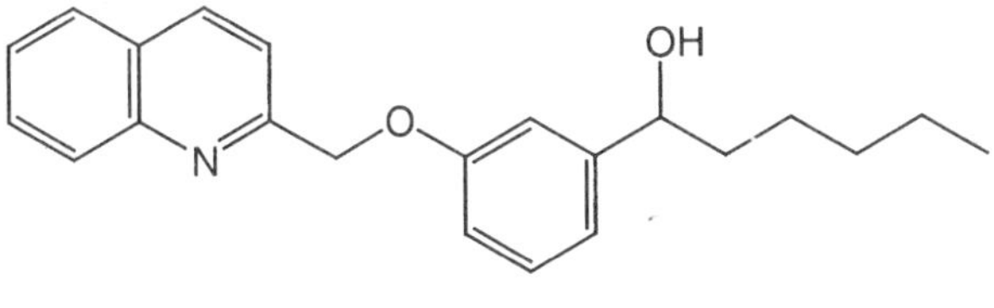

Fig. 8   The structure of REV 5901.

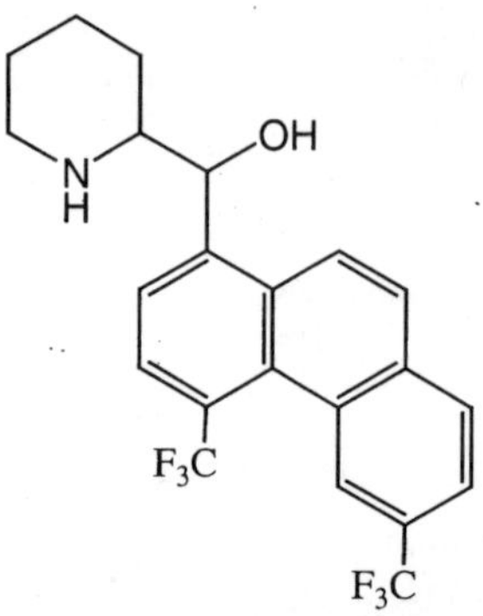

Fig. 9 The structure of α-(2-piperidyl)-3,6-bis(trifluoromethyl)-9-phenanthrene-methanol.

drochloride (Fig. 9). The melting point and solubility data are shown in Table 3 (13). Overall, a substantial decrease in solubility was observed with the increase in melting point of the salts. It should also be noted that the solubility of salts can be affected not only by changing the lattice energy (melting point), but also by enhancing water–drug interactions. The study of chlorhexidine (Fig. 10) showed that the solubility of this drug was significantly enhanced by increasing the number of hydroxyl groups on the conjugate acid (14).

The melting point of a drug substance salt can be greatly influenced by the counterion. For UK47880 (Fig. 11), a relationship was observed between the melting points of the salts and the corresponding conjugate acid (12). Salts pre-

Table 3 Melting-Point and Solubility Data for
α-(2-piperidyl)-3,6-bis(trifluoromethyl)-9-phenanthrenemethanol
hydrochloride (13)

Salt form	Melting point of salt (°C)	Aqueous solubility (mg/mL)
Free base	215	7.5
DL-Lactate	172	1850
L-Lactate	192	925
2-Hydroxyethane sulfonate	251	620
Sulfate	270	20
Mesylate	290	300
Hydrochloride	331	13

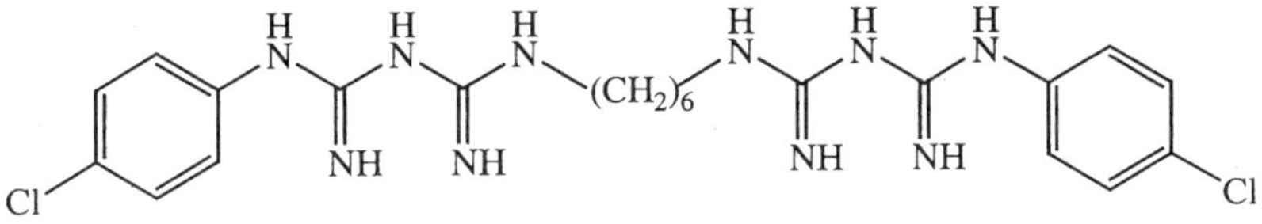

Fig. 10 The structure of chlorhexidine.

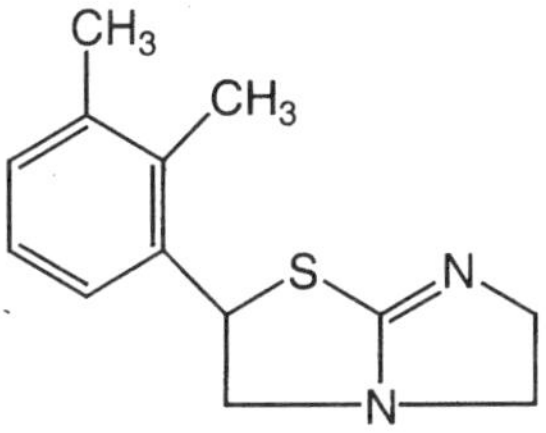

Fig. 11 The structure of UK47880.

pared from high-melting aromatic acids exhibited higher melting points, whereas salts prepared from low-melting flexible aliphatic acids yielded oils. In the case of epinephrine (Fig. 12), the effect of hydrogen bonding on the melting points of the salts was apparent (12). Small acids prone to form hydrogen bonds (malonic and maleic) resulted in higher-melting salts. The bitartrate and fumarate salts were found to be lower-melting due to their size and possibly unfavorable symmetry, respectively.

A salt can also provide improved chemical stability compared to the parent drug substance. An example of this was reported for xilobam, whose structure is shown in Fig. 13 (15). In order to protect xilobam from the effects of high temperature and humidities without decreasing the dissolution rate, three arylsulfonic acid salts (tosylate, 1-napsylate, and 2-napsylate), as well as the saccharate salt, were prepared. The 1-napsylate was found to be the most chemically stable form at 70°C and 74% RH after 7 days. Dissolution data from compressed tablets

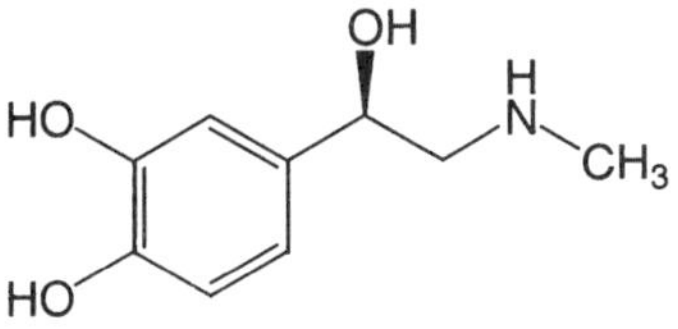

Fig. 12 The structure of epinephrine.

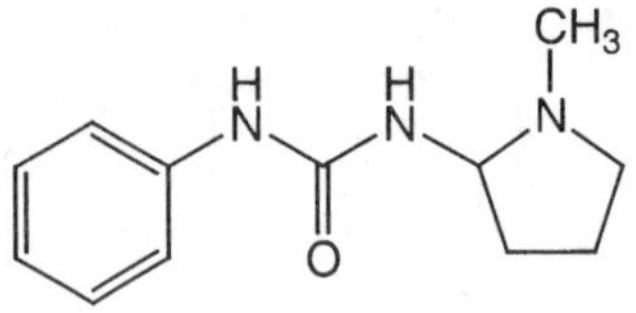

Fig. 13   The structure of xilobam.

showed that the 1-napsylate salt released xilobam at a faster rate than the free base. This work demonstrated that a strong acid with an aryl group protected the easily hydrolyzed base from the effects of high temperature and humidity.

A choice of salts can also expand the formulation options for a material. The antimalarial agent α-(2-piperidyl)-3,6-bis(trifluoromethyl)-9-phenanthrene-methanol hydrochloride (Fig. 9) exhibited poor solubility, was delivered as an oral formulation, and required a single dosing of 750 mg (13). Seven salts and the free base were evaluated. The lactate salt was found to be 200 times as soluble as the hydrochloride salt (Table 3). This enhanced solubility would make it possible to reduce the oral dose to achieve the same therapeutic response as well as develop a parenteral formulation for the treatment of malaria. However, the case of lidocaine hydrochloride (Fig. 14) demonstrates that a compound limited to parenteral and topical formulations can be expanded to oral administration by changing to a salt form with acceptable physical properties (16). The hydrochloride salt was hygroscopic, difficult to prepare, and hard to handle. Six salts were evaluated for salt formation, solubility, and hygroscopicity. Other salts, such as phosphate, exhibited properties acceptable for dry pharmaceutical dosage forms.

Many other examples can be found in the literature that demonstrate the applicability of examining a number of salts to obtain the necessary properties needed for development and marketing of the drug substance. Excellent reviews on salts (9,12) discuss many of the issues involved in targeting salt forms of drug substances.

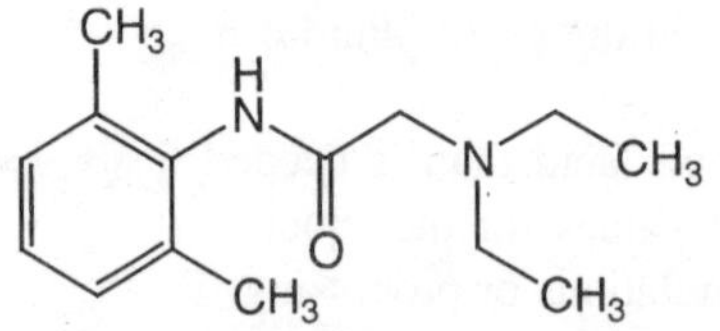

Fig. 14   The structure of lidocaine.

B. Counterions

Salt formation involves proton transfer from an acid to a base. In theory, any compound that exhibits acidic or basic characteristics can form salts. The major consideration is the relative acidity and/or basicity of the chemical species involved. To form a salt, the pK_a of the acidic partner must be less than the pK_a of the conjugate acid of the basic partner. These pK_a values need to be about two units apart for total proton transfer to occur, otherwise an equilibrium mixture of all components (acid, base, and salt) is likely to result. Even so, equilibrium mixtures of this type can often be used to prepare salts if a driving force is present, such as the crystallization of the salt from solution.

Another consideration is the toxicity of the counterion. A large number of anions and cations are available for pharmaceutical compounds, and tabulations of those approved by the FDA have appeared in the literature (9,12). An expanded but not comprehensive list of acceptable ions is presented in Table 4. In general, ions related to normal metabolic chemicals or present in food or drink are usually regarded as suitable candidates for preparing salts.

Target salts are chosen by considering a number of factors. The structure and pK_a of the drug substance are important values to determine initially. Available literature on structurally related compounds can result in excellent leads for target salts. The chemical stability of the drug substance, especially as related to pH stability, will also play a role. The ease of large-scale preparation of the salt, as well as the cost of the counterion and processing, will need to be considered to determine if the salt is a feasible choice. The type of drug product and anticipated loading of the drug substance in the drug product can also influence the choice of salts. For high drug loadings, a large, bulky counterion, which adds substantial mass to the loading, may not be the best choice. Anions that irritate the gastrointestinal tract should be avoided for certain drugs, such as anti-inflammatories. The relative acid/base strength of the resulting salt and the tendency to disproportionate should be considered when using basic excipients in a formulation.

A common salt choice for basic drug substances is the hydrochloride, because of its availability. However, a number of issues also need to be considered when using this salt. Reports have shown that hydrochloride salts do not always increase the solubility of poorly soluble basic drugs (1,13,17,18), due to the common-ion effect. The presence of chloride ions in the gastric fluid can result in a lower solubility for the hydrochloride salt.

The hydrochloride salt is often a stronger acid than is needed for many drug substances, which can result in low pH values for the aqueous solutions. This can lead to limitations in parenteral formulations or processing. The highly polar nature of hydrochloride salts can also lead to excessive hygroscopicity of the resulting salt. Dihydrochlorides are also found to be hygroscopic and may lose

Table 4 Summary of Acceptable Salts for Pharmaceutical Drug Substances (adapted from Ref. 9)

Anions—organic	Anions—organic	Anions—organic
Acetate/diacetate	Fumarate	Napsylate
Adipate[a]	Gluceptate (glucoheptonate)	Oxalate[a]
Alginate[a]	Gluconate/digluconate	Palmitate
Aminosalicylate[a]	Glucuronate	Pamoate (embonate)
Anhydromethylenecitrate[a]	Glutamate	Pantothenate
Arecoline[a]	Glycerophosphate[a]	Pectinate[a]
Arginine	Glycollylarsanilate (p-glycollamidophenyl-arsonate)	Phenylethylbarbiturate[a]
Ascorbate[a]		
Aspartate	Hexylresorcinate	Picrate[a]
Benzenesulfonate (besylate)	Hydrabamine (N,N'-di(dehydroabietyl)eth-ylenediamine)	Polygalacturonate
Benzoate	Hydroxynaphthoate	Propionate
Bicarbonate	Isethionate (2-hydroxyethanesulfonate)	Saccharate
Bitartrate	Lactate	Salicylate
Butylbromide[a]	Lactobionate	Stearate
Butyrate	Lysine	Subacetate
Calcium edetate	Malate	Succinate/disuccinate
Camphorate	Maleate	Tannate
Camsylate (camphorsulfonate)	Mandelate	Tartrate
Carbonate	Mesylate	Terephthalate
Citrate	Methylbromide	Teoclate (8-chlorotheophyllinate)
Edetate	Methylenebis(salicylate)[a]	Thiocyanate[a]
Edisylate (1,2-ethanedisulfonate)	Methylnitrate	Tosylate (toluenesulfonate)
Estolate (lauryl sulfate)	Methylsulfate	Triethiodide
Esylate (ethanesulfonate)	Mucate	Undecanoate[a]
	Napadisylate (1,5-naphthalenedisulfonate)[a]	Xinafoate (1-hydroxy-2-naphthalenecarbox-ylate)

Anions—inorganic	Cations—organic	Cations—inorganic
Bisulfate[a]	Benethamine (N-benzylphenethylamine)[a]	Aluminum
Bromide	Benzathine (N,N'-dibenzylethylenediamine)	Barium[a]
Chloride	Chloroprocaine	Bismuth
Hydrobromide/dihydrobromide	Chloline	Calcium
Hydrochloride/dihydrochloride	Clemizole (1-p-chlorobenzyl-2-pyrrolidin-1'-ylmethylbenzimidazole)[a]	Lithium
Hydrofluoride[a]	Diethanolamine	Magnesium
Hydroiodide[a]	Diethylamine[a]	Potassium
Iodide	Ethylenediamine	Sodium
Nitrate/dinitrate	Meglumine (N-methylglucamine)	Zinc
Persulfate[a]	N,N'-dibenzylethylenediamine	
Phosphate/diphosphate	Piperazine[a]	
Sulfate/disulfate/hemisulfate	Procaine	
	Tromethamine (tris(hydroxymethyl)amino-methane)	

[a]Drugs containing these counterions approved in other countries.

hydrogen chloride gas upon heating or under reduced pressure (lyophilization). Although hydrochloride is commonly used for salt formation, other salts may be better alternatives in the long run.

C.   Salt Preparation

Salts can be produced on a small scale using a variety of methods. Selected methods are described below.

1.   Salt Formation from Free Acid/Base

In salt formation from free acid or base, the free acid/base of the drug substance is combined with the base/acid containing the desired counterion in specific molar ratios in a suitable solvent system. There must be adequate solubility of each reactant in the solvent system chosen. The product can be isolated in different ways, often simply by evaporation of the solvent.

2.   Salt Formation by Salt Exchange

For salt formation by salt exchange, the salt of the drug substance is combined with a salt containing the desired counterion in specific molar ratios in a suitable solvent system. As described above, there must be adequate solubility of each reactant in the solvent system. If the desired salt of the drug substance is less soluble than the starting materials, it will precipitate out and can be isolated by filtration. If no precipitate is obtained, other isolation methods can be employed. A method that was described for iodide salts (19) involved precipitation of the unwanted counterion first. In this case silver salts were used for the counterions (silver sulfate, silver *ortho*-phosphate, silver lactate) and a silver iodide precipitate was isolated first by filtration. The desired salt of the drug substance was then precipitated from the filtrate by addition of an antisolvent.

3.   Other Reported Methods

Variations of the traditional methods for salt preparation described above have been reported. To produce the hydrochloride and hydrobromide salts of the anes-

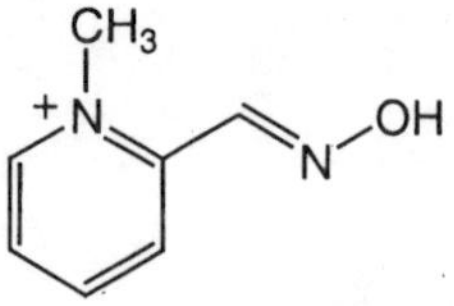

Fig. 15   The structure of *N*-methyl pyridinium-2-aldoxime.

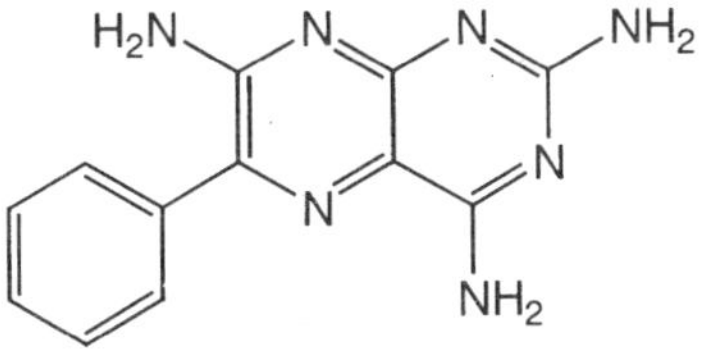

Fig. 16 The structure of triamterene.

thetic lidocaine (Fig. 14), dry hydrogen chloride or hydrogen bromide gas was bubbled into anhydrous ether solutions of the lidocaine base (16). Chloride and lactate salts of the cholinesterase reactivator N-methyl pyridinium-2-aldoxime (Fig. 15) were prepared using ion-exchange resins (19). Complex salts of the diuretic triamterene (Fig. 16) were produced from various acids (hydrochloric, nitric, sulfuric, phosphoric, and acetic) using the phase-solubility technique (20). Profiles of apparent solubility as a function of pH detected complex salt species containing both protonated and unprotonated triamterene. The stoichiometries of

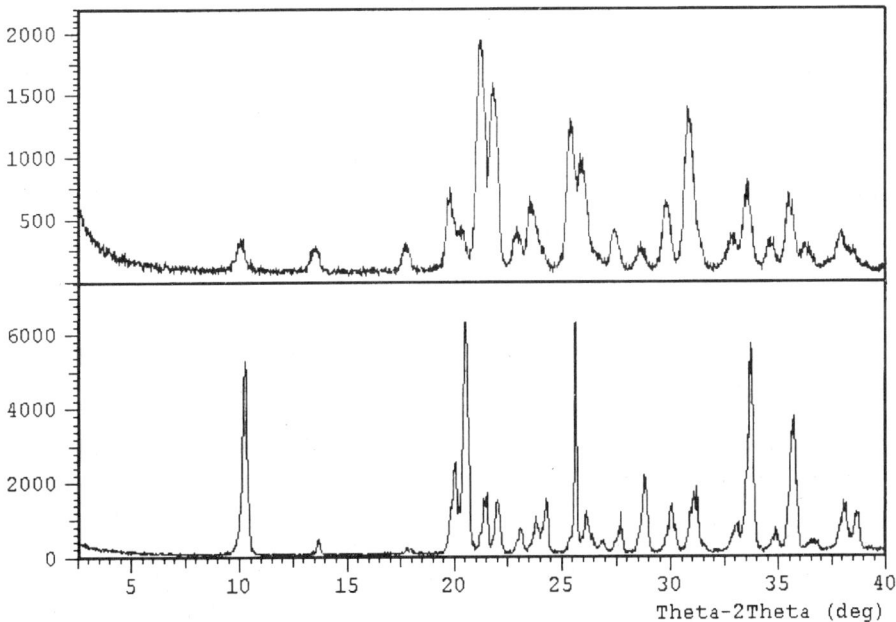

Fig. 17 XRPD patterns of the product obtained from an attempt to prepare a glutamate salt (top) and glutamic acid (bottom).

the complexes determined from the plots were confirmed by elemental analysis of the solids.

Once a salt is produced, it must be characterized. XRPD of the solid can confirm that the reaction occurred and starting materials were not recovered. This is illustrated in Fig. 17. Attempts to produce the glutamate salt from the free base resulted in the XRPD pattern of crystalline glutamic acid. If the drug substance salt is produced, the crystallinity can be determined by XRPD. Confirmation of the stoichiometry of the salt can be obtained by a variety of methods, including elemental analysis and solution NMR spectroscopy. The possibility of chemical degradation during salt formation can be determined using chromatographic (thin-layer, high-performance) or spectroscopic (IR, NMR) methods. The melting point can be obtained from melting-point measurements, hot-stage microscopy, or DSC analysis. The formation of hydrates or solvates can be investigated using Karl Fischer titration or TG analysis. Other information, such as solubility, hygroscopicity, and stability, is also useful. Because of all the information required, determining the best salt for development can be a complicated, time-consuming task.

D. Systematic Approach to Salt Selection

A variety of factors need to be considered when selecting the optimum chemical form of a new drug candidate. These include all physicochemical properties which would influence physical and chemical stability, processability under manufacturing conditions, dissolution rate, and bioavailability. Such selection of chemical form must be done at the initial stages of development, when material and time are limited. Often the medicinal and process chemists select salt forms based on a practical basis, such as previous experience with the salt type, ease of synthesis, reaction yield, etc. Pharmaceutical considerations such as stability, handleability, hygroscopicity, and suitability for a specific dosage form may be secondary considerations.

A salt selection process based on melting point, solubility, stability, wettability, and other properties has been proposed (12). However, in the absence of clear go/no-go decisions at any particular stage of the process, this approach would lead to the generation of extensive physicochemical data on all salt forms produced. A more rational approach to expedite the salt selection process using a tiered methodology was reported (21). The tiers were planned so that the least time-consuming experiments were conducted early and the progressively more time-consuming and labor-intensive experiments were conducted later, when fewer salts were in contention. In this way, many different salt forms could be screened with minimal experimental effort. An expanded version of this process will be described in this section.

The first step in salt selection involves preparation of various salts using the methods described above. Preparation methods should be chosen with eventual scale-up in mind whenever possible. Procedures which are feasible for small-scale production may not be practical for large-scale manufacture. On the other hand, it is sometimes desirable to produce salts by the most convenient method in the laboratory, so that the properties of many products can be evaluated quickly.

A number of parameters have been identified that are of primary importance in salt selection, including crystallinity, hygroscopicity, solubility, stability, polymorphism, and process control. A tiered approach for evaluation of these parame-

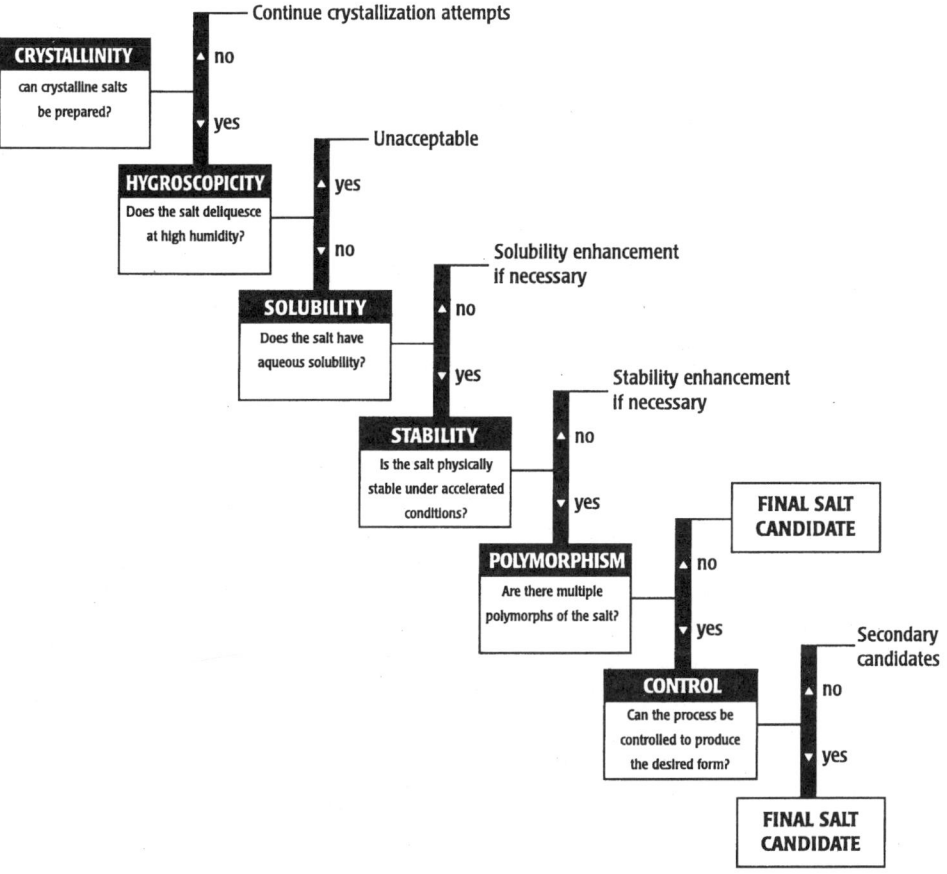

Fig. 18 A salt selection decision tree.

ters is illustrated by the flowchart in Fig. 18. Each step will be discussed in some detail below. It should be noted that the entire process can be carried out using between a few hundred milligrams and a few grams of drug substance.

The first step involves analyzing solid salts for crystallinity and melting point. Crystalline salts are usually the most desirable, because they exhibit superior processing, handling, stability, and purification properties. Low-melting salts may be relegated to a lower-priority status at this point.

In the second step, crystalline materials are evaluated for moisture sorption activity under elevated relative humidity conditions. A high degree of moisture sorption or desorption by salts under humidity conditions expected during manufacturing, handling, and storage may lead to problems. Batch variability in the potency of drug product may also be affected if the bulk drug substance is not maintained at the declared potency because of variations in water content. Based on hygroscopicity, salts that deliquesce or gain/lose excessive amounts of water are considered lower-priority than those that do not. Analyses of these materials after several days of exposure can also provide preliminary data related to hydrate formation.

At the next stage, equilibrium solubilities in the appropriate aqueous media are estimated and the pH values of solutions made in water are usually determined. This information can be used to assess any potential dissolution or bioavailability problems with the salts. These studies can also help determine if a solution dosage form is feasible. The selection of salts at this stage may be aided by the judgment of a drug development scientist, considering the type of dosage form and the expected dose of the compound. A salt with lower solubility that can still provide a good dissolution rate could be selected over one that is highly soluble but prone to crystal form changes. However, if the solubility is not high enough for a required oral or parenteral formulation, another salt with some propensity for crystal form changes at high humidity may be considered. Salts exhibiting appropriate solubilities are taken to the next step.

Physical stabilities are determined under accelerated conditions. Samples of each salt kept under appropriate conditions are periodically analyzed by the appropriate methods to ensure that their crystal forms are sufficiently stable. The appearance of new crystal forms suggests that polymorphic changes might occur during manufacturing or accelerated stability testing of bulk material or a solid dosage form. Determination of the hygroscopicity and solubility of any new forms found may be required. Abbreviated chemical stability, as well as compatibility screening with excipients, can also be monitored at this stage, depending on development timelines and material availability.

Salts that pass to the final stage are tested for their propensity to exist in polymorphic forms using an abbreviated screen, which is discussed in more detail in a later section. Salts which exist in a number of forms will require crystallization method development work to ensure that the manufacturing process is controlled and only the desired crystal form can be obtained reproducibly. Salts that appear to exist, or can be produced consistently, in one stable, crystalline polymorph are considered final salt candidates. As development proceeds and additional drug substance becomes available, these salt candidates can be prepared in larger quantities for comparison of other properties such as dissolution rate and excipient compatibility.

In the above scheme, the number of salt forms available and the physicochemical properties considered important for preparation of bulk drug substance, as well as stability and efficacy of the expected dosage form, will dictate how many steps will be necessary to select an appropriate salt. There may be situations in which all salts that make it to the final level are unacceptable for development. Additional salt forms or free acids/bases should be considered before reevaluating any salt that was dropped earlier in the salt selection. It should also be noted that the acceptance criteria for progression from one step to the next may depend on the physicochemical properties of the available salts. If all salts are found to be hygroscopic, it may be necessary to carry some to the next stage, with the realization that they may require special manufacturing and storage conditions if selected.

A real-world salt selection effort may not always allow strict adherence to the decision tree shown in Fig. 18. Steps may be removed, added, or performed in a different order as required by each specific situation. A multidisciplinary approach to salt selection with coordination and input from a variety of departments (pharmaceutics, chemical development, analytical, etc.) is essential for choosing the best salt for development.

E. Property Modification Using Salts

Many examples are reported in the literature describing the modification of drug substance properties using salts. A small sampling of these is shown in Table 5. Properties ranging from solubility to bitterness have been modified by producing salts.

A number of studies describe preformulation considerations during the salt selection process (10,11,15,22,28) and provide some comparative property data among salts. The integrated salt selection approach for BMS-180431 (Fig. 19) included screening of more than seven salts (sodium, potassium, calcium, zinc, magnesium, arginine, and lysine) in a 4- to 6-week period. Information on crystallinity, moisture content, hygroscopicity, crystal-form changes at various humidities, solubility, solid-state stability, and drug excipient interactions were collected on selected salts at various tiers of the salt selection process. The arginine and lysine salts were found to have comparable physicochemical properties. The

Table 5 Properties Modified by Salt Preparation

Property	Drug substance	Indication	Salts investigated	Ref.
Bioavailability	RS-82856	Positive inotropic agent	Chloride Hydrogen sulfate Potassium Sodium	10
	1-(2,3-Dihydro-5-methoxybenzo[b]furan-2-ylmethyl)-4-(o-methoxyphenyl)pipera-zine	Antihypertensive	Hydrochloride Dihydrochloride Disulfate	22
Bitterness	Erythromycin	Antibiotic	Cyclohexylsulfamate Ethyl phosphate Formate Lactate Lauryl sulfamate Lauryl sulfate Monostearyl phosphate Octylsulfamate Phosphate Stearate Stearyl sulfate Sulfamate	23
Hydrate stability	p-Aminosalicylic acid	Tuberculostatic agent	Calcium Magnesium Potassium Sodium	24
Processing	Ketoprofen	Antirheumatic	Sodium	25
Slow release	Albuterol	β_2-Adrenergic receptor	Adipate Stearate Sulfate	26
	9-[2-(Indol-3-yl)ethyl]-1-oxa-3-oxo-4,9-diazaspirol[5,5] undecane	Antihypertensive	Acetate Hydrochloride 3-Hydroxynaphthoate Methacrylic acid Methacrylate copolymer Napsylate	27

Property	Compound	Therapeutic category	Salt form	Ref.
Solid-state stability	Fenoprofen	Nonsteroidal anti-inflammatory, analgesic, and antipyretic	p-Hydroxybenzoate	28
			Sulfate	
			Tartrate	
	Vitamin A	Vitamin	Ammonium	29
			Benzylammonium	
			Calcium	
			Choline	
			Magnesium	
			Potassium	
			Sodium	
	Xilobam	Skeletal muscle relaxant	Acetate	15
			Nicotinate	
			1-Napsylate	
			2-Napsylate	
			Saccarinate	
			Tosylate	
Solubility	Methyl pyridinium-2-aldoxime iodide	Cholinesterase reactivator	Acetate	19
			Chloride	
			Dihydrogen phosphate	
			Fumarate	
			Hydrogen sulfate	
			Lactate	
			Nitrate	
			Tartrate	
	α-(2-Piperidyl)-3,6-bis(trifluoromethyl)-9-phenanthrenemethanol	Antimalarial	Hydrochloride	13
			2-Hydroxyethane-1-sulfonate	
			DL-Lactate	
			L-Lactate	
			Methanesulfonate	
			Sulfate	
	Oxazepam	Anxiolytic	Dihydrochloride	30
	Lorazepam		Hydrochloride	
			Maleate	
			Methanesulfonate	

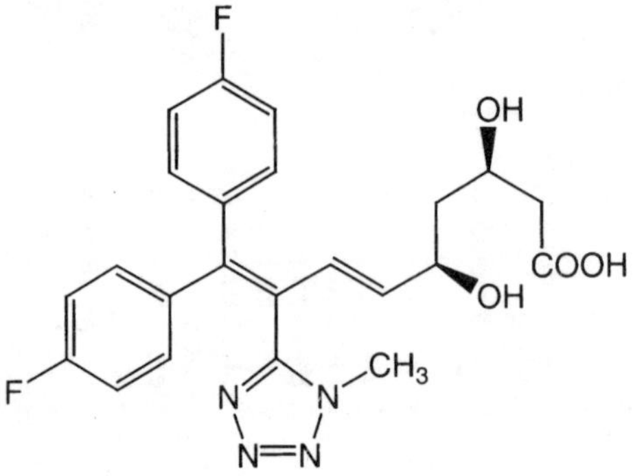

Fig. 19 The structure of BMS-180431.

arginine salt was chosen for development based on factors such as ease of synthesis, ease of analysis, experience with arginine salts, and marketing preferences.

Extensive studies of the various forms of the antiallergic agent nedocromil were reported. The free acid (Fig. 20) (31) as well as the magnesium (32), zinc (33), and calcium (34) salts were made. The commercially available form, nedocromil sodium trihydrate (35), converts to a heptahemihydrate above 80% RH. This situation leads to a possible problem when the drug is delivered by nasal inhaler and the drug substance particles enter the humid environment of the respiratory tract. Investigations of other salts in hopes of finding one that is more stable resulted in the discovery of multiple hydrated crystal forms of various salts, as summarized in Table 6.

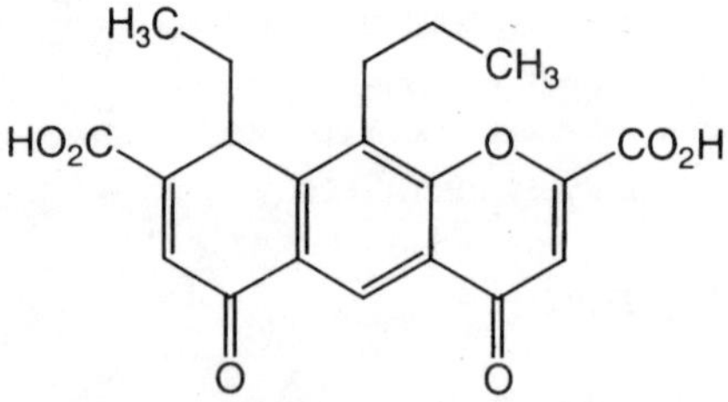

Fig. 20 The structure of nedocromil.

Table 6 Crystal Forms of Nedocromil

Salt form	Crystal forms	Ref.
Free acid	Unsolvated	31
Calcium	Pentahydrate	34
	8/3 Hydrate	
Magnesium	Pentahydrate	32
	Heptahydrate	
	Decahydrate	
Sodium	Trihydrate	35
	Heptahemihydrate	
Zinc	Pentahydrate A	33
	Pentahydrate B	
	Heptahydrate	
	Octahydrate	

IV. SOLID FORM SELECTION

It is obvious from the preceding discussion that salt and solid form selection are intertwined. The propensity of a compound, either neutral or a salt, to exist in different crystal forms is considered as part of the salt selection process. However, once selected for inclusion in drug product, the solid-state properties of a given compound must be evaluated in detail. The following section describes the solid-form selection process as it is carried out with a single chemical entity.

A. Solid Forms

The solid forms attained by organic compounds span a range of molecular order (Fig. 21). At one extreme is the amorphous state, characterized by no regular arrangement of molecules, as in a liquid. At the other is the crystalline state. In a crystal the molecules exist in fixed conformations and are packed against each other in a regular way. However, there are few if any "perfect" crystals. Imperfections in the packing arrangement during growth of a crystal can occur in many ways and, when present in sufficient number, provide a poor-quality crystal. Introduce enough packing dislocations to disrupt every intermolecular interaction, and the amorphous state results. Between amorphous and crystalline forms there can be states of partial order, as in liquid crystals.

Generally, organic molecules prefer to exist in crystalline form when solid. Amorphous material, even when isolable, is thermodynamically less stable than crystalline material. The practical consequence of this is that there is energetic

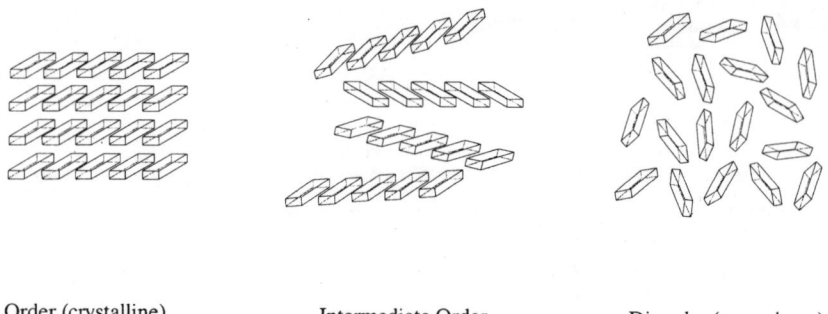

Order (crystalline)   Intermediate Order   Disorder (amorphous)

Fig. 21   An illustration of the concept of molecular order in solids.

pressure for an amorphous solid to crystallize, so selection of an amorphous form of a drug for development must be made with this in mind.

In addition to variations in the relative amount of molecular order in solids, there can also be variations in the nature of the order. Different crystalline arrangements of the same molecule can exist; this phenomenon is known as polymorphism. All types of substances exhibit this behavior, including elements, inorganic compounds, and organic compounds. Different crystal forms of elements are called allotropes, while different crystal forms of inorganic or organic compounds are called polymorphs. The propensity for polymorphism in organic compounds is great. In a crystalline organic solid, the forces holding the crystal together, the intermolecular bonds, are much weaker than those holding each molecule together, the intramolecular bonds. Typical intermolecular attractions consist of van der Waals and hydrogen bonds, which range in energy from <0.1 to 8 kcal/mole. Intramolecular covalent bonds range from about 50 to 200 kcal/mole. Thus, only small energy changes need be associated with changes in packing arrangements.

In many cases organic compounds incorporate water or solvents into their crystal lattice. These species are called hydrates or solvates, respectively. Crystals of this type are not strictly polymorphic, which is, by definition, different crystalline arrangements of a single substance. Practically, however, hydrates and solvates exhibit the same range of property differences as do polymorphs and must be considered as viable candidates in the form selection process.

In hydrates and solvates, the amount of water or solvent incorporation can vary. Often, stoichiometric amounts are found, but not always at a 1:1 ratio of water (solvent) to organic molecule. Some common hydrate ratios are shown in Table 7. Many compounds form hydrates or solvates, and examples are given

Table 7 Common Hydrates

Ratio of organic:water molecules	Hydrate type
2:1	Hemihydrate
2:3	Sesquihydrate
1:1	Monohydrate
1:2	Dihydrate
1:3	Trihydrate
1:4	Tetrahydrate
1:5	Pentahydrate

later in this chapter. An unusually complex crystal is the antibiotic doxycycline hydrochloride hyclate, which is a hemiethanolate hemihydrate (Fig. 22) (36).

Crystallographically, polymorphs differ from each other in packing arrangement and, at times, in molecular conformation. The antibiotic nitrofurantoin (Fig. 23) exists in two polymorphic forms, denoted α and β (37). In each polymorph the molecule adopts a planar conformation and forms extended sheets which are stacked to make a crystal. However, the hydrogen-bonding interactions

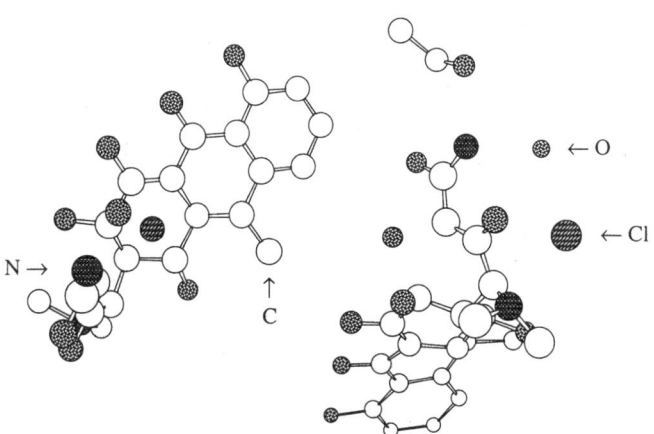

Fig. 22 The asymmetric unit of doxycycline hyclate, showing two water molecules, one ethanol molecule, and two independent conformations of doxycycline. Hydrogen atoms are omitted for clarity (36).

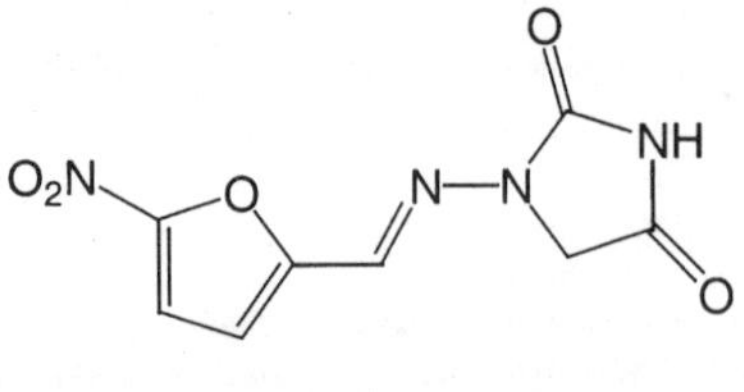

Fig. 23   The structure of nitrofurantoin.

that hold each sheet together are different (Fig. 24). Some of the data obtained from single-crystal X-ray structure determinations of both forms are shown in Table 8.

A compound whose polymorphic forms differ in both packing arrangement and molecular conformation is 5-methyl-2-[(2-nitrophenyl)amino]-3-thiophene-carbonitrile (Fig. 25) (38). This system is striking in that the conformations found in each of three polymorphic forms can be correlated to the color of the crystals

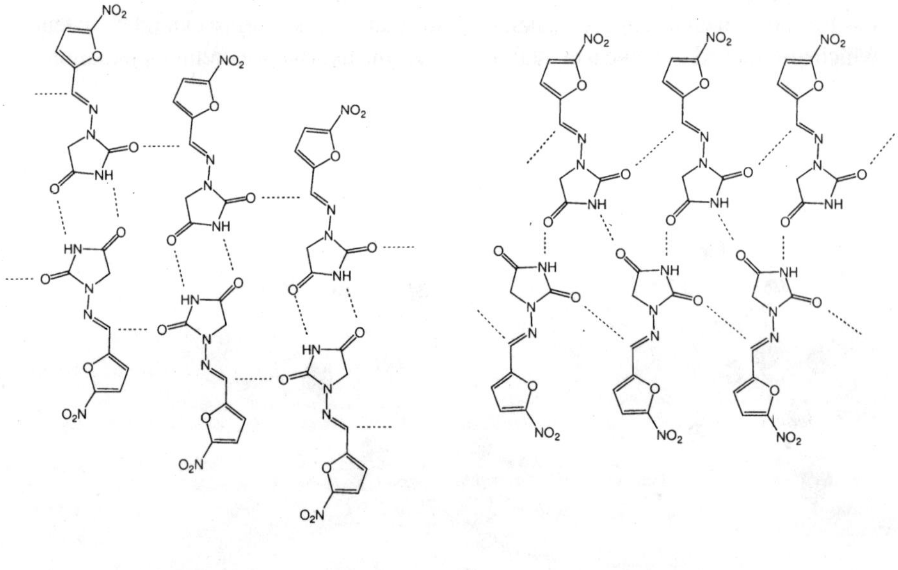

alpha form   beta form

Fig. 24   Depiction of the different hydrogen bonding patterns which hold molecular sheets together in the α- and β-polymorphs of nitrofurantoin. Hydrogen bonds are shown by dotted lines (37).

Table 8 Crystallographic Data for the α- and β-Polymorphs of Nitrofurantoin (37)

Form	Space group	Unit cell lengths (Å)			Unit cell angles (deg)			Z^a
		a	b	c	α	β	γ	
α	P1 bar	6.774(1)	7.795(1)	9.803(2)	106.68(1)	104.09(2)	92.29(1)	2
β	P2$_1$/n	7.840(5)	6.486(1)	18.911(6)	90	93.17(3)	90	4

[a] Z is the number of molecules in the unit cell.

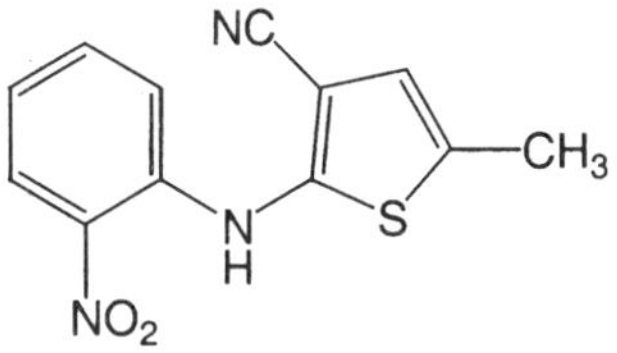

Fig. 25 The structure of 5-methyl-2-[(2-nitrophenyl)amino]-3-thiophenecarbonitrile.

(red, orange, and yellow). The packing arrangements in these forms differ, and the conformations of individual molecules in each form differ also. Rotations around the single bonds joining the aromatic rings result in more or less overlap of the pi electrons in these groups (Table 9, Fig. 26). In the red crystal the conformation attained provides the greatest amount of co-planarity, allowing for maximum pi overlap. The conformation in the yellow crystal is the least co-planar.

In some cases molecules adopt more than one conformation in a single crystalline arrangement. When this occurs, each conformer is part of the regular array from which the crystal is built. An interesting example of this phenomenon is the reverse transcriptase inhibitor lamivudine (Fig. 27). Two crystal forms of this material are known (39). Polymorphic form II is unspectacular, characterized by a highly symmetrical tetragonal lattice containing only one conformer of lami-

Table 9 Torsional Angles Found for Three Polymorphs of 5-Methyl-2-[(2-nitrophenyl)amino]-3-thiophenecarbonitrile (38)

Polymorph	Space group	Angle between aromatic rings (deg)
Red	P1 bar	46
Orange	P2$_1$/c	54
Yellow	P2$_1$/n	106

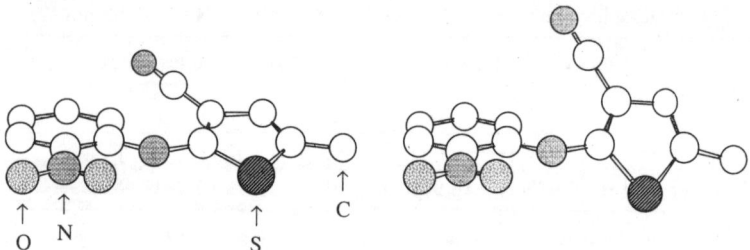

Fig. 26 Conformations of 5-methyl-2-[(2-nitrophenyl)amino]-3-thiophenecarbonitrile having more (left) and less (right) aromatic ring coplanarity. Hydrogen atoms are omitted for clarity.

vudine. Form I, on the other hand, is quite unusual in that it contains five different lamivudine conformers and one water molecule in the asymmetric unit (Fig. 28). Spectroscopic characterization of such crystals can be complicated, as the different conformations can give rise to different signals upon solid-state analysis. The solid-state ^{13}C-NMR spectrum of form II lamivudine exhibits resolved singlets for each of the eight carbon atoms, but the corresponding spectrum of form I consists of complex multiplets (39). This NMR feature, often referred to as crystallographic splitting, results from the fact that each conformer provides different, fixed local environments for each carbon atom in the molecule. Crystallographic splitting is evident in the solid-state NMR spectrum of a crystalline material which contains two independent conformations in the asymmetric unit, as shown in Fig. 29.

It is common to find multiple solid forms of a single organic compound. For example, the androgen dehydroepiandrosterone (DHEA, Fig. 30) exists in at least seven solid forms. Three polymorphic forms, three hydrates, and a methanol solvate were made and characterized (40). Single-crystal structure determinations were carried out on forms I, S1, S3, and S4 (41,42). Structural features of the various forms are compared in Table 10 and Fig. 31.

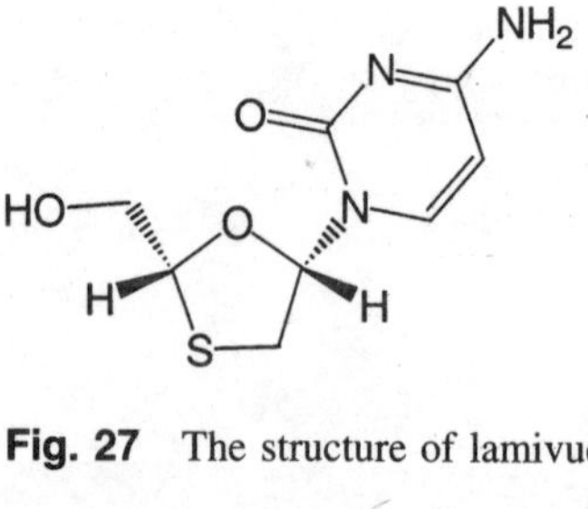

Fig. 27 The structure of lamivudine.

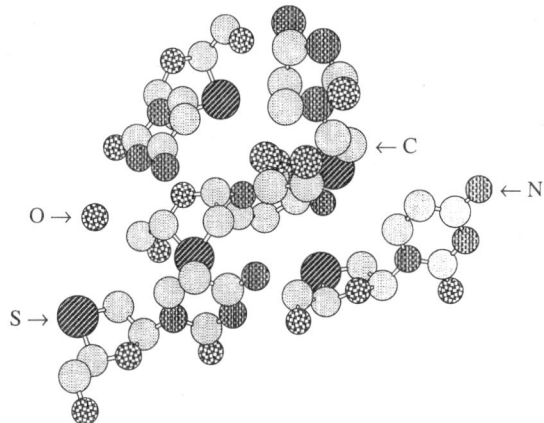

Fig. 28 The asymmetric unit of lamivudine form I, showing one water molecule and five independent conformations of lamivudine. Hydrogen atoms are omitted for clarity.

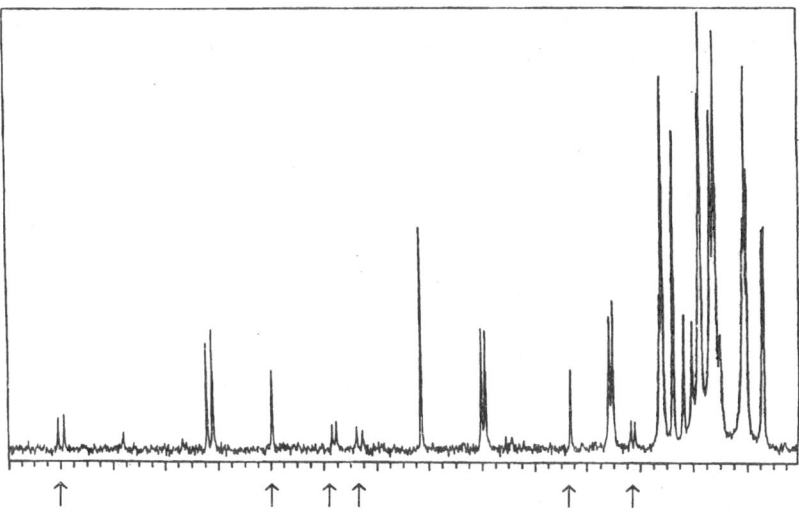

Fig. 29 The solid-state NMR spectrum of a crystalline compound having two conformations in the asymmetric unit. Note the splitting of many peaks into doublets. Arrows denote spinning side bands.

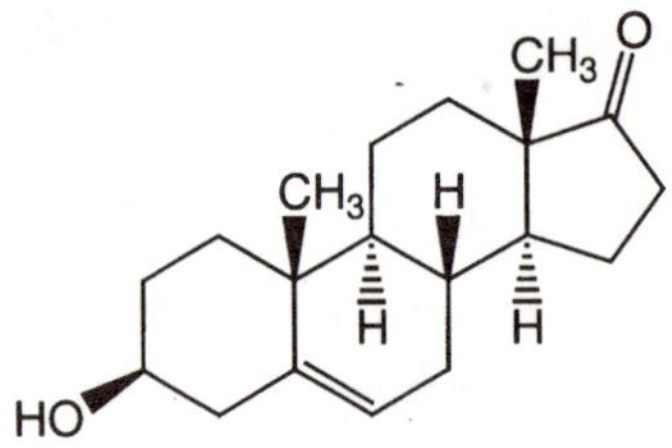

Fig. 30   The structure of DHEA.

So far we have concentrated on the crystalline state, but it is important to note that amorphous materials may also exist in various forms. An excellent review of the amorphous state, including a discussion of polymorphism, was published by Hancock and Zografi (43). It is well known that an amorphous substance behaves like a glass below and a rubber above its glass transition temperature. In addition, polyamorphism of glasses may be possible. For example, amorphous permethylated β-cyclodextrin (Fig. 32) was prepared by grinding and rapid cooling of the melt (quenching) (44). Calorimetric analyses of the products revealed that the ground material had twice the enthalpy of relaxation of the quenched material. Relaxation is the process of transformation from a higher-energy (less ordered) state to a lower-energy (more ordered) state. One explanation is that the amorphous materials prepared by different methods are solids having different degrees of order, or polyamorphic solids. In the amorphous condition the order under discussion is short-range, spanning fewer molecules than does the long-range order which is characteristic of a crystal.

More recently, samples of amorphous ursodeoxycholic acid (Fig. 33) were prepared by grinding and quenching (45). The products behaved differently upon exposure to ethanol vapor: the ground material crystallized, while the quenched

Table 10   Solid Forms of DHEA

Form	Type	DHEA : water (solvent) ratio	Number of conformers
I	Polymorph	—	2
II	Polymorph	—	Unknown
III	Polymorph	—	Unknown
S1	Hydrate	4:1	2
S2	Hydrate	1:1	Unknown
S3	Hydrate	1:1	2
S4	Methanolate	2:1	1

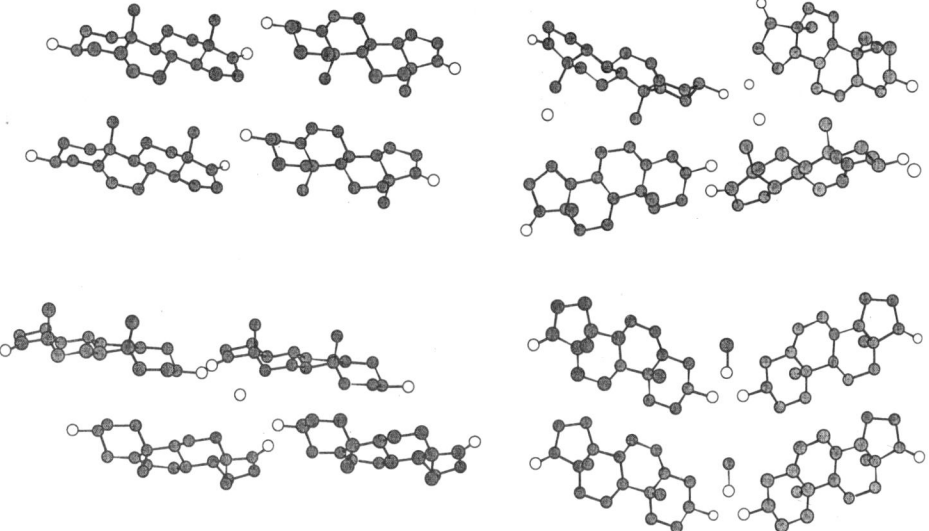

Fig. 31 Four solid forms of DHEA; polymorph I (top left), monohydrate S3 (top right), ¼ hydrate S1 (bottom left), and hemimethanolate S4 (bottom right). Solid circles are carbon atoms and open circles are oxygen atoms; hydrogen atoms are omitted for clarity (41,42).

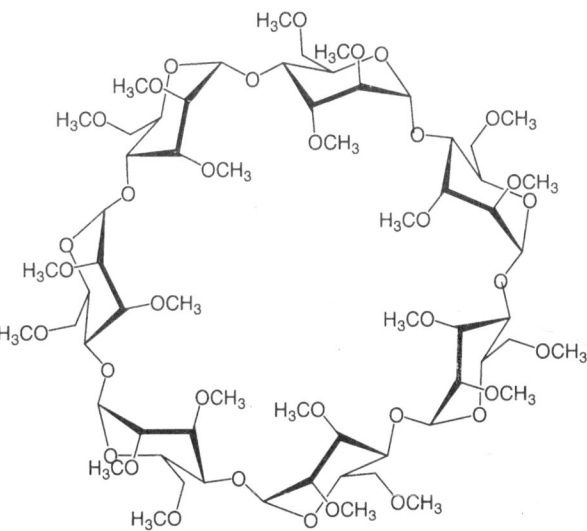

Fig. 32 The structure of permethylated β-cyclodextrin.

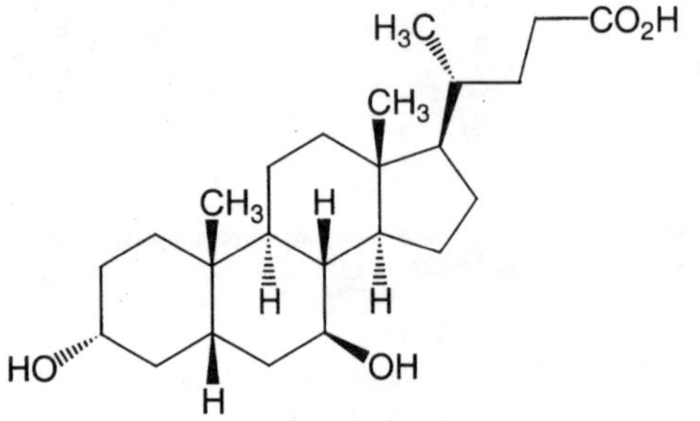

Fig. 33   The structure of ursodeoxycholic acid.

material did not. The authors attributed these behaviors to differences in the molecular states of the products, namely, polyamorphism.

B. Polymorph Screening

Given all the possible structures available to an organic solid, form selection can be a daunting task. The process typically begins once the molecular structure of the active has been selected, but can also accompany salt selection (see above). The first step in solid form selection is to determine if polymorphic and/or amorphous forms of the molecule of interest (drug substance) exist. This process is called polymorph screening (46). Once forms are identified, they must be characterized and their important properties determined. Only with such data in hand can a rational selection of final solid form be made.

Polymorph screening is an empirical process at present. While significant effort is being expended to develop algorithms that calculate crystal structure based on molecular formula, no programs useful for a wide range of compounds are available to date (47). One of the primary goals in a polymorph screen is to prepare as many solid samples of the drug substance under as many different conditions as possible. In this way, maximum opportunity is provided for the drug substance to organize into different forms. A more detailed description of the process follows.

A major consideration for a polymorph screen needs to be addressed at this stage. A polymorph screen, no matter how extensive, cannot guarantee that a new polymorphic form will not appear in the future. Stories of appearing and disappearing polymorphs are plentiful (48). Such occurrences are likely related to seeding. The first step in crystallization is formation of a seed, which involves collection of disordered molecules into an ordered array. Seed formation is the rate-determining step in a crystallization process. Now consider Ostwald's rule, which states that in passing from a less stable state (disordered) to a more stable state (crystalline), the product state is not the *most* stable state available, but is the nearest in energy to the starting state (49). The practical result is that a kinetically favored but metastable crystalline form of a new compound can exist alone for long periods of time (often years) because seeds of a more stable form are not present. However, once a seed of the more stable form arises, crystallization of that stable form can begin and all of the existing metastable material can, and eventually will, be converted to the stable material. Once a given polymorphic form has crystallized, its seeds are everywhere, and it can be difficult to remake original forms.

An example of problems caused by an appearing polymorph may be found in the case of Abbott Laboratories' protease inhibitor ritonavir (Fig. 34) (50). This compound was discovered, developed, manufactured, and marketed over a span of several years, so countless solid samples were generated during that time. One dosage form marketed was soft-gel capsules, which were prepared from crystalline material of the only solid form of drug substance known at that time (form I). After more than two years on the market, batches appeared which failed dissolution testing. These capsules were not released to market, but were studied to determine the cause of the failure. It was found that the capsules contained a new, more stable crystalline form of ritonavir (form II). Form II dissolves more slowly than does form I, and was the cause of the dissolution failures. Eventually, drug substance could no longer be manufactured as pure form I.

The human and financial costs resulting from the appearance of a new polymorphic form of ritonavir are likely to be significant. One group even suggested that there were likely to be shortages and an interruption in supply of the capsules to the estimated 60,000 to 70,000 patients with HIV who were taking

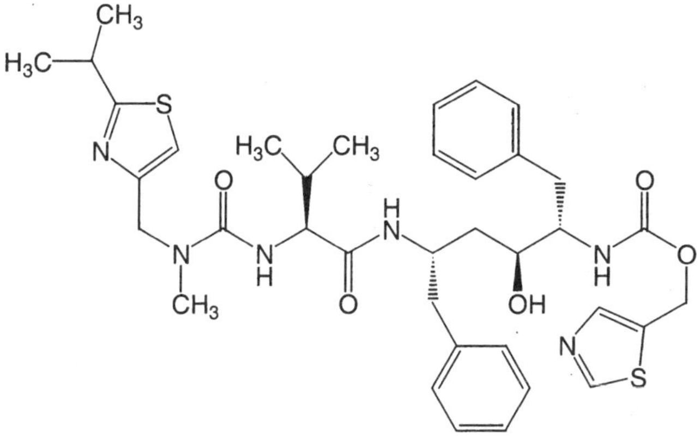

Fig. 34 The structure of ritonavir.

the drug (51). It was also predicted that other protease inhibitors will probably gain market share at the expense of ritonavir, and that the impact on Abbott's bottom line will likely depend on how long it takes to fix the problem (51).

Generation of samples for a polymorph screen can be carried out in many ways. Available methods were recently reviewed (52). A starting point usually involves crystallizations from a variety of solvents. Initial experiments should utilize solvents that provide a wide range of structural type and polarity, as well as solvents that are being or will be used in the drug substance manufacturing process. It is important to include water and/or water-containing solvent mixtures among the solvents selected to encourage formation of hydrates, and it is not wise to exclude solvents that might never be considered for manufacture; it is useful to think of polymorph screening as simply a hunt for seeds. In addition, some effort should be made to generate samples under conditions more favorable to the production of metastable forms.

Solvents often drive formation of particular solid forms. This is obvious in the case of solvates, where the presence of a solvent during crystallization is necessary for its inclusion in the crystal lattice. In some cases different anhydrous, unsolvated solid forms may be obtained at will, depending on the solvent of crystallization. Interactions between solute and solvent can affect the nature of the solute aggregation which leads to seed formation, and thus can control the crystallization process.

The production of various solid forms using the solvent methods described above may be illustrated by an investigation of the polymorphism of the anti-inflammatory sulindac (Fig. 35) (53). The forms were obtained as outlined in Table 11. Nonsolvated polymorph II resulted only from alcoholic solvents, suggesting that hydrogen-bonding associations between sulindac and the solvent

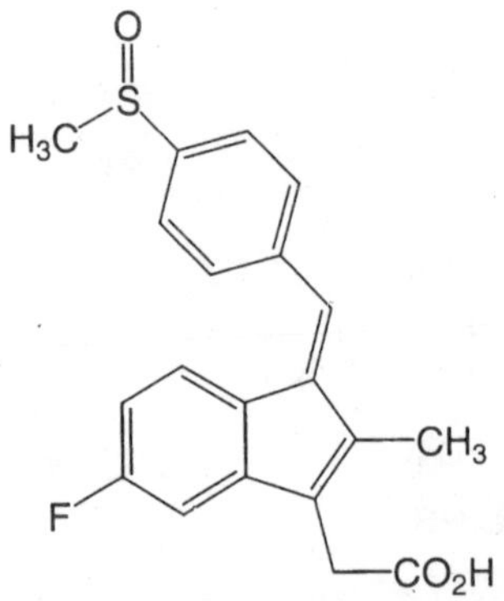

Fig. 35 The structure of sulindac.

Table 11 Production of the Solid Forms of Sulindac

Form	Production method
Polymorph I	Crystallization from chloroform at 5°C
Polymorph II	Crystallization from methanol or ethanol at 5°C
Polymorph III	Crystallization from chloroform at −20°C
Acetone solvate	Crystallization from acetone at 5°C
Chloroform solvate	Crystallization from chloroform at 25°C
Benzene solvate	Crystallization from benzene at 25°C

prompt organization of polymorph II crystals. Either polymorph I, polymorph III, or a chloroform solvate could be obtained from chloroform, depending on the crystallization temperature.

Once samples have been obtained by various crystallization methods, each should be analyzed by XRPD. If amorphous samples are generated, it will be immediately obvious from their XRPD patterns (Fig. 5). Comparison of patterns exhibiting reflections, which are indicative of crystalline material, usually allows organization of samples into groups based on similarities.

Experience in analysis of XRPD patterns is necessary to carry out the organization process. It is important to remember that XRPD patterns which differ in appearance do not necessarily represent different crystalline forms, since the patterns can be greatly affected by sample size and preparation, which cause preferred orientation. When the XRPD patterns have been organized, samples exhibiting each of the pattern types should be analyzed by other methods, such as DSC, TGA, IR spectroscopy, Raman spectroscopy, or solid-state NMR spectroscopy. The data resulting from these analyses can often be used to determine types of solid forms represented by samples exhibiting different XRPD patterns. It is also advisable to carry out additional analyses of several samples within an XRPD group to check for the presence of desolvated solvates. Desolvated solvates arise when loss of crystalline water or solvent is not accompanied by reorganization of the crystal lattice. For example, the antibiotic dirithromycin (Fig. 36) exists in a crystalline arrangement that can harbor six different solvents, ethanol, 1-propanol, 2-propanol, 1-butanol, acetone, and 2-butanone (54). The XRPD patterns of these isomorphous solvates are nearly identical.

The data in Fig. 37 are typical of those in hand at this stage of a polymorph screen. Two samples were generated which exhibit different XRPD patterns, and were thus tentatively classified as pattern A (sample 1) and pattern B (sample 2) material. The patterns could differ because of preferred orientation, as there are no peaks in either pattern which clearly have no counterpart (intensities ignored) in the other pattern. The thermal data in this case are very informative. Sample

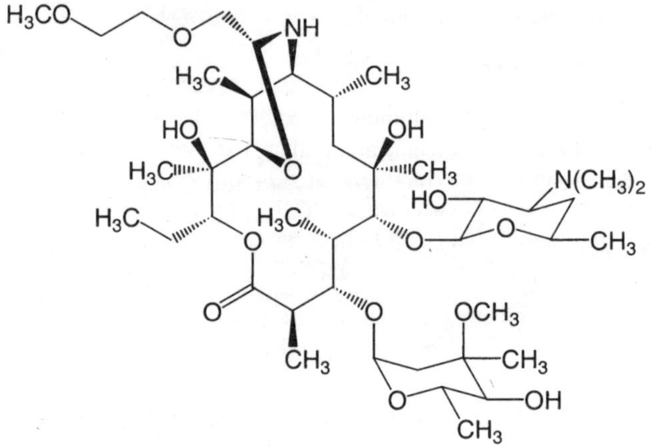

Fig. 36   The structure of dirithromycin.

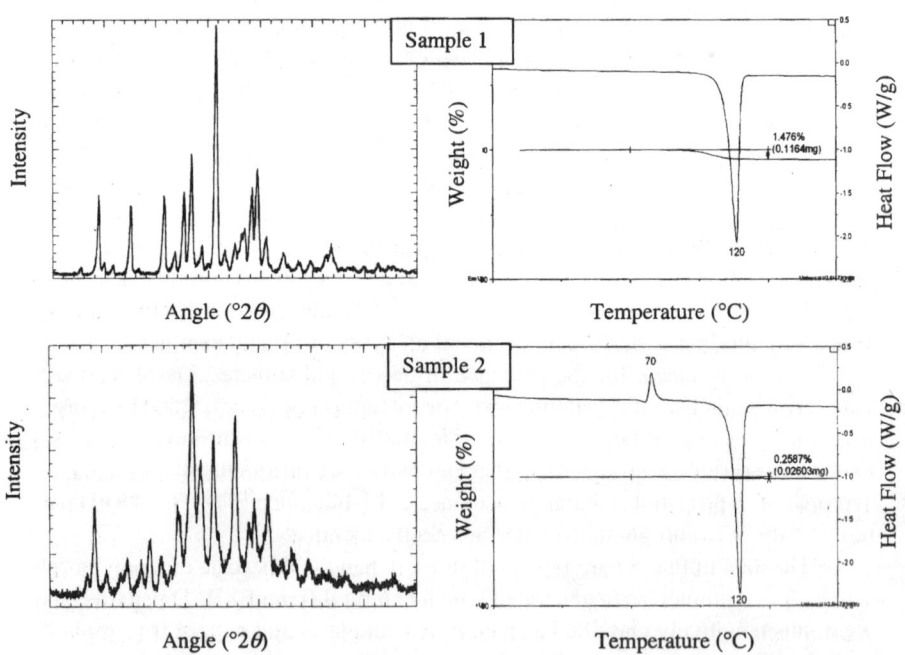

Fig. 37   XRPD, DSC, and TG data for sample 1 (top) and sample 2 (bottom) generated during a polymorph screen.

1 appears to contain volatile material (1.5% weight loss below 125°C by TG) and exhibits what is likely a melt (sharp endothermic event at 120°C by DSC). Sample 2 does not contain significant amounts of volatile components. Upon DSC analysis, sample 2 exhibits a small endothermic event followed immediately by an exothermic event (70°C), and finally a second endothermic event at 120°C. The latter data suggest that sample 2 undergoes a melt and immediate crystallization to a second form, which melts at the same temperature as does sample 1. Although each sample ultimately melts at 120°C, sample 1 loses 1.5% weight concurrent with the melt but sample 2 only loses 0.2%. Calculation of the weight loss expected for a monohydrate of this material (the molecular weight is 400) is 4.3%. Based on the data in Fig. 18, the following hypotheses might be generated.

> Samples 1 and 2 are different solid forms, call them forms I and II, which exhibit XRPD patterns A and B, respectively.
> Form I is a low-order hydrated or solvated form.
> Form II is an anhydrous, solvent-free form.
> A third form (form III) exists which melts at 120°C. The XRPD pattern of form III has yet to be observed.

Alternatively, one might hypothesize the following.

> Samples 1 and 2 are different solid forms, call them forms I and II, which exhibit XRPD patterns A and B, respectively.
> Each form is an anhydrous, solvent-free form which can sorb water or solvent.
> Form II has a higher melting point (120°C) than does form I (slightly below 70°C).

Clearly, more data are needed to distinguish the two possible situations. However, the combination of XRPD and thermal analyses has provided an excellent start, making selection of the next steps a matter of answering specific questions. For example, observation of the XRPD pattern of a sample as it is heated could be used to determine if a third form exists which melts at 120°C.

The next step of the polymorph screening process is to collect additional data which will allow identification and characterization of each solid form. Additional material is usually required. Attempts to scale up the production of each type of material should first involve simply repeating the procedures, at larger scale, that provided each material in the first place. If this is unsuccessful, other methods of production may need to be developed.

Once sufficient material of each type is in hand, many analytical methods can be brought to bear as needed. It is always worth the effort to attempt to grow single crystals which are suitable for X-ray structure determination, as this technique gives an unequivocal picture of the solid form. Waters or solvents of crystallization, as well as the number of independent conformations of the drug substance, are revealed by this method. However, it must be realized that only one crystal is selected for analysis in an X-ray structure determination. The crystallographer typically searches a batch of crystals to find one that has the best optical properties, and that crystal may not be representative of the batch. To understand the relationship of a single-crystal X-ray structure to the samples generated in the polymorph screen, a calculated XRPD pattern is beneficial. Since the information in an XRPD pattern is a subset of the information obtained in a single-crystal study, the XRPD pattern can be calculated from the single-crystal data. A variety of computer programs are available that perform this operation, at varying levels of sophistication. Comparison of a calculated pattern to experimentally determined patterns allows unequivocal form assignment. In Fig. 38 are shown XRPD patterns of two solid forms of a drug substance. Thermal analyses suggested that form I was a hydrate and form II was an anhydrate. A batch of form II material was characterized by XRPD and submitted for single-crystal X-ray analysis, but unexpectedly, the resulting structure contained a molecule of crystalline water. The calculated XRPD pattern clarified the situation: a crystal of form I was chosen by the crystallographer from among a much larger number of form II crystals, because the optical qualities of the former are always better than those of the latter.

It is not always possible to grow single crystals which are suitable for X-ray structure determination. In these cases spectroscopic methods can provide information critical to form characterization. Spectroscopic analysis of samples should be carried out on solid material that has been altered as little as possible. For example, an IR spectrum acquired in the diffuse reflectance mode is preferable to a transmission spectrum for which the sample was prepared as a mull or KBr disk. It is well established that solid-form interconversions can occur under

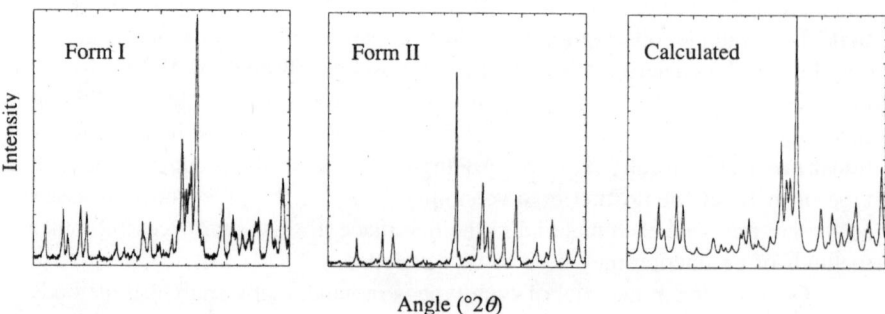

Fig. 38 Experimentally determined XRPD patterns for hydrated form I and anhydrous form II are compared to the pattern calculated from single-crystal X-ray data obtained from a crystal selected from a batch of form II material.

the pressures induced by either of these sample preparation techniques. In our laboratories, vibrational and NMR spectroscopic methods are used routinely. Water or solvents of crystallization are often detectable by either method. In addition, the environments around atoms are locked in a crystal and differ among different crystal forms. Thus, a given bond can vibrate at different frequencies or a given atom can resonate at different frequencies from one solid form to another. Sometimes spectroscopy can differentiate solid forms more clearly than XRPD.

More specialized techniques are often useful in form characterization. For example, TG–infrared spectroscopy or TG–mass spectroscopy combinations allow identification of volatile materials, making hydrate or solvate identification easier. Variable-temperature and variable-humidity sample chambers on XRPD or vibrational spectroscopy instruments provide the ability to watch crystal form changes associated with changing conditions. The decision to use such methods depends on the characteristics of the particular drug substance under study.

At this stage in the polymorph screen, enough data should be available to sort out the number and nature of solid forms obtained from crystallization experiments. As part of the characterization process, and to continue attempts to generate new forms, hot-stage microscopic and moisture sorption/desorption analyses should be carried out.

Hot-stage microscopy is the easiest way to determine if a substance sublimes. It has been estimated that two-thirds of all organic compounds sublime (7), and sublimation is a viable method for sample production in polymorph screening (52). Polymorphic forms may be missed in a typical screen without resorting to HSM. For example, a sample obtained by crystallization from solvent exhibited two endothermic events (at 160 and 180°C) by DSC analysis (Fig. 39). On initial HSM examination of bulk crystalline material no change was observed at 160°C, but the event at 180°C was shown to be melting. Cooling of the HSM sample afforded a crystalline film which, reheated on the hot stage, revealed a solid–solid transition at about 160°C. The transition was observed as a change in birefringence colors under crossed polarizers, and was found to be spontaneous and rapid in both directions. Variable-temperature XRPD analysis was then used to confirm that different crystalline forms existed above and below the transition temperature (Fig. 39). Without HSM analysis, the existence of the high-temperature solid form might have gone undetected.

Evaluation of drug substance by moisture sorption/desorption is a convenient method to search for hydrated forms. Formation of a hydrate is often accompanied by a well-defined weight gain, the magnitude of which is indicative of the order of the hydrate (that is, the molar ratio of water to drug substance). Some drug substances form multiple hydrates, all of which can sometimes be identified from the moisture sorption isotherm. For example, the sugar raffinose exists in several hydrated forms. The stepwise conversion of the trihydrate to the pentahydrate is clear from the sorption isotherm at 30°C (Fig. 40) (55). Information about

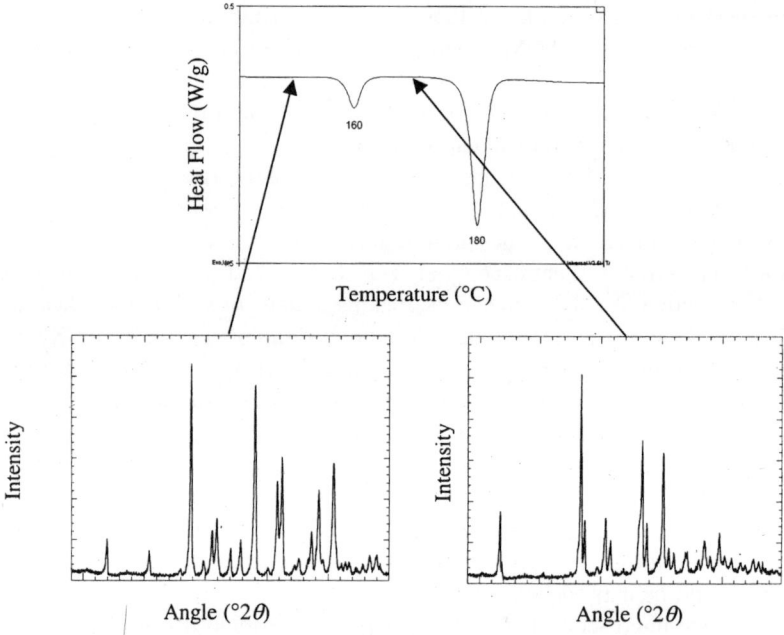

Fig. 39 XRPD patterns obtained before (left) and after (right) a solid-state phase transition (seen as the endothermic peak at 160°C in the DSC trace above).

the stability of both hydrated and anhydrous forms is also derivable from moisture sorption/desorption analysis.

The effects of pressure on a drug substance should be investigated as part of any polymorph screen. Common processing operations, such as milling, can cause solid form transformations. Sometimes grinding can be used to generate amorphous material, as noted above for permethylated β-cyclodextrin (Fig. 32) (44) and ursodeoxycholic acid (Fig. 33) (45). It is also possible to bring about crystalline form changes by grinding, including generation of a metastable form from a stable form in certain cases. In an interesting example, grinding of the antineoplastic cyclophosphamide monohydrate (Fig. 41) results in dehydration. Loss of the water occurs without a significant change in the crystal lattice, affording a metastable, anhydrous crystal form which undergoes a solid-state transformation to a more stable polymorph (56).

Having generated multiple solid samples and analyzed them using multiple techniques, enough data should be available to define the solid behavior of the drug substance. Initial XRPD pattern groups can be reorganized into a set of solid forms, each of which is known to be unique and is well characterized. At

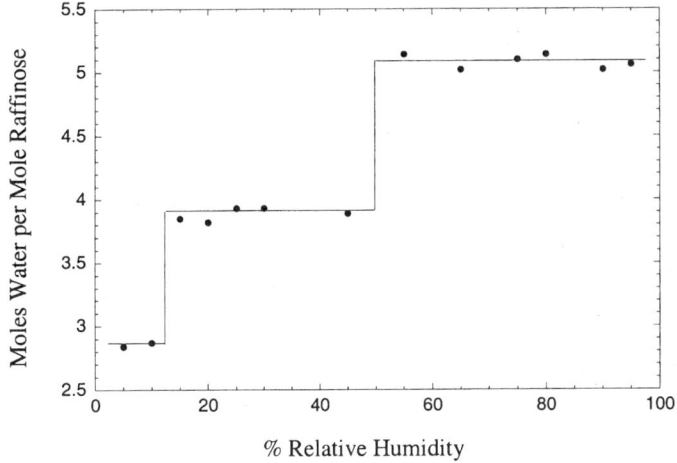

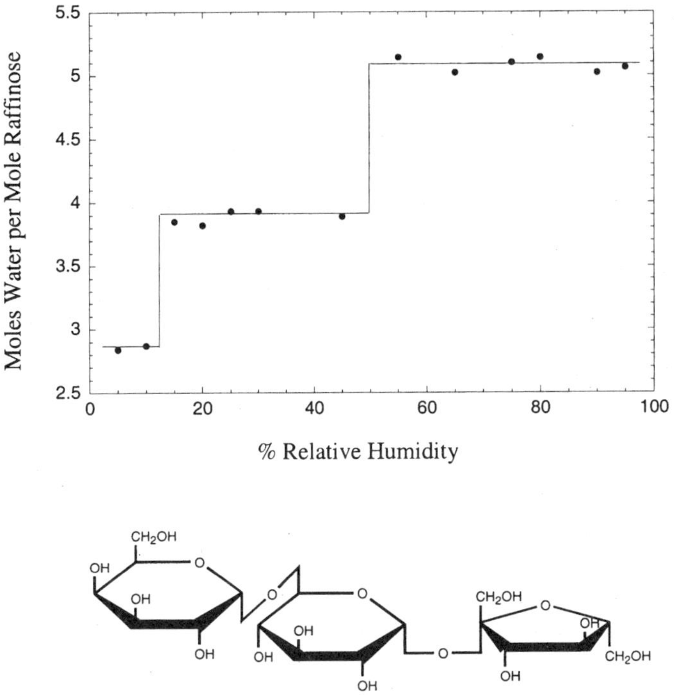

Fig. 40 Moisture sorption isotherm at 30°C showing the conversion of raffinose trihydrate to pentahydrate. This graph was constructed using data from Ref. 22.

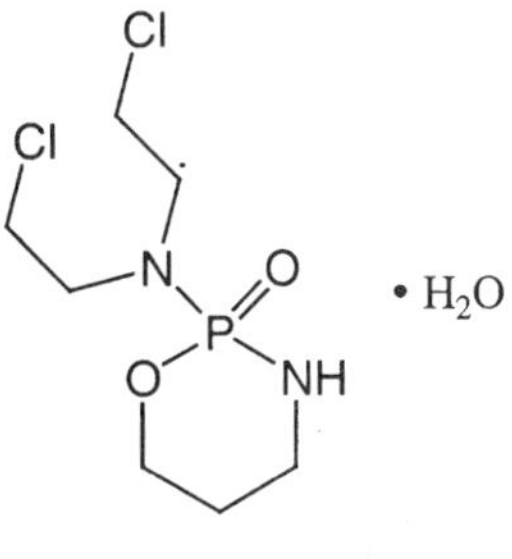

Fig. 41 The structure of cyclophosphamide monohydrate.

this point the screening process is completed, but evaluation of relative properties remains to be done (see below).

Perhaps the most important guidelines to remember in polymorph screening are to (a) generate as many samples as possible and (b) utilize multiple analytical techniques. The latter is critical in our view: be aware that sometimes people who develop an expertise in a single analytical technique tend to oversell its capabilities. Assembling all of the data typically acquired during a polymorph screen into a coherent description of the drug substance is often challenging and, like any endeavor, made more efficient by experience. It must also be stressed again that a polymorph screen, no matter how extensive, cannot guarantee that a new polymorphic form will not appear in the future.

C. Property Evaluation

When the various solid forms exhibited by a drug substance have been found and characterized, important properties of each form should be determined and compared. Some properties, such as melting point and hygroscopicity, will have been obtained in the form characterization process. Investigations of other properties will require directed research efforts. We will limit our discussion to selected properties that reflect internal arrangement (thermodynamic stability, solubility, and dissolution rate), and will not cover bulk properties which may be greatly affected by both solid form and particle morphology (flowability, particle size, etc.).

Knowing the relative thermodynamic stability of drug substance forms is essential to form selection. As in any chemical transformation, the conversion of one solid form to another is dependent on the relative free energies of the forms as well as the energetic barrier to conversion. Under a given set of conditions, the relative thermodynamic stabilities (free energies) of a set of forms is fixed. Any metastable forms will convert, at some rate, to the most stable form. The conversion rates depend on a number of factors and may be so slow as to be negligible, as illustrated by the stability of diamond, which is a metastable form of carbon under ambient conditions. However, choice of a metastable drug substance form for development must be made with the realization that transformation will occur if a low-energy pathway presents itself.

It is advisable to determine which of many solid forms is the most stable. Relative stability orders can be established either qualitatively or quantitatively. The thermodynamic relationship of forms is valid only at the temperature and pressure conditions under which the experiments were carried out. Stability orders may differ under different conditions. Since typical processing and storage conditions are more likely to vary in temperature rather than pressure, it is important to understand the thermodynamic relationships of forms over a range of temperatures.

In general, two energetic relationships are possible between two solid forms of the same drug substance at various temperatures. The first is monotropic, in which the free energies of each form remain constant up to the melting point. The second is enantiotropic, in which there is a reverse in relative stabilities at some temperature below the melting point. The easiest way to visualize these relationships is with energy–temperature diagrams, as shown in Fig. 42. These are plots of energy versus temperature for hypothetical solid forms (1) and (2), in which both enthalpies (H) and free energies (G) are shown. The enthalpic relationship of (1), (2), and the liquid compound remain unchanged from 0 K to the melting points; $\Delta H_1 < \Delta H_2 < \Delta H_L$. However, since the entropic contribution to free energy is temperature-dependent ($\Delta G = \Delta H - T_{Tr} \Delta S$), the free-energy relationships may vary.

In the monotropic system (Fig. 42, upper diagram) the free energy of (1) remains less than the free energy of (2) at all temperatures. Where the free-energy curves cross the free-energy curve of the liquid defines the melting points (mp_1 and mp_2). The heats of fusion are represented by the distance between the solid and liquid enthalpy curves at the melting points (ΔH_{f1} and ΔH_{f2}). Notice in the monotropic system that the more stable solid form (1) has the higher melting point and the higher heat of fusion.

The enantiotropic system is more complex (Fig. 42, lower diagram). The relative free energies of (1) and (2) reverse at some temperature below the melting points, which is called the transition temperature (T_{Tr}). The heat of transition (ΔH_{Tr}) is given by the distance between the enthalpy curves of (1) and (2) at T_{Tr}. Again, the intersection of the free-energy curves of the solids and the liquid define the melting points. In this case, however, (1) has the lower melting point and the higher heat of fusion, since it is the less stable at its melting temperature.

Various types of data may be used to understand energetic relationships. The Burger-Ramburger rules are useful in this regard if the appropriate information is available (57,58). Three of these are summarized in Table 12. Melting points and heats of fusion and transition may be obtained from DSC data if the compound is well behaved (does not decompose at its melting point, for example). Densities are easily derived from structures established by single-crystal X-ray analysis, or may be determined experimentally. Solubility measurements can also provide information necessary to construct an energy–temperature diagram. Relative free energies are reflected in equilibrium solubilities; the more stable solid is the less soluble. A description of the application of thermal, density, and solubility measurements to energy–temperature diagram construction was published (59).

It is important to remember that the energy–temperature relationships discussed above are thermodynamic only. Metastable forms may readily convert to a more stable form in the solid state (Fig. 39) or may never convert on the human

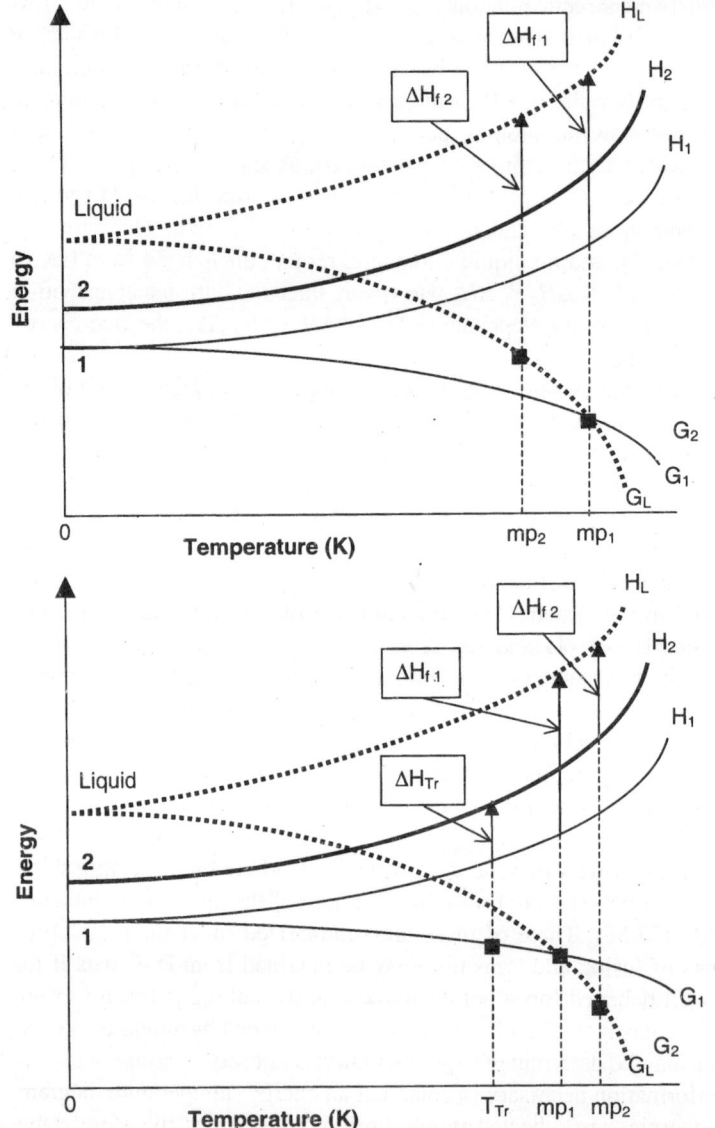

Fig. 42 Energy–temperature diagrams for a monotropic system (top) and an enantiotropic system (bottom).

Table 12   Three Burger-Ramburger Rules (57,58)

Rule	Description
Heat of transition rule	If an endothermic transition is observed at some temperature below the melting point, it may be assumed that there are two forms related enantiotropically. If an exothermic transition is observed below the melting point, it may be assumed that there are two forms related monotropically or the transition temperature is higher.
Heat of fusion rule	If the higher-melting form has the lower heat of fusion, the two forms are usually enantiotropic; otherwise they are monotropic.
Density rule	If one form has a lower density than another, the first may be assumed to be less stable at absolute zero.

time scale (diamond). The mechanisms of polymorphic transformations have been reviewed (60).

Understanding of the free-energy relationships of various solid forms can be important for various reasons. It is sometimes desirable to select a metastable, even amorphous, form of a drug substance for development, as such forms can offer bioavailability advantages. As mentioned above, this approach should be taken only with the realization that transformation to a more stable form will occur if a low-energy pathway presents itself. This pathway could be provided by relatively routine operations such as milling during processing or storage of drug product in a high-humidity environment. Certain formulations are more risky in this regard than others. Suspension formulations, for example, offer the opportunity for slurry interconversions to occur.

Enantiotropic systems whose transition temperatures fall within normal processing temperature ranges must be characterized in order to develop robust crystallization methods. The crystallization of a substance above its transition temperature can afford a form that is metastable under ambient conditions. Regardless of which form is desired, knowing the transition temperature is critical to planning the crystallization.

The properties of solubility and dissolution rate are key to the form selection process. Solubility in this discussion refers to equilibrium solubility, which is important to both process development and formulation activities. The rate of dissolution of drug substance in physiological media often correlates to the rate of attainment of therapeutic blood levels after administration of solid drug product. In such cases the appropriate choice of form is critical to product efficacy.

Equilibrium solubility is most easily determined by agitating a mixture of solid drug substance and solvent until equilibrium is reached, and then measuring the concentration of drug substance in solution. Any number of standard methods can be used to determine the concentration (high-performance liquid chromatography, ultraviolet/visible spectrophotometry, or simply evaporation of the solvent and weighing of the residue, for example).

The application of solubility measurements to crystallization method development is illustrated by the data shown in Fig. 43. Solubility curves were determined for two polymorphic forms of a drug substance, stable form 1 and metastable form 2. As expected, the stable form was the least soluble. A supersaturation limit curve was also determined for the drug substance. Either form could be reproducibly crystallized from a solution of drug substance in water by seeding while in a concentration/temperature regime below the supersaturation limit. For example, consider a 50-mg/mL solution. At 70°C this solution is supersaturated with respect to form 1 but not form 2, so seeding with form 1 and cooling at a slow enough rate to avoid primary nucleation afforded form 1 only. At 50°C a solution at the same concentration is supersaturated with respect to both forms; seeding with form 2, cooling as quickly as possible, and rapid harvest of the crystal crop afforded form 2 only.

Dissolution rates are typically determined by measuring the concentration of drug substance in a dissolution medium at various times. For comparison of solid-form dissolution rates, the experiments should be carried out with excess

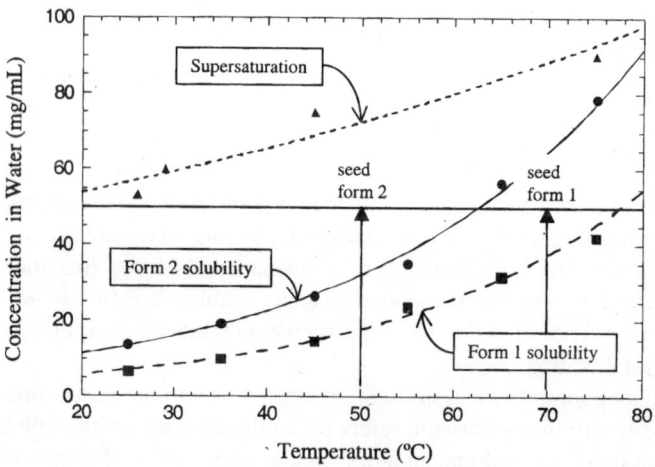

Fig. 43 Equilibrium solubility curves for stable form 1 (lower), metastable form 2 (middle), and the supersaturation curve (upper) for a drug substance in water. Either form could be reproducibly obtained by adding the appropriate seeds to a solution of 50 mg/mL of drug substance in water at the appropriate temperature, 70°C for form 1 and 50°C for form 2. Dissolution rates for these same forms are shown in Fig. 44.

solid of the appropriate form present throughout. If the goal is to relate dissolution rate to bioavailability, the medium should be aqueous-based. However, any medium providing a measurable rate may be used if, for example, relative thermodynamic stabilities are desired.

It is important to realize that particle size can have a dramatic effect on dissolution rates obtained using powders. Smaller particles, having a greater surface area, dissolve faster than larger ones. It is usually an acceptable practice to use powders for rate experiments as long as each sample is sieved to a common particle-size range. An advantage to this method is that dissolution rate and equilibrium solubility information can be obtained in a single experiment by analyzing samples until the concentration reaches a constant value. In some cases, however, different solid forms crystallize to give inherently different particle sizes. In such situations intrinsic dissolution measurements are called for. An intrinsic dissolution experiment involves compression of each sample into a tablet, which is retained in a specialized apparatus such that only one face of the tablet is exposed. The apparatus containing the tablet is then placed in the dissolution medium, allowing control of the exposed surface area of each sample. In this way dissolution rates of different solid forms can be obtained which are comparable without consideration of particle size effects. Remember when using the intrinsic dissolution method that crystal form changes can occur under pressure, so preface the dissolution experiment with analyses of solid before and after tablet formation.

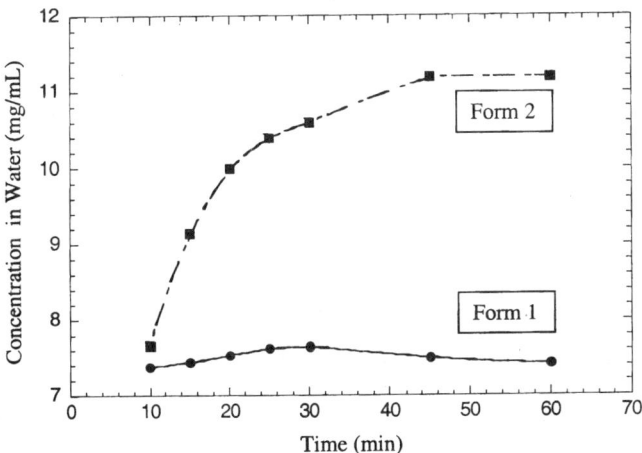

Fig. 44 Dissolution rate curves for stable form 1 and metastable form 2 for a drug substance in water at approximately 25°C. Equilibrium solubilities for these same forms are shown in Fig. 43.

An example of typical dissolution rate results is shown in Fig. 44. Stable form 1 and metastable form 2 are the same as those whose solubilities are plotted in Fig. 43. The data shown in Fig. 44 were obtained using sieved powders at approximately 25°C. Each form dissolves rapidly to a concentration just over 7 mg/mL, which is the equilibrium solubility of form 1. Form 2 continues to dissolve until it reaches its equilibrium concentration of just over 11 mg/mL. Based on the data in Figs. 43 and 44, one might decide to develop metastable form 2 into drug product because it is obtainable at will and it rapidly dissolves to give a higher concentration (compared to form 1) of drug substance in water. Of course, form 2 would have to be kinetically stable relative to conversion to the thermodynamically favored form 1 during processing and storage in drug product.

V. CONCLUSIONS

Selection of the appropriate solid form of a drug substance is critical to the stability and efficacy of the final drug product. Investigations necessary to make a rational solid form selection should be carried out as early in the drug development process as possible in order to (a) provide critical information to manufacturing and formulation development activities and (b) avoid having to change drug substance form late in the drug development process.

Drug substance solid form selection can be approached in a rational manner. The process consists primarily of salt selection and polymorph screening, both of which involve preparation and property evaluation of many samples. It is critical that multiple analytical techniques be employed during this work. The amount of information that needs to be collected is large, and evaluation of it requires experience and flexibility.

ACKNOWLEDGMENTS

The authors would like to thank Angela Thomas and Mark Andres for their help in preparing this manuscript.

REFERENCES

1. S Byrn, R Pfeiffer, M Ganey, C Hoiberg, G Poochikan. Pharmaceutical Solids: A Strategic Approach to Regulatory Considerations. Pharm Res 12(7):945, 1995.
2. HG Brittain, SJ Bogdanowich, DE Bugay, JD DeVincentis, G Lewen, AW Newman. Physical characterization of pharmaceutical solids. Pharm Res 8(8):963, 1991.

3. TL Threlfall. Analysis of organic polymorphs, a review. Analyst 120:2435, 1995.

4. KM Morris. Hydrates. In: J Swarbrick, JC Boylan, eds. Encyclopedia of Pharmaceutical Technology. New York: Marcel Dekker, 1993, vol. 7, p. 393.

5. HG Brittain, ed. Physical Characterization of Pharmaceutical Solids. New York: Marcel Dekker, 1995.

6. SR Byrn, RR Pfeiffer, JG Stowell. Solid-State Chemistry Drugs. West Lafayette, IN: SSCI, 1999.

7. M Kuhnert-Brandstätter. Thermomicroscopy in the Analysis of Pharmaceuticals. Oxford: Pergamon Press, 1971.

8. M Kuhnert-Brandstätter. Thermomicroscopy of organic compounds. In: G Svehla, ed. Comprehensive Analytical Chemistry. Amsterdam: Elsevier, 1982, vol. XVI, p. 329.

9. SM Berge, LD Bighely, DC Monkhouse. Pharmaceutical salts. J Pharm Sci 66(1): 1, 1977.

10. L Gu, O Huynh, A Beccker, S Peters, H Nguyen, N Chu. Preformulation selection of a proper salt for a weak acid-base (RS-82856)—A new positive inotropic agent. Drug Dev Ind Pharm 13(3):437, 1987.

11. ATM Serajuddin, PC Sheen, D Mufson, DF Bernstein, MA Augustine. Performulation study of a poorly water-soluble drug, α-pentyl-3-(2-quinolinylmethoxy)benzenemethanol: Selection of the base for dosage form design. J Pharm Sci 75(5):492, 1986.

12. PL Gould. Salt selection for basic drugs. Int J Pharm 33:201, 1986.

13. S Agharkar, S Lindenbaum, T Higuchi. Enhancement of solubility of drug salts by hydrophilic counterions: Properties of organic salts of an antimalarial drug. J Pharm Sci 65(5):747, 1976.

14. N Senior. Some observations on the formulation and properties of chlorhexidine. J Soc Cosmet Chem 24:259, 1973.

15. WD Walking, BE Reynolds, BJ Fegely, CA Janicki. Xilobam: Effect of salt form on pharmaceutical properties. Drug Dev Ind Pharm 9(5):809, 1983.

16. HM Koehler, JJ Hefferren. Mineral salts of lidocaine. J Pharm Sci 53(9):1126, 1964.

17. S Miyazaki, M Oshiba, T Nadai. Unusual solubility and dissolution behavior of pharmaceutical hydrochloride salts in chloride-containing media. Int J Pharm 6:77, 1980.

18. S Miyazaki, M Oshiba, T Nadai. Precaution on use of hydrochloride salts in pharmaceutical formulation. J Pharm Sci 70(6):594, 1981.

19. AA Kondritzer, RI Ellin, LJ Edberg. Investigation of methyl pyridinium-2-aldoxime salts. J Pharm Sci 50(2):109, 1961.

20. LW Dittert, T Higuchi, DR Reese. Phase solubility technique in studying the formation of complex salts of triamterene. J Pharm Sci 53(11):1325, 1964.

21. KR Morris, MG Fakes, AB Thakur, AW Newman, AK Singh, JJ Venit, CJ Spagnuolo, ATM Serajuddin. An integrated approach to the salt selection of optimal salt form for a new drug candidate. Int J Pharm 105:209, 1994.

22. S-L Lin, L Lachman, CJ Swartz, CF Huebner. Preformulation investigation I: Relation of salt forms and biological activity of an experimental antihypertensive. J Pharm Sci 61(9):1418, 1972.

23. PH Jones, EK Rowley, AL Weiss, DL Bishop, AHC Chun. Insoluble erythromycin salts. J Pharm Sci 58(3):337, 1969.

24. RT Forbes, P York, V Fawcett, L Shields. Physicochemical properties of salts of *p*-aminosalicylic acid. I. Correlation of crystal structure and hydrate stability. Pharm Res 9(11):1428, 1992.

25. G Hildebrand, CC Muller-Goymann. Ketoprofen sodium: Preparation and its formation of mixed crystals with ketoprofen. J Pharm Sci 86(7):855, 1997.

26. RN Jashnani, RN Dalby, PR Byron. Preparation, characterization, and dissolution kinetics of two novel albuterol salts. J Pharm Sci 82(6):613, 1993.

27. EJ Benjamin, L-H Lin. Preparation and *in vitro* evaluation of salts of an antihypertensive agent to obtain slow release. Drug Dev Ind Pharm 11(4):771, 1985.

28. CA Hirsch, RJ Messenger, JL Brannon. Fenoprofen: Drug form selection and preformulation stability studies. J Pharm Sci 67(2):231, 1978.

29. K Guillory, T Higuchi. Solid state stability of some crystalline vitamin A compounds. J Pharm Sci 51(2):100, 1962.

30. A Nudelman, RJ McCaully, SC Bell. Water-soluble derivatives of 3-oxy-substituted 1,4-benzodiazepines. J Pharm Sci 63(12):1880, 1974.

31. H-K Chan, I Gonda. Physicochemical characterization of a new respirable form of nedocromil. J Pharm Sci 84(6):692, 1995.

32. H Zhu, RK Khankari, BE Padden, EJ Munson, WB Gleason, DJW Grant. Physicochemical characterization of nedocromil bivalent metal salt hydrates. 1. Nedocromil magnesium. J Pharm Sci 85(10):1026, 1996.

33. H Zhu, BE Padden, EJ Munson, DJW Grant. Physicochemical characterization of nedocromil bivalent metal salt hydrates. 1. Nedocromil zinc. J Pharm Sci 86(4):418, 1997.

34. H Zhu, JA Halfen, VG Young Jr, BE Padden, EJ Munson, V Menon, DJW Grant. Physicochemical characterization of nedocromil bivalent metal salt hydrates. 1. Nedocromil calcium. J Pharm Sci 86(12):1439, 1997.

35. R Khankari, DJW Grant. Hydration-dehydration phenomena of nedocromil sodium trihydrate (PT 6195). Pharm Res 9(10):S-163, 1992.

36. JJ Stezowski. Chemical-structural properties of tetracycline antibiotics. 4. Ring A tautomerism involving the protonated amide substituent as observed in the crystal structure of α-6-deoxyoxytetracycline hydrohalides. J Am Chem Soc 99(4):1122, 1977.

37. EW Pienaar, MR Caira, AP Lötter. Polymorphs of nitrofurantoin. Preparation and X-ray crystal structures of two anhydrous forms of nitrofurantoin. J Crystallogr Spectrosc Res 23:785, 1993.

38. GA Stephenson, TB Borchardt, SR Byrn, J Bowyer, CA Bunnell, SV Snorek, L Yu. Conformational and color polymorphism of 5-methyl-2-[(2-nitrophenyl)amino]-3-thiophenecarbonitrile. J Pharm Sci 84:1385, 1995.

39. RK Harris, RR Yeung, RB Lamont, RW Lancaster, SM Lynn, SE Staniforth. "Polymorphism" in a novel anti-viral agent: Lamivudine. J Chem Soc Perkin Trans 2: 2653, 1997.

40. L-C Chang, MR Caira, JK Guillory. Solid state characterization of dehydroepiandrosterone. J Pharm Sci 84:1169, 1995.

41. MR Caira, JK Guillory, L-C Chang. Crystal and molecular structures of three modifications of the androgen dehydroepiandrosterone (DHEA). J Chem Crystallogr 25: 393, 1995.

42. PJ Cox, SM MacManus, BC Gibb, IW Nowell, RA Howie. Structure of 3β-hydroxy-5-androsten-17-one (DHEA) monohydrate. Acta Crystallogr C46:334, 1990.
43. BC Hancock, G Zografi. Characteristics and significance of the amorphous state in pharmaceutical systems. Pharm Sci 86:1, 1997.
44. I Tsukushi, O Yamamuro, H Suga. Heat capacities and glass transitions of ground amorphous solid and liquid-quenched glass of tri-O-methyl-β-cyclodextrin. J Non-Crystall Solids 175:187, 1994.
45. E Yonemochi, Y Inoue, G Buckton, A Moffat, T Oguchi, K Yamamoto. Differences in crystallization behavior between quenched and ground amorphous ursodeoxycholic acid. Pharm Res 16:835, 1999.
46. FDA flow chart 4 (Fig. 2).
47. GP Stahly, SR Byrn. The solid-state structure of chiral organic pharmaceuticals. In: AS Myerson, ed. Molecular Modeling Applications in Crystallization. Cambridge, U.K.: Cambridge University Press, 1999, chap. 6.
48. JD Dunitz, J Bernstein. Disappearing polymorphs. Accts Chem Res 28:193, 1995.
49. W Ostwald. Z Phys Chem 22:306, 1897.
50. SR Chemburkar, J Bauer, K Deming, H Spiwek, K Patel, J Morris, R Henry, S Spanton, W Dziki, W Porter, J Quick, P Bauer, J Donaubauer, BA Narayanan, M Soldani, D Riley, K McFarland. Dealing with the impact of ritonavir polymorphs on the late stages of bulk drug process development. Organic Process Research & Development 4:413, 2000.
51. Marketletter, August 3, 1998.
52. JK Guillory. Generation of polymorphs, hydrates, solvates, and amorphous solids. In HG Brittain, ed. Polymorphism in Pharmaceutical Solids. New York: Marcel Dekker, 1999, chap. 5.
53. MC Tros de Ilarduya, C Martín, MM Goni, MC Martínez-Ohárriz. Polymorphism of sulindac: Isolation and characterization of a new polymorph and three new solvates. J Pharm Sci 86:248, 1997.
54. GA Stephenson, JG Stowell, PH Toma, DE Domnan, GR Greene, SR Byrn. Solid-state analysis of polymorphic, isomorphic, and solvated forms of dirithromycin. J Am Chem Soc 116:5766, 1994.
55. A Saleki-Gerhardt, JG Stowell, SR Byrn, G Zografi. Hydration and dehydration of crystalline and amorphous forms of raffinose. J Pharm Sci 84:318, 1995.
56. J Ketolainen, A Poso, V Viitasaari, J Gynther, J Pirttimäki, E Laine, P Paronen. Changes in solid-state structure of cyclophosphamide monohydrate induced by mechanical treatment and storage. Pharm Res 12:299, 1995.
57. A Burger, R Ramburger. On the polymorphism of pharmaceuticals and other molecular crystals. I. Mikrochim Acta (Wien) II:259, 1979.
58. A Burger, R Ramburger. On the polymorphism of pharmaceuticals and other molecular crystals. II. Mikrochim Acta (Wien) II:273, 1979.
59. A Grunenberg, J-O Henck, HW Siesler. Theoretical derivation and practical application of energy/temperature diagrams as an instrument in preformulation studies of polymorphic drug substances. Int J Pharm 129:147, 1996.
60. WC McCrone. Polymorphism. In: D Fox, MM Labes, A Weissberger, eds. Physics and Chemistry of the Organic Solid State. New York: Interscience, 1965, chap. 8.

2

Preparation of Drug Samples for Analysis

David E. Nadig
The R. W. Johnson Pharmaceutical Research Institute, Raritan, New Jersey

I. INTRODUCTION

In this chapter the topic is the techniques for preparing pharmaceutical samples for analysis. Sample preparation of pharmaceutical products as well as biological samples is discussed. The scope of this chapter includes only sample preparation for chromatographic analysis in which the chromatographic system, not the sample preparation, achieves the ultimate isolation of the analyte. Isolation and purification of relatively pure substances for spectral analysis, e.g., infrared spectroscopy, nuclear magnetic resonance, and mass spectrometry, will be discussed elsewhere. This chapter is organized into three groups of discussions.

The first part of the chapter reviews some types of samples that a pharmaceutical analytical chemist would address. The second part deals with some fundamental chemical principles that are critical to the development of rugged sample preparation methods. The third part contains detailed discussions of specific techniques that a pharmaceutical analytical chemist should have in his or her arsenal. Each discussion will attempt to provide some practical examples of the technique as well as how the technique would fit into the overall analytical strategy.

The most common pharmaceutical analysis is the quantitative measurement of the active ingredient and related compounds in the pharmaceutical product. These determinations require the highest accuracy, precision, and reliability because of the intended use of the data: manufacturing control, stability evaluation, and shelf-life prediction. Determination of drugs and their metabolites in biological samples, generally plasma or urine, is important in elucidation of drug metabolism pathways as well as comparing bioavailability of different formulas.

II. PHARMACEUTICAL SAMPLES

Table 1 lists types of samples that are typically found in the pharmaceutical analytical chemistry lab. Different dosage forms and sample types represent different types of chemical mixtures and offer a different set of sample preparation problems. Pharmaceuticals are necessarily formulated as chemical mixtures to achieve the desired physiological effect. The purposes of inactive ingredients are varied: see Table 2.

The presence of the inactive ingredients is an important consideration when developing sample preparation methods. The product of the sample preparation steps must produce a solution with the following properties:

1. The analyte concentration is in the measurable range of the instrument.
2. The purity complements the chromatographic system.
3. The recovery is quantitative or reproducible from sample to sample.
4. The analyte is stable until the sample is analyzed.
5. The sample is compatible with the chromatographic system.

A. Range

The instrumental method must be developed before the sample preparation method. The concentration of the prepared sample must be within the working concentration range of the instrumental method. This will determine the volume of dilution for the sample; if the sample is too dilute, then the sample must be

Table 1 Types of Pharmaceutical Samples

Sample type	Explanation
Plasma sample	Metabolism or bioavailability
Tablet, capsule	Solid mixture
Packaging material, rubber gasket, bottle	Solid mixture
Transdermal patch, topical gel	Semisolid solution
Elixir	Solution
Aerosol	Solid in gas
Oral suspension, topical lotion	Solid suspended in liquid or
Oral suspension, topical cream	liquid–liquid suspension (emulsion)
Candy lozenge	Solid–Solid suspension

Table 2   Pharmaceutical Excipients

Excipient	Function
Starch	Dilution, improve compressibility characteristics
Antioxidants	Stability
Salts of glycolated starch	Tablet disintegration
Natural or artificial flavors, sweeteners	Improve taste
Gelatin	Contain drug, improve swallowability
Oils	Plasticizer
Carbohydrate or acrylic polymers	Delay or sustain release
Alcohol	Solubility
Magnesium stearate	Granulation flow improvement

concentrated using an appropriate sample preparation technique. For example, trace analysis methods often include a step to evaporate the sample dissolved in an organic solvent, with subsequent reconstitution in a smaller volume.

B. Selectivity

The sample preparation method must not only deliver a measurable amount of sample but the compounds accompanying the analyte must not interfere with the analysis. As an illustration, consider a gravimetric assay in which the detection step has no discriminating power and the sample preparation provides all of the specificity. In contrast, an enzyme assay can be performed on a very complex sample without any sample preparation or isolation of the analyte, because the changes in substrate concentration can be linked directly to the activity of the enzyme.

Similarly, the sample preparation must be complementary to the instrumental method. Chen and Pollack (1) compared four sample preparation techniques for a capillary electrophoresis assay of a peptide in human plasma. Table 3 is a

Table 3   Comparison of Sample Preparation Techniques for Biological Samples (1)

Preparation	Recovery	Clean-up
Ultrafiltration	Poor	Good
Microcentrifuge	Poor	Poor
Protein precipitation (acetonitrile)	Excellent	Poor
Solid-phase extraction	Excellent	Poor

summary of Chen and Pollack's observations of all four sample preparations and is an excellent example of the method development process. None of the investigated techniques alone satisfied both sample clean-up and analyte recovery criteria. The authors decided to combine the two techniques that offered excellent recovery, acetonitrile deproteination followed by solid-phase extraction. This combination offered a complement of selectivity that provided for adequate sample clean-up by removing the compounds that interfered with the instrumental method.

C. Recovery

The recovery of a sample preparation must be assessed, because the recovery determines the accuracy of the analysis. For drug substance and drug product analysis, recovery of 100% is generally required to maintain required levels of accuracy and precision. For biological samples, less than quantitative recovery is generally acceptable if the recovery is reproducible.

D. Stability

After the sample is prepared and is awaiting instrumental analysis, the analyte must be stable for a reasonable amount of time. If necessary, the sample may require one additional step, such as pH adjustment. This is particularly important for trace impurity or degradant methods. If the method is used to report degradants, then the sample must resist degradation so that the method can discriminate between degradation of the test solution and degradation of the test article.

E. Compatibility with the Instrumental Method

The first concern in the selection of the sample preparation solvent is to optimize recovery. However, a secondary consideration is the sample solvent's effect on the analysis. This is true whether the analytical technique is ultraviolet spectroscopy (UV), high-performance liquid chromatography (HPLC), or gas chromatography (GC). The method development sequence can be described as: (a) development of the chromatographic separation, (b) development of the sample preparation method, and then (c) evaluation and optimization of the interaction of the sample preparation with the instrumental method.

For UV analyses, the sample and the calibration standard must be dissolved in the same solvent to eliminate solvent effects on the UV absorbance. The sample and standard must also have similar pH. For many organic compounds, the UV absorbance maximum undergoes major shifts depending on the solvent and pH conditions. When the UV spectrum shifts to longer wavelengths it is termed a bathochromic shift, and likewise when the maximum shifts to shorter wavelengths it is termed a hypsochromic shift.

This solvent effect has also been attributed to errors in HPLC analysis, in which the sample solvent used for injection can alter the analyte's absorptivity enough to alter the accuracy of the analysis (2), although there is disagreement over the mechanism of this error (3). This problem can occur only if the slug of injection solvent has not dissipated prior to the analyte reaching the UV detector. Although this can be controlled by using identical solvents for calibration and measurement, it is better to keep the HPLC sample stream consistent throughout the analysis by dissolving the sample in a mobile phase or by designing a long enough HPLC separation so that the analyte is dissolved in the mobile phase when it reaches the detector. For reasons of solubility or stability, it sometimes is not possible to use the mobile phase as the injection solvent.

A larger concern in selection of HPLC injection solvent is chromatographic effects. Distortion of chromatographic peaks caused by the sample solvent is a well-known phenomenon in most method development laboratories and has been reported extensively (2,4,8,9). It is easily demonstrated by injection of a large volume of analyte dissolved in a strong solvent, e.g., acetonitrile for a reversed-phase system. Unless the mobile phase itself is almost all organic in composition, the analyte peak shape will be sharper when dissolved in the mobile phase than when dissolved in the strong solvent. The same effect is observed in a normal-phase system when the injection solvent is too strong (4). This effect results in changes in retention volume as well as in peak distortion such as tailing, fronting, and splitting (5,6).

Ng and Ng (4) have designed a computer simulation to model this HPLC chromatographic effect. According to this model, the sample enters the column as "slug" or "slice" and peak shapes can be calculated from retention data of the solute in the mobile phase and retention data of the solute if the sample solvent is used as mobile phase. If there is a large difference in solvent strength between the mobile phase and the sample solvent, then the peak will be asymmetrical. This simulation also proposes two other significant factors: injection volume and retention volume. A larger injection volume provides for longer slices of injection solvent, thus the solute will spend more time in the sample solvent before it exchanges to the mobile phase. If the retention volume is small, the solute will spend less time in the mobile phase relative to the time it spends in the sample volume. Dolan recommended that HPLC injections be made with sample solvents of mobile phase or a solvent that is one-half that of the mobile-phase solvent strength (7). Use of a sample solvent that is weaker is generally acceptable because the sample will lag behind only until the stronger mobile phase behind the solute reaches it, and then the separation will continue as designed.

Published examples include mostly cases where differences in organic solvent concentration and ionic strength (8,9) accounted for HPLC injection effects.

Ionic strength is particularly important for ion-exchange HPLC methods. Other factors affecting mobile phase, such as pH, must also be controlled. If the injected sample overwhelms the buffering capacity, then retention behavior will be affected if it is dependent on pH.

Injection effects may be minimized by dissolving the sample in a weak solvent; however, even in the absence of injection solvent effects, the volume of injection should be optimized depending on the goals and objectives of the assay. Bristow (10) recommends an injection volume less than one-tenth of the peak volume to maintain resolution and injection volume up to one-fourth of the peak volume if sensitivity concerns override resolution concerns. The peak volume is the volume of mobile phase that contains the peak at the detector (product of flow rate in volume per minute and peak width in minutes). If the solute's retention is increased, the peak is broadened, then the injection volume may be increased (10). This is consistant with the model developed by Ng and Ng (4).

Sample solvent effects on gas chromatographic analyses involve a very different set of considerations. For capillary GC, the effect of the injection solvent has been shown to affect injection precision caused by the expansion volume of the vaporized injection solvent. If the solvent evaporates too fast to a volume larger than the injector volume, then the rapid pressure increase in the injector can cause the sample to leak out through the septum, leading to poor injection precision (11,12). Different solvents have different expansion volumes and expansion rates, therefore proper solvent selection can overcome this problem. Other solutions are smaller injection volume, slower injection speed, or lower injector temperature (11).

F. Pharmaceutical Solids

The most reliable solid sample preparation is to completely dissolve the entire sample because the opportunities for recovery loss are minimized. This is possible for the analysis of drug substance samples for potency or impurities. However, drug products such as tablets or capsules are solid mixtures that may be formulated with insoluble excipients such as polymers, silicas, and starches, requiring a liquid–solid sample preparation. The dissolution of a drug in this situation requires that a solution be made in the presence of undissolved solids. At other times it is desired to leave behind solids to prevent interference with the chromatographic analysis. In these situations the resulting sample is not a clear solution, so the solvent must be chosen to maximize recovery. Poor recovery can be the result of undissolved analyte or analyte that is adsorbed or entrapped by the solids. Adsorption can be reduced by adjustment of sample pH or solvent concentration, depending on the nature of the interaction. Analyte entrapment can occur if the test article was manufactured with the matrix in the liquid state or for a residual solvent loosely included in the drug's crystal lattice (13,14). For residual solvent methods, it is best to completely dissolve the matrix to access the analyte

(11). It is often possible to find a solvent which dissolves both the matrix and the analyte; however, if this is not possible, a liquid–liquid extraction subsequent to dissolution of the matrix may be necessary (14).

A typical sample preparation method for a solid dose is to add the sample or a triturated dose to a suitable solvent. Dissolution of the compounds of interest can be accelerated by stirring, shaking, or sonication. These methods allow the precise control of volume and temperature. Soxhlet extractors (see Section V.A) are designed for liquid–solid extraction, but are rarely used.

Many recovery problems such as entrapment or adsorption are not revealed by the typical standard addition experiments because they may be a result of the manufacturing process. To assess the recovery, ruggedness experiments are required in which critical sample preparation parameters such as pH, solvent composition, or extraction time are varied. Alternatively, recovery can be compared to a second method, such as soxhlet extraction.

Automation of sample preparation of pharmaceutical solids has been quite successful. A variety of instrumentation is commercially available from several vendors. The instruments are capable of handling a dosage form, placing it into a vessel, adding solvent, and mechanically homogenizing the sample until the analyte is dissolved. Typically, the instrument can then dilute to the desired concentration. Some models are also capable of transferring the prepared solution to a HPLC injector or an ultraviolet spectrophotometer for measurement. These instruments use accurate balances to control the accuracy of solvent additions and dilutions.

G. Pharmaceutical Liquids

Pharmaceutical liquids require very minimal sample preparation: they can be injected directly, diluted in mobile phase or other suitable diluent, or extracted into an organic solvent. A diluent is chosen to maintain solubility of the analyte as well as compatibility with the chromatographic system. Pharmaceutical suspensions must be pretreated to dissolve the drug prior to analysis. Another liquid dosage form is an emulsion, in which the liquid is suspended in a second immiscible liquid such as polydimethylsilicone. The sample preparation must break the emulsion. In the analysis of a polydimethylsilicone suspension, a liquid–liquid extraction is generally used to break the emulsion and prepare the sample for simethicone measurement.

H. Biological Samples

Biological samples represent a very different set of challenging analytical problems. The concentration of the drug is generally very low, micrograms per milliliter or less, and the matrix is very complex. Additionally, the drug is often bound by protein or even covalently conjugated to form a more water-soluble metabolite. These problems require special techniques, many times in series. The individual techniques are discussed below. Wong has also reviewed many of the techniques used for analysis of biological samples (15).

Many of the more complex sample preparation techniques are used for analysis of biological samples. Liquid–liquid (16) or solid-phase extraction (17) are often used to isolate the compound from the sample, and sometimes it is necessary to apply both techniques in sequence to achieve adequate clean-up (18). Some plasma samples can be analyzed after a quick treatment to remove the protein (67).

III. FUNDAMENTAL THEORIES CONTROLLING PREPARATION TECHNIQUES

Before we begin the discussion of specific sample preparation techniques, it is necessary to review some of the fundamental theories that control these separation techniques (see Table 4). Phase equilibrium theories, phase contact, and countercurrent distributions provide the basis for the extraction techniques, e.g., liquid–liquid extractions as well as the various solid-phase extraction techniques. Solubility theories provide the basis for the preparation and dissolution of solid samples. Finally, understanding of the basic physicochemical theories that control intermolecular interactions is critical for successful development of sample preparation methods.

A. Physicochemical Interactions

Several types of intermolecular physicochemical interactions are critical to the development of sample preparation methods. For the purpose of discussing sample preparation techniques, the interactions can be divided into four types: ionic, dipole–dipole, hydrogen bonding, and hydrophobic interactions.

Ionic interactions are interactions of two charged species of opposite charge. Examples of sample preparation techniques using ionic interactions are ion-exchange or ion-pairing chromatography. The strength of these interactions depends on the ionic concentration of the solutions. As the ionic strength increases, the charge–charge interaction decreases. The presence of competing charged species (high ionic strength) results in weaker attractions of the species of interest. The pH of the solution plays a critical role, since most charged pharmaceutical compounds are weak acids or bases that can be neutralized by the addition of base or acid, respectively.

Dipole–dipole interactions occur when partial charges due to two separate dipoles of opposite charge come together, such as the adsorption of a solute onto the stationary phase in normal-phase chromatography. Hydrogen bonding occurs

Table 4 Summary of Sample Preparation Techniques

Technique	Application	Principles	Requirements
Liquid–liquid extraction	Isolation from liquid sample	Partitioning	Commonly available equipment and reagents
Solid phase extraction	Isolation from liquid sample	Step-gradient liquid chromatography mechanisms	Supplies are commonly available, including automated devices
Column-switching	In-line biological prep	Combines various liquid chromatography selectivities	Complex computer-controlled valve system
Solid-phase microextraction	Isolation from liquid samples, direct immersion or headspace	Partition or adsorption from solution or gas	Inexpensive segments of coated fused silica
Protein precipitation	Protein removal for plasma sample analysis	Decreases protein solubility using chaotropic agents	Commonly available equipment and reagents
Ultrafiltration	Protein removal for plasma sample analysis	Molecular-weight-selective membranes	Uses special filtration and low-speed centrifuge
Dialysis	Protein removal for plasma sample analysis	Molecular-weight-selective membranes combined with osmotic pressure	Dialysis membranes
Drug conjugates hydrolysis	Chemical hydrolysis of drug metabolites	Hydrolysis	Commonly available equipment and reagents
Direct HPLC injection	Analysis of plasma samples with no sample preparation	Achieves molecular-weight selectivity using specially designed bonded silica or maintains protein solubility with surfactants	Some methods require expensive columns
Derivatization	Chemical reaction to improve separation or detection	Treatment of active hydrogen functional group with electrophilic reagent	Small volumes of reactive derivatization reagents

when a hydrogen of one molecule interacts with an electronegative dipole of another molecule or functional group. Hydrogen-bonding interactions behave similarly to dipole interactions. These forces are very dependent on the amount of water present and the polarity of the solvent. The strength is maximized in nonpolar solvents. Dipole–dipole interactions can affect recovery if the analyte adsorbs onto an undissolved solid in a solid–liquid sample preparation.

Hydrophobic interactions are important in liquid–liquid extractions as well as in reversed-phase chromatography. This interaction is driven entropically by the exclusion of water from the nonpolar species: the inclusion of water in a nonpolar phase requires the molecular ordering of the water.

B. Solubility

To prepare solid samples for analysis, information about the solubility of the analyte must be collected. The solubility of a solute is the concentration of the solute when it is at equilibrium with the solid substance, that is, the solution is saturated. For the preparation of a solid drug substance or product, the final drug concentration should be considerably less than the drug's solubility, to assure reliable recovery from day to day.

Many factors affect solubility and must be considered. For example, pH, ionic strength, and temperature can significantly affect solubility. For example, the aqueous solubility of a carboxylic acid can be orders of magnitude higher at a pH above the pK_a than below the pK_a. This is due simply to changes in the polarity of the molecule. Conversely, the solubility of weak bases such as amines is higher when the pH is below the pK_a of the base. For aqueous sample preparations, addition of a water-miscible solvent such as acetonitrile or alcohol can be used to enhance solubility. For example, the solubility of acetaminophen in water is approximately 11 mg/mL, but the solubility is doubled by adding 2% ethanol (19).

For methods to measure trace impurity, the sample can be saturated with respect to the active ingredient because the solubility of the impurity is independent of the solubility of the drug. This technique, sometimes called the ''swish technique,'' allows for the use of small diluent volume and is an excellent technique for analysis of traces impurities such as degradants or residual solvents. However, recovery experiments are critical to demonstrate that the analyte is not adsorbed or entrapped by the solid drug.

C. Phase Equilibrium

Phase equilibrium theory is the fundamental basis for many of the separations techniques used for sample preparation, including liquid–liquid extraction, solid-phase extraction, solid-phase microextraction, and HPLC.

When a solute X is exposed to two immiscible phases, the partition equilibrium can be described by the following equation:

$$X1 : X2$$

and the partition coefficient is defined as follows:

$$K_p = \frac{[X]_1}{[X]_2}$$

where

K_p = partition coefficient
X_1 = concentration of X in phase 1
X_2 = concentration of X in phase 2

The partition coefficient is a very useful concept for understanding phase equilibrium theory and for developing analytical methods; however, partition coefficients are prone to so many contemporaneous variables, they cannot be cataloged and indexed. This becomes clear when we look at the derivation of the partition coefficient equation. For the partitioning process, the free-energy change is described as

$$\Delta G = \Delta G_0 + RT \ln(aX_1) - RT \ln(aX_2)$$

where

ΔG = free-energy change
ΔG_0 = standard free-energy change
R = gas constant
T = temperature
aX_1 = activity of X in phase 1
aX_2 = activity of X in phase 2

at equilibrium, $\Delta G = 0$, so the equation can be simplified:

$$\Delta G_0 = -RT \ln \frac{(ax_1)}{(ax_2)}$$

$$K_p = \frac{(aX_1)}{(aX_2)} = e - \frac{\Delta G}{RT}$$

Thus the constant K_p is determined by the temperature and ΔG_0. Furthermore, ΔG_0 for a liquid–liquid extraction is influenced by many other variables, such as pH, temperature, ionic strength of the aqueous phase, and the volumes of the two phases. Therefore, method development efforts must include identification and control of these critical variables that will influence the extraction. It should also be noted that for very high concentrations of two solvents that are not completely immiscible, the partition model is not valid, although many of the same concepts apply.

IV. VALIDATION

Sample preparation is often the most complex development step in a pharmaceutical analysis. The objective of sample preparation is to reproducibly provide 100% of the analyte in a solution that is ready for analysis. The objective of the validation experiments is to determine if the method is suitable for its intended purpose by running the method and comparing the results to predetermined criteria which will assess accuracy, precision, linearity, specificity, range, and ruggedness. Validation experiments and criteria are periodically suggested by regulatory authorities such as the U.S. Food and Drug Administration, the International Committee on Harmonization and the U.S. Pharmacopeia for drug substance and drug substance methods (20–22), as well as bioanalytical methods (23,24). Sample preparation is a critical part of the analytical method validation.

Validation of accuracy is done by applying the method to samples for which the amount of analyte is known. This can be done using a series of samples where the analyte is added to a pharmaceutical placebo or to blank plasma. Some manufacturing processes, such as wet granulation, can alter the physical form of the drug product, thus affecting sample preparation effectiveness. In this case it is difficult to validate the accuracy of the method using spiked placebo experiments, because recovery of the analyte from a spiked placebo experiment may not be representative of recovery of an analyte from a manufactured product. Conversely, preparing the product using the manufacturing process cannot provide samples with sufficiently reliable drug content to validate a method. Since it is not possible to validate the sample preparation procedure, the method development should include some robustness experiments which test the sample preparation variables. For example, mixing-time studies should be performed in which the time is increased until recovery is maximized.

Validation of bioanalytical methods requires experiments to assess accuracy, precision, limit of detection, limit of quantitation, range, linearity, selectivity, and sample stability (23). The sample preparation step has a great impact on all these parameters. The extraction efficiency must be experimentally determined by comparison of extracted samples to unextracted standards (24,25). The unextracted standards, because of interference, must be diluted in saline or the mobile phase rather than plasma.

V. SPECIFIC SAMPLE PREPARATION TECHNIQUES

A. Soxhlet Extraction

A classical liquid–solid sample preparation technique is Soxhlet extraction. The apparatus consists of a pot of extracting solvent which is refluxed to provide fresh hot solvent to drip through the solid sample. Solubility is maximized by using the clean hot solvent for dissolution. The sample is contained in a porous thimble held between the solvent flask and the reflux condenser. The Soxhlet technique is not often used in routine pharmaceutical sample preparations for two reasons: the high temperature can cause degradation which is unacceptable for stability-indicating assays, and the technique is difficult to perform on multiple samples because of time and space requirements.

B. Liquid–Liquid Extraction

Liquid–liquid extraction is generally reserved for more complex samples because it offers poorer precision than other techniques. It is most commonly used for the preparation of biological samples in which less precise methods can be tolerated. Occasionally, however, an extraction is necessary for the determination of a water-insoluble compound in a water-soluble matrix, such as the analysis of fat-soluble vitamins in tablets or menthol in pharmaceutical lozenges. In these cases, the water-soluble matrix must be treated with water to gain access to the analytes, but the solvent cannot be made sufficiently nonpolar to dissolve the analytes by adding a water-miscible solvent.

As discussed previously, in partitioning systems, there are some critical parameters that must be controlled. Selectivity and recovery can be manipulated by solvent selection (see Table 5). For ionizable drugs, the pH must be controlled to optimize the recovery. Generally, when the drug is in the ionized form, it is more soluble in the aqueous phase. Likewise, when the drug is in the uncharged state, it will generally partition into the organic layer, which is often exploited in sample clean-up steps called back-extractions. After the analyte is transferred to the organic solvent, a back-extraction technique can be used to take the sample back to an aqueous phase by adjusting the pH of the aqueous phase. By using sequential extractions and back-extractions, a high degree of selectivity can be achieved. However, care must be taken to maintain reproducible, high recovery.

To overcome poor extraction efficiencies, organic fraction from sequential extractions can be pooled (26). Liquid extractions are also used for trace enrichment of the analyte by evaporating the organic layer to dryness and then reconstituting into a smaller volume of mobile phase.

The addition of specific enhancers has been used to effect liquid–liquid extractions. Valenta et al. (79) have improved the extraction of phenobarbitol by using specific binding compounds to enhance the partition coefficient. By adding

Table 5 Typical Liquid–Liquid
Extraction Solvents (in order of
polarity)

Solvent	Dielectric constant (78)
Water	78.54
Methyl ethyl ketone	18.5
Isobutyl alcohol	15.8
Methylene chloride	9.8
Ethyl acetate	6.02
Chloroform	4.806
Diethyl ether	4.335
Toluene	2.379
Carbon tetrachloride	2.238
Benzene	2.284
Cyclohexane	2.023
Hexane	1.890

a soluble artificial receptor to the chloroform phase, the chloroform partition coefficient was improved enough to alter the liquid-phase ratios. The chloroform-to-serum sample ratio was reduced to 1:2; without the receptor, a ratio of 10:1 was required to achieve the same extraction efficiency. A more simple case is the use of the cation methylene blue to form an ion pair with alkylsulfonates (27). The ion-pair complex is partitioned into chloroform, where the amount of the alkylsulfonate can be measured spectrophotometrically in the visible spectrum.

A limitation to liquid–liquid extractions is that they are very difficult to automate. However, two approaches have been reported. Hsieh et al. reported the automated liquid–liquid extraction of hydrochlorothiazide from plasma and urine that simply mimics the actions of the analytical chemist (28). They made use of a robotic system that was programmed to combine the sample, internal standard solution, buffer, and solvent. The robot then mixed the sample, transferred the extract to a new tube, evaporated the extract, and injected it onto an HPLC.

A different automation approach, called solvent extraction-flow injection, was originally described by Karlberg and Thelander (29). This is a continuous-flow technique in which the aqueous sample is injected as segments into a stream of organic solvent. The sample flows through a series of mixing coils until the analyte reaches equilibrium. The measurement is then done on the separated organic phase. Lucy and Yeung (30) applied improvements to the separation step and applied it to the assay of dimenhydrinate with a precision of 2.5–4% and a throughput of 12 samples per hour.

C. Solid-Phase Extraction

Solid-phase extraction is a sample preparation technique that has been derived from liquid chromatography technology and has been applied extensively to the analysis of biological samples as well as pharmaceutical products. It is a step gradient technique in which the analyte, dissolved in a weak solvent, is retained on a stationary phase and subsequent additions of various moving phases of increasing solvent strength results in selective and controlled elution of the interferences and analytes.

In its most common mode, a small (about 0.1–1 g) open-bed column of octadecyl-bonded silica is used to retain an analyte from an aqueous sample while allowing interferences to pass through. Interferences are washed from the column using an appropriate solvent mixture such that the analyte remains on column. The analyte is then eluted, typically using a common organic HPLC solvent such as acetonitrile or methanol. The final eluate is then analyzed directly or exposed to further treatment such as evaporation to concentrate the analyte or derivatization. Alternatively, solid-phase extraction can also be used in a reverse mode, that is, to retain (filter) an interference while passing a sample through the column. In this case the solvent strength of the sample must be maximized with respect to the analyte and minimized with respect to the interferences.

Solid-phase extraction is most commonly used with reversed-phase packings, but the technique also works using other packings, such as in normal-phase or ion-exchange methods, or in combination with other techniques such as liquid–liquid extraction (28), to achieve the necessary levels of sample clean-up.

A significant advantage that solid-phase extraction offers over liquid–liquid extractions is the potential for automation of a routine procedure. The steps of adding solvent and samples to a small column are simpler to automate than in liquid–liquid extraction, where mixing of tube contents and removing of solvent at a liquid–liquid interface is required. The transfer of the eluate from a solid-phase extraction to the chromatographic system can be automated (31). There are multiple vendors of off-the-shelf instrumentation for automation of solid-phase extraction. Another advantage of solid-phase extraction is that it generally uses smaller volumes of organic solvents compared to more conventional methods such as liquid–liquid extraction.

The cost of solid-phase extraction can be limited by reusing the columns. Chen et al. investigated solid-phase extraction column reuse and found that the extraction efficiency decreases only after two or three reuses (32).

A related technique, solid disk extraction, is very similar except in the geometry of the solid phase. This technique uses octadecyl-bonded silica imbedded into a network of polymer fibers. The flat disk shape allows for excellent flow properties; thus, most of the applications involve the isolation of trace organic compounds from a large volume of sample. Tang and Ho described the extraction of phenol from a 1-L water sample (33). The use of solid disk extraction requires quick and tight retention of the drug onto the solid phase, because the sample path through the solid phase is very short. The advantage of the short path is that the drug can be removed using a very small volume of elution solvent.

Method development of solid-phase or solid disk methods can be simply accomplished by the following system. Apply a model sample to the solid phase and collect the sample as it passes through the solid phase. Directly inject the collected sample and compare it to an injection of an untreated model sample. The model sample should simulate the sample in terms of pH and solvent strength but generally must be cleaner than the intended sample, such as plasma or urine. If the injection of the collected sample is void of the analyte, then the conditions are suitable. In this way the conditions may be modified to optimize the sample conditions. The same procedure can be carried out to optimize the wash and elution steps.

D. Column-Switching Techniques

Solid-phase extraction is an example of using liquid chromatographic principles offline to prepare a sample for analysis. Similarly, liquid chromatographic principles have been applied in-line with an HPLC system, using a series of columns coupled with multiport liquid valves. Column-switching techniques have been applied extensively to the HPLC analysis of drugs in biological fluids and other trace analysis in complex matrices, allowing the analytical chemist to inject the sample directly, without prior sample preparation.

The technique has many variations, which Koenigbauer and Majors (34) classified into four types: direct transfer, indirect transfer, reversed transfer, and loop transfer. The analyte fraction of the first column eluate can be diverted directly to a second HPLC column (direct transfer), or eluted from column 1 to column 2 with a step gradient to a stronger mobile phase (indirect transfer), or eluted into a loop injector prior to the second column (loop transfer). The combination of the two columns provides sufficient clean-up for a specific analysis. With reversed transfer, the analyte is concentrated on the head of the first column and, after the sample interferences are passed through the first column, the analyte is transferred to the second column by reversing the flow through the first column.

In each case, the solvent strength must be carefully controlled so that the analyte is retained on the expected part of the column. For example, the reversed transfer requires a weak mobile phase in column 1 so that the analyte remains at the head of column 1. For indirect transfer, the stronger eluting mobile phase for column 1 becomes the separating mobile phase for column 2. Direct transfer requires that the same mobile phase be used in columns 1 and 2.

E. Solid-Phase Microextraction

Solid-phase microextraction (SPME) is a technique that was first reported by Louch et al. in 1991 (35). This is a sample preparation technique that has been applied to trace analysis methods such as the analysis of flavor components, residual solvents, pesticides, leaching packaging components, or any other volatile organic compounds. It is limited to gas chromatography methods because the sample must be desorbed by thermal means. A fused silica fiber that was previously coated with a liquid polymer film is exposed to an aqueous sample. After adsorption of the analyte onto the coated fiber is allowed to come to equilibrium, the fiber is withdrawn from the sample and placed directly into the heated injection port of a gas chromatograph. The heat causes desorption of the analyte and other components from the fiber and the mixture is quantitatively or qualitatively analyzed by GC. This preparation technique allows for selective and solventless GC injections. Selectivity and time to equilibration can be altered by changing the characteristics of the film coat.

The use of this technique in the headspace mode was more recently shown by Zhang and Pawliszyn (37) to reduce equilibration time, reportedly because the equilibration of the analyte with a gas phase is about four orders of magnitude faster than equilibration with an aqueous phase. This is due to the faster diffusion rate of a molecule in a gas than in water.

Solid-phase microextraction is controlled by diffusion rates and partition effects. In typical quantitative analyses, for this technique to be reproducible, the extraction process should continue until the partitioning events reach equilibrium and all variables affecting the partitioning must be controlled. For a two-phase system, the extraction is dependent on the analyte's partition coefficient and the volumes of the solid phase and the water. In the headspace technique, equilibrium must be reached between all three phases: the water, the vapor, and the solid phase.

This relatively new technique has found applications in the analysis of drugs in biological fluids using direct immersion (36) as well as headspace (37,38) methods. Headspace methods offer tremendous selectivity advantages for biological fluid methods because only volatile compounds will be extracted. Ishii (38) and co-workers used the headspace technique to measure phencyclidine in whole blood. The addition of base to the sample was made to improve the basic drug's vapor pressure. After exposure of the capillary to the headspace at 90°C, recovery was only about 50% or less but offered excellent linear response. The technique offers advantages of excellent sample clean-up and concentrated adsorption of the drug onto the small capillary, allowing detection to 1 ng/mL, which was a several-fold sensitivity improvement over a solid-phase extraction method (39).

F. Protein Precipitation Methods

HPLC analysis of plasma samples requires the removal of plasma protein. Solvent and solid-phase extraction methods achieve protein removal. Proteins can also be removed by precipitation and centrifugation by adding reagents that reduce the solubility of the proteins. Typical additives are acetonitrile (40), methanol (41), or perchloric acid (42). The resulting solution is then injected directly into the HPLC system. The protein precipitation must also assure that the drug is released by the protein. Many drugs, particularly hydrophobic drugs, are bound by plasma albumin protein by as much as 99% or higher. Although this is not a covalent linkage, the hydrophobic forces are enough to form a very tight bond, in some cases to specific sites on the albumin protein (43). The addition of miscible solvents often allows for the removal of the drug from the protein.

G. Ultrafiltration

Membrane filters can also be used to separate protein from plasma samples. This method differs from the other techniques, however, in that protein-bound drug is removed along with the proteins. In precipitation methods, liquid–liquid extraction, or solid-phase extraction techniques the protein and the solvent conditions undergo changes that release the drug from the protein, thus resulting in total drug concentration (sum of bound and free). Ultrafiltration is therefore the most common choice if the free drug concentration is needed. This measurement in combination with total plasma concentration allows the study of protein–drug interactions (44).

H. Dialysis

Dialysis is considered the reference method when studying protein binding of drugs (45) but is also useful for protein removal prior to plasma analysis. Dialysis is a classical separation method that uses semipermeable membranes to separate compounds by molecular weight. The protein sample is exposed to a buffer solution separated by the membrane. After the system is allowed to come to equilibrium, the small molecules diffuse to similar concentrations on both sides, while the sample side contains all of the protein

I. Other Methods to Study Protein Binding

More modern methods for studying qualitative parameters of the drug–protein complex have been developed; they are not used as sample preparation methods. They were developed to be more rugged and faster than the older conventional methods. These methods include size-exclusion chromatography, frontal analysis of HPLC, and capillary electrophoresis injections, as well as spectral methods (45,46,47).

Vaes et al. have used solid-phase microextraction fibers (SPME, see Section V.E) to study protein–drug binding (48,49). The advantages of using SPME fibers is that the fiber's phase volume is so small that the binding equilibrium is not disrupted by the removal of drug into the fiber phase. Note that one of the reasons liquid–liquid and solid-phase extraction of plasma sample yield total drug concentration is that the liquid or solid-phase volume, i.e., drug capacity, is so much in excess of the sample's protein drug capacity that the equilibrium shifts away from the protein. Vaes et al. showed that the SPME technique gave similar results to the dialysis reference method and it appears to be a useful method for studying protein binding as well as routine measurement of free drug in plasma.

J. Sample Preparation of Drug Conjugates

Compounds with poor water solubility are often metabolized in humans to more water-soluble compounds to improve excretion rates. Acetaminophen, for example, is metabolized to form glucuronide, sulfate, cysteine, and mercapturate conjugates (41,42). Full characterization of a drug's metabolism often requires analysis of both bound and free drug, particularly in urine, requiring analyses that are performed after hydrolysis of the conjugates. The drug conjugate can be hydrolyzed by the use of acid, base, or enzyme catalysts (50,65,71,72). Typical base hydrolysis conditions are pH 12 for 1 h at 80°C. The analyte itself must, of course, be stable at these severe conditions. More gentle conditions are used for the enzyme methods. Typical conditions are about 1000–5000 units/mL of β-glucuronidase at pH 5 for 1–18 h (71,72).

K. Direct-Injection Techniques for Plasma Samples

A critical first goal of plasma sample preparation steps is the selective removal of plasma proteins. If a plasma sample is injected directly onto a reversed-phase HPLC system, the proteins will precipitate on the hydrophobic stationary phase, causing column plugging. Pinkerton (80) addressed this problem by development of a bi-modal HPLC column in which the reversed-phase layer is bonded only to the internal pores of the silica. The larger protein molecules are size-excluded and pass through the column with the solvent front. The outer silica surfaces are coated with a hydrophillic layer, allowing the proteins to pass through without retention while the drug and other small molecules enter the pores and undergo reversed-phase separation. This technique has gained limited popularity, but column advances are still underway (51). There are two advantages of this technique:

there is significant time savings and there is no opportunity for recovery loss due to extraction.

The addition of surfactants to mobile phases has also been used to prevent on-column precipitation of proteins. The surfactants serve to keep the proteins in solution, allowing the direct injection of plasma samples (52).

L. Derivatization Techniques

Chemical derivatization reactions in analytical chemistry have been used to enhance detection, improve volatility for GC, or enhance selectivity such as for chiral separations. Sample preparation derivatization reactions occur at an active hydrogen such as an alcohol, phenol, amine, or sulfhydryl. The reaction types are largely alkylations, acylations, silylations, and condensations. A comprehensive review of derivatization reaction is available (53).

Many of the derivatization applications were developed to improve a compound's volatility so that gas chromatography may be employed. The presence of highly polar functional groups, such as carboxylic acids, generally results in poor volatility for gas chromatography because of intermolecular hydrogen bonding. Derivatization techniques are used to mask these functional groups, improving volatility and allowing analysis by gas chromatography.

Derivatization reactions can improve detection limits in trace analysis by addition of halogenated acyl or silyl groups for subsequent analysis by gas chromatography with electron-capture detection (54) or by addition of a good UV chromophore or fluorophore for subsequent analysis by HPLC (55).

Selectivity enhancement for chiral analysis can be done by derivatization of racemic mixtures with a chiral agent. The resulting diastereomers can then be separated using conventional chromatographic methods (26) such as reversed-phase HPLC or GC, obviating the need to use chromatographic systems with an expensive chiral stationary phase.

Solid-phase derivatization is an improvement that allows for the quick removal of the reagent from the sample after the reaction is complete. Fisher and Bourque described a novel polymeric system for derivatizing amphetamine, a primary amine to the corresponding dinitrobenzamide for the purpose of enhanced HPLC–UV detection (56). The derivatization reagent is a polymeric system that contains a dinitrobenzoylbenzotriazole moiety. The benzotriazole functions as a good leaving group for the dinitrobenzoyl upon nucleophillic attack by the drug's amine group. The reagent is readily removed because it is attached to an insoluble polymer. Derivatization efficiency was 72% and the reaction rates were fast, allowing the derivatization to be completed in less than 10 seconds.

Other advances in derivatizations include the use of the high-temperature GC injection port as a reaction vessel. Nimz and Morgan (57) describe the derivatization of silyl esters by this technique. Belsner and Buchele describe the analysis of cardiac glycosides using an in-line postcolumn derivatization to fluorescent derivatives (58).

M. Residual-Solvent Sample Preparation for Gas Chromatography

There are four types of sample preparation methods for residual solvents: headspace sampling, extraction, direct injection, and solid-phase microextraction.

Headspace methods include static methods as well as purge-and-trap methods. Both can be performed on solid or aqueous samples. In the static headspace methods, the sample is heated in a closed vessel and, at equilibrium, a portion of the headspace is injected into the GC (59). "Purge and trap" is a technique that is popular in analysis of volatile organic compounds in environmental samples and is used extensively in registered Environmental Protection Agency (EPA) methods. The techniques have found less use in pharmaceutical analysis. A temperature-controlled sample is purged with a stream of carrier gas which carries the volatile residue to a resin trap. The trapped volatiles are then thermally desorbed and transferred to the GC column for analysis. Each step must be carefully controlled with respect to temperature and time. The advantage of the purge-and-trap techniques and the static headspace techniques is that only the volatile organic compounds enter the gas chromatograph and there is no large solvent peak or drug peak to interfere with the analysis. However, the recovery precision is very dependent on the sample and its tendency to retain the volatile solvents by adsorption or crystal inclusion.

The preferred sample preparation method for residual solvent analysis of pharmaceuticals is direct injection of the dissolved sample (11,60). With this technique, the recovery is most reliable because there is no opportunity for recovery loss due to adsorption or entrapment. The other techniques involve a separation of the volatiles before the GC injection and there is a risk that the volatile will be trapped. Typical solvents for this analysis are water, dimethyl sulfoxide, benzyl alcohol, and dimethylformamide (11,12,61). The three latter solvents are chosen because they are higher-boiling than commonly used pharmaceutical solvents and thus elute after them and do not interfere with the analysis. Water offers the advantage that it contributes little interference with a flame ionization detector.

N. Sample Preparation for Capillary Electrophoresis

Capillary electrophoresis (CE) is a much newer technique, and much is yet to be learned about optimizing sample preparation and sample solutions for CE analysis. CE separations are done using flow that is induced by applying an electric field across a fused silica capillary. In contrast to the loop injectors that are used in HPLC, CE requires that the front end of the capillary to be immersed into the sample. After the sample enters the capillary, the sample is then moved to a second vial containing the running buffer. The two common sample introduction methods for CE are hydrodynamic loading and electrokinetic loading. Hydrodynamic loading is done by the pressurization of the sample vial with the capillary inserted. Electrokinetic loading is done by applying the electrical potential with the capillary inserted in the sample vial. The mode of injection is important in designing the sample preparation.

Hydrodynamic loading is consistently reported to be the more precise injection method. Hettiarachchi and Cheung have concluded that hydrodynamic loading should always be used for high-precision quantitative applications because it is less dependent on the ionic concentration of the sample buffer (62), but running buffer is the recommended sample diluent.

Electrokinetic loading is less precise because variations of sample ionic strength will affect the amount of analytes loaded. This has been used as an advantage to boost sensitivity. Taylor and Reid studied the effects of sample diluent changes (63) and reported that the use of methanol as an injection solvent improved the detection limit by a factor of 30 compared to water and a factor of 3 compared to the use of a field-amplified sample injection (FASI). FASI is a sample stacking technique in which a slug of water is preinjected to focus the analytes into tight bands, thereby improving detection limits.

Injection of biological samples that are prepared in organic solvents also has advantages with hydrodynamic injection. Shihabi and Constantinescu (64) reported improved plate counts when they used acetonitrile. The ability to use water-miscible organic solvent as sample diluents is very useful, because the acetonitrile and methanol are used as precipitation agents as well as elution solvents for solid-phase extraction. It is convenient to inject these solutions directly into the system. These are also excellent solvents for dissolving pharmaceutical solids that may be insoluble in water or running buffer. Shihabi and co-workers (65–67) recently reported direct injections of acetonitrile precipitations using a hydrodynamic injection as well as limited success when injecting serum directly.

O. Calibration Methods

Three primary methods are used for calibration: external standard calibration, internal standard calibration, and standard addition. Figs. 1–3 show the typical response curves for each method.

1. External Standard Calibration

External standard calibration is the most simple calibration method. It can be used if the recovery of the analyte is 100% and can be expected to be similar

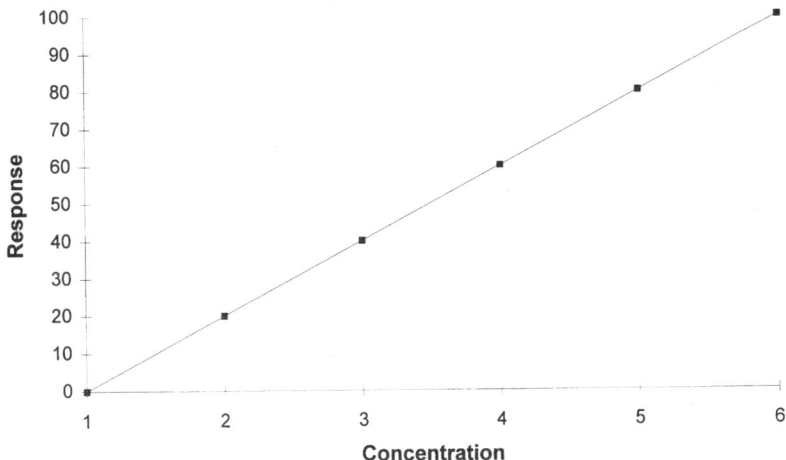

Fig. 1   External standard calibration.

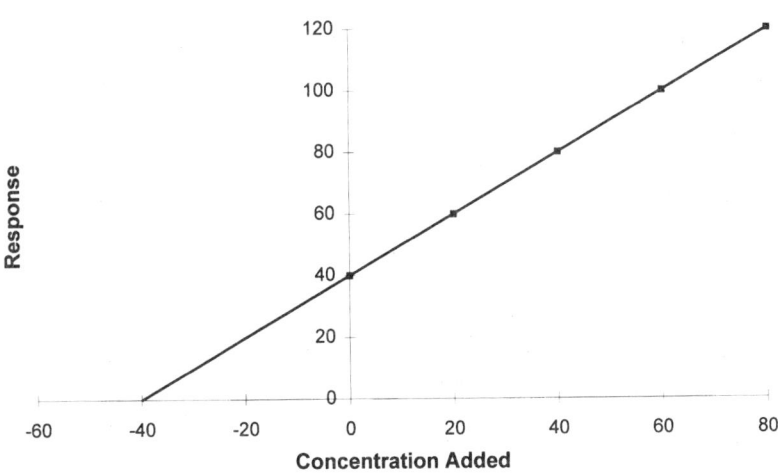

Fig. 2   Standard additional calibration.

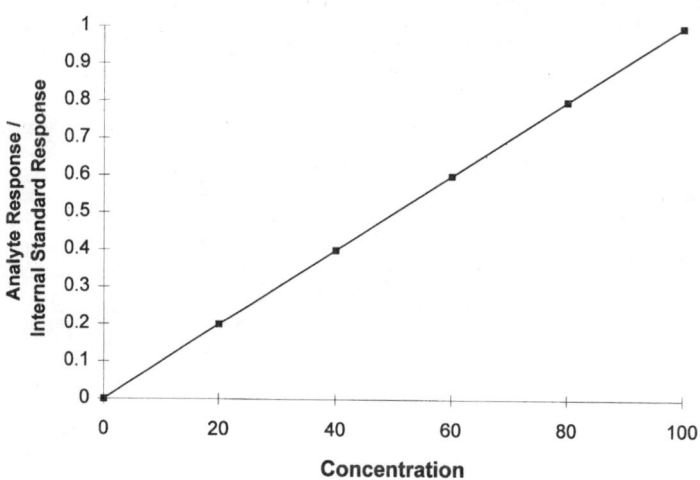

Fig. 3 Internal standard calibration.

for all expected samples. Consistent sample-to-sample recovery can be expected only for samples whose matrix is constant and predictable, such as in drug substances and products. This calibration method also requires reproducible dilution volumes and injection volumes. If the injection system of the chromatograph cannot deliver the same reliability, then internal standard calibration should be used.

2. Internal Standard Calibration

Internal standard calibration similarly requires the preparation of external standard solutions, but in addition a constant concentration of a second compound is added to each sample. The sample concentration is directly proportional to the ratio of the analyte to internal standard. This method is used for the analysis of biological samples and other more difficult analyses.

The internal standard can perform several functions. Its most common purpose is to correct for sample and standard injection volume. Since each sample and standard contains a constant internal standard amount, the injected amount is proportional to the internal standard concentration and the analyte concentration is proportional to the ratio of the analyte to internal standard signals. Likewise, the sample and standard dilution volume can be compensated for by use of an internal standard.

Proper selection of an internal standard also allows the internal standard to correct for extraction efficiency. The internal standard is chosen to be similar in structure, generally a structural analog to the analyte, so that they have similar partition coefficients in liquid extractions and extraction efficiencies in solid-phase extraction. The optimal internal standard is a deuterated or carbon-14 analog of the analyte (68,69), because they have identical chemical properties, resulting in identical extraction efficiencies and also identical reactivity for derivatization. Since deuterated analogs chromatographically co-elute with the analyte, they can only be used with mass-selective detectors because mass spectrometers allow for the mass discrimination of the co-eluting compounds.

3. Standard Addition

Standard addition is a useful calibration method when the sample matrix cannot be reproduced well enough to prepare standard solutions for calibration. The sample itself is used to prepare calibration standards by addition of known amounts of the analyte to three or more aliquots of the sample. The amount of added standard should cover a range of about 10–100% of the sample concentration. When the instrument response is plotted against the concentration added, the sample concentration is determined by calculating the absolute value of the x intercept. See Fig. 2. This technique is often used in conjunction with internal standards, where the instrument response is then the ratio of the analyte to internal standard response. This is a very good reference method to cross-check a primary method. It is rarely used for routine analysis of multiple samples because each test article requires multiple sample preparations and analyses.

VI. FUTURE TRENDS

As discussed previously, there have been many attempts to reduce or eliminate sample preparation steps for analysis of biological samples. These applications are methods to either automate the sample preparation or insert it in-line. The in-line development consists of plumbing improvements to allow for linking various techniques.

Another current trend that is well underway is the use of more specific analytical instrumentation that allows less extensive sample preparation. The development of mass spectrometric techniques, particularly tandem MS linked to a HPLC or flow injection system, has allowed the specific and sensitive analysis of simple extracts of biological samples (68,70–72). A similar HPLC with UV detection would require significantly more extensive sample preparation effort and, importantly, more method development time. Currently, the bulk of the HPLC–MS efforts have been applied to the analysis of drugs and metabolites in biological samples. Kristiansen et al. (73) have also applied flow-injection tandem mass spectrometry to measure sulfonamide antibiotics in meat and blood using a very simple ethyl acetate extraction step. This important technique will surely find many more applications in the future.

The ultimate development in the field of sample preparation is to eliminate it completely, that is, to make a chemical measurement directly without any sample pretreatment. This has been achieved with the application of chemometric near-infrared methods to direct analysis of pharmaceutical tablets and other pharmaceutical solids (74–77). Chemometrics is the use of mathematical and statistical correlation techniques to process instrumental data. Using these techniques, relatively raw analytical data can be converted to specific quantitative information. These methods have been most often used to treat near-infrared (NIR) data, but they can be applied to any instrumental measurement. Multiple linear regression or principal-component analysis is applied to direct absorbance spectra or to the mathematical derivatives of the spectra to define a calibration curve. These methods are considered secondary methods and must be calibrated using data from a primary method such as HPLC, and the calibration material must be manufactured using an equivalent process to the subject test material. However, once the calibration is done, it does not need to be repeated before each analysis.

One more trend that is worth mentioning is the miniaturization of sample preparation techniques. Solid phase microextraction is one good example of where very small samples are consumed and very small extracts are produced. Solid phase extractions can also be scaled down by reducing the bed volume or by use of coated membranes. Likewise liquid–liquid extractions can be scaled down conserving both sample and solvent.

REFERENCES

1. C Chen, GM Pollack. J Chromatogr B 681:363–373, 1996.
2. S Perman, JJ Kirschman. J Chromatogr 357:39–48, 1986.
3. KC Chan, ES Yeung. J Chromatogr 391:465–467, 1987.
4. T-L Ng, S Ng. J Chromatogr 329:13–24, 1985.
5. PK Tseng, LB Rogers. J Chromatogr Sci 16:436–438, 1978.
6. KJ Williams, ALW Po, WJ Irwin. J Chromatogr 194:217–223, 1980.
7. JW Dolan. LC-GC 3:18–21, 1985.
8. EL Inman, AM Maloney, EC Rickard. J Chromatogr 465:201–213, 1985.
9. JW Ho, JYF Candy. J Liq Chromatogr 17:549–558, 1994.
10. PA Bristow. Liquid Chromatography in Practice. Wilmslow, Handforth, England, 1976, p. 170.
11. MS Bergen, DW Foust. Pharmacopeial Forum 1963–1968, May–June 1991.
12. JA Krasowski, H Dinh, TJ O'Hanlon, RF Lindaur. Pharmacopeial Forum 1969–1972, May–June 1991.
13. T Kai, Y Akiyama, S Nomura, M Sato. Chem Pharm Bull 44:568–571, 1996.
14. M Shapiro, D Nadig. Abstr 118AAPS, Eastern Regional Meeting, June 6–7, 1994.
15. SHY Wong. J Pharm Biomed Anal 7:1011–1032, 1989.
16. J Fang, ЗB Baker, RT Coutts. J Chromatogr B 682:283–288, 1996.

17. D Cerretani, L Micheli, AI Fiaschi, G Giorgi. J Chromatog Biomed Appl 614:103–108, 1993.
18. JY-K Hsieh, C Lin, .K Matuszewski, MR Dobrinska. J Pharm Biomed Analy 12:1555–1562, 1994.
19. K Florey, ed. Analytical Profiles of Drug Substances. Vol. 3. New York, London: Academic Press, 1974.
20. J Garber, RF Lindaur, JA Sikes, BL Smith, M Thomas, RH Wendt, V Wotring, WE Zirk. Pharmacopeial Forum 22:2925.
21. US Pharmacopeia, Fourth Suppl., p. 3260, May 15, 1996.
22. GS Clarke, J Pharm Biomed Anal 12:643–652, 1994.
23. VP Shah, KK Midha, S Dighe, IJ McGilveray, JP Skelly, A Yacobi, T Layloff, CT Viswanathan, CE Cook, RD McDowall, KA Pittman, S Spector. Pharm Res 9:5488–5492, 1992.
24. JR Lang, S Bolton. J Pharm Biomed Anal 9:357–361, 1991.
25. AR Buick, MV Doig, SC Jeal, GS Land, RD Mcdowall. J Pharm Biomed Res 8:629–637, 1990.
26. M-J Zhao, C Peter, M-C Holtz, N Hugenell, J-C Koffel, L Jung. J Chromatogr 656:441–446, 1994.
27. ASTM Method D2330-88.
28. JY-K Hsieh, C Lin, BK Matuszewski, MR Dobrinska. J Pharm Biomed Anal 12:1555–1562, 1994.
29. B Karlberg, L Thelander. Anal Chim Acta 98:1–7, 1978.
30. CA Lucy, KK-C Yeung. Anal Chem 66:2220–2225, 1994.
31. PM Kabra, JH Wall, P Dimson. Clin Chem 33:2272–2274, 1987.
32. X-H Chen, J-P Franke, J Wijsbeek, RA deZeeuw. J Chromatogr Biomed Appl 619:137–142, 1993.
33. PH Tang, JS Ho. J High Res Chromatogr 17:509–518, 1994.
34. MJ Koenigbauer, R Majors. LC-GC 8:510–513, 1996.
35. D Louch, S Motlagh, J Pawliszyn. Anal Chem 64:1187–1199, 1992.
36. T Kumazawa, K Sato, H Seno, A Ishii, O Suzuki. Chromatographia 43:59–62, 1996.
37. Z Zhang, J Pawliszyn. Anal Chem 65:1943–1852, 1993.
38. A Ishii, H Seno, T Kumazawa, K Watanabe, H Hattori, O Suzuki. Chromatographia 43:331–333, 1996.
39. A Ishii, H Seno, T Kumazawa, M Nishikawa, K Watanabe, O Suzuki. Int J Legal Med 108:244–246, 1996.
40. C Coudray, C Mangournet, S Bouhadjeb, H Faure, A Favier. J Chromatogr Sci 34, 1996.
41. A Esteban, M Graells. J Chromatogr 573:121–126, 1992.
42. GSN Lau, JAJH Critchley. J Pharm Biomed Anal 12:1563–1572, 1994.
43. K Maruyama, H Nishigori, M Iwatsuru. Chem Pharm Bull 34:2989–2993, 1986.
44. ML Rosell-Rovira, L Pou-Clave, R Lopez-Galera, C Pascual-Mostaza. J Chromatogr B 675:89–92, 1996.
45. J Oravcova, B Bohs, W Linder. J Chromatogr B 677:1–28, 1996.
46. A Shibukawa, Y Yoshimoto, T Ohara, T Nakagawa. J Pharm Sci 83:616–619, 1994.
47. SF Sun, CL Hsiao. Chromatographia 37:329–335, 1993.

48. WH Vaes, C Hamwijk, EU Ramos, HJM Verhaar, JLM Hermans. Anal Chem 68: 4458–4462, 1996.
49. WH Vaes, EU Ramos, HJM Verhaar, W Seinen, JLM Hermans. Anal Chem 68: 4463–4467, 1996.
50. M Castillo, PC Smith. J Chromatogr 614:109–116, 1993.
51. D Song, JL-S Au. J Chromatogr B 676:165–168, 1996.
52. LJ Cline Love, L Zibas, J Noroski, M Arunyanart. J Pharm Biomed Anal 2:511–521, 1985.
53. DR Knapp. Handbook of Analytical Derivatization Reactions. New York: Wiley, 1979.
54. C-T Lai, ES Gordon, SH Kennedy, AN Bateson, RT Coutts, GB Baker. J Chromatogr B Biomed Sci Appl 749(2), 275–279, 2000.
55. DA Stead, RME Richards. J Chromatogr B 675:295–302, 1996.
56. DH Fisher, AJ Bourque. J Chromatogr 614:142–147, 1993.
57. EL Nimz, SL Morgan. J Chromatogr Sci 31:145–149, 1993.
58. K Belsner, B Buchele. J Chromatogr B 682:95–107, 1996.
59. KJ Mulligan, H McCauley. J Chromatogr Sci 33:49–54, 1995.
60. BS Kersten. J Chromatogr Sci 30:115–119, 1992.
61. ID Smith, DG Waters. Analyst 116:1327–1331, 1991.
62. K Hettiarachchi, AP Cheung. J Pharm Biomed Anal 11:1251–1259, 1993.
63. RB Taylor, RG Reid. J Pharm Biomed Anal 11:1289–1294, 1993.
64. ZK Shihabi, MS Constantinescu. Clin Chem 38:2117–2120, 1992.
65. ZK Shihabi, KS Oles. J Chromatogr B 683:119–123, 1996.
66. ZK Shihabi, TE Klute. J Chromatogr B 683:125–131, 1996.
67. ZK Shihabi, ME Hinsdale. J Chromatogr B 683:115–118, 1996.
68. YN Li, B Tatta, KF Brown, JP Searle. J Chromatogr B 683:259–268, 1996.
69. LJ Tulich, JL Randall, GR Kelm, KR Wehmeyer. J Chromatogr B 682:273–281, 1996.
70. GD Allen, R Griffiths, RW Abbott, S Bartlett, TA Brown, VA Lewis, M Nash, G Rhodes, JA Rontree. LC-GC 14(6):510–514, 1996.
71. DR Jones, JC Gorski, MA Hamman, SD Hall. J Chromatogr B 678:105–111, 1996.
72. J Ducharme, S Abdullah, IW Wainer. J Chromatogr B 678:113–128, 1996.
73. GK Kristiansen, R Brock, G Bojesen. Anal Chem 66:3253–3258, 1994.
74. BF Mcdonald, KA Prebble. J Pharm Biomed Anal 11:1077–1085, 1993.
75. MA Dempster, JA Jones, IR Last, BF MacDonald, KA Prebble. J Pharm Biomed Anal 11:1087–1092, 1993.
76. IR Last, KA Prebble. J Pharm Biomed Anal 11:1071–1076, 1993.
77. JA Jones, IR Last, BF Macdonald, KA Prebble. J Pharm Biomed Anal 11:1227–1231, 1993.
78. CRC Handbook of Chemistry and Physics, 77th ed., p. 8–98, 1996.
79. N Valenta, RP Dixon, AD Hamilton, SG Weber. Anal Chem 66(14), 2397–2403, 1994.
80. TC Pinkerton, TD Miller, SE Cook, JA Perry, JD Rateike, TJ Szczerba. BioChromatography 1(2), 96–105, 1986.

3
High-Performance Liquid Chromatography

Thomas H. Stout*
University of Cincinnati, Cincinnati, Ohio

John G. Dorsey
Florida State University, Tallahassee, Florida

I. INTRODUCTION

Chromatography is a general term applied to a wide variety of separation techniques based on the sample partitioning between a moving phase, which can be a gas, liquid, or supercritical fluid, and a stationary phase, which may be either a liquid or a solid. The discovery of chromatography is generally credited to Tswett (1), who in 1906 described his work on using a chalk column to separate pigments in green leaves. The term ''chromatography'' was coined by Tswett to describe the colored zones that moved down the column. The technique languished for years, with only periodic spurts of development following innovations such as partition and paper chromatography in the 1940s, gas and thin-layer chromatography in the 1950s, and various gel or size-exclusion methods in the early 1960s (2). Then in January 1969, the *Journal of Gas Chromatography* officially changed its name and became the *Journal of Chromatographic Science.* This change reflected the renewed interest in the technique of liquid chromatography (LC) and officially signaled the beginning of the era of modern LC. This renewed interest in the oldest of chromatographic techniques was brought about both because of the successes and because of the failures of gas chromatography (GC). On the one hand, GC provided a firm theoretical background on which

---

* Current affiliation: Eurand America, Inc., Vandalia, Ohio.

modern LC could build. However, this renewed interest in LC was being driven because of the inability of GC to handle thermally unstable or nonvolatile compounds. It has been estimated that fewer than 20% of organic compounds have sufficient volatility and thermal stability to traverse a GC column successfully. Admirers of LC also liked the selective interaction of its two chromatographic phases, and easy sample recovery because of its nondestructive detection methods and room-temperature operation.

In the following years the technique was known by many acronyms. It was at various times designated *high-pressure liquid chromatography*, *high-speed liquid chromatography*, *high-performance liquid chromatography*, *high-priced liquid chromatography* by cynical research directors, and finally *modern liquid chromatography*. The modifiers were all used to signify the difference in the technique brought about by small-particle-diameter stationary phases and the resulting high pressure needed to drive the mobile phase through the densely packed columns. These differences resulted in much faster separations with higher resolution than traditional column chromatography. Through the decade of the 1970s the technique grew with astounding speed, and for nine consecutive years from the middle 1970s to the early 1980s the technique led all other analytical instruments in growth rate in an annual survey conducted by *Industrial Research and Development* magazine.

A discussion of the history of liquid chromatography is beyond the scope of this volume; an excellent treatment has been published by Ettre (3). However, some perspective on the beginnings of modern liquid chromatography is useful for an understanding of the technique as now practiced. The first comprehensive publication on the technique was a book edited by Kirkland (4), which grew from a short course offered by the Chromatography Forum of the Delaware Valley. This then laid the foundation for *the* definitive book, now in its second edition, *Introduction to Modern Liquid Chromatography* (2).

During the development of modern liquid chromatography, advances have been driven by both instrumentation and chemistry. The technique as now practiced produces elegant separations that have been developed through chemical manipulation of the stationary and mobile phases, which are then effected and detected by modern instrumentation. This chapter is therefore divided into three broad sections: the first, on the fundamentals of chromatography, provides the necessary foundation for the second, on instrumentation, and the third, on separation modes.

II. FUNDAMENTALS OF CHROMATOGRAPHY

Chromatography is a separation method which employs two phases, one stationary and one mobile. As a mixture of analytes being carried through the system by the mobile phase passes over and through the stationary phase, the individual components of the mixture equilibrate or distribute between the two phases.

$$X_m \leftrightarrow X_s$$

The corresponding thermodynamic distribution coefficient K is defined as the concentration of component (X) in the stationary phase divided by its concentration in the mobile phase.

$$K = \frac{[X]_s}{[X]_m} \tag{1}$$

The use of a typical equilibrium constant K in chromatographic theory indicates that the system can be assumed to operate at equilibrium. As the analyte (X) proceeds through the system, it partitions between the two phases and is retained in proportion to its affinity for the stationary phase. At any given time, a particular analyte molecule is either in the mobile phase, moving at its velocity, or in the stationary phase and not moving at all. The individual properties of each analyte control its thermodynamic distribution and retention, and result in differential migration of the components in the mixture—the basis of the chromatographic separation. The effectiveness of the separation, however, is a function of *both* thermodynamics and kinetics.

A. Thermodynamic Considerations

In order to understand the thermodynamic considerations of the separation process, we need to look at the basic equations which describe the retention of a solute and relate the parameters of retention time, retention volume, and capacity factor to the thermodynamic solute distribution coefficient.

Figure 1 shows a model chromatogram for the separation of two compounds. The time required for the elution of a retained compound is given by t_r, the retention time, and is equal to the time the solute spends in the moving mobile phase plus the time spent in the stationary phase, not moving. If a solute is unretained and has no interaction with the stationary phase, it will elute at t_o, the hold-up time or dead time, which is the same time required for the mobile-phase solvent molecules to traverse the column.

Another fundamental measure of retention is retention volume, V_r. It is sometimes preferable to record values of V_r rather than t_r, since t_r varies with the flow rate F, while V_r is independent of F. If we wish to describe the volume of mobile phase required to elute a retained compound, then the retention volume is the product of the retention time and the mobile-phase flow rate.

$$V_r = t_r F \tag{2}$$

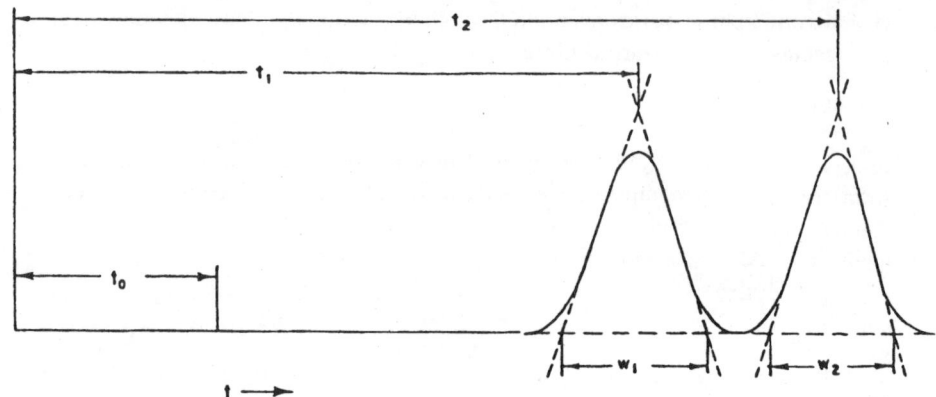

Fig. 1 A model chromatogram for the separation of two compounds.

V_o is a measure of the total volume of space available to the mobile phase in the system and correlates to t_o as the *volume* necessary to elute an unretained compound.

$$V_o = t_o F \tag{3}$$

The most commonly used retention parameter in high-performance liquid chromatography (HPLC) is the capacity factor, k'. While the distribution coefficient, K, describes the concentration ratios, the capacity factor, k', is the ratio of amounts of solute in each phase.

$$k' = \frac{\text{moles of X in stationary phase}}{\text{moles of X in mobile phase}} = \frac{[X]_s V_s}{[X]_m V_m} = \frac{KV_s}{V_m} \tag{4}$$

Equation (4) shows that the capacity factor is directly proportional to V_s, and so k' for each solute changes with the stationary-phase loading on the silica support.

The capacity factor is conveniently measured from retention parameters, since k' also describes the amount of time the solute spends in each phase.

$$k' = \frac{t_r - t_o}{t_o} = \frac{V_r - V_o}{V_o} \tag{5}$$

Rearrangement of Eq. (5) gives

$$V_r = V_o(1 + k') \tag{6}$$

The retention volume can then be related to the capacity factor by substituting Eq. (4) for k' in Eq. (6), where $V_m = V_o$.

$$V_r = V_o + KV_s \tag{7}$$

B. Kinetic Considerations

It is obvious from Fig. 1 that the chromatographic separation of components in a mixture is dependent on two factors: the difference in retention times of two adjacent peaks, or more precisely, the difference between peak maxima and the peak widths. It was shown in the preceding discussion that the retention of a solute is a thermodynamic process controlled by the distribution coefficient and the stationary-phase volume. The peak width, or band broadening, on the other hand, is a function of the kinetics of the system.

Early papers on chromatographic theory described the technique in terms similar to distillation or extraction. The Nobel Prize-winning work by Martin and Synge (5) in 1941 introduced liquid–liquid (or partition) chromatography and the accompanying theory that became known as the *plate theory*. Although the plate theory was useful in the development of chromatography, it contained several poor assumptions. The theory assumed that equilibration between phases was instantaneous, and that longitudinal diffusion did not occur. Furthermore, it did not consider dimensions of phases or flow rates. An alternative to the plate theory which came into prominence in the 1950s was the so-called *rate theory*. The paper which has had the greatest impact was published by the Dutch workers van Deemter, Zuiderweg, and Klinkenberg (6). They described the chromatographic process in terms of kinetics and examined diffusion and mass transfer. The popular van Deemter plot resulted, and the rate theory has become the backbone of chromatographic theory.

Several processes, which will be discussed shortly, contribute to the overall width of the band, and the contribution from each process can be described in terms of the variance (σ^2) of the chromatographic peak, which is the square of its standard deviation (σ).

The most common measure of efficiency of a chromatographic system is the plate number, N. Because the concept originated from the analogy with distillation, it was originally called the number of theoretical plates contained in a chromatographic column. This is not a useful analogy, since a chromatographic column does not contain "plates," but the terminology has become universally adopted and the original definition has persisted. The expression for the efficiency of a column is given by Eq. (8).

$$N \equiv \left(\frac{t_r}{\sigma_t}\right)^2$$

(8)

The parameters t_r and σ_t must be measured in the same units, so the number N is dimensionless.

For Gaussian peaks we can express σ in terms of peak width, where the width of the peak at the baseline (w_b) is equal to 4σ (see Fig. 2). N then becomes

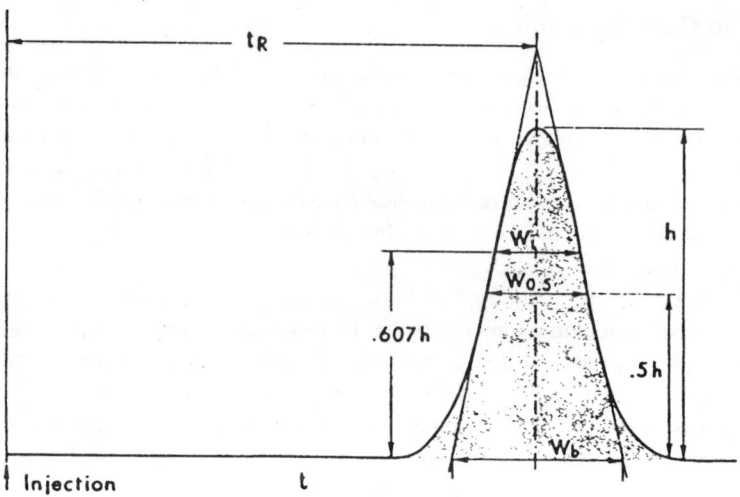

Fig. 2 Evaluation of a chromatographic peak for the calculation of efficiency.

$$N = \left(\frac{t_r}{\sigma_t}\right)^2 = 16\left(\frac{t_r}{w_b}\right)^2 \qquad (9)$$

Because measurements depending on w_b rely on the accurate positioning of tangents to the chromatographic peak, another common calculation of N uses the more easily measured peak width at half-height, where $w_{0.5} = 2.354\sigma$.

$$N = \left(\frac{t_r}{\sigma_t}\right)^2 = 5.54\left(\frac{t_r}{w_{0.5}}\right)^2 \qquad (10)$$

The assumption for the measurement of peak widths in Eqs. (8)–(10) is that the peak is Gaussian. Unfortunately, few peaks are truly Gaussian and, in general, for asymmetrical peaks, N, calculated by a Gaussian-based equation, increases the higher up on the peak the width is measured. Figure 3 shows the error in plate-count determinations for asymmetrical peaks as a function of the peak height at which the width is measured. A more accurate approach to efficiency measurement has been presented by Foley and Dorsey (7), which takes into account the peak asymmetry.

$$N = \frac{41.7\left(\dfrac{t_r}{w_{0.1}}\right)^2}{1.25 + B/A} \qquad (11)$$

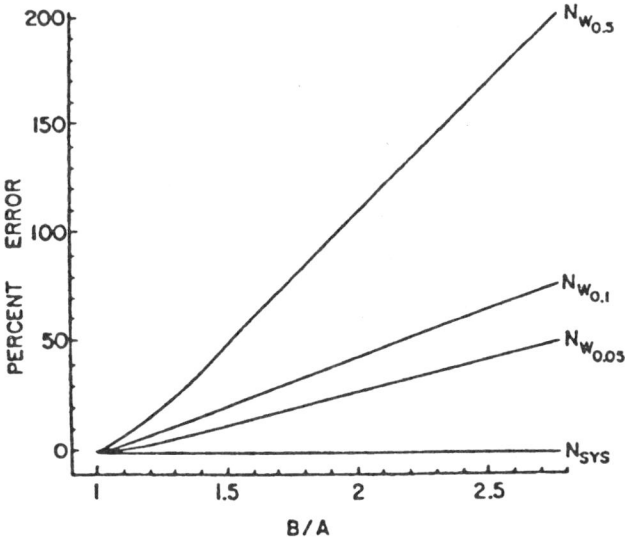

Fig. 3 Error in plate-count determinations for asymmetrical peaks.

Here the peak width is measured at 10% of peak height and the asymmetric ratio or tailing factor is calculated as the ratio of B/A (see Fig. 4). For a symmetrical peak, the B/A value will be 1, and the tailed peaks will have values greater than 1.

The plate number depends on the column length, making comparisons among columns difficult unless they are all of the same length. Another measure of efficiency which removes this dependency is given by the *plate height*, such that

$$H = L/N \qquad \text{where } L = \text{length of column} \qquad (12)$$

H can thus be thought of as the length of the column that contains one plate. For the highest efficiencies, the goal is to attain maximum N and small H values. We will see that H is proportional to the diameter of the stationary-phase particles, so a better measure of expressing efficiency is the *reduced plate height, h.*

$$h = \frac{H}{d_p} \qquad (13)$$

where d_p = stationary-phase particle diameter.

The reduced plate height is a dimensionless number (H and d_p being measured in the same units), and this calculation allows us to compare efficiencies

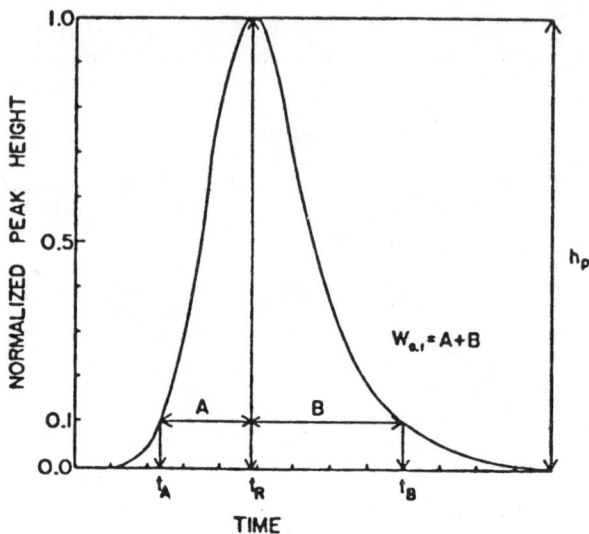

Fig. 4 Measurement of asymmetry factor.

of columns with different particle diameters. A well-operated chromatographic system should have h values of 2 to 5.

As mentioned earlier, the extent of band broadening, and thus the efficiency of a chromatographic column, is dependent on several contributing processes. The four important band broadening phenomena to be considered are (a) resistance to mass transfer, (b) eddy diffusion, (c) longitudinal diffusion, and (d) extra-column effects. Each of these independently would produce a Gaussian band. The variances of each band broadening process are additive to give the overall variance of the system.

$$\sigma_t^2 = \sigma_{rmt}^2 + \sigma_{ed}^2 + \sigma_{long}^2 + \sigma_{ex}^2 \tag{14}$$

1. Resistance to Mass Transfer

As solute molecules interact with the column packing, they continually transfer into and out of the stationary phase. Resistance to mass transfer relates to the rate at which the molecules exchange between phases and may be the dominant cause of band spreading.

If there were no flow through the column, there would be an equilibrium distribution of the analyte molecules between the mobile and stationary phases according to its partition coefficient. However, since there is flow, molecules in the mobile phase are swept downstream, creating a condition of nonequilibrium in the immediately adjacent stationary phase. In order to restore the system to equilibrium, some molecules left behind in the stationary phase must desorb and enter the mobile phase, and some that were carried ahead must leave the mobile phase and sorb onto the stationary phase. This mass transfer takes a finite time and results in band broadening. The faster the mass transfer occurs, the less will be the effect. Both the stationary and mobile phases play a role in determining the rate of mass transfer. A discussion of all the contributing factors is given by Giddings (8), but three of the most important are the diffusion coefficient of the solute in the mobile phase (D_m), the stationary-phase particle diameter (d_p), and mobile-phase flow velocity (V).

$$\text{RMT} \propto \frac{d_p^2 V}{D_m} \tag{15}$$

A molecule is in constant motion in solution. The larger the volume of mobile phase the molecule is in, the longer it will take to diffuse to a phase boundary. In chromatographic columns, the interstitial volume between particles is on the order of the same size as the particles themselves. Small-particle-diameter packings therefore result in a much higher rate of contact and a greater rate of mass transfer. A high rate of diffusion of the solute in the mobile phase also increases the rate of contact by decreasing the time for the molecule to move between stationary-phase particles. For this reason, many chromatographers choose low-viscosity solvents as eluents and/or operate the system at elevated temperatures to increase the diffusion coefficient of the solute in the mobile phase.

The contribution to band broadening by flow velocity is obvious, since the faster the mobile phase is flowing, the farther molecules will be swept downstream before they exchange phases.

2. Eddy Diffusion

The velocity of the liquid moving through a column may vary significantly across the diameter of the column, depending on the bed structure. The velocity of solutes being carried through the column by the mobile phase will therefore fluctuate between wide limits, and the total distance traveled by individual molecules will also vary. These variations are a result of different flow paths taken as a function of least resistance. Since the average velocity of the solute determines its retention time, these random fluctuations result in band broadening. There is still much disagreement about quantitation of this effect, but a generally accepted proportionality is given by Eq. (16).

$$\text{ED} \propto d_p V^{1/3} \tag{16}$$

The smaller the stationary-phase particle diameter, the closer will be the packing. This minimizes the variation in flow paths available to the molecules and reduces band broadening. High flow velocities affect the band broadening by causing molecules traveling through open pathways to move more rapidly than those in narrow pathways. The simple theory of eddy diffusion assumes that a particle will remain in a single flow path. In practice this is not the case, since there is nothing to stop it from diffusing laterally from one flow path to another. This process, called "coupling," averages out the two flow paths and reduces the band broadening so that the final band width, although still greater than the initial band width, is less than if coupling had not occurred. So, rather than a direct relationship to flow velocity, as theory would predict, eddy diffusion is more nearly proportional to velocity to the $1/3$ power.

3. Longitudinal Diffusion

As solute bands move through a column, diffusion of molecules into the surrounding solvent occurs from the region of higher concentration to the region of lower concentration in proportion to the diffusion coefficient, D_m, according to Fick's law. The faster the mobile phase moves, the less time the zone is in the column, the less time there is for diffusion, and the lower the band broadening due to diffusion.

$$D_{\text{long}} \propto \frac{D_m}{V} \tag{17}$$

In principle, longitudinal diffusion may occur in both mobile and stationary phases, but because the rate is so small in the stationary phase, this factor can be neglected.

Combination of terms 1–3 gives the well-known van Deemter equation,

$$H = A + B/V + CV \tag{18}$$

where the A term is due to eddy diffusion, the B term is due to longitudinal diffusion, and the C term to resistance to mass transfer. The familiar van Deemter plot is shown in Fig. 5.

4. Extra-Column Effects

Band broadening can occur in other parts of the chromatographic system as well as in the column. Contributions to this extra-column broadening may come from the injector, the detector flow cell, and the connecting tubing. Slow time constants of detectors and recorders may also contribute. These extra-column effects are more severe for early, narrow peaks in the chromatogram than for later, broader peaks. A more detailed discussion of these effects will be given in the section on instrumentation.

C. Resolution

We have seen that the effective separation of two peaks is a function of both thermodynamic and kinetic effects. A common measure of the separation, the

$$H = A + B/v + Cv$$

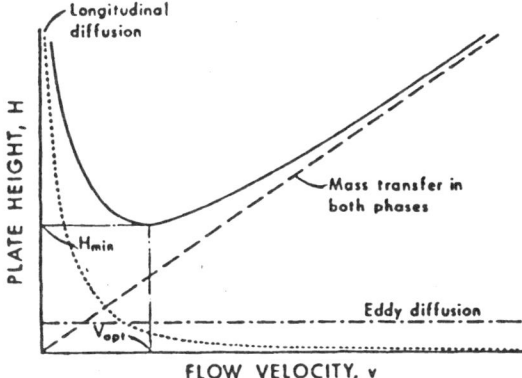

Fig. 5 Van Deemter plot.

resolution, is the ratio of the distance between peak maxima and the average peak width.

$$R_s = \frac{\text{peak separation}}{\text{avg. peak width}} \qquad \begin{matrix}\text{(thermodynamic)} \\ \text{(kinetic)}\end{matrix}$$

In order to obtain peak separation, one of the components in a mixture must be more selectively retained than another, or we may say the column showed selectivity toward the components in the mixture. Selectivity is measured by the separation factor or relative retention (α) and may be given by any of the following relationships:

$$\alpha_{2,1} = \frac{K_2}{K_1} = \frac{k_2'}{k_1'} = \frac{t'r_2}{t'r_1} = \frac{tr_2 - t_m}{tr_1 - t_m} \tag{19}$$

The relationship between the separation factor and the distribution coefficients K_2 and K_1 of the two components (2) and (1) emphasizes the thermodynamic basis of the separation.

Figure 6 illustrates the influence on resolution of peak separation and peak width. Figure 6a shows two components poorly resolved because of inadequate selectivity and large band widths. Figure 6b shows the components with the same band widths but good resolution as a result of improved selectivity. Figure 6c also demonstrates good resolution, but here narrow band widths are the contribut-

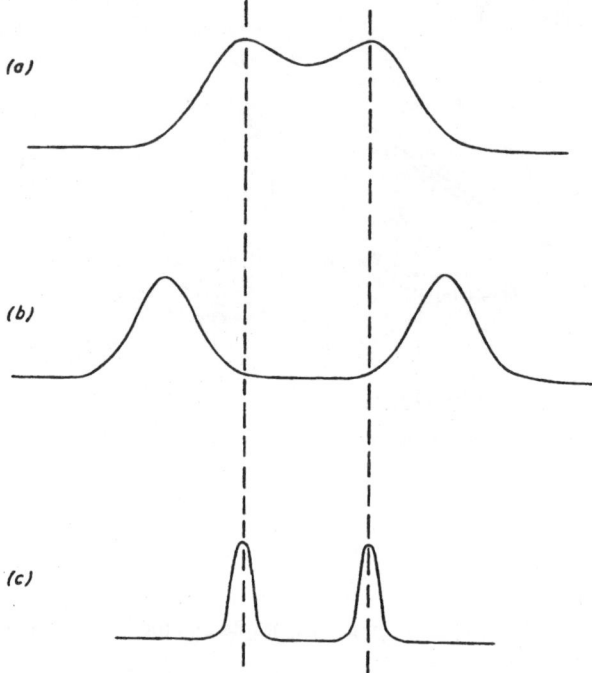

Fig. 6 Two-component chromatograms illustrating the importance of peak-to-peak separation and peak widths on separation.

ing factor even though the selectivity is unchanged. The quantitative measure of resolution is given by Eq. (19).

$$R_s = \frac{2(tr_2 - tr_1)}{W_1 + W_2} \tag{20}$$

In practice, it is seldom necessary to calculate the value of resolution. More frequently, the chromatographer simply looks at the shape of the peaks and estimates the R_s value from the peak shape. Of course, the concentrations and/or responses of the components in a mixture are rarely equal. Figure 7 illustrates the significance of the resolution value for two components with different relative concentrations.

The chromatographic control of resolution is a function of several factors. The fundamental resolution equation is

Rs   1/1   1/4   1/16

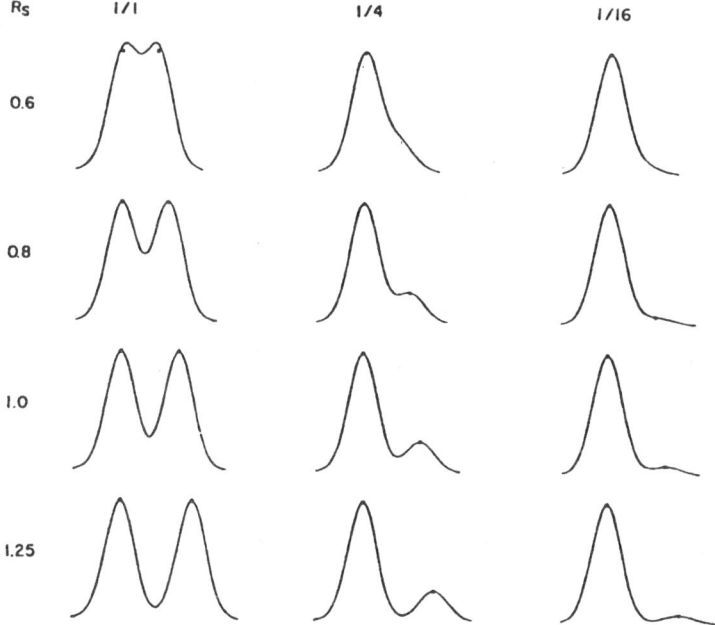

Fig. 7 Significance of the resolution value for two components with different relative concentrations.

$$R_s = \left(\frac{N^{1/2}}{4}\right)\left(\frac{\alpha - 1}{\alpha}\right)\left(\frac{k'}{k' + 1}\right) \tag{21}$$

where $\alpha = k_2'/k_1'$.

The plate number N (column factor) can be increased by lengthening the column, changing the packing particle diameter, or optimizing flow rate. However, improving resolution by increasing N is expensive in time. Doubling the column length doubles the elution time, solvent consumption, and pressure while only increasing R_s by 1.4. Likewise, reducing the particle diameter increases the resolution but may exceed the maximum allowable pressure.

The selectivity α (thermodynamic factor) can be increased by changing columns to a different stationary phase of by imposing secondary equilibria through changes in mobile-phase pH or the addition of complexing agents to the mobile phase, for example. In order to discuss the effect of selectivity changes, let us rearrange the fundamental resolution equation, solving for N.

$$N_{req} = 16R_s^2 \left(\frac{\alpha}{\alpha - 1}\right)^2 \left(\frac{k' + 1}{k'}\right)^2 \qquad (22)$$

If we assume $R_s = 1.0$ and $k' = 2$, then

$\qquad \alpha = 1.01 \qquad N_{req} = 367,000$

and

$\qquad \alpha = 1.05 \qquad N_{req} = 16,000$

Changing α to 1.05 gives us a 96% reduction in required column efficiency which translates to a 96% reduction in column length, analysis time, solvent consumption, and solvent disposal. Changing selectivity is certainly the most efficient way to improve resolution. However, the cost of changing α is in method development time.

Increasing the capacity factor, k', can also increase resolution, particularly for early-eluting compounds. This is done primarily by going to a weaker mobile phase. Looking at Eq. (21) again, for $R_s = 1.0$, and $\alpha = 1.05$,

for $\qquad k_2' = 0.1 \qquad N_{req} = 854,000$
$\qquad\qquad k_2' = 1.0 \qquad N_{req} = 28,000$
$\qquad\qquad k_2' = 10.0 \qquad N_{req} = 8,500$

While a 97% decrease in required column efficiency is calculated for changing k' from 0.1 to 1.0, another 70% decrease occurs for a k' change from 1.0 to 10.0. It is clear from Eq. (21) that as k' becomes large, the resolution is unaffected by small changes in k'. Changing k' to improve resolution thus has an upper limit of effectiveness at approximately $k' = 10$. This is illustrated in Fig. 8.

D. Time Considerations

While any of the above methods will improve resolution, they are far from equal in terms of their effects on time. Recalling that

$\qquad t_r = t_o(1 + k')$

and

$$H = \frac{L}{N}$$

and realizing that

$\qquad t_o = L/v \qquad$ where v = average flow velocity

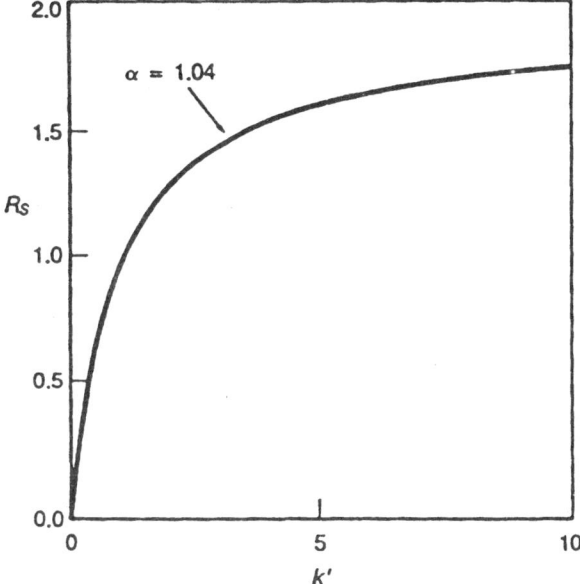

Fig. 8 Resolution of chromatographic peaks as a function of k'.

then

$$t_o = HN/v$$

so

$$t_r = (HN/v)(1 + k')$$

rearranging gives

$$N = \left(\frac{t_r v}{H}\right)\left(\frac{1}{1 + k'}\right) \tag{23}$$

then, setting Eq. (23) equal to Eq. (22) and solving for t_r gives

$$t_r = 16R_s^2\left(\frac{\alpha}{\alpha - 1}\right)^2\left[\frac{(k' + 1)^3}{k'^2}\right]H/v \tag{24}$$

We can see that to double the resolution by increasing N (holding α, k', H, and v constant) requires a $4\times$ increase in time.

In the previous section we saw how changes in selectivity can improve resolution while decreasing analysis time. If we assume $N = 10,000$ and $k' = 2$, and further if we could manipulate chromatographic conditions such that k'_2 would remain unchanged as we change selectivity, then by Eq. (21) a change in α from 1.01 to 1.05 would increase resolution from 0.2 to 0.8. We can see from Eq. (24), then, that with this improvement in resolution and the stated change in selectivity (holding H and v constant), that retention time would decrease by a factor of $t_{r2}/t_{r1} = 0.9$, or 10%.

Figure 9 shows that the $(k' + 1)^3/(k')^2$ term in Eq. (24) goes to infinity at both large and small values of k'. Thus, in the effective range of k' values from 1 to 10 (holding H and v constant, and assuming $\alpha = 1.05$ and $N = 10,000$), the resolution would improve by a factor of 1.8 with a reduction in analysis time of 66%.

We know we can reduce analysis time by shortening the column, but we have a required efficiency necessary to separate the components in a mixture with the desired resolution. Assuming that a well-operating chromatograph yielding a reduced plate height of 3, then 10,000 plates can be generated with 10-μm-diame-

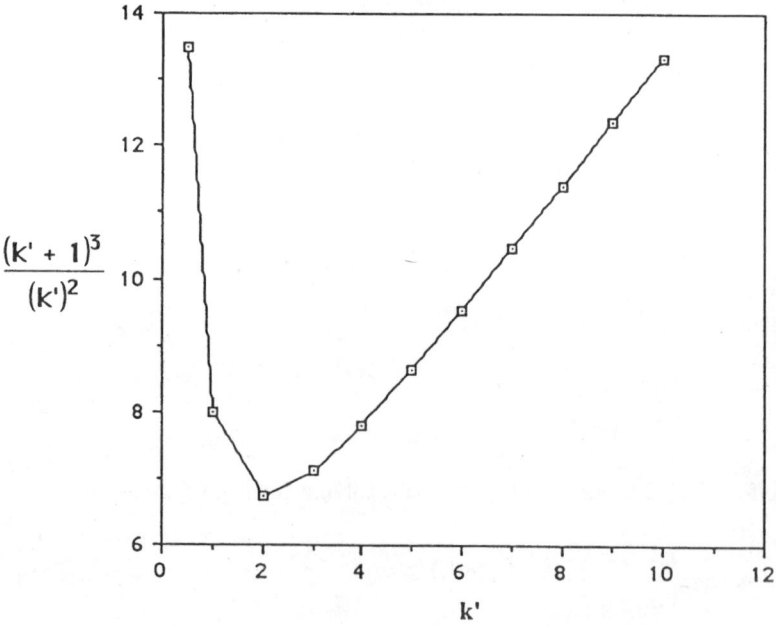

Fig. 9 Plot of capacity factor term versus k' for the modified fundamental resolution equation.

ter particles in 30 cm, with 5-μm particles in 15 cm, or with 3-μm particles in 9 cm. Assuming a solute with a capacity factor of 2, the retention volumes for these columns would be 9.0, 4.5, and 2.7 mL, respectively, and the peak base width would be 360, 180, and 108 μL, respectively.

E. Peak Capacity

Peak capacity is the maximum number of components in a mixture that can be resolved with a resolution of 1 between the inert peak and the last peak.

$$PC = 1 + \left(\frac{N^{1/2}}{4}\right) \ln(1 + k') \tag{25}$$

As samples become more complex, the ability of a particular separation method to resolve all components decreases. A statistical study of component overlap has shown that a chromatogram must be approximately 95% vacant to provide a 90% probability that a given component of interest will appear as an isolated peak (9). This is shown graphically in Fig. 10, where the probability of separation is plotted as a function of the system peak capacity for cases where the number

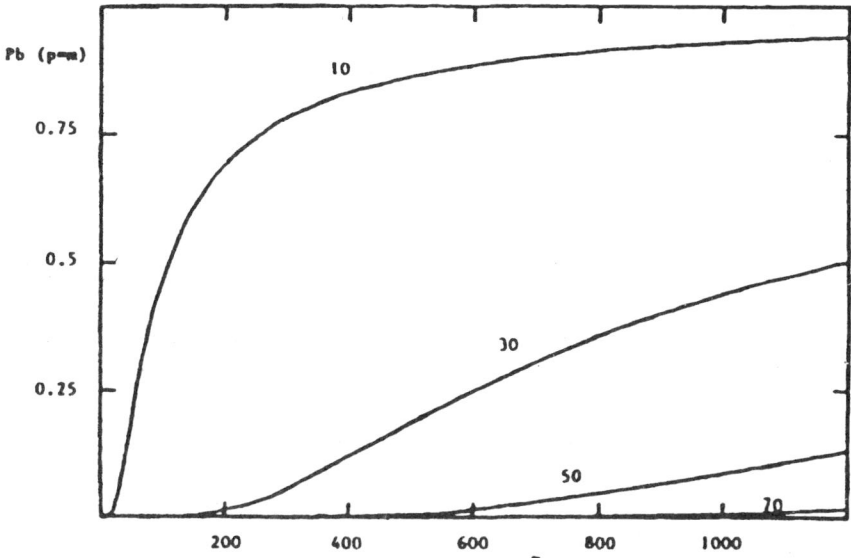

Fig. 10 Probability of separation as a function of system peak capacity. (From Ref. 10. Reprinted courtesy of American Chemical Society.)

of components varies from 10 to 70. A thorough discussion of resolving power is given by Martin et al. (10).

III. INSTRUMENTATION

The instrumental requirements of modern liquid chromatography were defined early. In 1963 Giddings noted that for separation efficiencies comparable to GC, "for the fastest analysis the particle diameter for LC will be about $1/30$ of that for the analogous GC system" (11)—thus the origin of high-pressure liquid chromatography! Cramers et al. (12) addressed flow processes in chromatography, and the integrated form of Darcy's law, which describes laminar flow conditions for liquid chromatography, shows that for a given flow velocity, the pressure drop required is inversely dependent on the square of the stationary-phase particle diameter—thus the first distinction between traditional and modern LC. High-pressure systems were developed to force the mobile phase through columns packed with small-diameter stationary phases, and now the velocity term in the van Deemter equation was easily controllable, in contrast to traditional gravity-flow column chromatography, allowing optimization of separations to be more readily accomplished.

The stationary-phase particle diameters have continually decreased, from the pellicular packings of the early 1970s to the totally porous 10-, 5-, and 3-μm diameters of the present. Research reports now are concerned with 1-μm packings. With this decrease in particle diameter has come more stringent requirements for virtually all aspects of the instrumentation, including injection techniques, pumping systems, and detection. Along with the greater pressure drops required for small-diameter packings, the peak volumes can be extremely small, which mandates careful attention to extra-column variance contributions. At various times in the development of modern liquid chromatography, the column technology has been ahead of the instrumentation, and vice versa. Hopefully, with the present maturity of the technique, advances will come more in parallel. Figure 11 shows a block diagram of a typical LC instrument.

A. Pumps

The high-pressure pumping system is at the heart of modern liquid chromatography. The requirements are clear. The pump should be able to reproducibly deliver virtually pulseless flow over a wide range of volume flow rates at pressures up to 6000 psi; there should be a very small delay for programmed solvent changes and the solvent contact points should be inert against pH, salt, and organic solvent variations. An excellent chapter on liquid chromatography equipment has appeared which gives a detailed discussion of the various pumping principles and

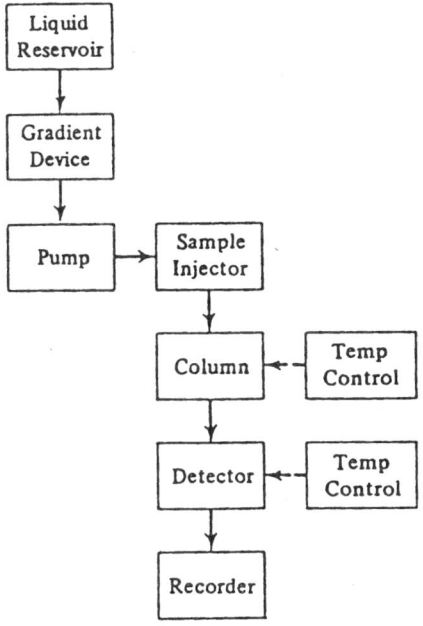

Fig. 11   Block diagram of a liquid chromatograph.

includes an extensive listing of commercially available pumps and pertinent technical specifications (13). While constant-pressure pumps were used in the early development of modern LC, constant-flow pumps are now virtually the only pumps used for analytical liquid chromatography. The constant-pressure pumps used flow rate as a variable and made the comparison of retention data difficult at best; constant-flow pumps allow pressure to vary while maintaining a constant flow rate, and retention volume is then exact and easily calculable.

Three popular types of pumps dominate liquid chromatography as practiced today. The syringe pump was one of the first pumps used in the development of modern LC, and has had a rebirth of interest which has been driven by the interest in micro-LC and supercritical-fluid chromatography. Syringe pumps contain the solvent in a large reservoir and the solvent is displaced by the action of a stepping motor such that the reservoir is pressurized and the solvent is delivered to the column at the defined flow rate. The primary disadvantage of this type of pump is the limited volume of the syringe (solvent reservoir), which requires periodic refilling and subsequent cessation of pumping. This problem becomes much less objectionable with small-diameter columns, and these pumps are now often the best choice for use with micro-LC, where they provide extremely stable flow rates and the syringe volume may be sufficient for an entire day of operation. These pumps will deliver virtually pulseless flow and are ideal for detectors which are flow-sensitive, such as electrochemical or mass spectrometric detectors. The elimination of pulse noise serves to lower the limit of detection, and these pumps are often the best choice for trace analysis.

Single-head reciprocating pumps are the simplest and among the most popular pumps currently used. They are generally the least expensive pumps available, and they are used heavily for quality control and routine applications. Here a piston is driven reciprocally in a small stainless steel chamber and a check valve serves to open the chamber to the solvent reservoir during the piston's "fill" stroke, and close the path to the solvent reservoir during the "pump" stroke. Solvent is then drawn into the chamber by suction, and forced into the column under pump pressure. Clearly, the single worst feature of the single-head pump is the discontinuous pump cycle, causing serious solvent pulsations to be delivered to the column. These pulsations are generally damped with a pulse dampener which, while reasonably effective, adds a significant precolumn volume which slows solvent changeover and gradient formation.

Dual-head reciprocating pumps offer lower solvent pulsation at the expense of mechanical complexity. Here two pistons fill and pump 180° out of phase and in theory provide "pulseless" flow. Dual-head pumps are more expensive and have either two check valves (series heads) or four check valves (parallel heads). Both designs generally provide for some type of pressure or flow feedback control to further compensate for minor flow variations during switching from one head to the next. A detailed discussion of the mechanics of these and other pumping systems is provided elsewhere (13).

The mechanism of solvent mixing is also pertinent to the discussion of pumping systems. Even for isocratic separations, it is generally not advisable to premix the solvents manually and pump the resulting solvent mixture. Slight variations in the preparation of the solvent mixture from one time to the next can lead to slight variations in solvent strength, and selective solvent evaporation over the duration of usage of a single solvent preparation can also cause retention time "creep." There are two popular methods for solvent mixing and gradient formation, generally classified as *low-pressure mixing* or *high-pressure mixing*. In a low-pressure mixing system, a proportioning valve, operated on a time base, proportions the desired ratio of solvents, which are mixed by a static or dynamic mixer *before* they are pumped. Some commercial systems are capable of controlling three or four different solvents, providing ease of solvent selection for method development. Alternatively, high-pressure mixing systems use a separate pump for each solvent, and the solvents are mixed under high pressure. Subtle advantages and disadvantages are apparent with each system. The most obvious problem of high-pressure mixing systems is expense. Providing a separate pump for each solvent adds rapidly to the price of a chromatographic instrument. An often frustrating problem of low-pressure mixing systems is their greater susceptibility to bubble formation. The individual mobile-phase components must be thoroughly degassed before mixing, to prevent bubbles from forming at the point of mixing and subsequently becoming trapped in the pump head. They also generally have a larger volume between the point of mixing and the top of the column. Care must be exercised with both types of solvent mixing when proportioning less than about 10% of one solvent. Here the proportioning valve may be operating at such a low time rate for the minor solvent as to cause irreproducible mixing. An equivalent problem for high-pressure mixing systems may arise when one pump is required to pump at a flow rate lower than that required for reproducible flow. A simple test of the accuracy and delay time of gradient formation is to spike one solvent with a UV-absorbing compound, such as acetone, and run a slow 0–100% gradient, following the actual formation with a UV detector.

B. Sample Introduction

An important factor in obtaining both good performance and accurate quantitative analysis is the reproducible introduction of sample onto the top of the column.

The injection process plays an important role in determining total peak width, as it is one of several important sources of extra-column variance. Kirkland et al. (14) and Colin et al. (15) have addressed this problem in detail. Sternberg (16) has shown that the variance contribution from the injection process cannot be less than $(V_{inj})^2/12$, where V_{inj} is the volume injected. In reality, the contribution will always be greater than this, as this is the theoretical limit representing a rectangular plug injection. While both syringe and valve injection methodologies exist, syringe injection is rarely used except during studies concerning the injection process itself.

Direct syringe injection through a self-sealing septum was the first method of sample introduction used in the development of modern liquid chromatography, but this method has several serious drawbacks. In contrast to GC, where direct syringe injection is the most common method, in LC the high pressures present at the top of the column make this method of sample introduction difficult. Generally, mobile-phase flow must be stopped and the pressure allowed to fall to ambient before the injection can be made. Along with the obvious inconvenience comes other problems. Small pieces of septum are occasionally torn loose and collect at the top of the column, causing pressure buildup. There is an inherent loss of quantitative precision through the addition of "syringe error," the imprecision of reproducibly injecting microliter quantities of sample; and finally, this injection method does not lend itself to automation, making it impractical for routine users of liquid chromatography.

By far the most common method of sample injection is through the use of a rotary valve. These are typically four- or six-port devices which allow the sample to be loaded into an external loop, by use of a syringe at atmospheric pressure, and subsequent activation of the valve shifts the solvent flow stream such that the sample loop is incorporated into the flow, and the sample is delivered onto the top of the column under pressure. A diagram of a typical injection valve is shown in Fig. 12. These devices have a number of advantages which account for their great popularity. The quality of the injection profile is not dependent on operator skill as with direct syringe injection, and if the external loop is overfilled, syringe error is eliminated. This greatly increases the precision of quantitative analysis, and the reproducibility of the injection process is generally quoted at ≤0.2%. The external loops may also be only partially filled, giving added flexibility in terms of injection volume. It is here, however, that a serious mistake is often made by those unfamiliar with the details of LC theory. The most common injection loop volumes are from 5 to 20 µL, and analysts often fail to utilize the variable of injection volume to improve the limit of detection. Very often much more sample can be loaded onto the top of the column than is delivered by the fixed-volume loops. This will be discussed in more detail in a later section. Coq et al. (17) have investigated extra-column variance contributions from the use of very large sampling loops, and have shown that volumes of up to several milliliters can be injected with variance of as little as 1.01 times the minimum if the sampling loop is packed with glass beads. This study was concerning the injection process *only*, and does not address the sample loadability of the chromatographic column. A major advantage of the valve injectors is their ease of automation. Microprocessor-controlled autoinjectors are now used routinely, which allows unattended or overnight operation of the chromatograph. The rate and volume

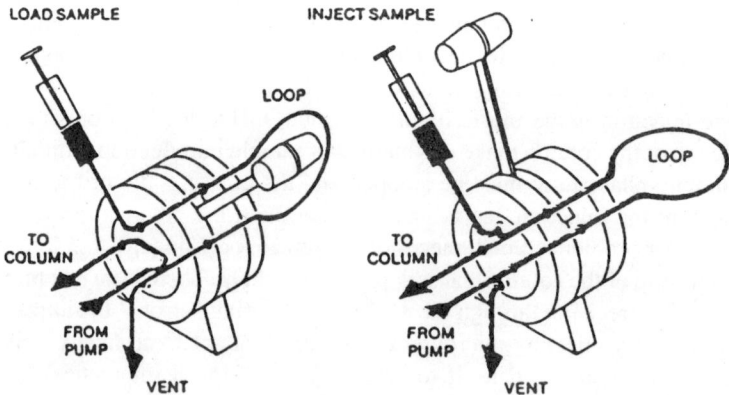

Fig. 12 Flow diagram of a typical LC injection valve.

of the injection process can be programmed, including multiple injections from the same sample vial.

C. Column Hardware

The chemistry of the separation process will be dealt with in a later section; here we look at column configurations and dimensions. By far the most common column material is stainless steel. It is generally inert (except to acid halides) and is able to withstand the high operating pressures necessary for the densely packed columns. At least one column manufacturer advocates glass-lined stainless steel columns, although the benefits of this seem small. Similarly, one manufacturer offers radially compressed columns, which are soft-walled and are operated in a radial compression module, where hydraulic pressure forces glycerol around the sides of the column at a constant applied pressure. This concept reduces wall effects and channels, thereby increasing chromatographic efficiency. This improvement is small, however, unless the columns are operated at the optimum flow velocity. At velocities greater than optimum, resistance to mass transfer is the dominant band broadening process, and the reduction in the wall effects is inconsequential (18).

The inside diameter and length of the columns vary greatly. Commercially available analytical columns range from 1.0 to about 8 mm in inside diameter, and from 3 to over 30 cm in length. A survey conducted in 1994 is very illustrative of column usage (19). From a sampling of 171 respondents, 61.2% reported using columns with diameters of either 4.0 or 4.6 mm, and 25-cm column lengths were the most popular for the 4.6-mm-i.d. category. Furthermore, over 71% reported they used columns packed with 4- to 5-μm stationary-phase particles. An interesting observation is that the percentage of respondents reporting using columns with internal diameters of 2.0–2.1 or 3.0 mm almost doubled to 24.2% from 1991 to 1994.

The internal diameter of the column affects several chromatographic aspects. In the beginnings of modern liquid chromatography there was much discussion of the "infinite diameter" effect (20–22). Due to slow radial mass transfer, for certain combinations of particle diameter and column diameter, solute injected directly onto the center of a column will traverse the length of the column without ever approaching the column walls. For poorly packed columns this significantly increases column efficiency, by eliminating wall effects. However, for well-packed columns the effect is rather small. The practical utilization of this phenomenon also requires specialized injection apparatus and decreased column sample capacity. For these reasons, this concept is now little discussed.

The internal diameter directly affects mobile-phase usage, sample loading capacity, sample dilution, and peak volumes. A 2.0-mm-i.d. column will use 80%

less solvent than one of 4.6 mm i.d., and is generally compatible with existing equipment. A 1.0-mm-i.d. column will use 95% less solvent than a 4.6-mm-i.d. one, but will likely require new capital equipment. This small saving in solvent consumption in going to the 1.0-mm-i.d. column is illustrated in Fig. 13. The fundamental equation describing retention in a chromatographic process, which was introduced in Sec. II, is

$$V_r = V_m + KV_s$$

where V_r is retention volume, V_m is the mobile phase or void volume, K is the thermodynamic distribution coefficient, and V_s is the stationary-phase volume.

It is clear that as the internal diameter of the column increases, so does the mobile-phase volume, and then the volume of mobile phase that is necessary for any given chromatographic analysis. This affects the cost of the analysis, both directly in increased usage of chromatographic solvents, and also through increased solvent disposal cost. This cost factor is one of the primary reasons driving the interest in small-diameter columns, which will be discussed in more detail shortly.

Sample loading is also directly affected by column diameter, as it can be shown that (23)

$$V_{\text{inj,max}} \propto \frac{V_r}{N^{1/2}} \tag{26}$$

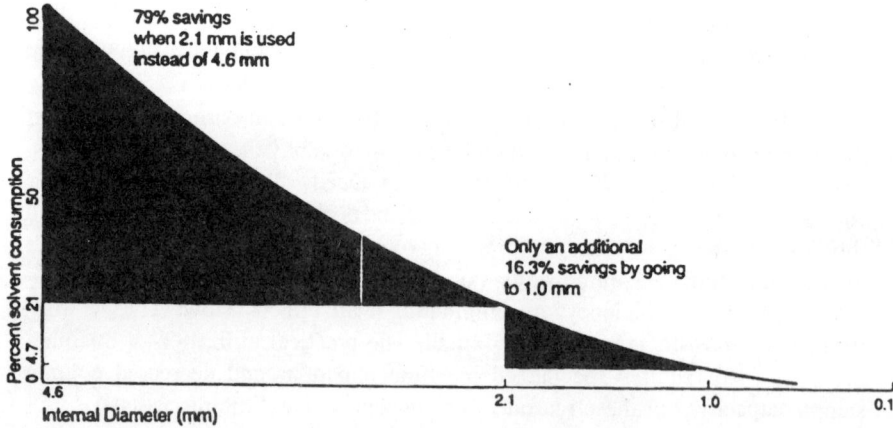

Fig. 13 Effect of column diameter on solvent consumption.

where $V_{inj,max}$ represents the maximum injection volume for a given tolerable loss in resolution, and N is the column efficiency. Likewise, sample dilution and peak volume are also affected (23),

$$V_p = \frac{4V_m(1 + k')}{N^{1/2}} \tag{27}$$

and

$$h_p = \frac{q_{inj}N^{1/2}}{(2\pi)^{1/2}V_r} \tag{28}$$

where V_p and h_p are peak volume and peak height, respectively, k' is capacity factor, and q_{inj} is the amount injected.

The peak volume is also an important consideration when determining the importance of extra-column band broadening processes. It is well known that

$$\sigma_t^2 = \sigma_{intra}^2 + \sigma_{extra}^2 \tag{29}$$

where σ^2 represents the variance of the chromatographic peak, with subscripts referring to total, intra-column and extra-column processes, respectively. The extra-column contribution is a constant amount determined by the injector, connecting tubing, the detector flow cell, and electronics. Clearly, the extra-column contribution becomes more important as the peak volume becomes small. This places severe restrictions on the design of injectors, detectors, and connections as the chromatographic process is miniaturized. For this reason, the use of short columns packed with 3-μm particles may be best accomplished by columns of wide diameter. Stout et al. (24) have discussed the use of 8×0.62 cm columns packed with 3-μm particles which provide high-speed separations with a minimum of concern for extra-column contributions.

A too-often-overlooked consideration in extra-column contributions is the diameter of the connecting tubing. Dolan (25) has thoroughly discussed the effect of tubing on the operation and maintenance of an efficient chromatographic system. Tables 1 and 2 show the volume of the tubing, in milliliters per centimeters and the maximum length allowable for a 5% increase in bandwidth for several typical dimensions.

As was shown in the survey of column usage (19), the most popular length for analytical columns is still 25 cm, with a significant migration to 15-cm lengths. The hesitancy to go to shorter columns is somewhat surprising. Smaller stationary-phase particle diameters and shorter columns reduce analysis times, solvent consumption, disposal costs, and peak volume, which translates into higher sensitivity and lower limits of detection. A discussion of the effects of varying stationary-phase particle diameter and column length on retention and

Table 1 Tubing Conversions

Internal diameter		Volume
(in.)	(mm)[a]	(μL/cm)[b]
0.007	0.18	0.25
0.010	0.25	0.51
0.020	0.50	2.03
0.030	0.75	4.56
0.040	1.00	8.11
0.046	1.20	10.72

[a] Nominal value.
[b] Calculated from English internal diameter measurements.
From Ref. 25.

peak volumes was given in Sec. II. A thorough comparison of the performance and practical limitations of 3- and 5-μm-diameter column packings has been performed by Cooke et al. (26). They showed that, for most practical applications, 5-μm stationary-phase particles are the best compromise, as going to short columns with 3-μm particles places severe restrictions on extra-column volumes, detector cells, and electronics. Going to shorter columns while maintaining the

Table 2 Guide to Tubing Length and Inner Diameter

Column characteristics				Maximum length (cm) for 5% increase in band width		
L (mm)	d_c (mm)	d_p (μm)	N	0.007 in.	0.010 in.	0.020 in.
33	4.6	3	4,400	22	9	*
50	4.6	3	6,677	33	14	*
100	4.6	3	13,333	67	27	*
150	4.6	5	12,000	167	68	*
250	4.6	10	10,000	556	228	14
250	4.6	5	20,000	278	114	*
250[a]	2.0	5	20,000	50	20	*
250[b]	1.0	5	20,000	12	*	*

L = column length, d_c = column internal diameter, d_p = particle diameter, N = column plate number; flow rate = 1 mL/min, except: [a]0.2 mL/min; [b] 0.05 mL/min.
* Less than 8 cm.
From Ref. 25.

same stationary-phase diameter is often advantageous as well; shortening the column by a factor of 2 will halve the analysis time and solvent consumption, and will only reduce the resolution by a factor of 1.4.

The interest in small-diameter columns for liquid chromatography, with inside diameters <1.0 mm, has waned somewhat since the early 1980s. The initial interest in small-diameter columns was driven by the tremendous improvement in efficiency found in gas chromatography upon going from packed to open tubular columns. In liquid chromatography, however, many problems have prevented the practical realization of the goals. Guiochon has shown that a packed column and an open tubular column of the same length give approximately the same efficiency and the same analysis time if the diameter of the particles in the packed column is half the diameter of the capillary column and if both are operated at the optimum flow velocity (27). This means that open tubular capillary columns must be only a few micrometers in diameter just to be competitive with conventional packed columns. He also showed that the speed advantage found with open tubular columns in gas chromatography does not exist in liquid chromatography (27). Because the liquid-phase compressibility is negligible compared to that of the gas phase, the much larger permeability of the open tubular column has no consequence for the average velocity at which the column is operated. For those few separations where more than about 100,000 plates are necessary, open tubular columns for LC do offer a permeability advantage over the linking of several traditional packed columns. However, this technology is still far from being routine or commercially available. An excellent overview of the advantages and technological barriers for microcolumn liquid chromatography is given by Novotny (28).

An often-cited advantage of microcolumn LC is that of enhanced detection. However, careful examination of these claims reveals that most often comparisons are made between micro- and traditional LC columns with a fixed sample injection volume. Here there will certainly be enhanced sensitivity with the small-diameter column, as sample dilution is proportional to the square of the inside diameter of the column. However, if the injection volumes onto different columns of different diameters are scaled proportionally to the square of their diameters, then the dilution of the two samples will be equivalent (29). This means that microcolumn LC will only offer enhanced detection sensitivity when the available sample volume is limited. A critical comparison of micro- and standard LC columns in terms of sample detectability using UV absorbance detectors has been made by Cooke et al. (30).

With columns of any diameter, a common problem is contamination of the top of the column by strongly retained compounds present in the sample being analyzed. Repeated injection of these samples can affect column performance through changes in plate count, capacity factor, and occasionally solute selectivity. For these reasons it is advantageous to use a "guard" column when dealing with real samples. Most often the guard column is a short column, typically 1–5 cm in length, packed with the same packing as the analytical column, and is located between the injector and analytical column. The guard column is then discarded and replaced at intervals dependent on the contamination level present in the sample being analyzed. While its use slows degradation of the analytical column, it represents a further site for sample dilution and dispersion. Kirkland et al. have shown, with both variance calculations and experiments, that it should be possible to use pellicular packing in the guard column without significantly degrading the separation from the analytical column, as long as the sample is retained to the same extent or less on the guard column compared to the analytical column (14). A second problem which can occur with modern siliceous stationary phases is the slow dissolution of the silica in the predominantly aqueous mobile phases used in reversed-phase liquid chromatography. This problem can be overcome by the use of a "saturator" column, located between the pump and injector (31). The saturator column is filled with bare silica, of any particle diameter, and serves to saturate the mobile phase with dissolved silicates. This greatly slows dissolution of the analytical column and, by virtue of its placement, adds no extra-column volume to the separation system. Guard and "scavenger" columns, including expected lifetimes, dimensions, suggested packings, and dispersion characteristics, have been thoroughly discussed by Dong et al. (32).

A too-often-overlooked aspect of liquid chromatography is temperature control. While liquid chromatography is (generally) a room-temperature technique, it is a *constant*-room-temperature technique. As the partitioning of a solute between the stationary and mobile phases is thermodynamically driven, changes in temperature will affect that partitioning. This leads to changes in retention time and peak height and possible misidentification of solute peaks. Gilpin and Sisco (33) performed a careful study on the effects of temperature on precision of retention measurements in both normal and reversed phase LC. They showed that a 1°C change in temperature can result in up to a 5% change in retention in reversed-phase systems, and in some cases 25% or more for normal-phase systems. Chmielowiec and Sawatzky have also shown with the separation of polyaromatic hydrocarbons that *peak inversion* can occur over a 5° change in temperature (34)! Changes in solute selectivity are also possible with changes in temperature, and this will be addressed in a later section. Poppe et al. have addressed temperature gradients which arise from viscous heat dissipation and the effect of this on the efficiency of modern LC columns (35,36). Perchalski and Wilder have also noted the importance of *preheating* the mobile phase to the column temperature before it enters the column (37).

The method of thermostatting the column is also of interest. There are three popular methods, each of which has disadvantages. Water jackets and circulators are likely the best way to ensure temperature control and homogeneity, yet this is the most inconvenient due to the necessity of locating a bulky water bath next to the chromatograph and running hoses to the column. Water does, however, provide excellent heat transfer and has a high heat capacity. The mobile phase can also be conveniently preheated by thermostatting the guard column and/or saturator column. Column block heaters are popular commercially, yet these are more expensive and have been reported to suffer from "hot spots" or uneven heating. Block heaters also require the use of certain length columns to fit properly, without the use of lengthy connecting tubing which increases extra-column band broadening. Finally, air baths or ovens have been popular with some manufacturers, yet these exhibit poor heat transfer characteristics and require lengthy precolumn tubing to equilibrate the mobile phase to the column temperature.

D. Detectors

Detection in liquid chromatography has long been considered one of the weakest aspects of the technique. Low concentrations of a solute dissolved in a liquid modify the properties of the liquid to a much smaller extent than low concentrations of a solute in a gas. For this reason there is no sensitive universal, or quasi-universal, detector such as the flame ionization or thermal conductivity detectors for GC. A comprehensive review of detectors has been published by Fielden (38), as well as two recent books by Scott (39) and Patonay (40). The Fundamental Review issue of *Analytical Chemistry*, published in even-numbered years, contains a comprehensive review of developments in instrumentation for LC, including detection techniques.

Detectors can be broadly classified into bulk property and solute property detectors. Bulk property detectors continuously monitor some property of the mobile phase, such as refractive index, conductance, or dielectric constant, which changes as solute is added to the mobile phase. Bulk property detectors have a finite signal in the absence of a solute, and this results in two serious limitations of these detectors. First, the addition of a low concentration of solute will add only a small increment to what may already be a large background signal; as a result, these detectors generally have poor limits of detection and are in general not suitable for trace analysis. Second, as they also respond to the mobile phase, the signal changes with changes in mobile-phase conditions, and these detectors are largely incompatible with gradient elution techniques. Solute property detectors respond to some specific property of certain compounds, such as the ultraviolet absorbance detector. These detectors generally have much lower limits of detection, but are applicable only to those compounds showing that specific property. While the search for a sensitive, universal detector continues, it is likely that this elusive detection scheme would create more problems than it would solve! Selective detectors can be used advantageously to simplify the chromatography in many instances; two solutes need not be separated if the detector responds only to one of them.

There are also two classifications for the response observed; some detectors give a concentration-dependent signal, which is proportional to the concentration of solute in the mobile phase and independent of the mobile-phase flow rate. If the mobile-phase flow rate is stopped, the signal decreases to zero with a time constant approximately that of the detector response time. Electrochemical detectors and mass spectrometers belong to this category.

1. Figures of Merit

Before discussing the individual types of detectors, there are certain figures of merit that are relevant to all detection techniques.

a. Sensitivity

Sensitivity is likely *the* most often misused word in analytical chemistry. Sensitivity refers to the response per unit concentration, *not* the lowest amount of solute detectable. Sensitivity is usually determined from the slope of the calibration curve, and is correctly reported in units of signal/amount of solute.

b. Noise

There are two types of noise relevant to chromatographic-determinations. Detector, or electrical, noise is the random fluctuation of the baseline signal in the presence of mobile-phase flow. Noise values can be reported as either peak-to-peak values, or as a root-mean-square (rms) value. The rms value can be estimated easily as $1/5$ the peak-to-peak value. This estimation follows from statistical considerations; noise is a randomly occurring phenomenon, and as such the values should follow Gaussian statistics. Ninety-nine percent of the values should then fall within the mean value ± 2.5 standard deviations. It has been recommended that the baseline region measured be sufficiently wide as to encompass at least 20 base widths of the analyte peak (23). The measurement of noise in chromatographic systems has been addressed in detail (23).

Chemical noise is also present in most chemical analyses. This arises from other solutes which give rise to a *nonrandom* signal at the same retention time as the solute of interest. This noise is much more difficult to discuss and quantitate, as the accurate estimation of this noise would require a true sample blank containing the *exact* matrix minus the solute of interest.

c. Detection Limit

The limit of detection (LOD) is that amount of solute that can be distinguished with some level of certainty from the baseline noise. For spectrochemical analysis, the IUPAC recommends the LOD be defined as that amount of solute giving a signal three times the standard deviation of the blank. This can be easily calculated from

$$\mathrm{LOD} = \frac{3S_b}{S} \tag{30}$$

where S_b is the standard deviation of the blank and S is the analytical sensitivity. A tutorial on the statistical basis of this measurement has been published by Long and Winefordner (41). Many different definitions of the LOD are found in the chromatographic literature, which, when coupled with the chromatographic variables, makes evaluation of reported LODs extremely difficult. As well as the inherent noise and sensitivity of the detector, for a true chromatographic LOD, the column void volume, solute capacity factor, column efficiency, and injection volume all play an important role, as seen in Eq. (27). The difficulties of assessing chromatographic LOD values have been discussed by Karger et al. (29) and by Foley and Dorsey (23).

d. Linearity

To be useful for quantitative purposes, the detector response should be linear with amount of solute over a reasonable range of solute concentrations. Most often the logarithm of the detector response is plotted versus the logarithm of the amount injected, and the linear range is taken as that concentration range over which the slope of the corresponding line is 1.00 ± 0.03. However, as Colin et al. have pointed out (13), such a plot dampens fluctuations and may be misleading. They have recommended that a plot of the ratio of detector signal to amount injected versus the logarithm of amount injected is much more instructive.

e. Response Time

Response time refers to the finite time a detector takes to respond to an instantaneous change in solute concentration in the detector cell. At least two methods of reporting this figure of merit exist, so the analyst must compare numbers carefully. The *time constant* is the time necessary for a device to respond to 0.632 of maximum peak height after application of a step change, while the *response time* is generally taken to be the time necessary for a device to respond to 0.90 of maximum peak height after application of a step change. The response time of the detector and subsequent recording device can modify both peak shape and peak height, causing an *apparent* decrease in both column efficiency and detector sensitivity. Both theoretical and experimental treatments of this effect have been reported (16,40,42). It is generally assumed that the detector response time should modify the peak by <5%, and Scott (40) has shown that to meet this requirement the maximum value of the detector time constant should be no greater than 32% of the time standard deviation of the earliest eluting peak. For a 10-cm column, packed with 3-μm-diameter stationary phase operating with a reduced plate height of 2, a solute capacity factor of 2, and a void time of 50 s, this means that the maximum response time of the detector should be 0.37 s. Modern detectors usually have a user-variable time constant, with a minimum value of 0.05 or 0.1 s; however, many older fixed-wavelength detectors have a *fixed* time constant, which often exceeds 1 s. Alternatively, if the time constant is too short, the filter is not eliminating as much noise as is desirable, and the limit of detection may be adversely affected.

f. Cell Volume

The dimensions of the detector cell are also important for both maximum detector sensitivity and to prevent loss of peak resolution obtained from the column. For optical detectors it is desirable to have a long optical path length, and many commercial detectors have a 1-cm path length while still maintaining a total cell volume of only 8 μL. It is generally assumed that the volume of the detector cell should be <25% of the volume of the fastest-eluting peak. For traditional 4.6-mm-i.d. columns of 10 cm or longer, this is not difficult to achieve. However, for short columns packed with small-diameter stationary-phase material, or for very-small-i.d. columns, this can be a severe restriction. A good perspective on the design of optical flow cells for traditional and small-diameter columns has been given by Cooke et al. (30).

g. Quantitation

There is much debate over whether peak height or peak area should be used for quantitation. Both have their uses, and selection depends on the type of analysis being performed. McCoy et al. (43) published the results of a cooperative study comparing the precision of peak height and peak area measurements in liquid chromatography. In general, for *well-behaved* peaks, precision of area measurements is usually better than height measurements, although peak height is less dependent on flow rate than is peak area. One of the problems affecting peak areas is that neighboring peaks can cause band broadening, resulting in poor resolution from other components or the solvent front. In addition, noisy baselines can contribute to inaccurate measurements. On the other hand, variations in retention time as a result of temperature or mobile-phase composition changes will affect peak height measurements, as peaks become broader and shorter with increasing k' values. Another problem causing peak height variation occurs when columns degrade and the plate count decreases.

Another issue concerning quantitation of chromatographic peaks is which method of integration should be used. Papas and Delaney (44,45) have evaluated chromatographic data systems, and Papas and Tougas (46) have examined the accuracy of peak deconvolution algorithms within chromatographic integrators. The most reliable and simple peak area deconvolution methods found were the tangent skim (TS) method and the perpendicular drop (PD) method, even though *both are inherently inaccurate*. When the resolution of two overlapping tailed peaks decreases, the PD method becomes very inaccurate for the smaller peak; dropping a perpendicular makes the smaller peak grossly overestimated, whereas the larger peak is underestimated. Criteria for deciding whether to use either tangent skim or perpendicular drop are not well defined. A general rule of thumb employs the peak height ratio, and a value of 10:1 has been suggested as the decision point. Very real and large differences in decision criteria exist among manufacturers, leading to large differences in peak quantitation. Any given example of overlapping peaks would be deconvoluted differently by different integration systems, leading to potentially large discrepancies not attributed to the physical separation.

2. Ultraviolet-Visible Absorbance Detectors

UV-visible detectors were among the first detectors utilized for liquid chromatography and remain by far the most popular. There are now three popular types of UV-visible detectors: fixed-wavelength detectors which operate from a discrete source, variable-wavelength detectors which utilize a continuum source and a monochromator, and rapid-scanning detectors which are generally based on a linear photodiode-array light detector. A survey (47) reported that in 1985, 78% of respondents reported using a variable-wavelength detector, and 18% reported the use of a diode array, although both of these figures are almost assuredly much higher now. Unfortunately, this was the last survey run by *LC-GC*, and another would be very revealing.

A fixed-wavelength detector will generally offer lower limits of detection than a variable-wavelength detector operated at the same wavelength. The 254-nm detector is almost ubiquitous; the use of a low-pressure mercury lamp gives a very strong source at 253.6 nm, and with the use of a silicon photodiode light detector gives a very inexpensive detector with high signal-to-noise ratio. The inherent sensitivity of this detector makes it useful for virtually all organic compounds with any degree of conjugation or other chromophore. This detector can also be used at other wavelengths by filtering the emission source to give other lines, or even using phosphor screens to give lines not available from mercury. These other wavelengths will have significantly lower signal-to-noise ratios, however. The American Society for Testing and Materials (ASTM) has developed a standard practice for testing fixed-wavelength detectors (ASTM E685) (48). However, not all important criteria are included. Larkins and Westcott have compared several commercially available UV detectors using an otherwise identical chromatographic system and found significant differences in apparent column efficiency due to differences in detector design which contribute to extra-column band broadening (49). Results of a cooperative study on the precision of LC measurements at low signal-to-noise ratios have shown that the best overall precision is noted for peak area measurements (as compared to peak height) with fixed-wavelength detectors (as compared to variable wavelength) (50).

The popularity of variable-wavelength detectors has increased greatly, both due to improved sensitivity from operating at the wavelength of maximum absorbance of the solute of interest, and also due to the ability to operate at wavelengths where other solutes will *not* absorb. Perhaps the greatest operating difficulty with these detectors is the somewhat limited lamp lifetime, especially when compared to the low-pressure mercury lamp used in the fixed-wavelength 254-nm detector.

The most information rich of these detectors is the photodiode array. Here entire spectra are collected simultaneously over some desired wavelength range, which offers several advantages. First, every compound can be quantitated at its wavelength of maximum absorbance, which can offer advantages for trace analysis. Second, compounds not resolved chromatographically can *sometimes* be resolved spectrally; arguably, no area has benefited from the application of chemometric techniques as much as has LC with photodiode array detection. Finally, the molecular absorption spectrum can be used for peak identification and peak tracking. These detectors have been increasing rapidly in popularity, not because of improvements in the detectors themselves, but rather because of improvements in the software available for post-run data manipulation. Castledine and Fell published a review of various strategies for peak-purity assessment in LC using photo-diode array detection (51). Figure 14 shows a typical three-dimensional chromatogram, absorbance versus time, as generated by a diode array detector.

3.   Refractive Index Detectors

Refractive index (RI) detectors were reported used by 37% of the respondents to the detector survey (47). These are clearly the most popular of the truly universal detectors. Dark has published a review of the evolution of both UV-visible and refractive index detectors (52). RI detectors are an excellent example of the problems facing universal detectors. They respond to changes in the refractive index of the mobile phase, and this changes not only with solute concentration, but also with temperature, pressure, dissolved gases, and changes in mobile-phase conditions. Therefore, they are generally applicable only for isocratic separations, and to be operated at maximum sensitivity they need careful temperature and flow control. Modern detectors can respond to changes in the index of refraction as low as 1×10^{-8} refractive index units, which should be compared with a change of 4×10^{-4} for a temperature change of 1°C and 4×10^{-5} for a pressure change of 1 atm. Clearly, this can be the cause of great frustration! Three general designs are employed in commercial instruments, and a good description of each is given by Yeung and Synovec in a tutorial on advances in LC detection (53).

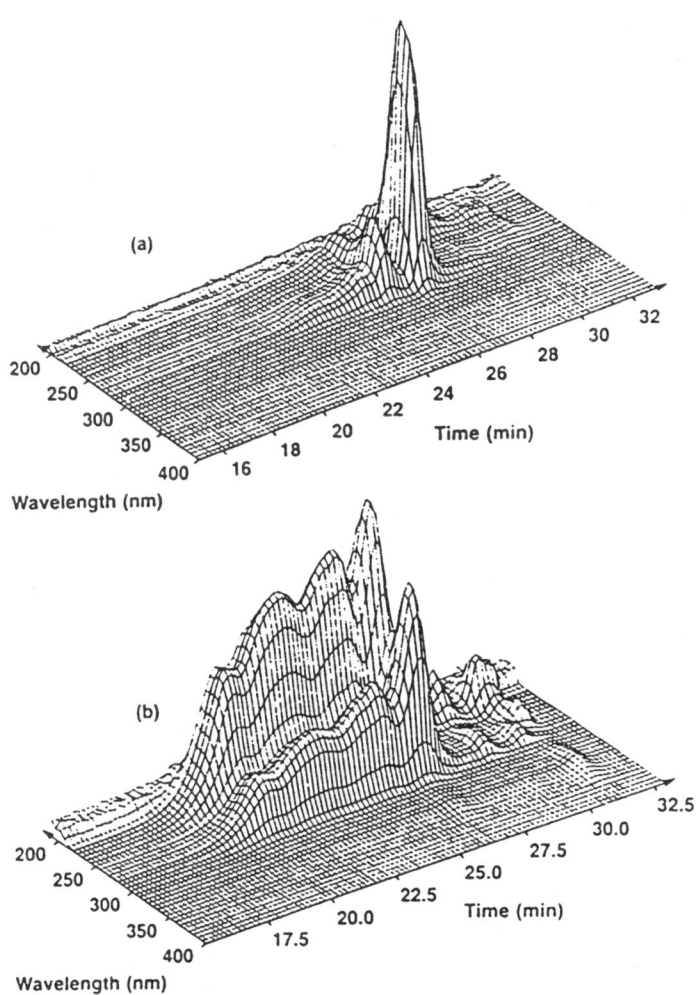

Fig. 14 3-D chromatogram from a diode-array UV absorption detector: (a) blended vegetable oil; (b) the blended vegetable oil after reaction with a resole phenolic resin. (From N Hessefort, M Hedstrom, W Greive. LC-GC 7:130, 1989. Reprinted courtesy of Aster Publishing Company.)

RI detectors are highly useful for preparative chromatography, where they can be operated at low sensitivity, and for polymer or macromolecular separations, where the change in refractive index from that of the mobile phase will be great, even for low concentrations of solute.

4. Fluorescence Detectors

Fluorescence detectors were reported used by 31% of the respondents to the survey (47). The fluorescence detector is a near-ideal detector *for those solutes that exhibit molecular fluorescence*. It exhibits excellent limits of detection because of near-zero signal in the absence of a fluorescing solute, it is highly selective, and, through the use of a photodiode-array light detector, it can give additional structural information about the solute. Commercial fluorescence detectors generally provide detection limits about 100 times lower than commercial absorption detectors. Many biologically active compounds, pharmaceutical products, and environmental contaminants are naturally fluorescing. In addition, many derivatization schemes exist to add a fluorophore to a nonfluorescing solute. A review of luminescence detection, including chemiluminescence and postcolumn derivatization techniques, has been published by Brinkman et al. (54).

The design of commercial fluorescence detectors generally follows one of three directions. The simplest and least expensive instrument uses a low-pressure mercury lamp as an excitation source, and the total luminescence is then collected at a right angle to the line of excitation. More complicated (and expensive) instruments add either one or two monochromators to select the excitation and emission wavelengths. With one monochromator, the excitation wavelength is chosen, and the emission wavelengths are selected with the use of a cutoff filter. Alternatively, two monochromators can be used to select both the excitation and emission wavelengths. Clearly, added selectivity comes with additional control of excitation and emission wavelengths. Some commercial detectors allow programming of the emission wavelength to select the appropriate wavelength for each eluting solute.

Laser excitation for fluorescence detection has received much research interest, but as of yet there is no commercially available instrument. Fluorescence intensity increases with excitation intensity, and it is generally assumed that laser excitation would then offer improved limits of detection. However, as Yeung and Synovec have shown, various types of light scattering, luminescence from the flow cell walls, and emission from impurities in the solvent all increase with source intensity as well, yielding no net improvement in signal-to-noise ratio (53). Where laser excited fluorescence may prove useful is for the design of fluorescence detectors for microbore packed and open tubular LC columns, where the laser source can be focused to a small illuminated volume for on-column detection.

Spectroscopists have shown many powerful variations on the fluorescence experiment, which can generate additional selectivity and structural information. These include two-photon excitation (55); supersonic jet expansion (56); constant-energy synchronous fluorescence, where the excitation and emission monochromators are scanned synchronously to maintain a constant energy difference (57); and pulsed excitation for the real-time measurement of fluorescence lifetimes (58). However, it is far from certain that these techniques will ever be useful for the practicing analyst.

5. Electrochemical Detectors

Electrochemical detectors were reported used by 21% of the respondents to the detector survey (47). Electron transfer processes offer highly sensitive and selective methods for detection of solutes. Various techniques have been devised for this measurement process, with the most popular being based on the application of a fixed potential to a solid electrode. Potential pulse techniques, scanning techniques, and multiple electrode techniques have all been employed and can offer certain advantages. Two excellent reviews of electrochemical detection in flowing streams have appeared (59,60), as well as a comprehensive chapter in a series on liquid chromatography (61).

This technique has increased rapidly in popularity over the past several years. In certain situations an electrochemical detector can offer picogram limits of detection. Furthermore, it is one of the few detectors that is easily adaptable for use with microcolumns. White et al. have shown the feasibility of using a *single carbon fiber* as the working electrode inserted into the end of a 15-μm-i.d. capillary column (62,63). Slais has reviewed the use of electrochemical detectors with low-dispersion (microbore) columns (64).

The early development of electrochemical detectors was driven by the need for increased sensitivity for certain neurotransmitters. The commercialization of the technique was an entrepreneurial effort by Peter Kissinger, one of the early developers, who was quoted as saying, "We were essentially willing to give the technology away for free to any company that would commercialize it, but at the time nobody wanted it" (65).

By far the most popular of the electrochemical detection techniques is *amperometric* detection. Here a fixed potential is applied to the electrode, most often glassy carbon, and a solute which will oxidize (or reduce) at that potential yields an output current. Very little of the solute species, often less than 10%, is involved in the actual electron transfer process. A second method is *coulometric* detection. Here 100% of the solute species is converted, which offers advantages of no mobile-phase flow dependence on the signal and absolute quantitation through Faraday's law, but a large-area electrode must be used. This then makes the electrode much more susceptible to fouling, and offers no improvement in signal-to-noise ratio. While more of the solute species is converted, the larger electrode area contributes proportionally more noise as well.

The efficient use of an electrochemical detector is somewhat more complicated than the popular spectroscopic detectors. Knowledge of the oxidation (or reduction) potential of the solute(s) of interest is necessary, but this is best estimated from the chromatographic system. Generally a hydrodynamic voltammogram is prepared, a plot of peak current versus potential, and the potential is set at the *lowest* potential that will yield an acceptable signal. Lower operating potentials give lower noise and better selectivity against other solutes present. Oxidative electrochemistry is the most convenient, and by far the most popular. Reductive electrochemistry adds complications, as trace amounts of metal ions present from the chromatographic system, hydrogen ion, and dissolved oxygen are all easily reducible, and contribute to background noise. For this reason, to operate near the limits of detection, it is necessary to replace all Teflon tubing in the chromatographic system with stainless steel and vigorously deaerate the mobile-phase supply.

There are two popular designs for the flow cells, a thin-layer design and a wall-jet design, both shown in Fig. 15. The potentiostats are generally of the

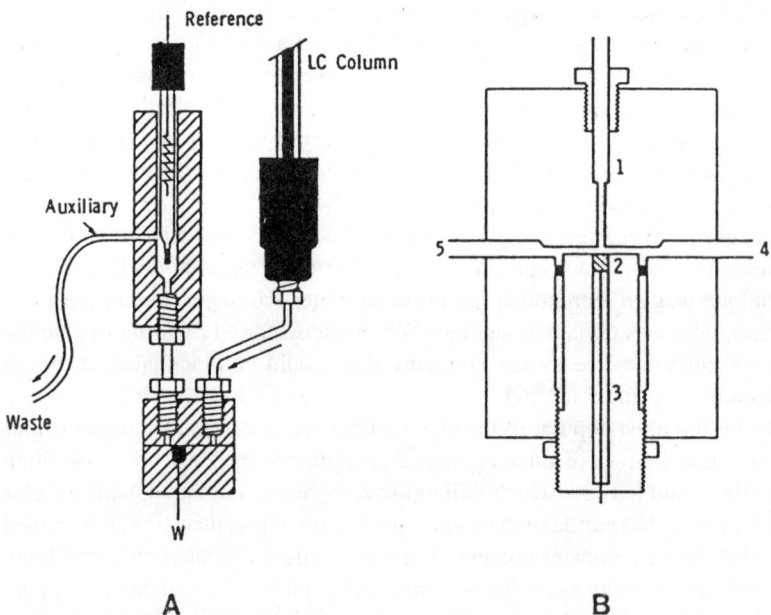

Fig. 15 Amperopetric electrochemical detector cells: (a) thin-layer design; (b) wall-jet design. Legend: (1) inlet from column; (2) working electrode; (3) cell block; (4) passage to reference electrode; (5) to counter electrode.

three-electrode design to ensure a constant potential on the working electrode. As background noise is dependent on mobile-phase conditions, it is difficult to utilize these detectors with gradient elution separations.

6. Conductance Detectors

Conductance detectors were reported used by 15% of the respondents to the detector survey (47). The measurement of electrical conductance is a subset of electrochemical detectors, although it is generally considered separately since it is a *non-Faradaic* electrochemistry; that is, no electron transfer reaction takes place. Conductance is one of the oldest of the LC detection techniques and is a universal technique for ionic solutes. Conductance detection went through a period of low popularity because of the difficulty of detecting a low concentration of an ionic solute in the presence of a highly conducting mobile phase. The popularity of the technique has increased dramatically since the introduction in 1975 of a postcolumn method of removing the background conductance from the mobile phase (66). This technique has since become known as "ion chromatography," and will be discussed in a later section. Electrochemical detectors for ion chromatography, including conductance detectors, have been reviewed by Jandik et al. (67). This reference also includes an excellent discussion of cell designs, development of electronic measurement circuitry, and temperature effects.

7. Postcolumn Reaction Detection

Postcolumn reaction detection was reported used by 9% of the respondents to the detector survey (47). The most popular LC detectors are *solute* property detectors. From a cursory glance at the popular detection techniques already discussed, it is apparent there are many classes of important compounds for which there is no sensitive solute-property detector. For this reason, many types of postcolumn chemistries have been devised to derivatize separated solutes to form a detectable species. Postcolumn reaction detection has been thoroughly reviewed (68,69).

Derivatization reactions can be performed either pre- or postcolumn. As outlined by Brinkman, there are important advantages to using the postcolumn techniques whenever possible (68). First, the analytes can be separated in their original form, which often permits the adoption of published separation procedures. Second, artifact formation is generally not a serious problem, in contrast to precolumn derivatization, where it increases the separation difficulty and causes problems with quantitation. Third, the reaction does not need to be complete and the reaction products need not be stable; the only requirement is reproducibility. Several reaction principles have been extensively applied. These include true chemical derivatization such as with dansyl chloride or *o*-phthalaldehyde; UV irradiation, which can convert the analyte of interest into a more easily detectable species; solid-phase reactions, including catalytic reactions such as with the use of immobilized enzymes; and chemiluminescence techniques.

8. Hyphenated Techniques

Hyphenated detection schemes generally involve the coupling of a spectrometric technique which can offer structural elucidation information. Easily the most heavily applied of this group is LC coupled with mass spectrometry. LC–MS has proven to be a much greater challenge than was originally thought. There are at least four different commercially available schemes for interfacing these two seemingly incompatible instruments; unfortunately, none of them seems to be universally applicable to all separation processes. Garcia and Barcelo have provided a good discussion of the possible interfaces (70). Other molecular spectroscopic techniques that have been extensively utilized are FT–IR (71) and NMR (72). While the information content of these combinations is exceedingly high, the coupling of the instruments is not trivial. Generally, compromises must be made in the performance of either the chromatographic system or the spectroscopic system. LC–MS is the subject of the following chapter in this volume.

9. Other Detection Techniques

Virtually every imaginable (and even some unimaginable!) instrument has been utilized as a detector for liquid chromatography. Most of these have been nothing more than research curiosities; many have not exhibited the low limits of detection or the small cell volume necessary for a popular, useful detection technique. The reader is encouraged to consult the latest issue of the Fundamental Review issue of *Analytical Chemistry*, published in June of even-numbered years, for a thorough review of detection schemes.

IV. SEPARATION MODES

The heart of liquid chromatography lies in the highly selective chemical interactions that occur in both the mobile and stationary phases. It is now possible to rapidly separate compounds whose difference in free energy of transfer between the mobile and stationary phases is only a few calories per mole. Columns exhibiting virtually every type of possible selectivity exist—from shape selectivity to charge selectivity to size selectivity to enantio-selectivity. It is also possible to generate additional selectivity through clever manipulations of the mobile phase; additives that interact with the solute in the mobile phase can create unique selectivities in columns that do not show that type of selectivity.

　　　Most LC stationary phases are now either bare of surface derivatized microsporous silica particles of 10, 5, or 3 μm in diameter. These small-diameter packings provide very high efficiencies due to rapid mass transfer, in contrast to the much larger pellicular packings of the 1970s. In a survey of column usage taken in 1994, 24.6% of respondents reported using 1–3 μm materials; 71.3% reported using 4–5 μm materials, and 26.3% reported using 6–10 μm materials (19). One-micrometer nonporous packings have been reported in the literature, although there are many practical barriers to their routine use, including extremely small peak volumes, and viscous heating of the stationary phase from the mobile-phase flow (73).

A. Adsorption Chromatography

Adsorption, or liquid–solid chromatography, is the oldest of the chromatographic separation techniques, and was the workhorse of chromatography until the development of derivatized silicas. In the survey of column usage, adsorption chromatography was reported used by only 8.8% of respondents to the 1994 survey (19). The two most common adsorbents are silica gel and alumina, with silica being by far the most popular. Porous carbon also enjoys some popularity, but remains largely the research object of a few academic groups. Chromatography on silica and alumina stationary phases has been well treated by Engelhardt and Elgass (74), chromatography on carbon has been reviewed by Unger (75), and the use of alumina in liquid chromatography has been discussed by Billiet et al. (76). A comprehensive book on porous silica and its properties and use as a chromatographic stationary phase has been published by Unger (77).

This mode of chromatography with polar stationary phases and nonpolar mobile phases is known as "normal-phase" chromatography, and is so named because it was the chromatographic technique originally described by Tswett. It is best suited for the separation of polar compounds, but has fallen into disfavor because of several rather serious drawbacks. First, irreversible adsorption is an unpredictable but often encountered event. The adsorption sites on silica and alumina are not energetically homogeneous, and adsorption of solute onto the strongest of these sites can lead to extreme peak tailing, or even irreversible adsorption. Second, day-to-day reproducibility of retention times is often poor. Since the stationary phase is polar, the strongest mobile phase is a polar solvent. This means that any variation in water content of the mobile phase can cause dramatic differences in retention, and as the water content of typical organic solvents can vary even with room humidity, it is extremely difficult to control. Along with this change in solvent strength comes an even more serious problem. The additional water content of the mobile phase can also lead to a change in the activity of the silica or alumina stationary phase. Caude and Rosset et al. have extensively studied the influence of water on retention data in liquid–solid chromatographic systems and found in one instance that capacity factors changed by up to one order of magnitude as the water content of the mobile phase changed from 100 to 600 ppm (78–80).

Nevertheless, there are still applications where adsorption chromatography will perform better than other separation techniques. These include class separations, isomeric separations, and especially preparative-scale separations. Here, the economy of the silica or alumina packings is a very powerful advantage.

Retention mechanisms of adsorption chromatography have been extensively studied. There are two popular models for this process. The "displacement model," originally proposed by Snyder, treats the distribution of solute between a surface phase, usually assumed to be a monolayer, and a mobile phase as a result of a competitive solute and solvent adsorption. A treatment of this model, including the significance of predictions of solvent strength and selectivity in terms of mobile-phase optimization strategies, has been published by Snyder (81).

Alternatively, Jaroniec and Martire have described liquid–solid chromatography in terms of classical thermodynamics (82). They show that a rigorous consideration of solute and solvent competitive adsorption in systems with a nonideal mobile phase and a surface-influenced nonideal stationary phase leads to a general equation for the distribution coefficient of a solute involving concurrent adsorption and partition effects. This equation is phrased in terms of interaction parameters and activity coefficients, which would need to be evaluated or estimated in actual applications.

B. Ion-Exchange Chromatography

Ion-exchange chromatography was reported used by 12.1% of the respondents to the 1994 survey (19). This method of separation has seen a large resurgence of interest since the development of postcolumn methods for removing the background electrolyte, allowing sensitive detection of trace levels of analytes (66). This "new" technique of ion chromatography has been reviewed by Dasgupta (83) and covered extensively in a revised book (84). Ion exchange was first recognized as a means of chemical analysis in 1947. At the National Meeting of the American Chemical Society in New York City there was a symposium in which reports were given of the separation of the rare-earth elements on columns of ion-exchange resins, research that had been done under wartime secrecy in the Manhattan Project. The fission of uranium produced all of the rare-earth elements, and they had to be separated; G. E. Boyd and his associates suggested ion exchange for this purpose (85). This was an especially challenging task, as the lanthanide contraction gives these elements highly similar charge-to-size ratios. With modern ion-exchangers, this separation can be performed in less than 30 min. Figure 16 shows a separation in less than 20 min which originally took hours to perform (86).

Ion-exchange chromatography involves the reversible exchange of ions between mobile and stationary phases. The stationary phase has charge bearing functional groups, with the most common mechanism being simple ion exchange. The sample ion is in competition with the mobile-phase ion for the ionic sites on the ion exchanger. Simple ion-exchange separations are based on the different

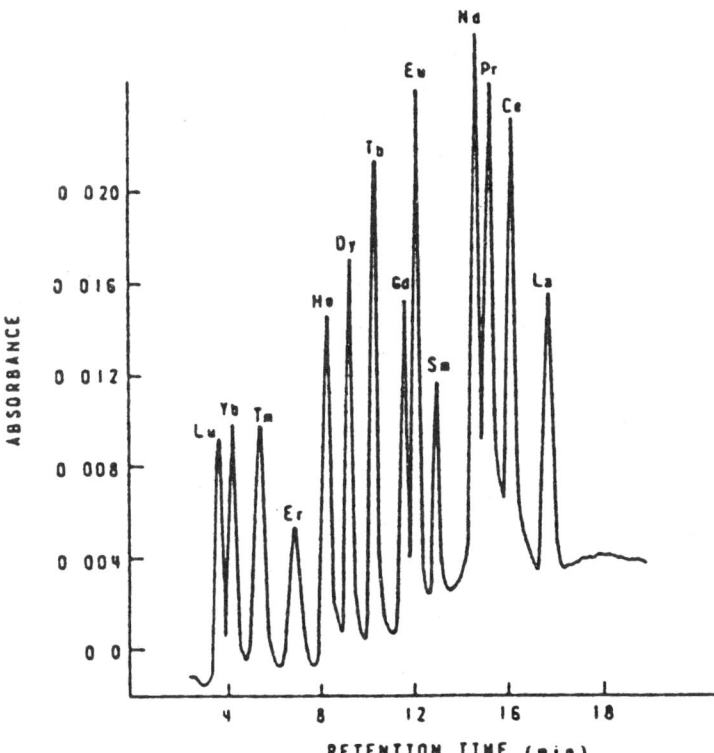

Fig. 16 Separation of the lanthanide ions on 5-μm Nucleosil SCX. (From Ref. 86. Reprinted courtesy of American Chemical Society.)

strengths of the solute-ion/resin-ion pair interactions. Traditional ion exchangers are polystyrene resins which have been cross-linked with divinylbenzene and then sulfonated to form strong acid cation exchangers, or quaternized to form strong base anion exchangers. Silica gel which has been surface derivatized to give exchangeable groups is also highly popular as a stationary-phase material. These will be discussed in more detail in the section on bonded-phase chromatography. Other support materials, such as alumina, agarose, and polymethacrylate, have all received recent research interest, as there are large differences in the ion-exchange properties of macromolecules, such as proteins, on sorbents based on these supports.

The technique of "ion chromatography" was originally developed by Small et al. (66) to allow the use of conductivity detection at high sensitivity. Conductivity is a universal detection technique for ions, but the large concentration of ions present in the mobile phase made it difficult to detect low concentrations of an ionic solute. Using a "stripper column" that was opposite in charge to the analytical column, they developed separation chemistries in which the mobile-phase ion would exchange onto the stripper column and the ion released would form water. For example, for cation analysis, HCl could be used as the mobile phase, and the stripper column would be an anion exchanger in the OH⁻ form. Then the pertinent reaction would be

$$HCl + resin\ OH^- \rightarrow resin\ Cl^- + HOH \qquad (31)$$

While the original technique required occasional regeneration of the stripper column, new membrane suppressers allow continuous regeneration. Suppressed-ion chromatography has been patented; therefore other companies have developed highly sensitive methods of conductivity detection that are not based on removal of the background electrolyte. These rely on both electronic suppression of the background conductivity and on the development of very-low-capacity ion-exchange resins which permit the use of dilute electrolyte solutions as the mobile phase. These systems still exhibit somewhat poorer limits of detection than the suppressed methods.

Ion-exchange packings can also be used to separate neutral and charged species by mechanisms other than simple ion exchange. Furthermore, since ion exchange can be performed under conditions approximating physiological conditions, it has become an important technique in the purification of biological macromolecules. The reader is encouraged to consult the latest Fundamental Review issue of *Analytical Chemistry* for the most recent advances in these areas.

C. Size-Exclusion Chromatography

Size-exclusion chromatography (SEC) was reported used by only 8.5% of the respondents to the 1994 survey (19). This is somewhat surprising in view of the large polymer and biotechnology industries, and may reflect the fact that many practitioners of this technique do not consider themselves "chromatographers," and either did not respond or were not included in the survey mailings. An excellent if somewhat dated book is available on SEC (87).

Size-exclusion chromatography is fundamentally the easiest mode of chromatography to understand and perform. The technique is known by many names, gel permeation, gel filtration, and steric exclusion to name a few, and is applicable to a wide range of materials covering both high and low molecular weight. It is unique among all of the LC techniques in that separation is from purely entropic forces. There should be no enthalpic contribution to retention. In other words, interactions such as adsorption, ion exchange, and partitioning should be absent in the ideal size-exclusion system.

Exclusion columns contain porous particles with different pore diameters. A solute injected onto such a column will diffuse into those pores that have a diameter greater than the effective diameter of the solute, and the effective diameters of the various solutes control elution order. The component with the largest effective diameter will elute first, since as this diameter gets larger the number of pores that it can fit into decreases. Thus, if a molecule is of such a diameter that it cannot diffuse into any of the pores, it is defined as totally excluded and will be unretained, so it will elute with the void volume of the column. A solute that is capable of diffusing into all the pores is said to totally permeate the packing. A solute of this type will require a much larger volume of solvent to achieve elution. Separation of components may be achieved if they elute with a retention volume between V_0 and the total permeation volume.

The selection of the pore size of the packing generally depends on the size of the solute molecules to be separated. Each size-exclusion packing of different pore size will have its own calibration curve, or plot of molecular size versus elution volume. Neither the exclusion limit nor the permeation limit is sharply defined, because the pores of the packing do not have a narrow distribution of sizes around the limits. If the pore distribution is wide, the curve will have a steep slope, large-molecular-weight operating range, but poor resolution for species close in size. If the pore distribution is narrow, the curve will be flatter, with good resolution of molecules of closely related size, but a small-molecular-weight operating range.

One of the most serious disadvantages of SEC is limited peak capacity; that is, only a few separated bands can be accommodated within the total chromatogram, as all peaks must elute between the total exclusion volume and the total permeation volume.

There are two popular stationary-phase materials. Polystyrene–divinylbenzene resins are useful materials, and the pore size can be easily controlled by the amount of divinylbenzene crosslinking. Silica and derivatized silica are also heavily used, and provide better mechanical rigidity. The problem with these and all packings used for SEC is to find a material that will be truly inert to all chemical interactions with the solutes.

D. Bonded-Phase Chromatography

Early in the development of modern liquid chromatography, attempts were made to adjust the stationary-phase chemistry in a fashion analogous to gas chromatography; that is, viscous organic liquids were physically coated onto an inert support. This proved highly frustrating for routine users and researchers alike. While there were many commercially available liquids for use as phases, the inability to keep a constant amount of the liquid coated on the support led to extremely poor reproducibility. Even when the mobile phase was presaturated with the organic liquid to slow dissolution, the shear forces of the liquid mobile phase rushing past the immobile supports were enough to physically strip the liquid from the support. Liquid–liquid chromatography is now practiced by only a few research groups.

In the late 1960s the technology was developed to chemically bond functional groups onto spherical silica particles, and these "bonded phases" provided the efficient, highly reproducible stationary phases needed for liquid chromatography to gain acceptance as a routine method of analysis. Interesting perspectives on the early development of bonded phases can be found in a book devoted to the history of liquid chromatography (88). A recent tutorial on bonded phases, and the remaining challenges in both normal and reversed-phase bonded-phase chromatography was published by Dorsey and Cooper (89).

The synthetic technology for preparing these bonded phases has advanced greatly since their first introduction. Early problems with a lack of reproducibility from column to column, even from the same manufacturer, have largely been solved. This irreproducibility was a result of many factors, including poor control of the physical and chemical properties of the initial silica as well as the bonding reaction. Realization of the importance of the starting silica material has led many column manufacturers now to make their own starting silica. It is now generally assumed that commercial columns from the same manufacturer should give retention variations of <5%. Variation from one manufacturer to another, however, may be dramatic. Differences in synthetic methodology and in starting silica both play a role in the retention properties.

It is not within the scope of this chapter to provide a comprehensive discussion of silica gel chemistry. An excellent treatise is available (77). The parameters that most significantly affect bonding chemistries and solute retention properties are surface area, pore volume, pore diameter, trace metal impurities, and thermal pretreatments. Both Sander and Wise (90) and Sands et al. (91) have studied the effect of pore diameter and surface treatment of the silica on bonding reactions. Boudreau and Cooper (92) have studied the effects of thermal pretreatments at 180, 400, and 840°C on the subsequent chemical modification of silica gel, and showed that thermal pretreatment at temperatures >200°C can produce more homogeneous distribution of "active" silanols which are available for subsequent derivatization.

An extremely thorough review of bonded phase has been published by Sander and Wise (93). By far the most popular synthetic scheme for the preparation of bonded phases involves the aptly named "monomeric reaction." Here a functionalized silane with a single leaving group is reacted with silica to form a siloxane bridge. The generalized reaction is depicted in Fig. 17. The primary advantage of monomeric stationary phases is that they provide a very well-defined single layer of coverage of the silica surface. With careful control of the reaction conditions, the end product is very reproducible in terms of bonding

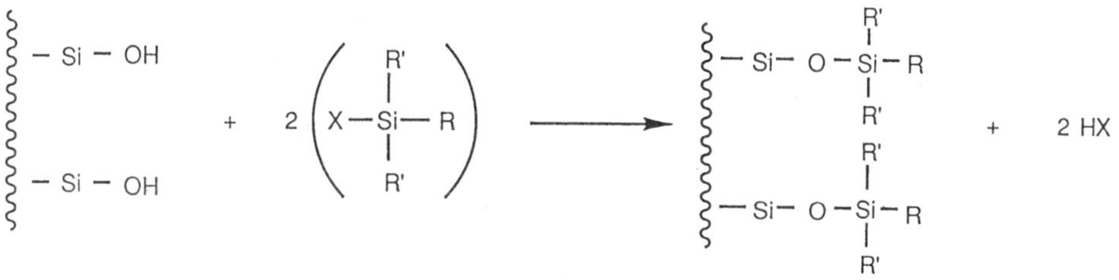

Fig. 17 Generalized bonding reaction for derivatization of silica surface by alkylsilanes. X = leaving group. R' and R are any desired functionalities; R' is typically methyl, and R is C_8, C_{18}, etc.

density of the grafted groups. The most popular leaving group is chloride, although methoxy, ethoxy, and dimethylamino groups, among others, have also been used. The most common reaction involves slurrying the starting silica with a functionalized dimethylchlorosilane in a suitable solvent such as toluene along with an "activator" or scavenger base and refluxing for several hours. Some manufacturers, for some bonded phases, then use a second silanization reaction with a trimethylchlorosilane to react with as many remaining hydroxyl sites as possible. This "end-capping" reaction is to prevent hydrogen-bonding solutes from interacting with these remaining hydroxyls and causing a mixed retention mechanism.

If a di- or trireactive silane is used in the synthetic scheme, a more complex surface chemistry results. These phases are generally referred to as "polymeric phases." In this case, there are a number of possibilities for reaction sites. The silane may simply anchor at two (or three) silica surface sites and still yield only a single layer of surface coverage. More likely, however, one or more of the leaving groups on the silane will hydrolyze and then react with other leaving groups to form a polymeric network extending out from the silica surface. This polymerization reaction may occur in solution before bonding to the silica surface, after the silane has already been bonded to the surface, or both. Both the extent of cross-linking and the amount of silane bonded to the surface are very sensitive to the reaction conditions. Early work with polymeric phases led to a belief that they were irreproducible and generally showed poor stationary-phase mass transfer characteristics. The reactions are, however, simpler to run, and for this reason polymeric phases have continued to be commercially available. While the polymeric phases do result in a higher carbon content, or more bonded "mass," they are still not free from the secondary interaction of solutes with residual hydroxyl sites. While the silica surface may be totally protected, hydroxyl sites will often occur on the "last" silanes in the polymeric network from hydrolysis of the leaving groups. Sander and Wise (94) have shown that polymeric phases can be made reproducibly when reaction conditions are carefully controlled. Using a single lot of silica and carefully controlling the water content in the reaction mixture, they reported the preparation of a polymeric phase with a relative standard deviation of only 0.96% in surface coverage over four trials.

· This bonding technology has the advantage of being able to "tailor" the surface chemistry of small-diameter silica particles to give virtually any desired interaction. There are reports in the literature of virtually every conceivable functional group being bonded to the surface, both alone and in mixtures to give mixed interaction phases. While this seems ideal, the bonded phases suffer from one serious flaw: they have only a limited pH stability. At pH < 2.5, the Si–C bond is cleaved, and at pH >7.5 soluble silicates are formed, and the column material actually begins to dissolve! High pH can be tolerated, if absolutely necessary, by placing a silica "saturator" column between the pump and injector. This sacrificial column serves to saturate the mobile phase with soluble silicates, so that the analytical column is preserved.

1. Normal-Phase Chromatography

Bonded-phase columns in the normal mode were reported used by 14.1% of the respondents to the 1994 survey (19). The term "normal-phase" chromatography, sometimes referred to as "straight phase," defines the chromatographic condition that retention of solutes *decreases* with *increasing* solvent polarity. That is, the stationary phase is (usually) more polar than the mobile phase. The types of solutes amenable to this separation mode are generally polar solutes, with the order of retention and types of separations being quite similar to those of adsorption chromatography. The stationary phases are typically formed with a propyl spacer away from the siloxane bridge, terminating in an appropriate polar functionality. The most common terminating groups are $-CN$, $-NH_2$, and $-(OH)_2$. An overview of retention on bonded-phase, normal-phase packings has been provided by Snyder (95).

Normal-phase, bonded-phase columns offer a number of advantages compared with traditional adsorption chromatography. These all arise from the energetic nonhomogeneity of the hydroxyl sites on bare silica. The most "active" sites on bare silica can lead to irreversible adsorption, peak tailing, and slow equilibration with changing solvent systems in adsorption chromatography. However, during a surface derivatization, it is these highly energetic hydroxyl sites which react *first* with the derivatizing agent. This means that the remaining unreacted hydroxyls are the weakest, and cause little interference with the separation. Irreversible adsorption is seldom a problem with bonded-phase columns. Control of water content is not as crucial; while water will still modify the mobile-phase strength, it will not cause deactivation of the stationary phase as with bare silica or alumina. Finally, solvent reequilibration is much more rapid, which means that gradient elution techniques can be used advantageously.

A wide choice of solvents and columns is available to effect separation. The selectivity of both the mobile and stationary phases have been approached from a logical perspective. Snyder first proposed classifying solvents into a "solvent selectivity triangle," based on their proton donating, proton accepting, and dipole interaction abilities (96), shown in Table 3 and Fig. 18. To effect a change in selectivity on a particular column, the goal is to investigate solvents from as near the apices of the solvent selectivity triangle as possible. Using hexane as the carrier, or weakest solvent, Glajch et al. (97) recommended chloroform as the best proton-donating solvent, methylene chloride as the best dipole interaction solvent, and ethyl ether, or methyl-*tert*-butyl ether as the best proton-accepting solvent. Figure 19 shows the location of the recommended solvents for both normal- and reversed-phase chromatography.

The specific chemical interactions from the stationary phase have been approached in the same manner. Snyder (95) first proposed that the three types of bonded phases should provide maximum differences in selectivity. He assigned the amino-phase to group I, the cyano- phase to group VI, and the diol-phase to group IV. Cooper and Smith (98,99) have extensively studied the three common types of normal-phase, bonded-phase columns and, using extended solubility parameters, have experimentally located the three columns on a stationary-phase selectivity triangle, shown in Fig. 20. Both the amino-phase and the cyano-phase fall near the predictions; however, the diol-phase shows significantly less

Table 3 Classification of Solvent Selectivity

Group	Solvent
I	Aliphatic ethers, tetramethylguanidine, hexamethyl phosphoric acid amide (trialkyl amines)
II	Aliphatic alcohols
III	Pyridine derivatives, tetrahydrofuran, amides (except formamide), glycol ethers, sulfoxides
IV	Glycols, benzyl alcohol, acetic acid, formamide
V	Methylene chloride, ethylene chloride
VI	(a) Tricresyl phosphate, aliphatic ketones and esters, polyethers, dioxane
	(b) Sulfones, nitriles; propylene carbonate
VII	Aromatic hydrocarbons, halo-substituted aromatic hydrocarbons, nitro compounds, aromatic ethers
VIII	Fluoroalkanols, *m*-cresol, water (chloroform)

From Ref. 96.

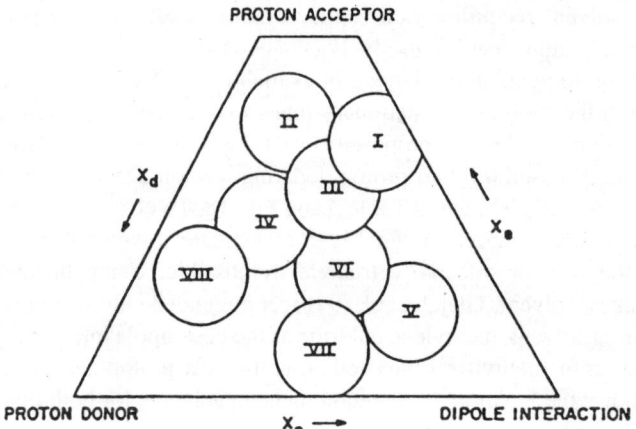

Fig. 18 Snyder solvent selectivity triangle for solvents of Table 3.

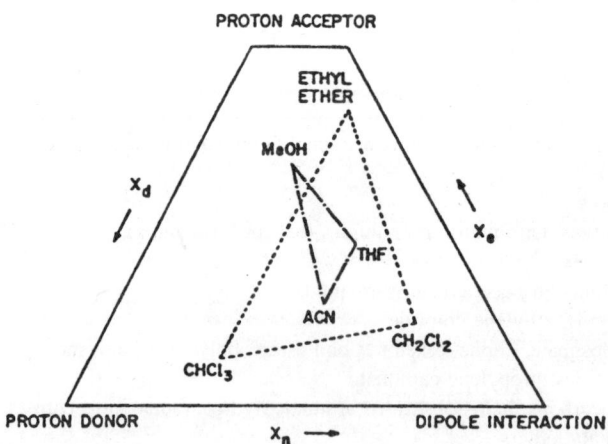

Fig. 19 Selectivity triangles for preferred solvents in reversed-phase and normal-phase chromatography: ·······, reversed phase; -----------, normal phase. (From Ref. 97. Reprinted courtesy of Elsevier Scientific Publishers.)

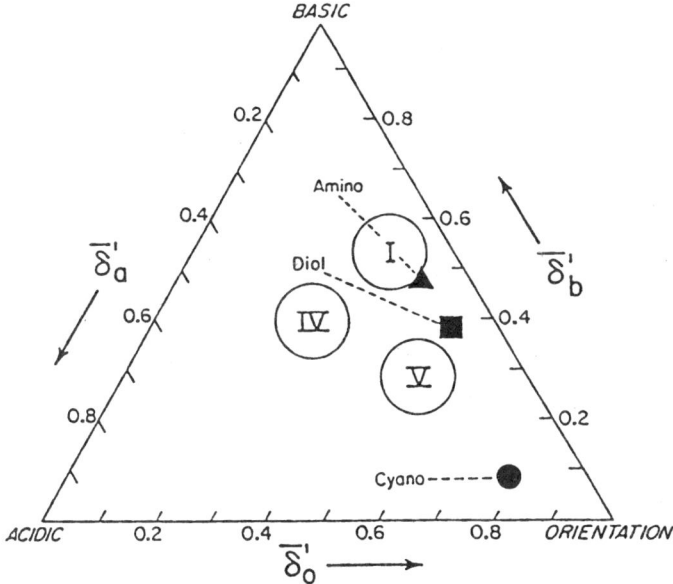

Fig. 20 Stationary-phase selectivity triangle. Circles represent expected locations based on solvent selectivities; I, amino (pure proton donors); IV, diol (alcohols); V, cyano (pure dipole interactions). (From Ref. 99. Reprinted courtesy of Friedr, Vieweg & Sohn.)

acidic character than predicted. Smith and Cooper (99) attribute this to the particular diol-phase studied, which contained an ether linkage, which they suggested was acting to neutralize to some extent the acidity of the hydroxyl groups through hydrogen bonding. It would be highly useful to study other diol-columns to see if this phenomenon is unique to this particular linkage group. If not, columns with other functional groups in the acidic apex of the selectivity triangle would be highly desirable.

Normal-phase, bonded-phase columns are likely underutilized for separations where they should be the method of choice. This is due both to the ease of use of reversed-phase, bonded-phase columns, discussed next, and also to the many problems inherent in the use of bare silica and alumina. Very straightforward method development in normal-phase chromatography can be performed by combining the solvent and stationary-phase selectivity triangles. The three columns, each used with the three recommended modifiers, should provide the maximum difference in selectivity available. These nine experiments, used in conjunction with chemometric optimization schemes, should then provide a rational method for methods development in normal-phase chromatography. A tutorial on chemometric solvent optimization has been provided by Glajch et al. (100).

2. Reversed-Phase Chromatography

Reversed-phase chromatography was reported used by 50.9% of the respondents to the 1994 column survey (19). The term "reversed-phase" chromatography at first seems inappropriate for what is by far the most popular mode of modern liquid chromatography. The term itself can be traced to Howard and Martin (101) in 1950. In attempting the separation of long-chain fatty acids, they realized that the "normal" mode of chromatography, using a polar stationary phase and a nonpolar mobile phase, would not work, as the hydrophobic compounds had too little retention to effect a separation. They were able to treat Kieselguhr with dimethyldichlorosilane vapor and then coat this hydrophobic support with a nonpolar liquid stationary phase. Both the polarity of the phases and the respective elution order of solutes were reversed from traditional chromatographic systems, and they christened the technique "reversed-phase" partition chromatography.

The popularity of the reversed-phase mode is the result of several advantages over normal-phase chromatography. As the stationary phase is nonpolar, the weakest mobile phase is water, and the strength is adjusted by adding polar organic solvents. These aqueous mobile phases provide optically transparent mobile phases at very low solvent costs, compatibility with many biological samples, compatibility with electrochemical detectors, and the ability to control secondary chemical equilibria as a further means of modulating separational selectivity. Furthermore, the interactions of the solute with the stationary phase are weak, which provides rapid mass transfer with little chance of irreversible absorption, and provides rapid equilibration with solvent changes, making the technique ideal for gradient elution methods.

Because of the popularity of the technique, much work has been devoted to understanding the retention mechanism of reversed-phase chromatography. The conditions of reversed-phase chromatography require a nonpolar stationary phase, but this condition can be met by many different ligands. In fact, there are commercially available columns of at least C_1, C_2, C_4, C_8, C_{18}, C_{30}, phenyl, and cyano functionalities, where the carbon numbers refer to the length of a fully saturated hydrocarbon chain. While the cyano phases are not highly nonpolar, they can behave in a reversed-phase manner. The question then arises as to how these different phases affect retention and the retention mechanism.

Although many statements exist in the literature about correlations between chromatographic properties and carbon content of the bonded phase, this is *not* the relevant parameter. As Unger et al. (102) pointed out early in the history of bonded phases, carbon content alone is often misleading because of differences in the surface areas of the original silica, which results in different surface densities of the bonded alkyl groups. As the surface area of various chromatographic silicas ranges from about 60 to several hundred square meters per gram, the actual surface density of the bonded alkyl chains must be calculated for relevant comparisons among different stationary phases. This surface density is most often described in units of $\mu mol/m^2$, but careful consideration is also necessary here before accepting literature values. The relevant surface areas should be the area of the *underivatized* silica, as this is where the bonding reaction occurs and is the point of attachment of the alkyl chains. This then gives the relevant chain density. Some workers have reported the area of the derivatized silica, which may be 30–70% lower and which will give highly inflated surface densities.

One popular model of retention has been the "solvophobic theory," which relates retention to the surface tension of the mobile-phase solvents (103). As important as the solvophobic theory has been to the development of modern LC, it is based on an incorrect model of the relevant solution processes. It supposes that retention can be modeled in terms of the association of two solute molecules in a single solvent rather than on the transfer of a solute from one solvent to another. Hence the solvophobic theory does not take cognizance of the interactions of the solute with the second "solvent," the cavity in the stationary phase; it takes into account only the cavity in the mobile phase.

The actual partitioning process involves (a) the creation of a solute-sized cavity in the stationary phase, (b) the transfer of the solute from the mobile to the stationary phase, and (c) the closing of a solute-sized cavity in the mobile phase. Furthermore, the alkyl chains of the stationary phase cannot be completely bulklike, since they are constrained by the interface. Chains in the bulk state are defined as those that have the freedom to explore all possible conformations. But when chains are grafted to an interface, such as silica, two constraints prevent access to all possible conformations. The first is the boundary condition imposed by the interface; certain configurations are prohibited by the requirement that the chain cannot penetrate the solid interface to which it is grafted. The second constraint, which applies only at sufficiently high surface densities of the grafted chains, arises from lateral interactions among neighboring chains. Both constraints cause interfacially grafted chains to be more "ordered" than bulk chains. In the present context, ordering refers to the partial alignment of the chains normal to the interface. Such interfacially constrained systems of chains have been referred to as "interphases."

Dill has described mean-field statistical mechanical theory which accounts for the interactions of the solute in both the mobile and stationary phases (104,105), and Dorsey and Dill have reviewed the theory and experimental verification of theory of retention in reversed-phase liquid chromatography (106). Partitioning is strongly dependent on the surface density of the grafted chains, increasing with surface coverage of the silica by hydrocarbon, until it reaches a point at which lateral packing constraints among neighboring chains give rise to chain ordering. Beyond that density, further increases in surface density lead to entropic expulsion of solute (106,107). The chain anisotropy at these higher densities also leads to higher solute selectivities (106,108). Hence the stationary phase plays a role of fundamental importance in retention and selectivity in reversed-phase liquid chromatography.

This statistical mechanical theory has dealt only with small molecules, and only with pure alkyl chain phases. There is still much that is not understood, including effects from the variation of pore diameter of the silica, the effect of other functional groups such as phenyl and cyano, and retention processes involving larger molecules, including both synthetic and biopolymers.

Antle et al. (109,110) have studied variations in retention and selectivity among different reversed-phase columns. They showed that differences in solute retentivity were correlated with three effects: (a) the effective phase ratio of the column as measured by the average retention of all solutes; (b) the "polarity" of the bonded phase; and (c) the dispersion solubility parameter of the bonded phase. The phase ratio of the column is a function of the chain length, the bonding density, and the surface area of the silica, and is almost never reported. This alone may account for many discrepancies in the literature concerning the effects of chain length.

The choice of functional group is of interest to both those interested in the fundamental mechanisms of retention as well as to practicing chromatographers. Antle et al. (109,110) have suggested the use of C_8, cyano, and phenyl columns for maximum change in column selectivity, or as apices of the "column selectivity triangle." These columns have independently been found to provide a wide range of selectivity for the separation of PTH-amino acids (100). Cooper and Lin (111) used solutes and solvents at the apices of Snyder's solvent selectivity triangle to systematically characterize retention on these three types of columns, and concluded that differences in retention are due primarily to differences in basic group selectivities of the three phases. These differences in basic group selectivities were found not to affect the retention of nonbasic solutes. The three solvents which give maximum difference in retention are methanol, acetonitrile, and tetrahydrofuran. Their location in the Snyder solvent triangle is shown in Fig. 19.

E. Secondary Equilibria

Certainly one of the major advantages of the aqueous-based mobile phases used in reversed-phase liquid chromatography is the ability to control secondary chemical equilibria. In liquid chromatography the primary equilibrium is the distribution of the solute between the stationary and mobile phases. Any other equilibrium involving the solute in the mobile or stationary phases is considered "secondary." These secondary equilibrium processes change the chemical form of the solute, and can be used advantageously to change retention of solutes that will enter into the secondary equilibrium. Typical equilibria that are employed are ionization control of weak acids and bases, ion pairing of strong electrolytes with a hydrophobic counterion, complexation of a metal ion or ligand, and solute–micelle association. Both Karger et al. (112) and Foley (113) have thoroughly reviewed the use of secondary equilibria in liquid chromatography. Foley and May have addressed the optimization of these types of separations, and have derived a general equation to predict the optimum mobile-phase conditions (114,115). They showed that the selectivity of secondary chemical equilibria-based separations had been substantially underestimated because nonoptimum mobile-phase conditions were employed in previous selectivity estimates.

Ionization control and ion pairing are the two most popular techniques used. Ionization control gives the ability to separate weak acids and bases based on differences in the pK values. Figure 21 is a dramatic example of the effect of pH on the separation of two weak acids with pK_a values of 3.68 and 3.85, and shows the value and utility of the optimization theory of Foley and May (114,115).

Ion pairing extends reversed-phase chromatography to the separation of ionic solutes, which are generally unretained by hydrophobic stationary phases. A hydrophobic counterion, such as octyl sulfonate or tetrabutyl ammonium ion, is added to solution and forms a coulombic ion pair with oppositely charged

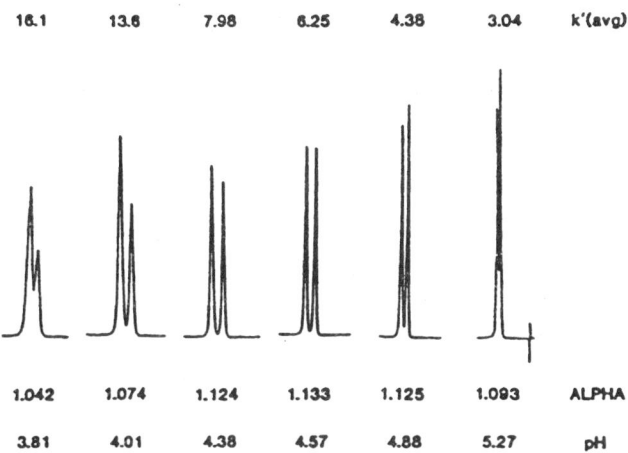

16.1	13.6	7.98	6.25	4.38	3.04	k'(avg)

1.042	1.074	1.124	1.133	1.125	1.093	ALPHA

3.81	4.01	4.38	4.57	4.88	5.27	pH

Fig. 21 Separation of 3- and 4-chlorobenzoic acid; pK_a 3.68 and 3.85, respectively. Optimum pH predicted from theory of Ref. 114 was 4.58, optimum found was 4.57 ± 0.05. Changes in relative peak heights with pH are due to changing molar absorptivities, not a reversal in elution. (From Ref. 114. Reprinted courtesy of American Chemical Society.)

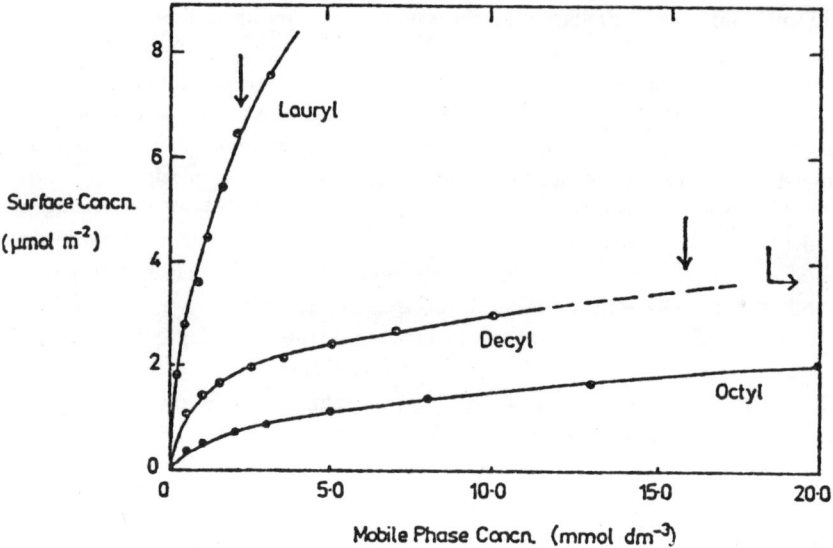

Fig. 22 Adsorption isotherms of alkyl sulfates on ODS Hypersil from standard mobile phase of water–methanol, 80:20, made 0.02 M in phosphate buffer pH 6.0 and 0.05 M in Na$^+$. (From Ref. 116. Reprinted courtesy of Elsevier Scientific Publishers.)

solutes. The mechanism of retention of these systems is still somewhat subject to debate. What is clear is that these separation systems are not very robust; that is, the separations are extremely sensitive to changes in temperature, organic modifier content, salt content, concentration of counterion, etc. The hydrophobic counterions are in general surface active, and adsorb on the stationary phase. Ion-pairing chromatography is performed at counterion concentrations on the steep slope of the adsorption isotherm of the counterion on the stationary phase, making the technique very sensitive to changing conditions. Figure 22 shows the adsorption isotherms of several typical ion-pairing agents on a C$_{18}$ stationary phase (116).

F. Gradient Elution

Gradient elution, or solvent programming, provides the most general solution to the general elution problem in liquid chromatography. The general elution problem is common with complex mixtures, and can be characterized by poor separation of components with small k' values, excessive separation times for components with large k' values, and broad elution bands that are difficult to detect for components with large k' values. Gradient elution involves a continuous change in mobile-phase conditions to increase the strength of the mobile phase. Gradient elution separations generally provide faster separations with improved limits of detection and less tailing for most compounds present in the sample. Gradient elution theory has been largely developed by Snyder, and an excellent treatment is available (117).

The first consideration in developing an appropriate gradient is the shape of the gradient, usually described as the change in volume composition of the "B" solvent, or strong solvent. A linear solvent strength (LSS) gradient is the desired form, as this provides approximately constant peak variance, equal resolution for both early- and late-eluting peaks, and regular band spacing. The actual change of solvent composition then depends on the retention processes of the type of chromatography being utilized. It is generally assumed that a linear solvent composition change with time provides a LSS gradient for reversed-phase, normal-bonded phase, and ion-exchange chromatography with pH gradients. A concave solvent composition change with time provides an approximately LSS gradient for liquid–solid chromatography and ion exchange with salt gradients.

As well as gradient shape, gradient steepness is important. This is generally described as

$$\varphi' = \frac{\text{change in volume fraction B}}{\text{time}} \tag{32}$$

Well-retained compounds essentially do not move until $k'_{\text{instantaneous}} < 10$, where $k'_{\text{instantaneous}}$ is the capacity factor of a solute under the mobile-phase conditions *at that instant at the top of the column*. This means that injected solutes remain at the top of the column until the gradient has progressed to the point that $k'_{\text{instantaneous}}$ is approximately 10. If the gradient is too steep, there will be poor resolution; and if the gradient is too shallow, broad peaks will result and time will be wasted. Just as there are optimal values for k' for isocratic separations, there are also optimum values for gradient steepness. These are based on the same considerations of the opposing figures of merit of resolution and peak height. Table 4 shows approximate optimum values of φ' for reversed-phase columns of differing void times for seven typical "B" solvents (118).

Just as important as steepness is consideration of the concentration of the beginning and ending solvent. If the beginning, or "A" solvent, is too strong, or the ending, or "B" solvent is too weak, the quality of the separation will be degraded. In the first case there will be poor resolution of the first-eluting peaks, and in the second case compounds will continue to elute after the end of the gradient, affecting detectability and time of analysis. If the A solvent is too weak, or the B solvent too strong, all peaks will occupy one part of the chromatogram and time will be wasted. Ideally, the first peak should have $k' \leq 1$, and the last peak should elute just as the final solvent reaches the column exit.

Table 4 Optimal Gradient Steepness (for maximal resolution) in Reversed-Phase LC

Solvent B	φ' (%v/min change in B)			
	$t_o = 10$ s	$t_o = 30$ s	$t_o = 1$ min	$t_o = 2$ min
Methanol	40	13	6.7	3.3
Acetonitrile	39	13	6.5	3.2
Acetone	35	12	5.9	3.0
Dioxane	34	11	5.7	2.8
Ethanol	33	11	5.6	2.8
Isopropanol	29	10	4.8	2.4
Tetrahydrofuran	27	9	4.5	2.3

From Ref. 118.

There are also several practical considerations in designing a gradient elution separation. First, in reversed-phase chromatography, water quality is crucial. Any trace organic contaminants will collect at the top of the column during the aqueous-rich portion of the gradient, and will elute as ghost peaks as the solvent strength increases. A blank gradient, consisting of the exact program with no sample injection, should always be run to check for ghost peaks. Second, there may be reproducibility problems at extremes of either the A or B solvent. Here the proportioning valve for a low-pressure mixing system is working at an extremely low time base, or with a high-pressure mixing system the pump may be pumping at less than its certified minimum flow rate.

Perhaps the worst problem of gradient elution separations is the need to reequilibrate the column with the initial solvent before a second sample can be run. An often-quoted rule of thumb is that up to 20 column volumes of the initial solvent may be necessary for this reequilibration process. The best test of reequilibration is the elution time of a weakly retained solute. These solutes will be greatly affected by an incompletely equilibrated stationary phase, and the retention time will vary. Cole and Dorsey have described a simple and convenient method for the reduction of column reequilibration time following gradient elution reversed-phase chromatography (119). Their method utilizes the addition of a constant 3% 1-propanol to the mobile phase throughout the solvent gradient to provide consistent solvation of the stationary phase. They noted reductions in reequilibration times of up to 78%!

G. Physicochemical Measurements

As well as for chemical analysis, liquid chromatography can be used to obtain physicochemical information. A recent book (120) and two reviews have dealt with these techniques (121,122). While diffusion coefficients, kinetic parameters, critical micelle concentrations of surfactants, and other information have been estimated from chromatographic data, by far the most common application is the estimation of hydrophobic parameters, especially as models of partitioning processes. These estimates are based on linear free-energy relationships and are then used as estimates of bioavailability, bioaccumulation, soil sorption and transport, etc.

An entire area of chemistry, often referred to as quantitative structure activity relationships (QSAR), has been built around estimation of partitioning, availability, and activity of compounds, and the most often used measure is the octanol–water partition coefficient. The *chromatographic* estimation of these values is often referred to as quantitative structure retention relationships (QSRR). Liquid chromatography has a number of advantages over traditional shake-flask methods, including the ability to measure several solutes simultaneously, access to a much wider dynamic range of the partition coefficient, and the ability to handle impure solutes. Unfortunately, there is much contradiction in the literature about the usefulness of the chromatographic correlations. It is not surprising that there is little agreement from one report to another. Liquid chromatography has long had a reputation for having little reproducibility of separation from columns of one manufacturer to another, which is a function of differences in both the base silica as well as in the density of the grafted alkyl chains. Illustrative of this problem is the comparison of two reports correlating chromatographic retention with octanol–water partition coefficients. Thus and Kraak (123) reported that a phenyl bonded phase gave significantly better correlations than an octadecyl bonded phase. Minick et al. (124), using the same type of phenyl column, but a different octadecyl column, reported significantly better correlations with the C_{18} column! Recent reports suggest that capillary electrophoresis in the micellar mode (MEKC) may be *much* more useful for the estimation of octanol–water partition coefficients than LC. Herbert and Dorsey described a universal calibration curve for over 100 solutes with widely varying functionality, covering in excess of 9 orders of magnitude in log K_{ow} and 4 orders of magnitude in log capacity factor (125). This method reduces the laboratory-to-laboratory variability and the long analysis time due to the multiple mobile phases necessary in current LC methods, while retaining many of the desired advantages of chromatographic techniques.

Hopefully, these problems will be resolved in the near future. DeYoung and Dill (126) have shown that the partition coefficient of benzene into bilayer membranes decreases with increasing chain density of the bilayer membranes, analogous to the decrease in partitioning of solutes to reversed-phase stationary phases of sufficiently high chain density (vide supra). This gives hope that reversed-phase stationary phases of significantly higher chain density than now commercially available will give true models of biological partitioning processes.

Hsieh and Dorsey have recently shown such promise (127), with a study using a high-density C_{18} stationary phase. They found that correlations of log k'_w (retention in 100% water) with bioavailability are equivalent to or better than correlations of bioavailability with the octanol–water partition coefficient.

V. CONCLUSIONS

Modern liquid chromatography is a mature technique, but is far from being static in terms of understanding and discoveries. Innovations and improvements in liquid chromatography are now coming from *chemistry* research, rather than instrumentation research. The future will bring more selective separations, improved understanding of retention processes, better estimates of physicochemical information, and untold hours of fun for those involved in making these discoveries.

REFERENCES

1. M Tswett. Ber Deut Botan Geo 24:385, 1906.
2. LR Snyder, JJ Kirkland. Introduction to Modern Liquid Chromatography. New York: Wiley-Interscience, 1979.
3. LS Ettre. Evolution of Liquid Chromatography: A Historical Overview. In Cs. Horvath, ed. High Performance Liquid Chromatography: Advances and Perspectives. Vol. 1. New York: Academic Press, 1980.
4. JJ Kirkland, ed. Modern Practice of Liquid Chromatography. New York: Wiley-Interscience, 1971.
5. AJP Martin, RLM Synge. Biochem J 35:1358, 1941.
6. JJ van Deemter, FJ Zuiderweg, A Klinkenberg. Chem Eng Sci 5:271, 1956.
7. JP Foley, JG Dorsey. Anal Chem 55:730, 1983.
8. JC Giddings. Unified Separation Science. New York: Wiley-Interscience, 1991.
9. JM Davis, JC Giddings. Anal Chem 55:418, 1983.
10. M Martin, DP Herman, G Guiochon. Anal Chem 58:2200, 1986.
11. JC Giddings. Anal Chem 35:2215, 1963.
12. CA Cramers, JA Rijks, CPM Schutjes. Chromatographia 14:439, 1981.
13. H Colin, G Guiochon, M Martin. In H Engelhardt, ed. Practice of High Performance Liquid Chromatography: Applications, Equipment, and Quantitative Analysis. Berlin: Springer-Verlag, 1986, chap. 1.
14. JJ Kirkland, WW Yau, HJ Stoklosa, CH Dilks Jr. J Chromatogr Sci 15:303, 1977.
15. H Colin, M Martin, G Guiochon. J Chromatogr 185:79, 1979.
16. JC Sternberg. Adv Chromatogr 2:205, 1966.
17. B Coq, G Creiter, JL Rocca, M Porthault. J Chromatogr Sci 19:1, 1981.
18. JS Landy, JL Ward, JG Dorsey. J Chromatogr Sci 21:49, 1983.
19. RE Majors. LC-GC 12:890, 1994.
20. DS Horne, JH Knox, L McLaren, Sep Sci 1:531, 1966.

21. JH Knox, JF Parcher. Anal Chem 41:1599, 1969.
22. JH Knox, GR Laird, PA Raven. J Chromatogr 122:129, 1976.
23. JP Foley, JG Dorsey. Chromatographia 18:503, 1984.
24. RW Stout, JJ DeStefano, LR Snyder. J Chromatogr 261:189, 1983.
25. JW Dolan. LC, Liq Chromatogr HPLC Mag 3:92, 1985.
26. NHC Cooke, BG Archer, K Olsen, A Berick. Anal Chem 54:2277, 1982.
27. G Guiochon. Anal Chem 53:1318, 1981.
28. M Novotny. Anal Chem 60:500A, 1988.
29. BL Karger, M Martin, G Guiochon. Anal Chem 46:1640, 1974.
30. NHC Cooke, K Olsen, BG Archer. LC, Liq Chromatogr HPLC Mag 2:514, 1984.
31. JG Atwood, GL Schmidt, W Slavin. J Chromatogr 171:109, 1979.
32. MW Dong, JR Gant, PA Perrone. LC, Liq Chromatogr HPLC Mag 3:787, 1985.
33. RK Gilpin, WR Sisco. J Chromatogr 194:285, 1980.
34. J Chmielowiec, H Sawatzky. J Chromatogr Sci 17:245, 1979.
35. H Poppe, JC Kraak, JFK Huber, JHM van den Berg. Chromatographia 14:515, 1981.
36. H Poppe, JC Kraak. J Chromatogr 282:399, 1983.
37. RJ Perchalski, BJ Wilder. Anal Chem 51:774, 1979.
38. PR Fielden. J Chromatogr Sci 30:45, 1992.
39. RPW Scott. Liquid Chromatographic Detectors. 2nd ed. Amsterdam: Elsevier, 1986.
40. G Patonay, ed. HPLC Detection. New York: VCH, 1992.
41. GL Long, JD Winefordner. Anal Chem 55:712A, 1983.
42. GKC Low, PR Haddad. J Chromatogr 198:235, 1980.
43. RM McCoy, RL Aiken, RE Pauls, ER Ziegel, T Wolf, GT Fritz, DM Marmion. J Chromatogr Sci 22:425, 1984; 24:273, 1986.
44. AN Papas, MF Delaney. Anal Chem 59:54A, 1987.
45. AN Papas. CRC Crit Rev Anal Chem 20:35, 1989.
46. AN Papas, TP Tougas. Anal Chem 62:234, 1990.
47. JW Dolan. LC-GC 4:526, 1986.
48. T Wolf, GT Fritz, LR Palmer. J Chromatogr Sci 19:387, 1981.
49. LA Larkins, SG Westcott. Anal Proc (London) 23:258, 1986.
50. RE Pauls, RW McCoy, ER Ziegel, GT Fritz, DM Marmion, DL Krieger. J Chromatogr Sci 26:489, 1988.
51. JB Castledine, AF Fell. J Pharm Biomed Anal 11:1, 1993.
52. WA Dark. J Chromatogr Sci 24:495, 1986.
53. ES Yeung, RE Synovec. Anal Chem 58:1237A, 1986.
54. UAT Brinkman, GJ De Jong, C Gooijer. Pure Appl Chem 59:625, 1987.
55. WD Pfeffer, ES Yeung. Anal Chem 58:2103, 1986.
56. T Imasaka, N Yamaga, N Ishibashi. Anal Chem 59:419, 1987.
57. MJ Kerkhoff, JD Winefordner. Anal Chim Acta 175:257, 1985.
58. DJ Desilets, PT Kissinger, FE Lytle. Anal Chem 59:1830, 1987.
59. DC Johnson, SG Weber, AM Bond, RM Wightman, RE Shoup, IS Krull. Anal Chim Acta 180:187, 1986.
60. K Stulik. Anal Chim Acta 273:435, 1993.
61. RE Shoup. Liquid chromatography/electrochemistry. In Cs. Horvath, ed. High Per-

formance Liquid Chromatography: Advances and Perspectives. Vol. 4. New York: Academic Press, 1986.

62. JG White, RL St. Claire III, JW Jorgenson. Anal Chem 58:293, 1986.
63. JG White, JW Jorgenson. Anal Chem 58:2992, 1986.
64. K Slais. J Chromatogr Sci 24:321, 1986.
65. SA Borman. Anal Chem 57:1124A, 1985.
66. H Small, TS Stevens, WC Bauman. Anal Chem 47:1801, 1975.
67. P Jandik, PR Haddad, PE Sturrock. CRC Crit Rev Anal Chem 20:1, 1983.
68. UAT Brinkman. Chromatographia 24:190, 1987.
69. WJ Bachman, JT Stewart. LC-GC 7:38, 1989.
70. JF Garcia, D Barcelo. J High Res Chromatogr 16:633, 1993.
71. PR Griffiths, SL Pentoney, A Giogetti, KH Shafer. Anal Chem 58:1349A, 1986.
72. DA Laude Jr, CL Wilkins. TrAC, Trends Anal Chem 5:230, 1986.
73. LF Colwell, RA Hartwick. J Liq Chromatogr 10:2721, 1987.
74. H Engelhardt, H Elgass. Liquid chromatography on silica and alumina as stationary phases. In Cs. Horvath, ed. High Performance Liquid Chromatography: Advances and Perspectives. Vol. 2. New York: Academic Press, 1980.
75. KK Unger. Anal Chem 55:361A, 1983.
76. H Billiet, C Laurent, L de Galan. TrAC, Trends Anal Chem 4:100, 1985.
77. KK Unger, Porous Silica: Its Properties and Use as Support in Column Liquid Chromatography. New York: Elsevier, 1979.
78. L Szepesy, C Combellas, M Caude, R Rosset. J Chromatogr 237:65, 1982.
79. C Souteyrand, M Thibert, M Caude, R Rosset. J Chromatogr 262:1, 1983.
80. C Souteyrand, M Thibert, M Caude, R Rosset. J Chromatogr 316:373, 1984.
81. LR Snyder. Mobile phase effects in liquid solid chromatography. In Cs. Horvath, ed. High Performance Liquid Chromatography: Advances and Perspectives. Vol. 3. New York: Academic Press, 1983.
82. M Jaroniec, DE Martire. J Chromatogr 351:1, 1986.
83. PK Dasgupta. Anal Chem 64:775A, 1992.
84. J Weiss. Ion Chromatography. 2nd ed. New York: VCH, 1995.
85. HA Laitinen, GW Ewing, eds. A History of Analytical Chemistry. Washington, DC: Division of Analytical Chemistry, American Chemical Society, 1977, pp. 309–310.
86. S Elchuk, RM Cassidy. Anal Chem 51:1434, 1979.
87. WW Yau, JJ Kirkland, DD Bly. Modern Size Exclusion Chromatography: Practice of Gel Permeation and Gel Filtration Chromatography. New York: Wiley, 1979.
88. LS Ettre, A Zlatkis, eds. 75 Years of Chromatography: A Historical Dialog. Amsterdam: Elsevier, 1979.
89. JG Dorsey, WT Cooper. Anal Chem 66:857A, 1994.
90. LC Sander, SA Wise. J Chromatogr 316:163, 1984.
91. BW Sands, YS Kim, JL Bass. J Chromatogr. 360:353, 1986.
92. SP Boudreau, WT Cooper. Anal Chem 61:41, 1989.
93. LC Sander, SA Wise. CRC Crit Rev Anal Chem 18:299, 1987.
94. LC Sander, SA Wise. Anal Chem 56:504, 1984.
95. LR Snyder. LC, Liq Chromatogr HPLC Mag 1:478, 1983.
96. LR Snyder. J Chromatogr Sci 16:223, 1978.

97. JL Glajch, JJ Kirkland, KM Squire, JM Minor. J Chromatogr 199:57, 1980.
98. WT Cooper, PL Smith. J Chromatogr 355:57, 1986.
99. PL Smith, WT Cooper. Chromatographia 25:55, 1988.
100. JL Glajch, JJ Kirkland. Anal Chem 55:319A, 1983.
101. GA Howard, AJP Martin. Biochem J 46:532, 1950.
102. KK Unger, N Becker, P Roumeliotis. J Chromatogr 125:115, 1976.
103. W Melander, Cs. Horvath. Reversed phase chromatography. In Cs. Horvath, ed. High Performance Liquid Chromatography: Advances and Perspectives. Vol. 2. New York: Academic Press, 1980.
104. KA Dill. J Phys Chem 91:1980, 1987.
105. KA Dill, J Naghizadeh, JA Marqusee. Annu Rev Phys Chem 39:425, 1988.
106. JG Dorsey, KA Dill. Chem Rev 89:331, 1989.
107. KB Sentell, JG Dorsey. Anal Chem 61:930, 1989.
108. KB Sentell, JG Dorsey. J Chromatogr 461:193, 1989.
109. PE Antle, AP Goldberg, LR Snyder. J Chromatogr 321:1, 1985.
110. PE Antle, LR Snyder. LC Mag 2:840, 1984.
111. WT Cooper, LY Lin. Chromatographia 21:335, 1986.
112. BL Karger, JN LePage, N Tanaka. Secondary chemical equilibria. In Cs. Horvath, ed. High Performance Liquid Chromatography: Advances and Perspectives. Vol. 1. New York: Academic Press, 1980.
113. JP Foley. Chromatography 2(5):43, 1987.
114. JP Foley, WE May. Anal Chem 59:102, 1987.
115. JP Foley, WE May. Anal Chem 59:110, 1987.
116. JH Knox, RA Hartwick. J Chromatogr 204:3, 1981.
117. LR Snyder. Gradient elution. In Cs. Horvath, ed. High Performance Liquid Chromatography: Advances and Perspectives. Vol. 1. New York: Academic Press, 1980.
118. LR Snyder, JW Dolan, JR Gant. J Chromatogr 165:3, 1979.
119. LA Cole, JG Dorsey. Anal Chem 62:16, 1990.
120. R Kaliszan, Quantitative Structure-Chromatographic Retention Relationships. New York: Wiley, 1987.
121. R Kaliszan. CRC Crit Rev Anal Chem 16:323, 1986.
122. JG Dorsey, MG Khaledi. J Chromatogr 656:485, 1993.
123. JLG Thus, JC Kraak. J Chromatogr 320:271, 1985.
124. DJ Minick, JJ Sabatka, DA Brent. J Liq Chromatogr 10:2565, 1987.
125. BJ Herbert, JG Dorsey. Anal Chem 67:744, 1995.
126. LR DeYoung, KA Dill. Biochemistry 27:5281, 1988.
127. MM Hseih, JG Dorsey. Anal Chem 67:48, 1995.

4
Mass Spectrometry in Pharmaceutical Analysis

Frank J. Belas* and Ian A. Blair
*Center for Cancer Pharmacology, University of Pennsylvania,
Philadelphia, Pennsylvania*

I. INTRODUCTION

Mass spectrometry (MS) has become one of the most important analytical tools employed in the analysis of pharmaceuticals. This can most likely be attributed to the availability of new instrumentation and ionization techniques that can be used to help solve difficult bioanalytical problems associated with this field (1–8). Perhaps the best illustration of this occurrence is the development of electrospray (ESI) and related atmospheric-pressure ionization (API) techniques, ionspray (nebulizer-assisted API), turbo ionspray (thermally assisted API), and atmospheric pressure chemical ionization (APCI; nebulization coupled with corona discharge), for use in drug disposition studies. The terms ESI and ionspray tend to be used interchangeably in the literature. For the purpose of this review, the term API will be used to describe both ESI and ionspray. In recent years there has been an unprecedented explosion in the use of instrumentation dedicated to API/MS (4,6,8–14). API-based ionization techniques have now become the method of choice for the analysis of pharmaceuticals and their metabolites. This has made thermospray (TSP), the predominant LC/MS technique during the 1980s, obsolete (15). Numerous reports describing the utility of API/MS for pharmaceutical analysis have appeared in the literature over the last decade (7). The

* Current affiliation: Lilly Research Laboratories, Lilly Corporate Center, Indianapolis, Indiana.

increased sensitivity that can be attained with API-based methodology, together with the availability of robust instrumentation, has resulted in TSP being phased out as a practical technique in pharmaceutical analysis. Thus, methodology based on TSP ionization will not be discussed in detail in this review. The pharmaceutical industry has certainly been at the forefront of implementing API/MS, and this, perhaps, reflects the utility of this technique for the analysis of pharmaceuticals. Several reviews have appeared recently concerning the use of MS in drug disposition and pharmacokinetics (6,16–18).

The use of stable isotope analogs has assumed increasing importance in drug disposition studies (18–20). The isotope cluster technique is particularly useful. This method allows for the recognition of metabolites in complex biological matrices such as urine and plasma (21,22). Numerous methods have been developed for the deconvolution of isotope clusters from stable isotope analogs so that mass spectral identification can be facilitated (20,23). In the field of pharmacokinetics, stable isotope analogs are used as internal standards for quantitative purposes (24), to determine bioequivalence (25,26) and in chiral drug disposition (27,28). In studies regarding chiral drugs, the use of a pseudo-racemate, where one enantiomer is labeled and the other is unlabeled, provides the basis for a robust method that can be used to determine pharmacokinetic parameters of individual enantiomers and their metabolites (27–31). An exciting innovative approach to the use of stable isotopes was pioneered by Abramson and his co-workers (32), who introduced the chemical reaction interface (CRI) mass spectrometer for isotope-independent quantitative analysis. This type of approach could ultimately provide an alternative to utilizing radioisotopes in drug disposition studies (33,34). There is a genuine need for this type of methodology in view of the increasing cost of disposal of radioactive waste, the cost of radiolabeled compounds, and the increasing reluctance to dose normal subjects with radiolabeled compounds.

Tandem MS (MS/MS) is used in combination with a number of different ionization techniques for pharmaceutical analysis (7,35). In fact, a major factor that has ensured the rapid acceptance of API-based methodology is the ability to conduct collision-induced dissociation (CID) of molecular ions and MS/MS analysis. The resultant product ion spectra (36), constant neutral loss (CNL) spectra (37–40), or precursor ion spectra (41) can be obtained with extremely high sensitivity. The combination of CID with a soft ionization process that produces abundant molecular species provides very powerful methodology for use in structural studies. It is also possible in some instances to obtain significant structural information using API techniques without employing a tandem mass spectrometer, by performing CID in the API source of a single-stage mass spectrometer (in-source collisions) (42). Thus, even relatively simple single-quadrupole instrumentation can often provide rich structural information. The enormous literature that has accrued over the last decade illustrates the importance of MS in the field of pharmaceutical analysis. The present review will focus on literature published since 1991 and will use selected examples rather than attempting to cite all the references that have accrued in the field. This should provide the reader with sufficient information to allow consultation of all the relevant background sources.

II. INTRODUCTION TO IONIZATION TECHNIQUES

Electron ionization (EI) is the oldest available technique in MS and is still useful, particularly when compounds can be derivatized to improve their EI characteristics (18). Although numerous soft ionization techniques are available for the analysis of drug conjugates, in some suitable cases derivatization can be performed in order to facilitate their analysis by EI/MS (43). The advantage of EI is that extensive fragmentation can occur in the source of the mass spectrometer, and this can provide a wealth of structural information. Furthermore, the fragmentation processes that occur are very reproducible. Thus, it is relatively easy to reproduce spectra obtained in other laboratories. Because it is a relatively soft ionization technique, positive chemical ionization (PCI) is often used to complement data obtained with EI. Fragmentation processes are normally much reduced compared with EI, and protonated molecular or adduct ions are detected, depending on the type of gas employed (18). This can be extremely useful, both in structural studies and in quantitative studies, where inadequate specificity is obtained using EI.

The exquisite sensitivity of electron-capture negative chemical ionization (ECNCI)/MS has facilitated quantitative determinations of drugs (18,27,44,45–51) at femtomole and attomole concentrations. ECNCI/MS is the method of choice when pharmacokinetic determinations require a very high level of sensitivity. The method normally requires that analyte molecules are derivatized so that they can capture thermal electrons generated in the source of the mass spectrometer. Because it is relatively easy to prepare in high yield under very mild conditions, the pentafluorobenzyl (PFB) derivative has found the widest applicability (52). The derivative is extremely efficient in capturing gas-phase thermal electrons, but paradoxically, it is lost during ionization through a homolytic cleavage to produce the molecular negative ion (M) − PFB. This process is known as dissociative electron capture and occurs with almost all PFB derivatives, independent of whether they are attached to oxygen or nitrogen. Dissociative electron capture provides an anion that contains the intact analyte molecule and virtually no fragmentation is observed. Therefore, assays using this technique have high specificity in addition to high sensitivity.

Liquid secondary-ion (LSI) MS has the highest sensitivity when using Cs^+ ionization, but more studies have been conducted using fast-atom bombardment (FAB) with Xe atoms, probably because of the wider availability of this technique

(18). Most studies have employed the use of off-line LC purification of metabolites followed by LSI analysis using FAB/MS (53). One problem that is often encountered in LSI/MS is the interference from material present in organic solvents used to extract drugs and their metabolites from biological fluids. Even HPLC-grade solvents contain significant amounts of impurities. A recent study has advocated the use of reversed-phase extraction cartridges in the final stage of sample purification to remove such impurities (54).

FAB/MS has made an enormous contribution to the structural characterization of drug conjugates. This technique has been responsible for the characterization of structures that would have previously required a formidable effort. Conjugates tend to show prominent ions from the protonated molecule in the positive-ion mode or ions from molecules that have lost a proton in the negative-ion mode. Polar metabolites that have been identified by FAB and LSI/MS include sulfates (55), glutathione adducts (39), ether-linked glucuronides (56,57), ester-linked glucuronides (58), and N-glucuronides (59–61). Significant improvements in sensitivity have also been obtained using flow techniques (62). However, there are few reports concerning the use of flow LSI/MS for trace analysis, presumably because of the reluctance to use fused-silica capillary columns. Most reports on the use of flow techniques either employ direct infusions or sample splitting from a conventional column. Although TSP ionization has played an important role in the past, it has largely been superseded by API methods for the analysis of pharmaceutical compounds. For a thorough discussion of this methodology, one should refer to recent reviews (6,18).

III. ATMOSPHERIC-PRESSURE IONIZATION (API) TECHNIQUES

API techniques have experienced tremendous growth in the past few years, and are unquestionably the most rapidly evolving techniques currently being used in the analysis of pharmaceuticals. The tremendous interest in API and APCI stems from the recognition that they are powerful techniques for both structural and quantitative analyses of drugs and their metabolites (5,10,13,63–68). More traditional ionization techniques, such as particle-beam (PB), TSP, continuous-flow (CF) LSI, and CF-FAB, have now been overshadowed by the utility of API technology (7,8,14). The relatively low cost and high sensitivity of quadrupole ion trap (IT) instruments has further increased the availability and utility of API/MS (69,70). However, CF-LSI/MS is still being employed in selected applications, particularly with magnetic sector instruments (71,72). Other novel techniques, such as LC coupled with time-of-flight (TOF)/MS (73) and capillary electrophoresis (CE)/MS, are maturing to the point where they are having a significant impact in the area of pharmaceutical analysis (74).

For API-based ionization techniques, LC is typically conducted using narrow-bore (2.1-mm inside diameter) or conventional columns (4.6-mm inside diameter). The narrow-bore columns are ideally suited for LC/MS coupling because the optimal flow rates of these columns (between 100 and 500 μL/min) can be accommodated quite easily by current API interfaces. If sufficient quantities of material are available, it is possible to split the LC effluent from the column into the API source. The effluent that does not pass into the API source can be collected and stored for use in additional analytical studies. The mechanism by which API results in the conversion of charged droplets to ions in the gas phase is not completely understood (9,63,75). However, it is known that the charged droplets contain both positive and negative ions, and the predominant charge is dependent on the polarity of the induced potential. Heat exchange with a counter-current flow of gas causes rapid size reduction of the droplets until the coulombic forces overcome attractive forces, and the droplets explode into smaller droplets. These droplets undergo further ion evaporation so that ions are transferred from the condensed phase to the gaseous phase and are then analyzed by the mass spectrometer (9). Ionization of compounds using API-based techniques appears to be much less sample-dependent than TSP ionization. The interface is not heated; therefore, the thermal degradation of samples, often observed using TSP, is eliminated.

API has been particularly useful in the analysis of platinum-containing drugs and their metabolites (76). An impressive study on the metabolism of the HMG Co-A reductase inhibitor pravastatin used API (77). Biotransformation pathways that were elucidated included: isomerization, ring hydroxylation, ω-1 oxidation of the ester side chain, β-oxidation of the carboxy side chain, ring oxidation followed by aromatization, oxidation of a hydroxyl group, and conjugation. Structural assignments were confirmed by ^{1}H-NMR spectroscopy. API has been used in the analysis of sulfonamides (78) and a series of β-agonists (79). In one study (67), LC/API/MS was employed to investigate the metabolism of the dopamine D_2 agonist N-0923 by using an in-vitro isolated liver perfusion system. This method did not require the use of radioactivity or extensive sample clean-up procedures. The parent drug was found to be metabolized to at least 15 different metabolites, nine of which were new. All metabolites were exclusively excreted into the bile, except for the despropyl metabolite, which was also detectable in the perfusate. 5-O-Glucuronidation and N-depropylation followed by 5-O-glucuronidation were the most important metabolic routes. N-dealkylation of the thienylethyl group followed by 5-O-glucuronidation and sulfation was a second major metabolic pathway.

LC/MS has also proved useful in obtaining information concerning stereochemical factors in drug disposition studies (80). LC/MS was used to determine the identity of the metabolites excreted in bile after isolated rat liver perfusions with the quaternary ammonium derivatives of the enantiomeric drugs dextrorphan and levorphanol. The drugs were labeled with deuterium and mixed with unlabeled drugs to create an artificial isotope pattern in the mass spectrum. This aided in the recognition of unknown metabolites. In mass spectra that were recorded under normal conditions, fragmentation was absent and metabolites of N-methyl dextrorphan and N-methyl levorphanol were visible as parent "doublets." The metabolite structures were further confirmed using CID experiments. For N-methyl dextrorphan, the glucuronide, the glutathione conjugate, and the glucuronide of the N-demethylated metabolite were found in bile. For N-methyl levorphanol, the glucuronide, the glutathione conjugate, the sulfate conjugate, and the glucuronide of a hydroxylated N^4-methyl levorphanol were excreted in bile. Thus, LC/MS allowed for the demonstration of the remarkable selectivity occurring in the metabolism of these quaternary ammonium compounds in the rat liver.

LC/MS was also used to study the deposition of cyclosporine and its metabolites in needle biopsy samples from kidney and liver, where sample size was limited (66). The limit of detection in the single-ion monitoring mode was 500 fg (450 amol). Several other thermally labile drugs and their metabolites have been analyzed by API-based techniques, including metabolites of the neuroleptic agent haloperidol (81), and a potentially neurotoxic pyridinium haloperidol metabolite (82), the polyether antibiotic semduramicin (83), the PGE_2 antagonists SC-42867 and SC-51089 (84), the H_2-receptor antagonist famotidine (68), metabolites of the antitumor agent 5,6-dimethylxanthenone (53), and the immunosuppressant cyclosporine A (85). API/MS has also been used to identify the presence of over 40 metabolites of the antiulcer agent omeprazole (22).

IV. COLLISION-INDUCED DISSOCIATION (CID)

CID requires the tandem pairing of two mass analyzers (tandem mass spectrometers). In this technique, an ion of one particular mass is allowed to pass through the first analyzer into a field-free region, where it collides with an inert gas to form product (product) ions. The product ions are then separated in the second analyzer, which in turn transmits them to the detector (4,35). The type of instrumentation used in these studies can be a triple-stage quadrupole mass spectrometer, a four-sector mass spectrometer, or a hybrid of sector and quadrupole mass spectrometers. In the triple-stage quadrupole instrument, the first and third quadrupoles act as the first and second analyzers, respectively. CID is conducted within the second analyzer stage (quadrupole or octapole), where only the Rf is varied. The tandem four-sector instruments employ an electrostatic and magnetic sector as the first analyzer and an electrostatic and magnetic sector as the second analyzer; a collision cell is located between the two analyzers. CNL scans (conducted using quadrupole instruments) can be employed to screen for particular metabolites that may be present in a complex biological matrix. In this mode of operation, the first analyzer is scanned through the desired mass range and the second analyzer is scanned at a constant mass lower than the first analyzer. This makes it possible to characterize molecules with particular structural features such as glucuronides or sulfates. Metabolites can then be characterized based on their full CID spectra (37,86–90). Another useful mode of operation involves scanning for parents of selected product ions. In this mode, the second mass analyzer is fixed to transmit the product ions of interest, and the first mass analyzer is scanned over a wide mass range to detect the parent ions for this product ion. A signal is obtained on the detector when a true parent ion is allowed to pass through the first mass analyzer, and the appropriate product ion is transmitted through the second mass analyzer. This can be particularly useful when it is difficult to detect a molecular ion because of interfering substances that are present in the biological matrix. A combination of CNL and parent-ion scans can be used to classify the number and type of primary metabolites and polar drug conjugates that are present in a complex sample matrix (4,91–93). The resulting information can be used to assess the overall biotransformation routes that are available to a drug or other xenobiotic substance.

LC/API/MS/MS is rapidly becoming the method of choice for the structural characterization of drug metabolites. For example, novel glutathione conjugates of disulfiram and diethyldithiocarbamate (94) and valproic acid derivatives (92) were identified using this technique. This methodology has also proved useful in the analysis of chemotherapeutic agents. One example of this technique is the first direct structural elucidation of glucuronide conjugates of tamoxifen in human urine samples using on-line LC/API/MS/MS (95). A minor N-oxide metabolite was also identified in plasma extracts using CID. In another study (96), the mechanism of cytotoxicity of the antineoplastic mitoxantrone was determined using LC/MS/MS techniques. The identification in bile and urine of glutathione and N-acetylcysteine conjugates of 2-chloroethyl isocyanate as metabolites of the antitumor agent N,N'-bis(2-chloroethyl)-N-nitrosourea have also been reported using LC/API/MS/MS methods (97). CID in combination with API and MS/MS provided structural information, allowing for the identification of common fragmentation pathways and the differentiation of isomeric sulfonamides. In another study (98), following intraperitoneal administration of Compound A (a breakdown product of the volatile anesthetic, sevoflurane), the presence in bile of two types of Compound A–glutathione conjugates, and the urinary excretion of two types of Compound A–mercapturic acid conjugates, was demonstrated by using API–MS/MS. MS/MS has also been used for quantitative purposes (99,100). A particularly innovative study involved the cross-validation of a radioimmunoassay for the class II antiarrhythmic agent MK-0499 in human plasma and urine using LC/API/MS/MS (101). Another study (99) quantitated dihydrocodeine and its metabolite, dihydromorphine, in human serum using GC–MS/MS (in multiple reaction monitoring mode) after one simple extraction step and derivatization to their respective pentafluoropropionic esters. The sensitivity of the method was excellent and allowed for the reproducible quantification of dihydrocodeine and dihydromorphine with limits of quantification of 2 ng/mL and 40 pg/mL of serum, respectively.

FAB is a particularly powerful structural tool when used in combination with CID and MS/MS. FAB MS/MS has been used in numerous studies to characterize oligonucleotide adducts of the antitumor drug cisplatin (102); show that an impurity in the oxytocin antagonist atosiban contained 5-aminovaleric acid instead of a proline (103); identify the in-vitro metabolites of cyclosporine A (85); analyze 2-amino-1-benzylbenzimidazole and its metabolites (88); characterize metabolites of the H_2-receptor antagonist mifentidine (104); analyze metabolites from the antimuscarinic agent cimetropium bromide (37,54); identify S-oxidized metabolites of the investigative calcium channel blocker, AJ-2615, in rat plasma (105); and identify N-acetylcysteine conjugates of 1,2-dibromo-3-chloropropane (106). These reports illustrate the diverse applications of CID in combination with FAB and MS/MS that have appeared in recent years. It is anticipated that API techniques will make an equally valuable contribution to the field of drug disposition in the next few years.

An interesting application of CID involves the use of on-line CF dialysis TSP coupled with MS/MS for quantitative screening of drugs in plasma (107). The potential utility of the method was demonstrated by the quantitative analysis of the anticancer drug rogletimide in the plasma of patients after treatment. In-vivo microdialysis and TSP/MS/MS of the dopamine uptake blocker 1-[2-[*bis*(4-fluorophenyl)methoxy]ethyl]-4′-(3-phenylpropyl)-piperazine (GBR-12909) was conducted in the rat (108). The maximum concentration of GBR-12909 in the brain for a dose of 100 mg/kg i.p. was determined to be 250 nmol/L, with the maximal concentration occurring approximately 2 h post injection. This represents a 40-fold lower concentration of GBR-12909 in the brain as compared to cocaine concentrations obtained at a dose of 30 mg/kg. The authors suggested that this could explain the discrepancy between relative in-vivo and in-vitro potencies of the two drugs. A combination of microdialysis and FAB/MS/MS was used to follow the pharmacokinetics of penicillin G directly in the bloodstream of a live rat (109). After intramuscular injection of the antibiotic, the blood dialysate was allowed to flow into the mass spectrometer via the continuous-flow/FAB interface.

V. ATMOSPHERIC-PRESSURE CHEMICAL IONIZATION (APCI)

APCI is becoming increasingly popular for the routine analysis of drugs and their metabolites. The ionization of analytes by APCI results from the effect of a corona discharge rather than through ion evaporation, which occurs with other API techniques (7,63). This provides several advantages over other API techniques for the accurate and precise analysis of drugs and their metabolites in biological fluids. Thus, APCI can be employed in combination with LC flow rates of up to 1 mL/min, making it possible to use conventional (4.6-mm) LC columns. These columns are more rugged (and permit faster reequilibration after gradient elution) than the microbore (2.0-mm-i.d.) columns normally used at lower flow rates. In addition, greater volumes can be injected on-column, which makes it easier to filter and transfer aqueous solutions containing the analytes. Perhaps the major advantage of APCI is that, in contrast to conventional API methodology, little suppression of ionization by the constituents of the biological matrix is observed (110). For assays that use minimal extraction, when there is likely to be substantial contamination from the biological matrix, APCI is clearly the method of choice. Mobile-phase additives are often required either to improve chromatography or to improve the sensitivity of APCI (111). Assays based on LC/MS tend to perform minimal chromatography and rely on the power of the mass spectrometer to provide specificity so that large numbers of samples can be analyzed. However, great care has to be taken to ensure that none of the metabolites or constituents from the biological matrix interferes in the analysis (7,110). In order to eliminate such potential problems, care must be taken to ensure that potentially interfering substances are well separated. The use of stable-isotope internal standards eliminates any problems that could arise through suppression of the internal standard signal by endogenous contaminants or by metabolites that may be present in different plasma samples.

APCI was used to determine the metabolic fate of the antineoplastic drug 1-(2-chloroethyl)-3-cyclohexyl-1-nitrosourea (CCNU) (112). The identification of carbamoylated thiol conjugates as products of CCNU in rats and humans was a novel finding, in that it represents for the first time structurally characterized metabolic species that provide evidence for the in-vivo carbamoylating activity of chloroethylnitrosoureas. In rats, 4-hydroxycyclohexyl, 3-hydroxycyclohexyl, and cyclohexyl isocyanate were trapped and identified as glutathione conjugates in the bile and as N-acetyl-L-cysteine conjugates in urine. In the case of the patient, the N-acetyl-L-cysteine conjugates of 4-hydroxycyclohexyl and 3-hydroxy-cyclohexyl isocyanate were identified in the urine. LC/APCI/MS/MS has also been employed for monitoring human urine extracts for the conjugated metabolites of 4-hydroxyandrost-4-ene-3,17-dione, an anticancer drug (113). This provided the first evidence of the presence of the sulfate conjugates and analogs in patients' urine. The same technique of APCI/MS/MS has allowed the determination of the β-adrenergic blocker timolol in plasma after ocular administration to volunteers. Using multiple reaction monitoring, timolol and its deuterated internal standard were detected quantitatively (114). This technique was also used recently in a quantitative assay for the antibiotic azithromycin in human serum (115).

LC/APCI/MS/MS in the negative-ion mode was employed for quantitation of the glucuronide metabolites of the new pharmaceutical agent BW 1370U87 (116). In an excellent study, Matuszeweski et al. (110) demonstrated that assays based on LC/MS/MS can sometimes provide erroneous data. They demonstrated that conventional API/MS techniques such as ESI, ionspray, and turboionspray may be adversely affected by lack of specificity and selectivity. This could arise through ion suppression caused by the sample matrix or result from interference by an unknown metabolite. LC/APCI/MS/MS proved to be the technique that was least affected by such problems. However, the use of conventional chromatographic methodology was advocated until assay specificity could be ascertained. By paying attention to such details, high sensitivity and reproducibility was attained for analysis of the human 5α-reductase inhibitor, finasteride, at concentrations in the pg/mL range (110).

VI.   ADDITIONAL TECHNIQUES

A.   Chemical Reaction Interface Mass Spectrometry (CRIMS)

CRIMS is a relatively new technique that has been shown to be useful in detecting stable isotopes (117–119). The CRI completely decomposes analytes to individual atoms in a helium flow that comes directly from a GC or through an LC interface. The atoms are then allowed to recombine with atoms from a reagent gas such as oxygen or sulfur dioxide. The elemental and isotopic characteristics of these newly formed molecules (usually NO or CO_2) are monitored by a conventional mass spectrometer. A number of studies have been conducted using this technique (118–122). The accuracy of the LC/CRIMS method was evaluated in two studies by comparing the LC/CRIMS data with on-line radioactivity detection (33,34). The results of these studies indicated that the LC/CRIMS technique is a stable-isotope-sensitive and compound-independent detection tool that is especially suitable for drug metabolism studies. An excellent correlation between the CRIMS and the radioactive methods was observed. The use of CRIMS in combination with continuous-flow isotope ratio MS (123) could provide both mass balance data on total metabolites as well as quantitative information on each individual metabolite.

B.   Capillary Electrophoresis Mass Spectrometry (CE/MS)

Another interfacing technique that has stimulated significant interest in recent years involves the interfacing of CE with API/MS (124–130). In this technique, charged compounds are separated under the influence of a high electric field in small-diameter capillary tubes filled with buffer. Efficient separations, short analysis times, and low sample consumption make CE a very attractive separation technique. Developments in CE/MS interfacing have been based either on API at atmospheric pressure or on continuous-flow (CF) FAB. The CE/MS interface is needed to facilitate coupling of the low-CE buffer flow to the mass spectrometer or to introduce the matrix required for the CF-FAB process. A make-up liquid is introduced either by a liquid junction or as a sheath flow arrangement. The buffer in the liquid junction is typically the same as the separation buffer for CE and provides a suitable make-up buffer flow to sustain a stable spray to the API mass spectrometer. The coupling of CE with API requires some compromise to be reached with the electrophoretic buffer in order to obtain a high ion current response. Involatile buffers such as sodium citrate, phosphate, and borate, commonly used in CE separations, are generally not appropriate in conjunction with MS, because the ion evaporation mechanism which produces gas-phase ions under these conditions operates best with volatile buffers at low concentrations. On-line CE in combination with API/MS was employed for the analysis of metabolites of the antiestrogen, tamoxifen (128). An SDS concentration of 7 mM lowered the API/MS signal response of N-desmethyltamoxifen by a factor of approximately 3. However, separation of tamoxifen metabolites using 7 mM SDS was augmented relative to the unadulterated methanol electrolyte. This enabled the separation of α-hydroxytamoxifen and 4-hydroxytamoxifen, which were not resolvable in methanol electrolyte devoid of SDS. A recent study has demonstrated that CE can be interfaced with an IT and that quantitative measurements can be performed in biological fluids (74).

VII. FUTURE DIRECTIONS IN LC/MS INTERFACING

A. Electrospray Ionization and Sector Instruments

Interfacing API with sector instruments makes it possible to perform mass measurements with relatively high accuracy. This can provide a direct means to determine the charge state of a particular peak in an ion envelope, which can be particularly important in structure elucidation of macromolecular biomolecules. A number of groups have initiated research in this area (131–133). In one study (132), accurate mass measurements were obtained for a number of compounds with molecular weights in the 500–16,000 Da range. Mass accuracies in the low-parts-per-million range were obtained for positive and negative ions. These mass measurements were sufficiently accurate to permit the determination of elemental compositions for smaller molecules, and they provided confidence in the mass assignments for larger biomolecules. The API mass spectrum of the lysozyme $[M + 9H]^{9+}$ species measured at a resolution of 10,000 permitted the determination of the masses of individual isotopes to within 0.15 Da (12 ppm) of the theoretical value.

B. Electrospray Ionization and Fourier-Transform Ion Cyclotron Resonance Mass Spectrometry (FT–ICR/MS)

Considerable effort has also been invested into the coupling of API and FT–ICR/MS for the analysis of drugs and biomolecules (134–138). High-resolution measurements to obtain accurate molecular masses and improvements in absolute detection limits by signal averaging and other techniques are the goals of these studies. One interesting study (139) combined CE–API/FT–ICR/MS for the direct analysis of proteins. The on-line acquisition of high-resolution mass spectra (average resolution ≥45,000) was determined for both the α- and β-chains of hemoglobin acquired from the injection of 10 human erythrocytes (corresponding to 4.5 fmol of hemoglobin). Several groups have also coupled MALDI with FT–ICR/MS (140,141). This technique is particularly useful for high-sensitivity analysis of high-molecular-weight biomolecules. FT–ICR/MS is becoming a more general and routinely applicable instrument with potential utility in pharmaceutical analyses.

C. Electrospray Ionization and Ion Trap Mass Spectrometry (IT/MS)

The interfacing of API with the IT is another rapidly emerging area that offers new analytical opportunities for pharmaceutical laboratories (142). ITs combine advantages of quadrupole filters such as ease of operation and unit-mass resolution with some advantages of FT–ICR/MS instruments such as trapping and selection of ions, multistage MS/MS, and high-resolution measurements. The excellent sensitivity of the ITs and their low cost relative to conventional quadrupole filters suggests that bench-top IT instruments for LC/MS/MS will eventually compete with triple-stage quadrupole instruments for pharmaceutical analysis. On-line LC/MS with ITs has been described for thermospray (143), MALDI (144), particle-beam (145,146), and API interfaces (126,142,147–149). In addition, a new method of selected ion monitoring has recently been described that is capable of unit-mass isolation throughout the operation mass range of an IT mass spectrometer (150).

D. Electrospray Ionization and Time-of-Flight Mass Spectrometry (TOF/MS)

Another area of recent interest is the interfacing of API techniques with TOF mass spectrometers. TOF/MS instruments combine ease of operation, relatively low cost, excellent ion transmission, and virtually unlimited mass range. The only significant disadvantage with respect to other mass spectrometers is the limited mass resolution. A tremendous effort in development and performance optimization of TOF instruments is stimulated by MALDI as a powerful ionization technique for biological macromolecules (4,151,152). Other groups have described API in conjunction with TOF instruments (153–155). In one of these studies (155), it was discovered that by orienting the ion source perpendicular to the field-free drift region, the longitudinal energy spread of the ion packet was substantially reduced. This allowed a mass resolving power of over 1000 to be achieved for both low-mass and high-mass ions of biological interest. Also, the sensitivity of the instrument allowed for the routine detection of low-picomole and sub-picomole quantities of large, multiply charged species such as cytochrome c. The potential utility of this instrument for conducting rapid off-line screening of chromatographic effluents is evident in light of its simplicity, rapid scanning speed, and high sensitivity. A combination of quadrupole or IT and TOF analyzers will permit relatively high-resolution MS/MS studies, making it possible to characterize product ions derived from unknown metabolites more accurately (73).

VIII. METABOLISM STUDIES

A. In-Vitro Studies

Polar conjugates of drugs and their metabolites can be prepared by synthesis or from in-vitro incubations. Unusual N-glucuronides of the calcium channel antagonists gallopamil and verapamil were synthesized in an elegant study by Mutlib and Nelson (156). FAB mass spectra of the conjugates were obtained and characteristic ions were found in the FAB mass spectra of glucuronides isolated from the bile of rats dosed with gallopamil and verapamil. Fortunately, glucuronides can also be readily prepared in vitro. For example, glucuronides of the antiepileptic agent lamotrigine (157) the neuroleptic fluphenazine (56), and the combined α- and β-adrenoceptor antagonist labetalol (158) were prepared by UDPGA-fortified microsomal preparations. The lamotrigine glucuronide was unusual in that glucuronidation occurred at N^2 of the triazine ring, which led to the formation of a quaternary glucuronide. Glutathione conjugates can be prepared either synthetically or by the use of immobilized glutathione transferases. The availability of these conjugates has allowed their mass spectral characteristics to be examined so that they can then be identified from in-vivo metabolism studies. In addition to the preparation of conjugates, in-vitro studies can be useful for helping to establish the primary routes of metabolism of a particular drug. Perfused rat liver preparations have provided valuable information concerning the metabolism of the dopamine D_2 agonist N-0923 (67); the morphinan analogs N-methyl dextrorphan and N-methyl levorphanol (80); the antitumor agent ET18-OME (159); the inotropic agent ethimizol (160); and the antitumor agent 5,6-dimethylxanthenone (53).

Hepatocytes and microsomes can also provide information concerning pathways of metabolism that occur in vivo (81,86,96,128,161–165). In one of these studies (162), metabolism of the antimalarial agent WR 238605 was investigated using rat liver microsomes. Metabolism involved O-demethylation, N-dealkylation, N-oxidation, and oxidative deamination. In addition, C-hydroxylation involving the 8-aminoalkylamino side chain, which was previously unknown for 8-aminoquinoline analogs, was found to be an important metabolic pathway. In another study (164), it was shown that the oxidation of the angiotensin II receptor antagonist losartan to its active carboxylic acid metabolite E3174 in human liver microsomes was catalyzed by CYP2C9 and CYP3A4. The immunosuppressive agent FK506, a 23-member macrolide, was shown to produce two metabolites when incubated with human microsomes (166). However, these were eventually shown to be different conformers with the same structure. The negative FAB mass spectrum of the metabolite showed a molecular ion at m/z 766, indicating that it was simply a demethylation product, although it was not possible to determine which of the seven methyl groups had been lost. Hepatic microsomes from rat, monkey, and humans were used to study the oxidative metabolism of L-746,530 and L-739,010, two potent and specific 5-lipoxygenase inhibitors (163). These experiments have shown that L-739,010 and L-746,530 can form intermediates that irreversibly bind to the hepatic microsomal proteins and that the dioxabicyclo portion of the molecule is a reactive moiety. Covalent binding is an important issue because of its potential to mediate toxic responses, such as idiosyncratic and immunoallergic reactions. Differences in the in-vitro metabolism of the neuroleptic drug haloperidol were demonstrated by using mouse and guinea pig hepatic microsomes (81). In another study, rat hepatocytes were used to examine metabolism of the NSAID ibuprofen. Cellular extracts and the media were analyzed for ibuprofen using a stereoselective GC/MS assay (167). When (R)-(−)-ibuprofen was incubated with the hepatocytes, its concentration declined in an apparent first-order manner with the concomitant formation of metabolites. (R)-(−)-ibuprofen was also shown to undergo a chiral inversion to the (S)-(+)-enantiomer. However, the (S)-(+)-enantiomer was not converted to the (R)-(−)-enantiomer, indicating that chiral inversion was unidirectional in these cells. Rat and human hepatocyte cultures were also used to elucidate the metabolism of the PGE$_2$ antagonists SC-42867 and SC-51089 (84). In a recent study (168), incubations of primary cultures of rat hepatocytes with CI-937 resulted in the formation of three glutathione conjugates and one glucuronic acid conjugate. The structures of the glutathione conjugates were established by reference synthesis with activated horseradish peroxidase, LC/MS/MS, and two-dimensional NMR measurements. The glucuronic acid derivative of CI-937 was identified by MS.

Microsomal incubations coupled with mass spectral analysis of products has also been conducted for the H$_2$-receptor antagonist mifentidine (104), the immunosuppressant tacrolimus (169), the antipsychotic drug tiospirone (170), and the antifertility drug norgestimate (171). Primary metabolites of the important immunosuppressive drug cyclosporine isolated from human urine were shown to undergo further biotransformations when incubated with human liver microsomes (172). Five of the 14 new metabolites were characterized by FAB/MS.

In-vitro studies have also been conducted in order to provide mechanistic support for proposed pathways of metabolism. Three studies have examined the metabolism of valproate by rat liver mitochondria. In two of these studies, it was demonstrated that valproic acid undergoes metabolism through β-oxidation (173,174). An unidentified metabolite observed in these studies was later shown by LSI/MS/MS to be valproyl-AMP formed during the activation of valproic acid in rat liver (175). Aldehyde oxidase isolated from mouse liver cytosol was shown by in-vitro studies to be involved in the oxidative metabolism of the anti-schistosomal agent niridazole (176). Metabolism of the LTD_4 antagonist verlukast was investigated using rat liver and kidney cytosols (72). In rat liver cytosol, the metabolite was a 1,4 Michael addition product in which GSH had added to position 12 of the styryl quinoline of verlukast. Incubation with kidney cytosol produced the GSH, cysteinylglycine, and cysteine conjugates of verlukast. In bile collected from rats dosed intravenously with 50 mg/kg of verlukast, approximately 80% of the dose was recovered up to 4 h post dose. The GSH conjugate accounted for 16.5% of the dose. The cysteinylglycine, cysteine, and N-acetylcysteine conjugates were observed and together accounted for 7.5% of the dose. Verlukast accounted for 14.5%, and the remainder of the metabolites (40.5%) were oxidation or acyl glucuronide metabolites.

The glucuronide metabolite of AZT was isolated from rat and human liver microsomal incubations (177), and the formation of a toxic metabolite of AZT was demonstrated in rat hepatocytes and liver microsomes (177). The metabolism of tamoxifen was examined in human liver homogenate and human Hep G2 cell line preparations by LC/API/MS (178). Several metabolites were detected in the human liver homogenate extracts, namely, N-didesmethyltamoxifen, α-hydroxytamoxifen, 4-hydroxytamoxifen, N-desmethyltamoxifen, and tamoxifen N-oxide. All of these metabolites, except the N-didesmethyltamoxifen, were observed in the samples after incubating tamoxifen with the human Hep G2 cell line. In-vitro studies have also used MS to examine potential drug–drug interactions. For example, (179) demonstrated that the α_2-agonist, dexmedetomidine, inhibited metabolism of the anesthetic alfentanil, whereas clonidine had no effect.

B. In-Vivo Studies in Animal Models

Drug metabolism studies in animal models provides important information concerning the structures of metabolites before embarking on costly human studies. MS is used to provide confirmatory information if synthetic metabolites are available and to help in the structural characterization of novel metabolites. Many studies have employed off-line HPLC purification of urinary metabolites followed by mass spectral analysis. Surprisingly, in view of the recognized utility of soft ionization techniques, a number of studies published over the last few years have employed EI for the structural characterization of metabolites. This probably reflects the availability of EI instrumentation coupled with the ease of operation in this mode, although studies which use API techniques are becoming much more prevalent.

There have been numerous studies involving in-vivo animal models to elucidate the metabolism of particular drugs (22,91,98,119,180–182). The nonsteroidal antiestrogen tamoxifen is subject to extensive biotransformation in humans and laboratory animals (183). In particular, the dimethylamino group of tamoxifen undergoes N-demethylation and formal replacement with a hydroxyl group, affording the major metabolites tamoxifen amine and tamoxifen alcohol, respectively. In this particular study in ovariectomized rats, tamoxifen was eliminated in part as metabolites arising from conversion of its basic side chain to an oxyacetic acid moiety. Thus, tamoxifen acid was characterized spectrally from the urine of rats after intraperitoneal administration of tamoxifen. It was not detected in fecal extracts. In contrast, a second metabolite, 4-hydroxy-tamoxifen, was detected and characterized only from fecal extracts, indicative of a qualitative difference in routes of elimination for tamoxifen and 4-hydroxy-tamoxifen. Fifteen metabolites of the anticonvulsant drug stiripentol were identified in the urine of rats treated with the drug (184). The major pathway of metabolism involved hydroxylation of the methylenedioxy ring. Yet another study involved the characterization and formation of the glutathione conjugate of clofibric acid (185). In-vivo studies indicated that, following an intravenous infusion of clofibric acid to rats (75 mg/kg), the concentration of clofibryl glutathione excreted in bile over 4 h was about 0.1% of the concurrent clofibryl glucuronide concentrations. Although these results indicated a minor role for glutathione-catalyzed reactions in clofibrate metabolism in vivo, they did define 1-O-acyl glucuronides as a new class of substrates for glutathione S-transferase. The tricyclic antidepressant, trimipramine, was shown to undergo extensive metabolism in the rat (186). In this extremely thorough study, 20 metabolites were identified in the urine after hydrolysis with β-glucuronidase. Four major routes of metabolism were identified: aliphatic and aromatic hydroxylation, and two different kinds of N-demethylation. In another study, four new rat urinary metabolites of the potent antihistamines, tripelennamine and pyrilamine, were identified (187). The metabolites arose by N-dealkylation pathways most likely as a consequence of oxidation α-to the nitrogen atom undergoing N-dealkylation. An interesting mechanism for the depyridination of tripelennamine and pyrilamine is proposed that involves the formation of an oxazine intermediate. In a recent study (98), following intraperitoneal administration of Compound A (a breakdown product of sevoflurane), the presence in rat bile of two types of Compound A–glutathione conjugates, and the urinary excretion of two types of Compound A–mercapturic acids conjugates, was demonstrated by API/MS/MS. In other studies, a new urinary hydroxylated metabolite of aprophen in the rat was identified (188), and the antiarrhythmic agent, bucromarone was shown to undergo both O- and N-dealkylation in rats and mice (189).

A thiazolidinedione hypoglycemic agent, CP-68,722, was shown to undergo biotransformation to seven metabolites in the rat (190). Five of the metabolites arose by hydroxylation and one by oxidation of the chromone. Chlorpheniramine, a potent antihistamine, underwent extensive metabolism in the rat and was shown to form primarily hydroxylated metabolites (21). The new antiatherosclerotic agent CI 976 was shown to undergo both β- and ω-oxidation in the rat in a manner analogous to long-chain fatty acids (191). A new *meta*-hydroxylated metabolite of the antiarrhythmic drug mexiletine was identified in the hydrolyzed urine of rats dosed with the drug (192). Three metabolites of the new anticonvulsant drug felbamate that arose through oxidation and hydrolysis were identified in the rat, rabbit, and dog (193). Unchanged drug and metabolites were excreted mainly in the urine. In an extension of earlier in-vitro studies, stable-isotope methodology was used to investigate the mechanism by which (R)-(−)-ibuprofen undergoes chiral inversion in vivo in the rat (194). The data were consistent with stereoselective formation of a CoA thioester of (R)-(−)-ibuprofen and conversion of this metabolite to an enolate tautomer. This resulted in the formation of a symmetrical intermediate through which racemization of ibuprofen occurred in vivo. Two novel N-oxide metabolites of metyrapone, a diagnostic drug used to test pituitary function, were identified in the urine of rats dosed with the drug (195). The N-oxides were surprisingly stable under EI conditions, and it was possible to observe molecular ions for both of the metabolites. A combination of EI and PCI/MS was used in the identification of urinary metabolites of imidapril, a new ACE inhibitor (196), in several animal models.

LSI/MS methodology has been employed in the analysis of peptide drugs, a new and important area for pharmaceutical research. For example, three truncated peptide metabolites of the synthetic decapeptide RS-26306 were characterized by FAB/MS (197). Initial FAB analyses were unsuccessful due to interfering substances derived from the biological matrix. However, hexyl ester formation allowed protonated molecular ions to be detected. The synthetic anticoagulant decapeptide MDL 28,050 underwent hydrolysis at four peptide bonds when administered to rats, and metabolites were excreted in the urine (198). Six resulting peptides, a *des*-alanine impurity present in the drug together with a metabolite derived from this impurity, were identified by CF–LSI/MS. Two biliary metabolites of irinotecan, a new antitumor agent, were identified in the rat (199). The first characterization of intact sulfate and glucuronide metabolites of the neuroleptic fluphenazine was conducted using FAB/MS (55). An interesting study concerning the metabolism of chlorpromazine N-oxide revealed that it could be reduced and excreted as an ether glucuronide in the bile of rats dosed with the drug (200). This is the first systematic study of N-oxide disposition that has been conducted using modern techniques of structural characterization. A combination of LSI and EI/MS was used in metabolism studies of the anticancer agent crisnatol (201). PCI/MS has been used primarily in quantitative studies and in combination with other ionization techniques, but there are two recent reports on its use in combination with MS/MS. Using this technique, it was possible to identify 17 hydroxylated urinary metabolites of praziquantel in the mouse (202). As noted above, numerous reports of API techniques for drug metabolism studies are appearing in the literature (22,203). A combination of API MS/MS and FAB/MS/MS was used to identify the major metabolites of the anticancer agent mitoxantrone in pig urine (204). In a recent elegant study, LC/API/MS/MS was used in combination with NMR spectroscopy to identify four glutathione conjugates (in rat bile) derived from a nephrotoxic degradation product of the anesthetic agent sevoflurane (205).

C. In-Vivo Studies in Humans

Human drug metabolism studies have made extensive use of EI methodology (206,207). The uricosuric drug benzbromarone was found to be extensively metabolized in humans, forming two major metabolites (206). The metabolites were 1'-hydroxybenzbromarone and 6-hydroxybenzbromarone. Benzbromarone was hydroxylated in vivo at the prochiral center C1' to 1'-hydroxybenzbromarone. Analysis of 1'-hydroxybenzbromarone from plasma and urine extracts by chiral HPLC revealed that two peaks were eluted which showed a mean enantiomeric ratio of 2:1 for plasma and 7:3 for urine. These data demonstrate that the formation and elimination of this metabolite is enantioselective. LSI/MS was used in studies that were focused upon the identification of polar conjugates (208,209). FAB/MS provided fascinating data regarding 12 new cyclosporine metabolites present in human bile from liver-grafted patients being treated with cyclosporine (172). One of the metabolites was a glucuronide conjugate and the others arose primarily through oxidative metabolism. The monohydroxylated, carboxylated metabolite was found to be elevated in patients with cholestasis. FAB was also used to identify a hydroxylated metabolite of diflunisal in human urine (210). FAB/MS/MS was used to determine the metabolic fate and disposition of taxol in cancer patients (211). Total urinary excretion was 14.3 ± 1.4% (SE) of the dose, with unchanged taxol and an unknown polar metabolite as the main excretion products. Total fecal excretion was 71.1 ± 8.2%, with 6α-hydroxytaxol being the largest component. Unchanged taxol and four other metabolites were also identified from fecal extracts.

Many of the published studies concerning human metabolism have used either EI or LSI techniques. However, as with animal studies, PCI has made a contribution in selected cases. In the metabolism of tiospirone it was found that sulfoxide and sulfone metabolites could be readily detected by PCI (170). Evidence for oxidative activation of the anticancer agent mitoxantrone was obtained by PCI MS/MS (212). GC/PCI/MS provided an efficient method for profiling metabolites derived from phenobarbitone, primidone, and their N-methyl and N-ethyl derivatives (213). The difficulty often encountered in obtaining sufficient quantities of metabolites for profiling studies in human subjects has stimulated a number of studies using highly sensitive ECNCI/MS methodology. Valproic acid and 14 of its metabolites were quantified using four internal standards in human urine (214). A carbomylglucuronide of the antiviral agent rimantadine was identified in human urine (215). In an important study (216), GC/ECNCI/MS was used to demonstrate that the mucolytic agent S-carboxymethyl-L-cysteine (CMC) was metabolized to thiodoglycolic acid, its sulfoxide, and (3-carboxymethylthio)lactic acid. Thus, CMC is not excreted as the sulfoxide or as its decarboxylation product. This careful study has provided compelling evidence that CMC does not undergo polymorphic sulfoxidation, as had been suggested previously.

There are a number of reports concerning the use of API-based techniques in human metabolism studies (64,113,178,217–222). Metabolism of the new H_2-receptor antagonist ebrotidine was investigated using LC/MS (221). Using APCI in both the positive and negative modes allowed for the identification of ebrotidine, 4-bromobenzenesulfonamide, and four S-oxidized metabolites in human urine. Employing LC MS/MS techniques (220), the antipsychotic agent iloperidone was found to be extensively metabolized to a number of metabolites by rats, dogs, and humans. It was shown that iloperidone was metabolized via O-dealkylation, oxidative N-dealkylation, reduction, and hydroxylation. Identification of some of the unknown metabolites in rat bile was successfully achieved by a combination of LC/NMR and LC/MS with a minimum amount of sample clean-up. The utility of coupling a semipreparative HPLC to an LC/MS instrument for further characterization of collected metabolites was also demonstrated. The major metabolites of stanozolol in human urine were identified using LC APCI and API/MS/MS (203). Conjugated metabolites were identified using the API technique. The identification of two human urinary metabolites of the antitumor agent biantrazole (CI-941) was determined using LC/API methodology (217). One metabolite was identified as an oxidation product of CI-941 with both side chains oxidized at the hydroxymethylene groups, and other metabolite was the analogous monooxidation product. The biotransformation of pravastatin in humans was studied using API/MS (77). The parent drug was the major drug-related material found in the urine, and at least 15 metabolites were also detected. None of the metabolites accounted for more than 6% of the dose; but careful use of API/MS and NMR spectroscopy made it possible to provide structures for virtually all of them. In another study (112), the metabolism of the antineoplastic agent 1-(2-chloroethyl)-3-cyclohexyl-1-nitrosourea (CCNU) was examined using LC/MS/MS techniques.

IX. SUMMARY

This review highlights the important role of MS in the analysis of pharmaceuticals. GC/EI/MS and GC/ECNCI/MS were used extensively for quantitative analyses until the early 1990s. Over the last few years there has been a dramatic shift to the use of methodology based on LC/API/MS (6,18). However, for assays that require the ultimate in sensitivity, GC/ECNCI/MS is still the method of choice. There have been several reports concerning the coupling of CE with API/MS for drug analysis (74,130). Whether this technique has practical utility for routine drug disposition studies remains to be seen. The power of MS/MS in combination with API ionization techniques has been aptly illustrated in the study of Weidolf and Covey (22), in which more than 40 metabolites were detected. The development of quantitative methodology for the analysis of pharmaceuticals and their metabolites is still heavily reliant on the availability of authentic standards. This can overcome the potential problem of analyzing individual metabolites that may have quite different ionization characteristics. The groundbreaking studies of Abramson and his colleagues on the development of CRIMS may eventually provide methodology for quantitative studies that are compound-independent (32). This will permit quantitative analyses to be performed without the need to synthesize authentic standards. The introduction of MALDI/MS (151,223) will affect significantly our ability to analyze macromolecular drugs such as proteins and targeted monoclonal antibodies (224). The wide availability of LC/API/IT methodology is making it a valuable technique for the identification of unknown drug metabolites (70,225). Some reports are emerging describing the use of LC/API/IT for routine quantitative determinations (69), although it is not yet fully accepted for such applications. LC/API/TOF and various hybrid variations including the quadrupole/TOF and IT/TOF (73) instruments afford higher resolution than can be obtained with triple-quadrupole instruments. Therefore, this methodology should also find a niche for drug metabolism studies. There are several groups using a combination LC/MS with LC/NMR (226) in order to facilitate metabolite identification, particularly for unstable compounds. This combination of techniques holds significant promise for the rapid metabolite identification. The search for improved sensitivity continues. LC/MS sensitivity will continue to improve through advances in instrumentation together with the development of novel API sources such as that described recently by Wang and Hacket

(227). It is anticipated that LC/MS methodology will eventually be used to the exclusion of all other techniques for the analysis of pharmaceuticals.

REFERENCES

1. FP Abramson. Mass spectrometry in pharmacology. Meth Biochem Anal 34:289, 1990.
2. AL Burlingame, TA Baillie, DH Russell. Mass spectrometry. Anal Chem 64:467R, 1992.
3. AL Burlingame, RK Boyd, SJ Gaskell. Mass spectrometry. Anal Chem 66:634R, 1994.
4. G Siuzdak. The emergence of mass spectrometry in biochemical research. Proc Natl Acad Sci (USA) 91:11290, 1994.
5. DS Ashton, CR Beddell, BN Green, RWA Oliver. Rapid validation of molecular structures of biological samples by electrospray-mass spectrometry. FEBS Lett 342: 1, 1994.
6. E Gelpi. Biomedical and biochemical applications of liquid chromatography-mass spectrometry. J Chromatogr A 703:59, 1995.
7. E Brewer, J Henion. Atmospheric pressure ionization LC/MS/MS techniques for drug disposition studies. J Pharm Sci 87:395, 1998.
8. WM Niessen. Advances in instrumentation in liquid chromatography-mass spectrometry and related liquid-introduction techniques. J Chromatogr A 794:407, 1998.
9. JB Fenn, M Mann, CK Meng, SF Wong, CM Whitehouse. Electrospray ionization—Principles and practice. Mass Spec Rev 9:37, 1990.
10. M Hamdan, O Curcuruto. Development of the electrospray ionisation technique. Int J Mass Spectrom Ion Processes 108:93, 1991.
11. EC Huang, JD Henion. Packed-capillary liquid chromatography/ion-spray tandem mass spectrometry determination of biomolecules. Anal Chem 63:732, 1991.
12. AP Bruins. Mass spectrometry with ion sources operating at atmospheric pressure. Mass Spec Rev 10:53, 1991.
13. T Wachs, JC Conboy, F Garcia, JD Henion. Liquid chromatography-mass spectrometry and related techniques via atmospheric pressure ionization. J Chromatogr Sci 29:357, 1991.
14. WMA Niessen, AP Tinke. Liquid chromatography-mass spectrometry. General principles and instrumentation. J Chromatogr A 703:37, 1995.
15. P Arpino. Combined liquid-chromatography mass spectrometry. Techniques and mechanisms of thermospray. Mass Spec Rev 9:631, 1990.
16. TA Baillie. Advances in the application of mass spectrometry to studies of drug metabolism, pharmacokinetics and toxicology. Int J Mass Spectrom Ion Processes 118/119:289, 1992.
17. S Naylor, M Kajbaf, JH Lamb, M Jahanshahi, JW Gorrod. An evaluation of tandem mass spectrometry in drug metabolism studies. Biol Mass Spectrom 22:388, 1993.

18. IA Blair. Mass-spectrometry in drug disposition and pharmacokinetics. In: PG Welling, LP Balant, eds. Pharmacokinetics of Drugs—Handbook of Experimental Pharmacology. Berlin: Springer-Verlag, 1994, p. 41.

19. PWN Lee, LO Byerley, EA Bergner. Mass isotopomer analysis: Theoretical and practical considerations. Biol Mass Spectrom 20:451, 1991.

20. MP Barbalas, WA Garland. A computer program for the deconvolution of mass spectral peak abundance data from experiments using stable isotopes. J Pharm Sci 80:922, 1996.

21. F Kasuya, K Igarashi, M Fukui. Metabolism of chlorpheniramine in rat and human by use of stable isotopes. Xenobiotica 21:97, 1991.

22. L Weidolf, TR Covey. Studies on the metabolism of omeprazole in the rat using liquid chromatography/ionspray mass spectrometry and the isotope cluster technique with [^{34}S]omeprazole. Rapid Commun Mass Spectrom 6:192, 1992.

23. K Korzekwa, WN Howard, WF Trager. The use of Brauman's least squares approach for the quantification of deuterated chlorophenols. Biomed Environ Mass Spectrom 19:211, 1990.

24. Y Shinohara, S Baba. Stable isotope methodology in the pharmacokinetic studies of androgenic steroids in humans. Steroids 55:170, 1990.

25. RL Wolen. The application of stable isotopes to studies of drug bioavailability and bioequivalence. J Clin Pharmacol 26:419, 1986.

26. MJ Avery, DY Mitchell, FC Falkner, HG Fouda. Simultaneous determination of tenidap and its stable isotope analog in serum by high-performance liquid chromatography/atmospheric pressure chemical ionization tandem mass spectrometry. Biol Mass Spectrom 21:353, 1992.

27. C Prakash, A Adedoyin, GR Wilkinson, IA Blair. Enantiospecific quantification of hexobarbital and its metabolites in biological fluids by gas chromatography/electron capture negative ion chemical ionization mass spectrometry. Biol Mass Spectrom 20:559, 1991.

28. C Fischer, F Schonberger, W Muck, K Heuck, M Eichelbaum. Simultaneous assessment of the intravenous and oral disposition of the enantiomers of racemic nimodipine by chiral stationary-phase high-performance liquid chromatography and gas chromatography/mass spectroscopy combined with a stable isotope technique. J Pharm Sci 82:244, 1993.

29. A Adedoyin, C Prakash, D O'Shea, IA Blair, GR Wilkinson. Stereoselective disposition of hexobarbital and its metabolites: Relationship to the S-mephenytoin polymorphism in Caucasian and Chinese subjects. Pharmacogenetics 4:27, 1994.

30. UG Eriksson, K-J Hoffman, R Simonsson, CG Regardh. Pharmacokinetics of the enantiomers of felodipine in the dog after oral and intravenous administration of a pseudoracemic mixture. Xenobiotica 21:75, 1991.

31. Y Shinohara, H Magara, S Baba. Stereoselective pharmacokinetics and inversion of suprofen enantiomers in humans. J Pharm Sci 80:1075, 1991.

32. FP Abramson. CRIMS—chemical reaction interface mass spectrometry. Adv Mass Spectrom 13:341, 1994.

33. CA Goldthwaite Jr, F-Y Hsieh, SW Womble, BJ Nobes, IA Blair. Liquid chromatography chemical reaction interface mass spectrometry (LC/CRI MS) as an alter-

native to radioisotopes for quantitative drug metabolism studies. Anal Chem 68: 2996, 1996.

34. FP Abramson, Y Teffera, J Kusmierz, RC Steenwyk, PG Pearson. Replacing C-14 with stable isotopes in drug metabolism studies. Drug Metab Disp 24:697, 1996.
35. C Fenselau. Tandem mass spectrometry: The competitive edge for pharmacology. In: AK Cho, ed. Annual Review of Pharmacology and Toxicology. Palo Alto, CA: Annual Reviews, Inc., 1992, p. 555.
36. DF Hunt, RA Henderson, J Shabonowitz, K Sakaguchi, H Michel, N Sevilir, AL Cox, E Appella, VH Engelhard. Characterization of peptides bound to class I MHC molecule HLA-A2.1 by mass spectrometry. Science 255:1261, 1992.
37. S Naylor, M Kajbaf, JH Lamb, M Jahanshahi, JW Gorrod. Rapid identification of cimetropium bromide metabolites using constant neutral loss tandem mass spectrometry. Biol Mass Spectrom 21:165, 1992.
38. JJ Vrbanac, IA O'Leary, L Baczynskyj. Utility of the parent-neutral loss scan screening technique: Partial characterization of urinary metabolites of U-78875 in monkey urine. Biol Mass Spectrom 21:517, 1992.
39. TA Baillie, MR Davis. Mass spectrometry in the analysis of glutathione conjugates. Biol Mass Spectrom 226:319, 1993.
40. TA Baillie. The role of LC-ionspray MS/MS in studies of drug metabolism and toxicology. 42nd ASMS Conf. Mass Spectrometry and Allied Topics, 1994, p. 862.
41. PE Mirkes, NA Brown, M Kajbaf, JH Lamb, PB Farmer, S Naylor. Identification of cyclophosphamide-DNA adducts in rat embryos exposed *in vitro* to 4-hydroperoxycyclophosphamide. Chem Res Toxicol 5:382, 1992.
42. RD Voyksner, T Pack. Investigation of collisional-activation decomposition process and spectra in the transport region of an electrospray single-quadrupole mass spectrometer. Rapid Commun Mass Spectrom 5:263, 1991.
43. WT Brashear, BR Kuhnert, R Wei. Structural determination of the conjugated metabolites of ritodrine. Drug Metab Dispos 18:488, 1990.
44. A Changchit, J Gal, JA Zirrolli. Stereospecific gas chromatographic/mass spectrometric assay of the chiral labetalol metabolite 3-amino-1-phenylbutane. Biol Mass Spectrom 20:751, 1991.
45. S Komatsu, S Murata, H Aoyama, T Zenki, N Ozawa, M Tateishi, JJ Vrbanac. Micro-quantitative determination of ciprostene in plasma by gas chromatography-mass spectrometry coupled with an antibody extraction. J Chromatogr 568:460, 1991.
46. HJ Leis, E Malle. Deuterium-labelling and quantitative measurement of ketotifen in human plasma by gas chromatography/negative ion chemical ionization mass spectrometry. Biol Mass Spectrom 20:467, 1991.
47. JM Neal, WN Howald, KL Kunze, RF Lawrence, WF Trager. Application of negative-ion chemical ionization isotope dilution gas chromatograph-mass spectrometry to single-dose bioavailability studies of mefloquine. J Chromatogr B 661:263, 1994.
48. CC Lang, CM Stein, RA Nelson, HB He, FJ Belas, IA Blair, M Wood, AJJ Wood. Sympathoinhibitory response to clonidine is blunted in patients with heart failure. Hypertension 30:392, 1997.
49. G Momerency, K Van Cauwenberghe, MS Highley, PG Harper, AT van Oosterom,

EA De Bruijin. Partitioning of ifosfamide and its metabolites between red blood cells and plasma. J Pharm Sci 85:262, 1996.

50. RL Fitzgerald, DA Herold. Serum total testosterone: Immunoassay compared with negative chemical ionization gas chromatography-mass spectrometry. Clin Chem 42:749, 1996.

51. NN Vachharajani, WC Shyu, DS Greene, RH Barbhaiya. The pharmacokinetics of butorphanol and its metabolites at steady state following nasal administration in humans. Biopharm Drug Dispos 18:191, 1997.

52. IA Blair. Electron capture negative ion chemical ionization mass spectrometry of lipid mediators. In: RC Murphy, FA Fitzpatrick, eds. Methods in Enzymology. San Diego, CA: Academic Press, 1990, p. 13.

53. LK Webster, AG Ellis, P Kestell, GW Rewcastle. Metabolism and elimination of 5,6-dimethylxanthenone-4-acetic acid in the isolated perfused rat liver. Drug Metab Dispos 23:363, 1995.

54. M Kajbaf, M Jahanshahi, K Pattichis, JW Gorrod, S Naylor. Rapid and efficient purification of cimetropium bromide and mifentidine drug metabolite mixtures derived from microsomal incubates for analysis by mass spectrometry. J Chromatogr 575:75, 1992.

55. C-JC Jackson, JW Hubbard, G McKay, JK Cooper, EM Hawes, KK Midha. Identification of phase-I and phase-II metabolites of fluphenazine in rat bile. Drug Metab Dispos 19:188, 1991.

56. C-JC Jackson, JW Hubbard, KK Midha. Biosynthesis and characterization for glucuronide metabolites of fluphenazine: 7-hydroxyfluphenazine glucuronide and fluphenazine glucuronide. Xenobiotica 21:383, 1991.

57. DC Mays, KF Dixon, A Balboa, LJ Pawluk, MR Bauer, S Nawoot, N Gerber. A nonprimate animal model applicable to zidovudine pharmacokinetics in humans: Inhibition of glucuronidation and renal excretion of zidovudine by probenecid in rats. J Pharm Exp Ther 259:1261, 1991.

58. M Tanaka, K Ono, H Hakusui, T Takegoshi, Y Watanabe, M Kanao. Identification of DP-1904 and its ester glucuronide in human urine and determination of their enantiomeric compositions by high-performance liquid chromatography with optical activity and ultraviolet detection. Drug Metab Dispos 18:698, 1990.

59. H Luo, EM Hawes, G McKay, ED Korchinski, KK Midha. The quaternary ammonium-linked glucuronide of doxepin: A major metabolite in depressed patients treated with doxepin. Drug Metab Dispos 19:722, 1991.

60. H Luo, G McKay, KK Midha. Identification of clozapine N(+)-glucuronide in the urine of patients treated with clozapine using electrospray mass spectrometry. Biol Mass Spectrom 23:147, 1994.

61. H Luo, EM Hawes, G McKay, ED Korchinski, KK Midha. N(+)-glucuronidation of aliphatic tertiary amines in human: Antidepressant versus antipsychotic drugs. Xenobiotica 25:291, 1995.

62. PTM Kenny, R Orlando. Tandem mass spectrometric analysis of peptides at the femtomole level. Anal Chem 64:957, 1992.

63. EC Huang, T Wachs, JJ Conboy, JD Henion. Atmospheric pressure ionization mass spectrometry. Anal Chem 62:713A, 1990.

64. D Wang-Iverson, ME Arnold, M Jemal, AI Cohen. Determination of SQ 33,600,

a phosphinic acid containing HMG CoA reductase inhibitor, in human serum by high-performance liquid chromatography combined with ionspray mass spectrometry. Biol Mass Spectrom 21:189, 1992.

65. GS Rule, JD Henion. Determination of drugs from urine by on-line immunoaffinity chromatography-high-performance liquid chromatography-mass spectrometry. J Chromatogr 582:103, 1992.

66. DA Whitman, V Abbott, K Fregien, LD Bowers. Recent advances in high-performance liquid chromatography/mass spectrometry and high-performance liquid chromatography/tandem mass spectrometry: Detection of cyclosporine and metabolites in kidney and liver tissue. Ther Drug Monitor 15:552, 1993.

67. PJ Swart, WEM Oelen, AP Bruins, PG Tepper, RA de Zeeuw. Determination of the dopamine D2 agonist N-0923 and its major metabolites in perfused rat livers by HPLC-UV-atmospheric pressure ionization mass spectrometry. J Anal Toxicol 18:71, 1994.

68. X-Z Qin, DP Ip, KHC Chang, PM Dradransky, MA Brooks, T Sakuma. Pharmaceutical application of LC-MS. 1—Characterization of a famotidine degradate in a package screening study by LC-APCI MS. J Pharm Biomed Anal 12:221, 1994.

69. R Wieboldt, DA Campbell, J Henion. Quantitative liquid chromatographic-tandem mass spectrometric determination of orlistat in plasma with a quadrupole ion trap. J Chromatogr B 708:121, 1998.

70. PR Tiller, AP Land, I Jardine, DM Murphy, R Sozio, A Ayrton, WH Schaefer. Application of liquid chromatography-mass spectrometry(n) analyses to the characterization of novel glyburide metabolites formed in vitro. J Chromatogr A 794:15, 1998.

71. N Chauret, D Nicoll-Griffith, R Friesen, C Li, L Trimble, D Dube, R Fortin, Y Girard, J Yergey. Microsomal metabolism of the 5-lipoxygenase inhibitors L-746,530 and L-739,010 to reactive intermediates thàt covalently bind to protein: The role of the 6,8-dioxabicyclo[3,2,1]octanyl moiety. Drug Metab Dispos 23: 1325, 1995.

72. DA Nicoll-Griffith, N Gupta, SP Twa, H Williams, LA Trimble, JA Yergey. Verlukast (MK-0679) conjugation with glutathione by rat liver and kidney cytosols and excretion in the bile. Drug Metab Dispos 23:1085, 1995.

73. JT Wu, MG Qian, MX Li, K Zheng, DH Huang, DM Lubman. On-line analysis by capillary separations interfaced to an ion trap storage/reflectron time-of-flight mass spectrometer. J Chromatogr A 794:377, 1998.

74. GA Bach, J Henion. Quantitative capillary electrophoresis-ion-trap mass spectrometry determination of methylphenidate in human urine. J Chromatogr B 707:275, 1998.

75. P Kebarle, L Tang. From ions in solution to ions in gas phase: The mechanism of electrospray mass spectrometry. Anal Chem 65:972A, 1993.

76. GK Poon, P Mistry, S Lewis. Electrospray ionization mass spectrometry of platinum anticancer agents. Biol Mass Spectrom 20:687, 1991.

77. DW Everett, TJ Chando, GC Didonato, SM Singhvi, HY Pan, SH Weinstein. Biotransformation of pravastatin sodium in humans. Drug Metab Dispos 19:740, 1991.

78. JR Perkins, CE Parker, KB Tomer. Nanoscale separations combined with electro-

spray ionization mass spectrometry: Sulfonamide determination. J Am Soc Mass Spectrom 3:139, 1992.

79. L Debrauwer, G Bories. Electrospray ionization mass spectrometry of some beta-agonists. Rapid Commun Mass Spectrom 6:382, 1992.

80. ABL Lanting, AP Bruins, BFH Drenth, K de Jonge, K Ensing, RA de Zeeuw, DKE Meijer. Identification with liquid chromatography-ionspray mass spectrometry of the metabolites of the enantiomers N-methyl dextrorphan and N-methyl levorphanol after rat liver perfusion. Biomed Mass Spectrom 22:226, 1993.

81. AJ Tomlinson, LM Benson, KL Johnson, S Naylor. Investigation of the metabolic fate of the neuroleptic drug haloperidol by capillary electrophoresis-electrospray ionization mass spectrometry. J Chromatogr Biomed Appl 621:239, 1993.

82. B Subramanyam, SM Pond, DW Eyles, HA Whiteford, HG Fouda, N Castagnoli Jr. Identification of potentially neurotoxic pyridinium metabolite in the urine of schizophrenic patients treated with haloperidol. Biochem Biophys Res Commun 181:573, 1991.

83. RP Schneider, MJ Lynch, JF Ericson, HG Fouda. Electrospray ionization mass spectrometry of semduramicin and other polyether ionophores. Anal Chem 63: 1789, 1991.

84. K Lee, Y Vanderberghe, M Herin, R Cavalier, D Beck, A Li, N Verbeke, M Lesne, J Roba. Comparative metabolism of SC-42867 and SC-51089, two PGE$_2$ antagonists, in rat and human hepatocyte cultures. Xenobiotica 24:25, 1994.

85. C Pham-Huy, N Sadeg, T Becue, C Martin, G Mahuzier, J-M Warnet, M Hamon, JR Claude. In vitro metabolism of cyclosporin A with rabbit renal or hepatic microsomes: Analysis of HPLC-FPIA and HPLC-MS. Arch Toxicol 69:346, 1995.

86. M Ulgen, M Kajbaf, JH Lamb, M Jahanshahi, JW Gorrod, S Naylor. Characterization of N-benzylcarbazole and its metabolites from microsomal mixtures by tandem mass spectrometry. Eur J Drug Metab Pharmacokinet 19:343, 1994.

87. B Yeung, P Vouros, M-L Siu-Caldera, GS Reddy. Characterization of the metabolic pathway of 1,25-dihydroxy-16-ene vitamin D$_3$ in rat kidney by on-line high performance liquid chromatography-electrospray tandem mass spectrometry. Biochem Pharmacol 49:1099, 1995.

88. O Ogunbiyi, M Kajbaf, JH Lamb, M Jahanshahi, JW Gorrod, S Naylor. Characterization of 2-amino-1-benzylbenzimidazole and its metabolites using tandem mass spectrometry. Toxicol Lett 78:25, 1995.

89. A Ding, P Zia-Amirhosseini, AF McDonagh, AL Burlingame, L Benet. Reactivity of tolmetin glucuronide with human serum albumin. Identification of binding sites and mechanisms of reaction by tandem mass spectrometry. Drug Metab Dispos 23(3):369, 1995.

90. A Ding, JC Ojingwa, AF McDonagh, AL Burlingame, LZ Benet. Evidence for covalent binding of acyl glucuronides to serum albumin via an imine mechanism as revealed by tandem mass spectrometry. Proc Natl Acad Sci (USA) 90:3797, 1993.

91. O van Tellingen, ALC Sonneveldt, JH Beijnen, WJ Nooijen, JJ Kettenes-van den Bosch, C Versluis, A Bult. Plasma pharmacokinetics, tissue disposition, excretion and metabolism of vinleucinol in mice as determined by high-performance liquid chromatography. Cancer Chemother Pharmacol 33:425, 1994.

92. K Kassahun, P Hu, MP Grillo, MR Davis, L Jin, TA Baillie. Metabolic activation of unsaturated derivatives of valproic acid. Identification of novel glutathione adducts formed through coenzyme A-dependent and -independent processes. Chemico-Biol Int 90:253, 1994.

93. W Tang, FS Abbott. Bioactivation of a toxic metabolite of valproic acid, (E)-2-propyl-2,4-pentadienoic acid, via glucuronidation. LC/MS/MS characterization of the GSH-glucuronide diconjugates. Chem Res Toxicol 9:517, 1996.

94. L Jin, MR Davis, P Hu, TA Baillie. Identification of novel glutathione conjugates of disulfiram and diethyldithiocarbamate in rat bile by liquid chromatography-tandem mass spectrometry. Evidence for metabolic activation of disulfiram *in vivo*. Chem Res Toxicol 7:526, 1994.

95. GK Poon, YC Chui, R McCague, PE Lonning, R Feng, MG Rowlands, M Jarman. Analysis of phase I and phase II metabolites of tamoxifen in breast cancer patients. Drug Metab Dispos 21:1119, 1993.

96. K Mewes, J Blanz, G Ehninger, R Gebhardt, K-P Zeller. Cytochrome P-450-induced cytotoxicity of mitoxantrone by formation of electrophilic intermediates. Cancer Res 53:5135, 1993.

97. MR Davis, K Kassahun, CM Jochheim, KM Brandt, TA Baillie. Glutathione and *N*-acetylcysteine conjugates of 2-chloroethyl isocyanate. Identification as metabolites of *N,N'*-bis(2-chloroethyl)-*N*-nitrosourea in the rat and inhibitory properties toward glutathione reductase *in vitro*. Chem Res Toxicol 6:376, 1993.

98. L Jin, TA Baillie, MR Davis, ED Kharasch. Nephrotoxicity of sevoflurane compound A [fluoromethyl-2,2-difluoro-1-(trifluoromethyl)vinyl ether] in rats: Evidence for glutathione and cysteine conjugate formation and the role of renal cysteine conjugate beta-lyase. Biochem Biophys Res Commun 210:498, 1995.

99. U Hofmann, MF Fromm, S Johnson, G Mikus. Simultaneous determination of dihydrocodeine and dihydromorphine in serum by gas chromatography-tandem mass spectrometry. J Chromatogr B 663:59, 1995.

100. G Engel, U Hofmann, M Eichelbaum. Highly sensitive and specific gas chromatographic-tandem mass spectrometric method for the determination of trace amounts of antipyrine metabolites in biological material. J Chromatogr B 666:111, 1995.

101. JD Gilbert, TF Greber, JD Ellis, A Barrish, TV Olah, C Fernandez-Metzler, AS Yuan, CJ Burke. The development and cross-validation of methods based on radioimmunoassay and LC/MS-MS for the quantification of the class III antiarrhythmic agent, MK-0499, in human plasma and urine. J Pharm Biomed Anal 13:937, 1995.

102. LBI Martin, AF Schreiner, RB van Breemen. Characterization of cisplatin adducts of oligonucleotides by fast atom bombardment mass spectrometry. Anal Biochem 193:6, 1991.

103. DJ Burinsky, R Dunphy, AR Oyler, CJ Shaw, ML Cotter. Characterization of a synthetic peptide impurity by fast-atom bombardment-tandem mass spectrometry and gas chromatography-mass spectrometry. J Pharm Sci 81:597, 1992.

104. M Kajbaf, JH Lamb, S Naylor, K Pattichis, JW Gorrod. Identification of metabolites derived from the H$_2$-receptor antagonist mifentidine using tandem mass spectrometry. Anal Chim Acta 247:151, 1991.

105. M Kurono, A Itogawa, K Yoshida, S Naruto, P Rudewicz, M Kanai. Identification

of AJ-2615 and its S-oxidized metabolites in rat plasma by use of tandem-mass spectrometry. Biol Mass Spectrom 21:17, 1992.

106. GL Weber, RC Steenwyk, SD Nelson, PG Pearson. Identification of N-acetylcys-teine conjugates of 1,2-dibromo-3-chloropropane: Evidence for cytochrome P450 and glutathione mediated bioactivation pathway. Chem Res Toxicol 8:560, 1995.

107. E van Bakergem, RA van der Hoeven, WM Niessen, UR Tjaden, J van der Greef, GK Poon, R McCague. On-line continuous-flow dialysis thermospray tandem mass spectrometry for quantitative screening of drugs in plasma: Rogletimide. J Chromatogr 598:189, 1992.

108. SD Menacherry, JB Justice Jr. *In vivo* microdialysis and thermospray tandem mass spectrometry of the dopamine uptake blocker 1-(2-[bis(4-fluorophenyl)-methoxy]ethyl)-4-(3-phenylpropyl)-pipe razine (GBR-12909). Anal Chem 62: 597, 1990.

109. RM Caprioli, S-N Lin. On-line analysis of penicillin blood levels in the live rat by combined microdialysis/fast-atom bombardment mass spectrometry. Proc Natl Acad Sci (USA) 87:240, 1990.

110. BK Matuszewski, ML Constanzer, and CM Chavez-Eng. Matrix effect in quantitative LC/MS/MS analyses of biological fluids: A method for determination of finasteride in human plasma at picogram per milliliter concentrations. Anal Chem 70:882, 1998.

111. WH Shaefer, F Dixon Jr. Effect of high-performance liquid chromatography mobile phase components on sensitivity in negative atmospheric pressure chemical ionization liquid chromatography-mass spectrometry. J Am Soc Mass Spectrom 7:1059, 1996.

112. AG Borel, FS Abbott. Identification of carbamoylated thiol conjugates as metabolites of the antineoplastic 1-(2-chloroethyl)-3-cyclohexyl-1-nitrosourea, in rats and humans. Drug Metab Dispos 21:889, 1993.

113. GK Poon, YC Chui, M Jarman, MG Rowlands, PS Kokkonen, WMA Niessen, JV van der Greef. Investigation of conjugated metabolites of 4-hydroxyandrost-4-ene-3,17-dione in patient urine by liquid chromatography-atmospheric pressure ionization mass spectrometry. Drug Metab Dispos 20:941, 1992.

114. TV Olah, JD Gilbert, A Barrish. Determination of the beta-adrenergic blocker timolol in plasma by liquid chromatography-atmospheric pressure chemical ionization mass spectrometry. J Pharm Biomed Anal 11:157, 1993.

115. HG Fouda, RP Schneider. Quantitative determination of the antibiotic azithromycin in human serum by high-performance liquid chromatography (HPLC)-atmospheric pressure chemical ionization mass spectrometry: Correlation with a standard HPLC-electrochemical method. Ther Drug Monitor 17:179, 1995.

116. LC Taylor, RL Johnson, L St John-Williams, T Johnson, SY Chang. The use of low-energy collisionally activated dissociation negative-ion tandem mass spectrometry for the characterization of dog and human urinary metabolites of the drug BW 1370U87. Rapid Commun Mass Spectrom 8:265, 1994.

117. AL Yergey, Y Teffera, NV Esteban, FP Abramson. Direct determination of human urinary cortisol metabolites by HPLC/CRIMS. Steroids 60:295, 1995.

118. Y Teffera, FP Abramson. Application of high-performance liquid chromatography/

chemical reaction interface mass spectrometry for the analysis of conjugated metabolites: A demonstration using deuterated acetaminophen. Biol Mass Spectrom 23: 776, 1994.

119. H Song, FP Abramson. Drug metabolism studies using "intrinsic" vs "extrinsic" labels: A demonstration using ^{15}N vs ^{35}Cl in midazolam. Drug Metab Dispos 21: 868, 1993.

120. H Song, FP Abramson. Selective detection of chlorine-containing compounds by gas chromatography/chemical reaction interface mass spectrometry. Anal Chem 65:447, 1993.

121. JJ Kusmierz, FP Abramson. Tracing ^{15}N with chemical reaction interface mass spectrometry: A demonstration using ^{15}N-labeled glutamine and asparagine substrates in cell culture. Biol Mass Spectrom 23:756, 1994.

122. F-Y Hsieh Jr, CA Goldthwaite, BJ Nobes, RF Mayol, LJ Klunk, IA Blair. Quantification of drug metabolism using a combination of chemical reaction interface mass spectrometry and NMR. Proc 42nd ASMS Conf on Mass Spectrometry and Allied Topics, 1994, p. 615.

123. TR Browne, GK Szabo, A Ajami, D Wagner. Performance of human mass balance/metabolite identification studies using stable isotope (^{13}C,^{15}N) labeling and continuous-flow isotope-ratio mass spectrometry as an alternative to radioactive labeling methods. J Clin Pharmacol 33:246, 1993.

124. IM Johansson, R Pavelka, JD Henion. Determination of small drug molecules by capillary electrophoresis-atmospheric pressure ionization mass spectrometry. J Chromatogr 559:515, 1991.

125. AJ Tomlinson, JM Benson, KL Johnson, S Naylor. Investigation of drug metabolism using capillary electrophoresis with photodiode array detection and online mass spectrometry equipped with an array detector. Electrophoresis 15:62, 1994.

126. RS Ramsey, DE Goeringer, SA McLuckey. Active chemical background and noise reduction in capillary electrophoresis/ion trap mass spectrometry. Anal Chem 65: 3521, 1993.

127. AJ Tomlinson, LM Benson, JW Gorrod, S Naylor. Investigation of the *in vitro* metabolism of the H$_2$-antagonist mifentidine by on-line capillary electrophoresis-mass spectrometry using non-aqueous separation conditions. J Chromatogr B 657: 373, 1994.

128. W Lu, GK Poon, PL Carmichael, RB Cole. Analysis of tamoxifen and its metabolites by on-line capillary electrophoresis-electrospray ionization mass spectrometry employing nonaqueous media containing surfactants. Anal Chem 68:668, 1996.

129. MH Lamoree, NJ Reinhoud, UR Tjaden, WMA Niessen, J van der Greef. On-capillary isotachophoresis for loadability enhancement in capillary zone electrophoresis/mass spectrometry of beta-agonists. Biol Mass Spectrom 23:339, 1994.

130. S Naylor, AJ Tomlinson, LM Benson, JW Gorrod. Capillary electrophoresis and capillary electrophoresis-mass spectrometry in drug and metabolite analysis. Eur J Drug Metab Pharmacokinet 19:235, 1994.

131. JR Chapman, RT Gallagher, EC Barton, JM Curtis, PJ Derrick. Advantages of high-resolution and high-mass range magnetic-sector mass spectrometry for electrospray ionization. Org Mass Spectrom 27:195, 1992.

132. RB Cody, J Tamura, BD Musselman. Electrospray ionization/magnetic sector mass

spectrometry: Calibration, resolution, and accurate mass measurements. Anal Chem 64:1561, 1992.

133. P Dobberstein, E Schroder. Accurate mass determination of a high molecular weight protein using electrospray ionization with a magnetic sector instrument. Rapid Commun Mass Spectrom 7:861, 1993.

134. SA Hofstadler, E Schmidt, Z Guan, DA Laude Jr. Concentric tube vacuum chamber for high magnetic field, high pressure ionization in a Fourier transform ion cyclotron resonance mass spectrometer. J Am Soc Mass Spectrom 4:168, 1993.

135. SC Beu, MW Senko, JP Quinn, FW McLafferty. Improved Fourier-transform ion-cyclotron-resonance mass spectrometry of large biomolecules. J Am Soc Mass Spectrom 4:190, 1993.

136. BE Winger, SA Hofstadler, JE Bruce, HR Udseth, RD Smith. High resolution mass measurements of biomolecules using a new electrospray ionization ion cyclotron resonance mass spectrometer. J Am Soc Mass Spectrom 4:566, 1993.

137. DP Little, JP Speir, MW Senko, PB O'Connor, FW McLafferty. Infrared multiphoton dissociation of large multiply charged ions for biomolecule sequencing. Anal Chem 66:2809, 1994.

138. MW Senko, JP Speir, FW McLafferty. Collisional activation of large multiply charged ions using Fourier transform mass spectrometry. Anal Chem 66:2801, 1994.

139. JH Wahl, SA Hofstadler, RD Smith, Direct electrospray ion current monitoring detection and its use with on-line capillary electrophoresis mass spectrometry. Anal Chem 67:462, 1995.

140. RT Coutts, F Jamali, F Malek, A Peliowski, NN Finer. Urinary metabolites of doxapram in premature neonates. Xenobiotica 21:1407, 1991.

141. T Solouki, JA Marto, FM White, S Guan, AG Marshall. Attomole biomolecule mass analysis by matrix-assisted laser desorption/ionization Fourier transform ion cyclotron resonance. Anal Chem 67:4139, 1995.

142. JD Henion, T Wachs, A Mordehai. Recent developments in electrospray mass spectrometry including implementation on an ion trap. J Pharm Biomed Anal 11:1049, 1993.

143. RE Kaiser Jr, JD Williams, SA Lammert, RG Cooks. Thermospray liquid chromatography-mass spectrometry with a quadrupole ion trap mass spectrometry. J Chromatogr 562:3, 1991.

144. L He, L Liang, DM Lubman. Continuous-flow MALDI mass spectrometry using an ion trap/reflection time-of-flight detector. Anal Chem 67:4127, 1995.

145. BL Kleintop, DM Eades, RA Yost. Operation of a quadrupole ion trap for particle beam LC/MS analyses. Anal Chem 65:1295, 1993.

146. ME Bier, PC Winkler, JR Herron. Coupling a particle beam interface directly into a quadrupole ion trap mass spectrometer. J Am Soc Mass Spectrom 4:38, 1993.

147. SA McLuckey, GJ Van Berkel, GL Glish, EC Huang, JD Henion. Ion spray liquid chromatography/ion tran mass spectrometry determination of biomolecules. Anal Chem 63:375, 1991.

148. H-Y Lin, RD Voyksner. Determination of environmental contaminants using an electrospray interface combined with an ion trap mass spectrometer. Anal Chem 65:451, 1993.

149. SA McLuckey, GJ Van Berkel, DE Goeringer, GL Glish. Ion trap mass spectrometry. Using high-pressure ionization. Anal Chem 66:737A, 1994.

150. G Wells, C Huston. High-resolution selected ion monitoring in a quadrupole ion trap mass spectrometer. Anal Chem 67:3650, 1995.

151. F Hillenkamp, M Karas, RC Beavis, BT Chait. Matrix-assisted laser desorption/ionization mass spectrometry of biopolymers. Anal Chem 63:1193A, 1991.

152. KL Walker, RW Chiu, CA Monnig, CL Wilkins. Off-line coupling of capillary electrophoresis and matrix-assisted laser desorption/ionization time-of-flight mass spectrometry. Anal Chem 67:4197, 1995.

153. JG Boyle, CM Whitehouse. Time-of-flight mass spectrometry with an electrospray ion beam. Anal Chem 64:2084, 1992.

154. SM Michael, BM Chien, DM Lubman. Detection of electrospray ionization using a quadrupole ion trap storage/reflection time-of-flight mass spectrometer. Anal Chem 65:2614, 1993.

155. AN Verentchikov, W Ens, KG Standing. Reflecting time-of-flight mass spectrometer with an electrospray ion source and orthogonal extraction. Anal Chem 66:126, 1994.

156. AE Mutlib, WL Nelson. Synthesis and identification of the N-glucuronides of nor-gallopamil and norverapamil unusual metabolites of gallopamil and verapamil. J Pharm Exp Ther 252:593, 1990.

157. J Magdalou, R Herber, R Bidfault, G Siest. *In vitro* N-glucuronidation of a novel antiepileptic drug, lamotrigine, by human liver microsomes. J Pharm Exp Ther 260: 1166, 1992.

158. NR Niemeijer, TK Gerding, RA DeZeeuw. Glucuronidation of labetalol at the two hydroxy positions by bovine liver microsomes. Isolation, purification, and structure elucidation of the glucuronides of labetalol. Drug Metab Dispos 19:20, 1991.

159. A Magistrelli, P Villa, E Benfenati, EJ Modest, M Salmona, MT Tacconi. Fate of 1-O-octadecyl-2-O-methyl-rac-glycero-3-phosphocholine (ET18-OME) in malignant cells, normal cells, and isolated and perfused rat liver. Drug Metab Dispos 23:113, 1995.

160. S Bezek, M Kukan, Z Kallay, T Trnovec, M Stefek, LB Piotrovskiy. Disposition of ethimizol, a xanthine-related inotropic drug, in perfused rat liver and isolated hepatocytes. Drug Metab Dispos 18:88, 1990.

161. HK Jajoo, IA Blair, LJ Klunk, RF Mayol. *In vitro* metabolism of the antianxiety drug buspirone as a predictor of its metabolism *in vivo*. Xenobiotica 20:779, 1990.

162. OR Idowu, JO Peggins, TG Brewer, C Kelley. Metabolism of a candidate 8-amino-quinoline antimalarial agent, WR 238605, by rat liver microsomes. Drug Metab Dispos 23:1, 1995.

163. N Chauret, C Li, Y Ducharme, LA Trimble, JA Yergey, C Ramachandran, DA Nicoll-Griffith. *In vitro* and *in vivo* biotransformations of the naphthalenic lignan lactone 5-lipoxygenase inhibitor, L-702,539. Drug Metab Dispos 23:65, 1995.

164. RA Stearns, PK Chakravarty, R Chen, S-HL Chiu. Biotransformation of losartan to its active carboxylic acid metabolite in human liver microsomes. Drug Metab Dispos 23:207, 1995.

165. LC Wienkers, RC Steenwyk, SA Mizsak, PG Pearson. *In vitro* metabolism of tirila-

zad mesylate in male and female rats. Contribution of cytochrome P4502C11 and delta⁴-5alpha-reductase. Drug Metab Dispos 23:383, 1995.

166. U Christians, H Radeke, R Kownatzki, HM Schiebel, R Schottmann, K-F Sewing. Isolation of an immunosuppressive metabolite of FK506 generated by human microsome preparations. Clin Biochem 24:271, 1991.

167. SM Sanins, WJ Adams, DG Kaiser, GW Halstead, TA Baillie. Studies on the metabolism and chiral inversion of ibuprofen in isolated rat hepatocytes. Drug Metab Dispos 18:527, 1990.

168. U Renner, J Blanz, S Freund, D Waidelich, G Ehninger, K-P Zeller. Biotransformation of CI-937 in primary cultures of rat hepatocytes. Formation of glutathione conjugates. Drug Metab Dispos 23:94, 1995.

169. A Lampen, U Christians, FP Guengerich, PB Watkins, JC Kolars, A Bader, A-K Gonschior, H Dralle, I Hackbarth, K-F Sewing. Metabolism of the immunosuppressant tacrolimus in the small intestine: Cytochrome P450, drug interactions, and interindividual variability. Drug Metab Dispos 23:1315, 1995.

170. HK Jajoo, IA Blair, LJ Klunk, RF Mayol. Characterization of in vitro metabolites of the antipsychotic drug tiospirone by mass spectrometry. Biomed Environ Mass Spectrom 19:281, 1990.

171. S Madden, DJ Back. Metabolism of norgestimate by human gastrointestinal mucosa and liver microsomes in vitro. J Steroid Biochem Mol Biol 38:497, 1991.

172. U Christians, S Strohmeyer, R Kownatzki, H-M Schiebel, J Bleck, J Greipel, K Kohlhaw, R Schottmann, K-F Sewing. Investigations on the metabolic pathways of cyclosporine. I. Excretion of cyclosporine and its metabolites in human bile-isolation of 12 new cyclosporine metabolites. Xenobiotica 21:1185, 1991.

173. J Li, DL Norwood, L-F Mao, H Schulz. Mitochondrial metabolism of valproic acid. Biochemistry 30:338, 1991.

174. SM Bjorge, TA Baillie. Studies on the beta-oxidation of valproic acid in rat liver mitochondrial preparations. Drug Metab Dispos 19:823, 1991.

175. L-F Mao, DS Millington, H Schulz. Formation of a free acid acyl adenylate during the activation of 2-propylpentanoic acid. J Biol Chem 267:3143, 1992.

176. JW Tracy, BA Catto, LT Webster Jr. Formation of N-(5-nitro-2-thiazolyl)-N'-carboxymethylurea from 5-hydroxyniridazole. Role of aldehyde dehydrogenase in the oxidative metabolism of niridazole. Drug Metab Dispos 19:508, 1991.

177. EM Cretton, DV Waterhous, R Bevan, J-P Sommadossi. Glucuronidation of 3'-azido-3'-deoxythymidine by rat and human liver microsomes. Drug Metab Dispos 18:369, 1990.

178. GK Poon, B Walter, PE Lonning, MN Horton, R McCague. Identification of tamoxifen metabolites in human Hep G2 cell line, human liver homogenate, and patients on long-term therapy for breast cancer. Drug Metab Dispos 23:377, 1995.

179. ED Kharasch, HF Hill, AC Eddy. Influence of dexmedetomidine and clonidine on human liver microsomal alfentanil metabolism. Anesthesiology 75:520, 1991.

180. MK Parkash, J Caldwell. Metabolism and excretion of [¹⁴C] furfural in the rat and mouse. Food Chem Toxicol 32:887, 1994.

181. M Asami, M Yamamura, W Takasaki, Y Tanaka. Quantitative determination of diol metabolites of CS-670, a new antiinflammatory agent, by capillary column gas chromatography-mass spectrometry. J Chromatogr B 665:107, 1995.

182. E Warner, JV Rajan, A Wynshaw-Boris, J Xu, GY Yin, KJ Abel, BL Weber, LA

Chodosh. The risk of breast cancer associated with mammographic parenchymal patterns: A meta analysis of the published literature to examine the effect of methods of classification. Cancer Detect Prevent 16:67, 1995.

183. PC Ruenitz, X Bai. Acidic metabolites of tamoxifen. Aspects of formation and fate in the female rat. Drug Metab Dispos 23:993, 1995.

184. K Zhang, F Lepage, G Cuvier, J Astoin, MS Rashed, TA Baillie. The metabolic fate of stiripentol in the rat. Studies on cytochrome P-450-mediated methylenedioxy ring cleavage and side chain isomerism. Drug Metab Dispos 18:794, 1990.

185. LJ Shore, C Fenselau, AR King, RG Dickinson. Characterization and formation of the glutathione conjugate of clofibric acid. Drug Metab Dispos 23:119, 1995.

186. RT Coutts, MS Hussain, RG Micetich, M Daneshtalab. The metabolism of trimipramine in the rat. Biomed Environ Mass Spectrom 19:793, 1990.

187. SY Yeh. N-depyridination and N-dedimethylaminoethylation of tripelennamine and pyrilamine in the rat. Drug Metab Dispos 18:453, 1990.

188. ND Brown, LR Phillips, H Leader, PK Chiang. Isolation and identification of beta-hydroxyethylaprophen: A urinary metabolite of aprophen in rats. J Chromatogr 563: 466, 1991.

189. JC Maurizis, C Nicolas, M Verny, M Ollier, M Faurie, M Payard, A Veyre. Biodistribution and metabolism in rats and mice of bucromarone. Drug Metab Dispos 19: 94, 1991.

190. HG Fouda, J Lukaszewicz, DA Clark, B Hulin. Metabolism of a new thiazolidinedione hypoglycemic agent CP-68,722 in rat: Metabolite identification by gas chromatography mass spectrometry. Xenobiotica 21:925, 1991.

191. TF Woolf, SM Bjorge, AE Black, A Holmes, T Chang. Metabolism of the acyl-CoA: Cholesterol acyltransferase inhibitor 2,2-dimethyl-N-(2,4,6-trimethoxyphenyl)dodecanamide in rat and monkey. Drug Metab Dispos 19:696, 1991.

192. O Grech-Belanger, J Turgeon, M Lalande, PM Belanger. Meta-hydroxymexiletine, a new metabolite of mexiletine. Isolation, characterization, and species differences in its formation. Drug Metab Dispos 19:458, 1991.

193. JT Yang, VE Adusumalli, KK Wong, N Kucharczyk, RD Sofia. Felbamate metabolism in the rat, rabbit, and dog. Drug Metab Dispos 19:1126, 1991.

194. SM Sanins, WJ Adams, DG Kaiser, GW Halstead, J Hosley, H Barnes, TA Baillie. Mechanistic studies on the metabolic chiral inversion of R-ibuprofen in the rat. Drug Metab Dispos 19:405, 1991.

195. JI Usansky, LA Damani. The urinary metabolic profile of metyrapone in the rat. Identification of two novel isomeric metyrapone N-oxide metabolites. Drug Metab Dispos 20:64, 1992.

196. Y Yamada, R Ohashi, Y Sugawara, M Otsuka, O Takaiti. Metabolic fate of the new angiotensin-converting enzyme inhibitor imidapril in animals. Arzneim Forsch/Drug Res 42:490, 1992.

197. RL Chan, SC Hsieh, PE Haroldsen, W Ho, JJ Nestor Jr. Disposition of RS-26306, a potent luteinizing hormone-releasing hormone antagonist, in monkeys and rats after single intravenous and subcutaneous administration. Drug Metab Dispos 19: 858, 1991.

198. MP Knadler, BL Ackermann, JE Coutant, GH Hurst. Metabolism of the anticoagulant peptide, MDL, 28,050, in rats. Drug Metab Dispos 20:89, 1992.

199. R Atsumi, W Suzuki, H Hakusui. Identification of the metabolites of irinotecan, a

new derivative of camptothecin, in rat bile and its biliary excretion. Xenobiotica 9:1159, 1991.

200. TJ Jaworski, EM Hawes, JW Hubbard, G McKay, KK Midha, The metabolites of chlorpromazine N-oxide in rat bile. Xenobiotica 21:1451, 1991.

201. DK Patel, JL Woolley Jr, JP Shockcor, RL Johnson, LC Taylor, CW Sigel. Disposition, metabolism, and excretion of the anticancer agent crisnatol in the rat. Drug Metab Dispos 19:491, 1991.

202. MH Ali, FP Abramson, DD Ferrerolf, VH Cohn. Metabolism studies of the antischistosomal drug praziquantel using tandem mass spectrometry: Distribution of parent drug and ten metabolites obtained from control and schistosome-infected mouse urine. Biomed Environ Mass Spectrom 19:186, 1990.

203. WM Muck, JD Henion. High-performance liquid chromatography/tandem mass spectrometry: Its use for the identification of stanozolol and its major metabolites in human and equine urine. Biomed Environ Mass Spectrom 19:37, 1990.

204. J Blanz, K Mewes, G Ehninger, B Proksch, B Greger, D Waidelich, K-P Zeller. Isolation and structure elucidation of urinary metabolites of mitoxantrone. Cancer Res 51:3427, 1991.

205. L Jin, MR Davis, ED Kharasch, GA Doss, TA Baillie. Identification in rat bile of glutathione conjugates of fluoromethyl 2,2-difluoro-1-(trifluoromethyl)vinyl ether, a nephrotoxic degradate of the anesthetic agent sevoflurane. Chem Res Toxicol 9: 555, 1996.

206. JX De Vries, I Walter-Sack, A Voss, W Forster, PI Pous, F Stoetzer, M Spraul, M Ackermann, G Moyna. Metabolism of benzbromarone in man: Structures of new oxidative metabolites, 6-hydroxy- and 1'-oxo-benzbromarone, and the enantioselective formation, and elimination of 1'-hydroxybenzbromarone. Xenobiotica 23: 1435, 1993.

207. J Kristinsson, I Snooradottir, M Johannsson. The metabolism of mebeverine in man: Identification of urinary metabolites by gas chromatography/mass spectrometry. Pharmacol Toxicol 74:174, 1994.

208. K Kassahun, K Farrell, F Abbott. Identification of characterization of the glutathione and N-acetylcysteine conjugates of (E)-2-propyl-2,4-pentadienoic acid, a toxic metabolite of valproic acid, in rats and humans. Drug Metab Dispos 19:525, 1991.

209. MW Sinz, RP Remmel. Isolation and characterization of a novel quaternary ammonium-linked glucuronide of lamotrigine. Drug Metab Dispos 19:149, 1991.

210. JI MacDonald, RG Dickinson, RS Reid, RW Edom, AR King, RK Verbeeck. Identification of a hydroxy metabolite of diflunisal in rat and human urine. Xenobiotica 21:1521, 1991.

211. T Walle, UK Walle, GN Kumar, KN Bhalla. Taxol metabolism and disposition in cancer patients. Drug Metab Dispos 23:506, 1995.

212. J Blanz, K Mewes, G Ehninger, B Proksch, D Waidelich, B. Greger, K-P Zeller. Evidence for oxidative activation of mitoxantrone in human, pig, and rat. Drug Metab Dispos 19:871, 1991.

213. AM Treston, WD Hooper. Urinary metabolites of phenobarbitone, primidone, and their N-methyl and N-ethyl derivatives in humans. Xenobiotica 22:385, 1992.

214. K Kassahun, K Farrell, J Zheng, F Abbott. Metabolic profiling of valproic acid in

patients using negative-ion chemical ionization gas chromatography-mass spectrometry. J Chromatogr 527:327, 1990.

215. SY Brown, WA Garland, EK Fukuda. Isolation and characterization of an unusual glucuronide conjugate of rimantadine. Drug Metab Dispos 18:546, 1990.

216. U Hofmann, M Eichelbaum, S Seefried, CO Meese. Identification of thiodiglycolic acid, thiodiglycolic acid sulfoxide, and (3-carboxymethylthio)lactic acid as major human biotransformation products of S-carboxymethyl-l-cysteine. Drug Metab Disp 19:222, 1991.

217. J Blanz, U Renner, K Schmeer, G Ehninger, KP Zeller. Detection and identification of human urinary metabolites of biantrazole (CI-941). Drug Metab Dispos 21:955, 1993.

218. SR Dueker, AD Jones, GM Smith, AJ Clifford. Stable isotope methods for the study of beta-carotene-d_8 metabolism in humans utilizing tandem mass spectrometry and high-performance liquid chromatography. Anal Chem 66:4177, 1994.

219. GK Poon, FI Raynaud, P Mistry, DE Odell, LR Kelland, KR Harrap, CFJ Barnard, BA Murrer. Metabolic studies of an orally active platinum anticancer drug by liquid chromatography-electrospray ionization-mass spectrometry. J Chromatogr A 712: 61, 1995.

220. AE Mutlib, JT Strupczewski, SM Chesson. Application of hyphenated LC/NMR and LC/MS techniques in rapid identification of *in vitro* and *in vivo* metabolites of iloperidone. Drug Metab Dispos 23:951, 1995.

221. E Rozman, MT Galceran, L Anglada, C Albet. Investigation of the metabolism of ebrotidine in human urine by liquid chromatography-atmospheric pressure chemical ionization mass spectrometry. Drug Metab Dispos 23:976, 1995.

222. MA Shirley, X Guan, DG Kaiser, GW Halstead, TA Baillie. Taurine conjugation of ibuprofen in humans and in rat liver *in vitro*. Relationship to metabolic chiral inversion. J Pharmacol Exp Ther 269:1166, 1994.

223. BT Chait, SBH Kent. Weighing naked proteins: Practical, high-accuracy mass measurement of peptides and proteins. Science 257:1885, 1992.

224. MM Siegel, IJ Hollander, PR Hamann, JP James, L Hinman, BJ Smith, APH Farnsworth, A Phipps, DJ King, M Karas, A Ingendoh, F Hillenkamp. Matrix-assisted UV-laser desorption/ionization mass spectrometric analysis of monoclonal antibodies for the determination of carbohydrate, conjugated chelator, and conjugated drug content. Anal Chem 63:2470, 1991.

225. RH Griffey, MJ Greig, HJ Gaus, K Liu, D Monteith, M Winniman, LL Cummins. Characterization of oligonucleotide metabolism *in vivo* via liquid chromatography/ electrospray tandem mass spectrometry with a quadrupole ion trap mass spectrometer. J Mass Spectrom 32:305, 1997.

226. JP Shockcor, SE Unger, ID Wilson, PJ Foxall, JK Nicholson, JC Lindon. Combined HPLC, NMR, spectroscopy, and ion-trap mass spectrometry with application to the detection and characterization of xenobiotic and endogenous metabolites in human urine. Anal Chem 68:4431, 1996.

227. H Wang, M Hackett. Ionization within a cylindrical capacitor: Electrospray without an externally applied high voltage. Anal Chem 70:205, 1998.

5

Ultraviolet–Visible Spectrophotometry

John H. Miyawa and Stephen G. Schulman
University of Florida, Gainesville, Florida

I. INTRODUCTION

The absorption of electromagnetic radiation of wavelengths between 200 and 800 nm by molecules which have π electrons or atoms possessing unshared electron pairs can be employed for both qualitative and quantitative analysis; as such, it is known as spectrophotometry. As a wide variety of pharmaceutical substances absorb radiation in the near-ultraviolet (200–380 nm) and visible (380–800 nm) regions of the electromagnetic spectrum, the technique is widely employed in pharmaceutical analysis.

The relationship between the concentration of analyte and the intensity of light absorbed is the basis of quantitative applications of spectrophotometry. In addition, features of absorption spectra such as the molar absorptivity, spectral position, and shape and breadth of the absorption band are related to molecular structure and environment and therefore can be used for qualitative analysis.

This chapter deals with the origin, nature, and measurement of the spectra of molecules arising from the absorption of near-ultraviolet and visible radiation and the dependence of the spectra on molecular structure, reactivity, and interactions with the environment. In addition, the instrumentation employed to obtain the spectra and the techniques used in absorption spectrophotometric analysis are discussed.

The absorption of near-ultraviolet or visible light by molecules occurs as a result of the interaction of the electric field associated with a light wave or photon with molecular electrons. The intensity, position in the spectrum, and appearance of the spectral band produced by this interaction depends on the energies of the molecular electrons and their dynamic characteristics with respect to the rest of the molecule. It is therefore appropriate to begin consideration of spectrophotometry with some of the details of the electronic structure of molecules and the nature of the interaction of the latter with the electric fields associated with electromagnetic radiations.

II. MOLECULAR ELECTRONIC STRUCTURE

Electronic absorption spectra originate with the excitation of the electrons which form the bonds holding the molecule together. A chemical bond originates from the overlap of occupied atomic orbitals. The geometry of the overlap classifies the type of chemical bonding, and the occupancy of the molecular orbital is governed by the Pauli exclusion principle (i.e., a maximum of two electrons can occupy one orbital). In σ bonds the overlap of two atomic orbitals occurs along the line joining the nuclei of the bonded atoms.

Because the electronic charge is localized between two atoms, electronic repulsion prevents the formation of more than one σ bond between any two atoms in the molecule. Electrons engaged in σ bonding are usually bound strongly, and considerable energy is required to promote these electrons to vacant molecular orbitals. This means that absorption spectra involving σ-electron excitations occur well into the vacuum ultraviolet (<200 nm) and are not observed in the spectral region encompassing the near-ultraviolet and the visible.

π-Bonds are formed by the overlap of two atomic orbitals at right angles to the line joining the nuclei of the bonded atoms. π-bonding is weaker than σ-bonding and consequently, π-electrons are higher in energy than σ-electrons. In a π-bond, the distribution of electronic charge is concentrated above and below the plane containing the σ- bond axis. While σ-electrons are strongly localized between the atoms they bind, π-electrons, not being concentrated immediately between the parent atoms, are more free to move within the molecule and are frequently distributed over several atoms. If several atoms are σ-bonded in series, and each has a p orbital with the proper spatial orientation to form a π-bond with the others, a set of π-orbitals is formed which is spread over the entire series of atoms. These π-orbitals are said to be delocalized or conjugated. In some cyclic organic molecules, π-delocalization may extend over the entire molecule. These compounds are said to be aromatic, and they comprise the largest group of substances of interest in absorption spectrophotometry.

Because π-electrons are not concentrated between the bonded atoms, they are not as tightly bound as σ-electrons. Hence, they can be excited to higher orbitals with smaller energies. For molecules containing isolated π-bonds, the transitions involving π-electrons are still in the vacuum ultraviolet or at the limit of the near-ultraviolet (e.g., ethylene absorbs at 180 nm). Molecules containing delocalized π-electrons usually have π-electron spectra in the near-ultraviolet (e.g., butadiene absorbs at 220 nm), while the superdelocalized π-systems, the aromatic molecules, have π-electron spectra which range from the near-ultraviolet for small molecules to the near-infrared for large ones.

In all atoms of the Periodic Table which have more than four electrons in the valence shell (e.g., nitrogen), there are electrons in the valence shell which are already paired. These electrons are unavailable for conventional covalent bonding and yet have energies comparable to other electrons in the same shell. Consequently, they are called nonbonding or n-electrons. Because the n-electrons are higher in energy than either the σ- or π-electrons, they must be considered as potential contributors to the spectral features of molecules possessing them.

While nonbonded electron pairs in molecules do not enter into covalent bonding in the usual sense, they may exhibit a secondary kind of valency by being transferred into vacant molecular orbitals in suitable acceptor molecules. This results in the transformation of a coordination complex in which the bond formed between the electron-pair donor and the acceptor is said to be a coordinate covalent or dative bond. Brønsted basicity is the simplest example of coordinate covalent bond formation. A Brønsted base donates a pair of nonbonded electrons to a vacant $1s$ orbital of a hydrogen ion to form the conjugate acid. The σ-bond formed between the base and the hydrogen ion results in the loss of identity of the nonbonded pair previously localized on the base. The formation of coordination complexes has significance in the interpretation of spectra of compounds having nonbonded electron pairs.

In addition to the occupied molecular orbitals which comprise the chemical bonds of molecules, each molecule has associated with it several higher-energy molecular orbitals which are normally unoccupied. These are called antibonding orbitals. Antibonding orbitals may have their electron density lying along the bonding axis and are then denoted as σ*-orbitals.

Electronic absorption entails the promotion of an electron, by the absorption of energy, from an originally occupied bonding or nonbonding orbital to an originally unoccupied molecular orbital (Fig. 1).

In all organic molecules except free radicals, the nonbonding orbitals are normally doubly occupied and the electrons in these orbitals have vectorially opposite spin angular momenta. However, when an electron is promoted to a higher orbital, the orbital originally occupied and the antibonding orbital occupied both become singly occupied it is then no longer required that the two electrons have opposite spins, and there is a finite probability that one electron will change spin.

The lowest possible electronic energy a molecule can have is referred to as its ground electronic state and corresponds to the state having the configuration in which all electrons are in the lowest energy orbitals available. Promotion of an electron from an orbital which is occupied in the ground state to one which

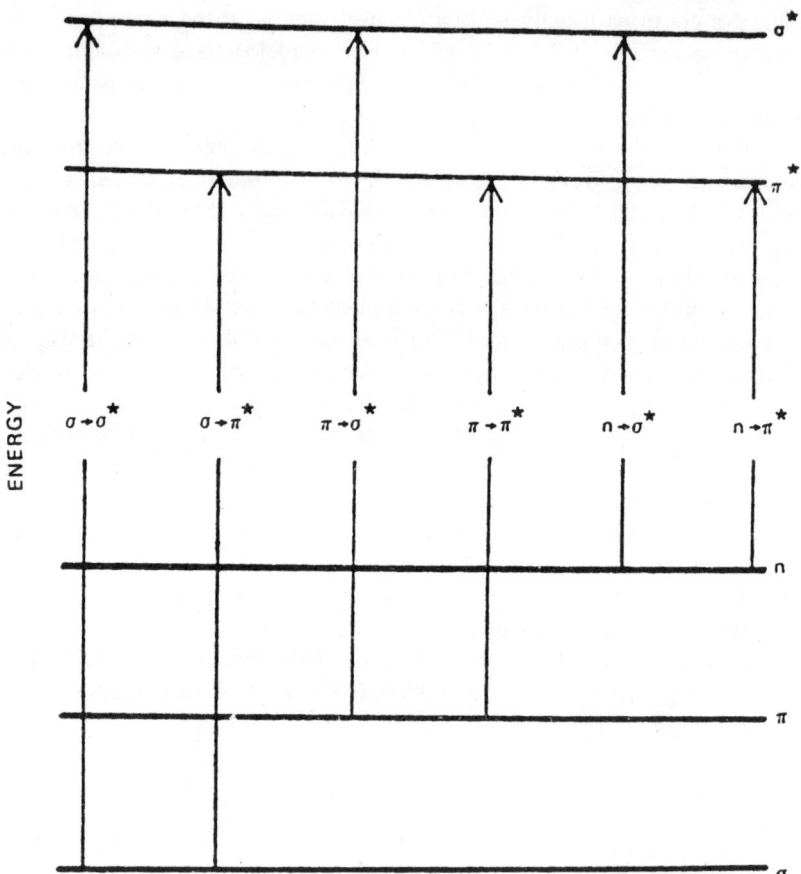

Fig. 1 The order of orbital energies and approximate order of electronic transition energies in a hypothetical unsaturated molecule containing a heteroatom with a nonbonded electron pair (n).

is normally unoccupied in the ground state is called an electronic transition and results in the formation of an electronically excited state of the molecule. Because there are several unoccupied orbitals in each molecule, several electronically excited states are possible. Each electronic state of a molecule is characterized by a particular distribution of electronic charge in the molecule. This means that the dipole moment of an electronically excited molecule will generally be different from that of the same molecule in the ground electronic state. In fact, a substantial change in the dipole moment accompanying electronic transition is a requirement for the occurrence of an intense band in the absorption spectrum. The electronically excited states of organic molecules which absorb in the near-ultraviolet and visible regions are created by the promotion of π-electrons to π^*- and n-electrons to π^*- or σ^*-orbitals. The states resulting from these promotions are called π,π^*-, n, π^*-, and n,σ^*-states, respectively. The formation of π,σ^*-states is also possible, but because of the relatively low energies of the π-orbitals and the higher energies of the σ^*-orbitals, transitions from the ground state to π,σ^*-states are invariably observable only in the vacuum ultraviolet (at wavelengths $<$ 200 nm). Generally, the production of n,σ^*-states is observable only in heteroatom-substituted, saturated compounds such as alkyl mercaptans. The n,σ^*-states and π,π^*-states of small conjugated olefinic molecules are produced in transitions giving rise to absorption bands in the near-ultraviolet, between 200 and 250 nm.

Electronic states are also designated by spin angular momentum. In most organic molecules the ground state has an even number of electrons which are paired in the lowest-lying molecular orbitals. There are two possible orientations of the spin angular momentum vactor of an electron. In the ground states of most organic molecules, the number of electrons with either spin vector orientation is the same and the net spin angular momentum is zero. States which have zero spin angular momentum (i.e., no unpaired electrons) are called singlet states.

If an electron is promoted from an occupied orbital to a higher, previously unoccupied orbital, the electron promoted will usually retain its original spin, in which case the excited state will also be a singlet state. If, however, the promoted electron does change spin, there will be two unpaired electrons in the molecule. In a molecule with two unpaired electrons there are three possible orientations of the resultant spin angular momentum vector in an externally applied magnetic field, each having a slightly different energy (in the same direction as the field vector, opposite to the field vector, or normal to the field vector). Thus, a molecule with two unpaired electrons is said to be in a triplet excited state.

The triplet states are of great importance in fluorescence and phosphorescence spectroscopy. However, the formation of a triplet excited state from a ground singlet state by direct absorption of radiation requires a change in spin angular momentum. A photon of light, however, has zero spin angular momentum. The production of the triplet state from the ground singlet, therefore, violates the law of conservation of angular momentum and thus is extremely improbable. This is manifested in absorption spectra in the form of vanishingly small band intensities. Consequently, singlet–triplet absorption spectra are not normally observed in typical absorption spectra.

In unsaturated molecules having nonbonded electron pairs, it is possible to have excited states formed by the promotion of an n-electron to a vacant orbital.

Excited states formed in this way are called n,π^*-states. While the bonding π orbitals are much lower in energy than the atomic orbitals from which they are formed, the n-electrons are only slightly lower in energy than atomic valence-shell electrons. As a result, the n-electrons are higher in energy than π-electrons in the same molecule, and the energy separation between the n orbitals and π^* orbitals is smaller than the separation between the π and π^* orbitals.

Hence, n,π^*-excited states generally lie lower than π,π^*-excited states involving the same π^* orbital. In aromatic molecules, n-orbitals are most often encountered in nitrogen heteroatoms and in carbonyl-substituent groups. The n-orbital is usually an sp^2 hybrid directed in the plane of the aromatic ring and therefore at right angles to the π- and π^*- orbitals.

Not all nonbonded electron pairs are projected perpendicular to the π- and π^*-orbitals. Some, such as those in exocyclic amino groups and hydroxy groups, lie almost parallel to the π-orbitals of the aromatic ring and in this case may participate in the π-electron structure of the aromatic system. The excited states formed by the promotion of these electrons (which will hereafter be called lone-pair or l electrons, to distinguish them from n-electrons) are generally found to behave more like π,π^*-states than n,π^*-states. These π,π^*-like states will hereafter be called intramolecular charge transfer states (or l,π^*-states), because they arise from the transfer of an electron from the exocyclic group to the aromatic ring. In n,π^*-states the promoted electron is localized near the group from which it originated, but in l,π^*-states the promoted electron may be displaced far from its site of origin.

Intramolecular charge-transfer excited states may also arise from the promotion of a π-electron from the aromatic ring to a vacant π^*-orbital localized on an exocyclic group. This phenomenon is often observed in aromatic carboxylic acids, aldehydes, and ketones, where the carbonyl group is conjugated with the aromatic ring.

The arrangement of the atoms comprising a molecule determines the molecular geometry. The atoms, however, are not rigidly fixed in space, but execute periodic motions with respect to one another and with respect to the center of mass of the molecule. These periodic motions of the molecular atoms are called normal vibrations. They result from the tendency of the positively charged nuclei to repel each other and the tendency of the bonding electrons to hold them together. Because of the mobility of the bonding electrons, the nuclei never come to rest but rather vibrate about an equilibrium position. The close interrelationship between electronic and vibrational structure arises from the dependence of both the electronic and vibrational properties on the electronic distribution of the molecule. The valence electrons may be thought of as "springs" which hold the vibrating molecule together. These "springs" will have restorative properties which vary with the electronic distribution in the molecule. As a result, the equilibrium positions of the component nuclei of the molecule may be different for different electronic states of the same molecule.

Just as it is possible to produce electronically excited states by the alteration of the electronic distribution of the molecule, it is possible to produce vibrationally excited states by displacing the atoms of the molecule. These states are vibrationally excited by comparison with the lowest-energy vibrational mode possible for the molecule, and their attainment requires the absorption of energy in steps or quanta, each of which is of the order of 10% of the magnitude of the energy required to promote an electron to a higher orbital. Because of differences in electronic distributions, each electronic state of a molecule has its own group of associated vibrational energy levels.

At room temperature (298 K), it may be assumed to a good degree of approximation that in any given electronic state of a molecular species, only the lowest vibrational level is populated if the system under consideration is in thermal equilibrium with its surroundings. This approximation will greatly simplify the consideration of the shapes and fine structures of electronic absorption bands.

III. THE ABSORPTION OF LIGHT BY MOLECULES

A. The Interaction of Light with Molecular Electronic Structure

A light wave is an electromagnetic disturbance traveling in a straight line with an *in-vacuo* speed (c) of 3.0×10^{10} cm/s. Perpendicular to the direction of travel of the wave there is an alternating electric field, and perpendicular to the direction of travel of the wave and to the plane of oscillation of the electric field vector there is an alternating magnetic field. The frequency of oscillation of the electric and magnetic field vectors is called the frequency of the light, ν. The distance traveled by the wave during the period of one complete cycle of the electric vector is called the wavelength of the light, λ. The speed, frequency, and wavelength of the light wave are related by the equation

$$c = \lambda\nu \tag{1}$$

Because of the electric field associated with light, an electron placed in the path of a light wave will experience a force and is capable of absorbing energy from the electric field of the light wave. If an electron belonging to a molecule in its ground electronic state absorbs energy from the electric field of a light wave, the electron will be promoted to an unoccupied orbital and will be transported from one site to another in the molecule. The net result will be that the molecule will have absorbed energy from the light and will be raised from the ground state to an electronically excited state. However, not all frequencies of light can be absorbed by a given molecule. Quantum theory tells us that the energy associated with one wavelength of light of frequency ν is

$$E = h\nu = \frac{hc}{\lambda} \tag{2}$$

where h is a proportionality constant known as Planck's constant ($h = 6.625 \times 10^{-27}$ erg/s). A necessary condition for light of frequency ν to be absorbed by a molecule in its ground state is that the energy gap between the ground state and the excited state to which excitation occurs is exactly equal to $h\nu$, or

$$E_e - E_g = h\nu \tag{3}$$

where E_g and E_e are the energies of the ground and excited states, respectively. If $E_e - E_g$ is not equal to $h\nu$, absorption will not occur and the molecule is said to be transparent to light of frequency ν. The absorption of light by a molecule causes an electronic transition. Because the distribution of molecular electronic charge changes when light is absorbed, the absorption of light by a molecule is also called an electronic dipole transition. The dipole moments associated with the various electronically excited states of a molecule are different, and the line along which the resultant dipole moment changes during the transition is called the direction of polarization of the electronic transition. The period or time of the absorption process is given, classically, by

$$T = \frac{1}{\nu} \tag{4}$$

Consequently, light of $\lambda = 300$ nm (3.0×10^{-5} cm) which has a frequency of 1×10^{15} s^{-1} will be absorbed in 1×10^{-15} s.

B. Electronic Absorption Spectra

The electronic absorption spectrum of a molecule is a graphical representation of the intensity of light absorbed in producing electronic transitions in the molecule as a function of the frequency of the light. In regions where the intensity of light absorbed is high, strong absorption bands are said to occur. In regions of frequency where the intensity of light absorbed is low, weak absorption bands are said to occur. Most absorption spectra recorded on commercial scanning spectrophotometers are represented as absorbance (absorption intensity) versus wavelength; this is a situation arising from electronic and optical convenience to the manufacturer of the instrument. A more physically meaningful spectrum would display absorbance versus frequency, because the frequency is linearly related to the energy gaps between the molecular electronic states represented in the spectral bands, while the wavelength is inversely related to energy.

Equations (2) and (3) suggest that absorption should occur at discrete values of λ or ν in the absorption spectrum and should therefore appear as narrow lines as is the case in atomic spectra. In molecules, however, each electronic state consists of several vibrational substates. In the ground electronic state nearly all absorbing molecules occupy the lowest vibrational sublevel. However, if the absorbing sample is scanned over a wide range of UV or visible wavelengths, excitation to some or all of the various vibrational sublevels of the terminal excited electronic state will occur. Since the vibrational sublevels of the excited state differ slightly in energy, the absorption will occur over a fairly broad wavelength region in the spectrum, giving rise to an absorption band rather than a line. Depending on the spectral region where absorption occurs and the rigidity of the molecule, absorption bands corresponding to a single electronic transition may range from about 50 to 100 nm in breadth.

C. The Spectral Positions of Electronic Absorption Bands

In the smaller aromatic molecules (derivatives of benzene, naphthalene, and phenanthrene) having no n-electrons, three absorption bands, corresponding to $\pi \rightarrow \pi^*$ transitions, are normally observed in the visible and near-ultraviolet regions. In benzene itself, the shortest wavelength of these bands actually lies in the far ultraviolet at ~ 184 nm. In larger aromatic molecules, e.g., anthracene derivatives, additional bands may be observed at the short-wavelength end of the ultraviolet, but these will not be of concern here. According to the free-electron model, the lowest π^*- and highest π-orbitals of an aromatic ring system become closer in energy as the number of fused rings increases (or, in general, as conjugation of the π-system is extended). As a result, the longest-wavelength absorption bands of anthracene lie at longer wavelengths (lower transition energy) than their counterparts in naphthalene, which in turn lie at longer wavelengths than those in benzene (Fig. 2).

The absorption spectra of substituted aromatic molecules bear remarkable similarities to the spectra of the parent hydrocarbons. Aromatic ring substituents exert their influences on the positions of spectral bands by electrostatically and electromerically altering the energies of the lowest excited states relative to that of the ground state. In this regard an aromatic ring is said to be a chromophore, because it is the part of the substituted molecule responsible for the general spectroscopic region where absorption takes place. The substituent is called an auxochrome because it affects the exact region where the absorbing molecule displays its absorption bands.

Auxochromes which do not conjugate with the chromophore or conjugate weakly, such as $-NH_3^+$, $-SO_3^-$, $-CH_3$, $-Cl$, $-Br$, and nitrogen as a heteroatom in molecules such as pyridine and quinoline, exert their influences on the spectral bands of the chromophore predominately by their electrostatic or polarizing ef-

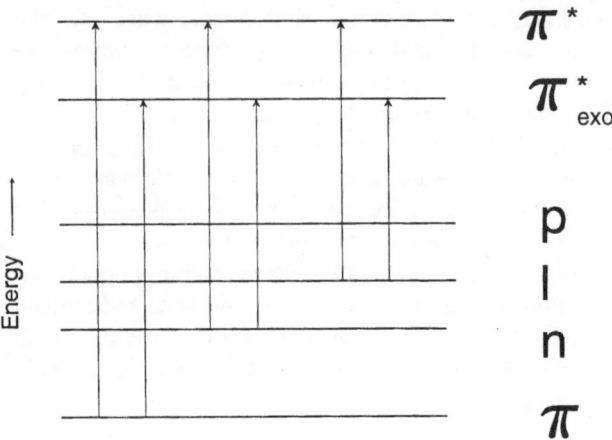

Fig. 2 The order of energies of the highest occupied π-orbital, the nonbonding (*n*)-and lone-pair (*l*) orbitals, and the lowest unoccupied π*-orbital of a typical substituted heteroaromatic molecule and the corresponding transitions. The relative position of the lowest π*-orbital, π^*_{exo}, results from the substitution of an exocyclic group having vacant π-orbitals (an electron acceptor) onto the aromatic system. The line denoted *p* represents the energy of an atomic *p*-orbital from which the *n*- or *l*-orbital arises.

fects on the charge distributions of the electronic states involved in the transitions giving rise to the spectral bands. Since the excited states are generally more polarizable than the ground state, the excited states are usually somewhat more stabilized than the ground state by the polarizing effect of the substituent, and small shifts to longer wavelengths (red shifts) of the spectral bands of the substituted molecules are usually observed. In the cases of $-SO_3^-$, where valence-shell expansion of the hexavalent sulfur atom (*d*-orbital participation), and $-CH_3$, where hyperconjugation may be involved, weak conjugative effects which are more pronounced in the excited state may also be responsible for the fact that molecules such as toluene absorb at wavelengths slightly longer than where benzene absorbs.

Substituents which conjugate strongly with aromatic rings donate electrons to or withdraw electrons from the π-system of the chromophore and thereby extend the size of the conjugated system. Those substituents which demonstrate this behavior in the ground electronic state usually behave in the same way, but to a much more exaggerated degree, in the excited states of the substituted molecules. The positions of the absorption bands of molecules containing strongly conjugating substituents vary from those of the parent hydrocarbons according to the difference between the degree of electron donation to, or electron withdrawal from, the chromophore in the ground and excited states. Groups such as

$-NH_2$, $-OH$, $-O^-$, $-SH$, and $-S^-$ have unshared pairs of electrons which can be transferred into π^*-orbitals belonging to the chromophore. These groups are best considered by treating the lone pair, in the ground state of the substituted molecule, as residing in a molecular orbital which is essentially localized on the exocyclic substituent group. The energy of this orbital is slightly lower than that of an atomic $2p$ orbital but considerably greater than that of the highest occupied π-orbital of the aromatic ring (Fig. 3). The energy gap between the highest occupied orbital (lone pair) and the lowest unoccupied orbital of the substituted molecule is thus lower than the corresponding gap of the unsubstituted molecule, and as a result, the absorption spectrum of the substituted molecule lies at longer wavelengths than that of the parent molecule.

Electronic absorption in the donor-substituted molecule entails transfer of an electron from the substituent group to the aromatic ring. Hence, the higher the energy of the lone-pair orbital, the lower is the energy required to cause the intramolecular charge-transfer process associated with absorption. Because nitrogen has a smaller atomic number than oxygen, its lone-pair electrons are less tightly bound by the exocyclic substituent and the energy of the nitrogen lone pair is higher than that of the oxygen lone pair. This is why β-naphthylamine

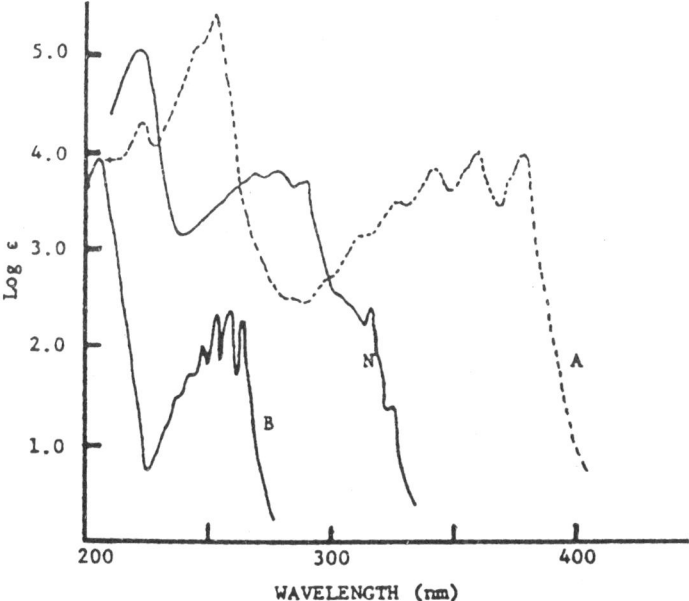

Fig. 3 Near-ultraviolet absorption spectrum of benzene (B), naphthalene (N), and anthracene (A).

absorbs at longer wavelengths than does β-naphthol. In general, amino-substituted aromatic rings will absorb at longer wavelengths than the corresponding hydroxy-substituted aromatic rings. Protonation of the nitrogen lone pair removes it entirely from the spectroscopic process because it is converted, by protonation, to a σ-bond. Hence, the spectrum of the β-naphthylammonium ion is almost the same as that of naphthalene (Fig. 4). Dissociation of β-naphthol to form the β-naphthoxide ion raises the energy of the oxygen lone pairs because of repulsions produced by the negative charge of the residual lone pair created as a result of dissociation. Consequently, absorption in the β-naphtholate anion occurs at lower energies or longer wavelengths than in β-naphthol. The lone pairs of sulfur atoms are higher in energy than the lone pairs of oxygen because of the greater radius

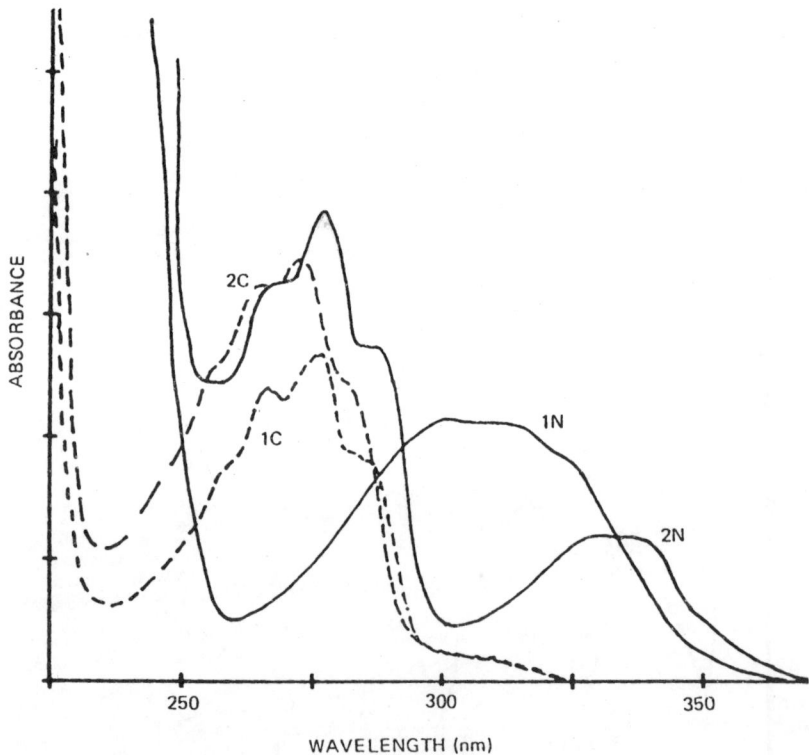

Fig. 4   The absorption spectrum of the cations (1C and 2C) and neutral molecules (1N and 2N) derived from 1-naphthylamine and 2-naphthylamine, respectively. Note the similarities of the spectra of the cations and their resemblance to the spectrum of naphthalene (Fig. 3).

of the former, which contributes to lower electrostatic attractions by the sulfur nucleus. Mercaptoaromatics, such as β-naphthylmercaptan, therefore, absorb at longer wavelengths than their oxygen counterparts, such as β-naphthol.

Electron acceptor groups such as $-NO_2$, $-COOH$, $-CHO$, and $-CN$ may be considered to interject vacant low-energy π^*-orbitals into the aromatic system between the highest π- and lowest π^*-orbitals of the unsubstituted chromophore (Fig. 3). Because the energy gap between the highest π- and lowest π^*-orbitals is smaller in the acceptor-substituted molecules, electronic absorption which entails transfer of an electron from the aromatic ring to the acceptor group occurs at longer wavelengths than in the parent hydrocarbon. Protonation of an acceptor group lowers the energy required to promote an electron from the aromatic ring to the acceptor group. This is reflected in the lowering of the π^*-orbitals of the substituted chromophore and causes the absorption spectrum of the protonated molecule to move to longer wavelengths relative to that of the unprotonated molecule. Hence, solutions of 9-anthraldehyde, which are yellow by virtue of absorption near 400 nm, turn red, indicating absorption near 500 nm, upon addition of concentrated sulfuric or perchloric acid. Similarly, dissociation of an acidic acceptor group such as $-COOH$ leaves a residual negative charge on the substituent which repels electrons of the aromatic ring. This raises the energy of the lowest π^*-orbitals of the carboxylic acid and causes the absorption spectrum to shift to shorter wavelengths (blue shift). Almost all aromatic carboxylic acids will show a blue shift upon dissociation to the corresponding aryl carboxylate.

In carbonyl compounds, the more strongly electron donating the substituents attached to the carbonyl group are, the higher the energies the π^*-orbitals of the compound will be and the shorter will be the wavelengths of absorption. Hence, the $\pi \rightarrow \pi^*$ absorption spectra of aromatic carbonyl compounds will fall into the following decreasing order of absorption wavelength:

$$
\underset{H}{\overset{O}{\underset{\|}{\diagup\!\diagdown}}} > \underset{CH_3}{\overset{O}{\underset{\|}{\diagup\!\diagdown}}} > \underset{OH}{\overset{O}{\underset{\|}{\diagup\!\diagdown}}} > \underset{NH_2}{\overset{O}{\underset{\|}{\diagup\!\diagdown}}} > \underset{O^-}{\overset{O}{\underset{\|}{\diagup\!\diagdown}}}
$$

It should be mentioned here that in dealing with carbonyl compounds only the $\pi \rightarrow \pi^*$ absorption spectra have been considered. These compounds may also show weak $n \rightarrow \pi^*$ absorption bands. In aldehydes and ketones the $n \rightarrow \pi^*$ absorption bands occur at wavelengths longer than the longest-wavelength $\pi \rightarrow \pi^*$ absorption bands. However, the $n \rightarrow \pi^*$ transitions are essentially localized on the carbonyl groups and entail electron transfer from the oxygen atom to the carbon atom. Electron-donating substituents such as $-OH$, $-OR$, and $-NH_2$ attached to the carbon atom of the carbonyl group repel the n-electron in the n,π^* excited state, considerably raising the energy of the latter. Hence, in carboxylic acids, esters, and amides, and in carboxylate anions, the $n \rightarrow \pi^*$ absorption bands are displaced to very short wavelengths (<250 nm), are well hidden under the intense shorter-wavelength $\pi \to \pi^*$ bands, and are rarely observed.

A heteroatom in the aromatic ring may be thought of as a special kind of substituent. Two types of heteroatoms may be distinguished. If the heteroatom contributes one π-electron to the aromatic system, it will be bonded to the adjacent atoms, very much like aromatic carbon. It will exert its effect on the electronic structure of the molecule by virtue of its possession of nonbonded electron pairs and by its ability, relative to that of aromatic carbon, to attract π-electrons. Nitrogen, as it occurs in pyridine, quinoline, isoquinoline, and acridine, is the only common member of this class of heteroatoms and will hereafter be referred to as pyridinic nitrogen.

If the heteroatom contributes two electrons to the π-electronic structure of the molecule (e.g., nitrogen in pyrrole or indole, oxygen in furan, sulfur in thiophene), it essentially represents a charge-transfer donor-type substituent. Its effect on the electronic structure and spectra will be dominated by its ability to donate electrons and to a lesser extent by considerations of electronegativity. Thus, the interactions of nitrogen, oxygen, and sulfur in, say, indole, benzofuran, and benzothiophene, with their aromatic systems, are similar to the interactions of amino, hydroxy, and mercapto exocyclic groups with their aromatic systems.

The principal difference between the nonbonded electron pair of a pyridinic nitrogen atom and the lone pair of a pyrrolic nitrogen atom is the orientation of the orbital accommodating the unshared pair with respect to the $p\pi$-orbitals of the aromatic system. The nonbonded orbitals of pyridinic nitrogen atoms are directed in the plane of the aromatic ring and perpendicular to the $p\pi$-orbitals. As a result, there is poor overlap between the two types of orbitals and therefore low transition probability. The lone pairs of pyrrolic nitrogen atoms are directed perpendicular to the plane of the aromatic ring and parallel to the $p \sim$ orbitals of the aromatic system, allowing for strong conjugation in the ground state and intense intramolecular charge transfer upon electronic absorption.

The pyridinic nitrogen atom is slightly more electronegative than the carbon atoms to which it is bonded. Thus the heteroatom has a slight polarizing effect on the electronic distribution of the heterocyclic ring. However, if no charge-transfer donor (e.g., $-NH_2$, $-OH$) substituents are substituted onto the aromatic ring, the effect of the pyridinic nitrogen atom on the ground and π,π^*-states of the aromatic ring is so nearly identical that the separation between these states in the heterocyclic ring is practically the same as in the corresponding homocyclic molecules. Thus, the absorption bands of the heterocyclic molecule are almost identical in spectral position with those of the parent homocyclic molecule. If, however, electron-donor substituents are also present, the excited states are so much more strongly polarized by the heteroatom that the absorption spectra lie at longer wavelengths than in the homocyclic system.

Because of the presence of the non–bonded electron pair, pyridinic nitrogen atoms have basic properties. Protonation at the pyridinic nitrogen atom has a substantial effect on the absorption spectrum. This arises from the dramatic increase in electronegativity of the heteroatom resulting from the acquisition of a formal positive charge. In this case the polarizing effect of the protonated nitrogen atom is so strong that its differential effects on the ground and excited state of the heterocycle are immediately obvious. The excited states, being more polarizable than the ground state, are more stabilized by the electrostatic attraction of the protonated nitrogen atom, and the spectrum shifts to the red relative to the spectrum of the unprotonated heterocycle.

The n-electrons associated with pyridinic N-heterocyclics are also capable of participating in $n \rightarrow \pi^*$ transitions. The absorption bands corresponding to these transitions are of low intensity and lie at wavelengths longer than the longest $\pi \rightarrow \pi^*$ bands. However, in polycyclic heterocyclics such as quinoline, isoquinoline, and acridine, they are not easily observed because they are buried under the long-wavelength tailing of the much more intense longest-wavelength $\pi \rightarrow \pi^*$ band. In pyridine the shoulder at ~270 nm on the longest-wavelength $\pi \rightarrow \pi^*$ band is believed to be an $n \rightarrow \pi^*$ band overlapped by the $\pi \rightarrow \pi^*$. In polyazaaromatics such as pyrazine, pryrimidine, pyridazine, triazine, and tetrazine, the $n \rightarrow \pi^*$ transitions tend to move to longer wavelengths with increasing numbers of nitrogen atoms in the ring. This no doubt is a result of repulsion between nonbonded electron pairs, which raises the energies of the n-orbitals and decreases the gap between the n-orbitals and the π^*-orbitals of the aromatic ring.

Protonation of a nonbonded electron pair will completely eliminate $n \rightarrow \pi^*$ absorption bands in molecules containing only a single pair of n-electrons. This is due to the conversion of the nonbonded pair to a σ-bond by protonation. In molecules having more than one nonbonded pair, protonation will shift the $n \rightarrow \pi^*$ transition to shorter wavelengths by diminishing repulsion between the electrons in the n-orbitals. The foregoing arguments are valid for substituent effects in substituted nonaromatic conjugated molecules as well as in substituted aromatics.

The introduction of a second substituent into an aromatic ring may have a very small or a very dramatic effect on the positions of the absorption bands. If both substituents are electron donors or electron acceptors, the effect of the second substituent will usually be rather small. If the two substituents are not identical, the electron donor with the highest-energy lone pair will dominate the spectrum. For example, the absorption spectrum of m-aminophenol will be very close to that of aniline. If the two substituents are electron acceptors, the one with the stronger electron-withdrawing influence will appear to be the dominating factor. The absorption spectrum of pyridine 3-aldehyde, for example, appears in very nearly the same place as the spectrum of benzaldehyde. If, however, one substituent is electron-withdrawing and one is electron-accepting, the spectral features will be much more red-shifted when both substituents are present in the aromatic ring than when either one is present alone. In essence, one may view the spectroscopic transition in the electron donor and acceptor disubstituted molecule as involving charge transfer from the lone pair of the donor group to the lowest π^*-orbital associated with the acceptor group.

There are certain special cases of substituent effects which limit the generality of the foregoing arguments. In 9-substituted anthracenes (such as 9-anthroic acid), when the substituent is bulky, steric hindrance between the substituent group and the peri hydrogen atoms in the 1- and 8-positions interfere with coplanarity of the substituent with the anthracene ring. In 9-anthroic acid the carboxyl group is perpendicular (or nearly so) to the anthracene ring, so the longest-wavelength transition does not contain an appreciable charge-transfer component. As a result, the absorption spectrum of 9-anthroic acid is almost identical to that of anthracene. Similarly, in certain *o*-substituted biphenyls, steric hindrance to coplanarity prevents coupling of the entirety of each aromatic system, so their spectra resemble those of substituted benzenes. In certain *o*-disubstituted compounds, such as salicylic acid, intramolecular hydrogen bonds may form between the substituents. In salicylic acid, the phenolic proton is hydrogen-bonded to the carboxyl group.

The intramolecular hydrogen bond leaves a partial negative charge on the hydroxyl group, making it a better electron donor, and leaves a partial positive charge on the carboxyl group, making it a better electron acceptor. As a result, the longest-wavelength absorption band of salicylic acid lies at slightly longer wavelengths than that of the electronically similar but non-hydrogen-bonded *o*-methoxybenzoic acid. However, when salicylic acid dissociates, the residual negative charge left on the carboxyl group which inhibits its electron-acceptor properties is partially offset by a strengthening of the intramolecular hydrogen bond. As a result, the anticipated blue shift of the longest-wavelength absorption band of salicylic acid is almost negligible, although *o*-methoxybenzoic acid demonstrates a substantial blue shift of its longest-wavelength absorption band upon dissociation, as is typical of carboxylic acids.

D. Intensities of Spectral Bands

The intensity of an absorption band (the absorbance) is determined by the rate of transition between the ground and excited states of the electronic transition giving rise to the band. The rate of transition is determined by the intensity of exciting radiation, the path length of exciting radiation through the sample, the concentration of potential absorbers in the sample, and the probability that an absorptive transition will occur from the ground to an excited state of the absorber. In quantitative analysis, it is the concentration of absorbers that is of interest. However, the concentration of absorbers, the optical depth of the sample, and the intensity of exciting radiation are experimentally controllable variables. Ultimately, the probability of absorptive transition, which is a function of molecular electronic structure and the nature of the transition, determines whether absorption will occur intensely enough to be analytically useful.

The probability of absorptive transition between two electronic states is described quantum mechanically by a term called the transition moment integral, R. Although the quantum mechanical derivation of R is beyond the scope of this treatment, it is appropriate to visualize R as the change in electronic dipole moment of the molecule which absorbs radiation (i.e., if e is the electronic charge and r is the vector along which the displacement of electronic charge in the atom occurs as a result of the transition, then $R = er$). The spectroscopic significance of R is that the molecular probability (or intensity) or radiative transition is proportional to R^2. The greater the value of the electronic dipole moment change in the electronic transition, the greater will be the inherent intensity of the transition.

For absorptive transition it can be shown that the measurable integrated inherent intensity (the molar absorptivity, ε) of an absorption band, expressed as absorbance A per unit concentration of absorber c and optical depth d of the sample, is given by

$$\varepsilon = \frac{A}{cd} = \int_0^\infty \varepsilon d\bar{v} = 1.085 \times 10^{38} \bar{v}_m |R|^2 \tag{5}$$

where $\bar{v}_m$, the reciprocal of the wavelength at the center of the band, is measured in cm^{-1}, ε', the absorptivity at any given value of $\bar{v}$ in the spectral band of interest, is usually expressed in terms of L/mole cm (where c is expressed as moles/L and d in cm), and R is expressed in esu cm. For Gaussian-shaped bands,

$$\int_0^\infty \varepsilon' d\bar{v} \approx \varepsilon_m \bar{v}_{1/2} \tag{6}$$

where ε_m is the absorptivity at the band maximum (center), and $\bar{v}_{1/2}$ is the band width at half-maximum absorption.

The absorbance is defined by the Beer-Lambert law,

$$A = -\log \frac{I}{I_0} = -\log T \tag{7}$$

where I and I_0 are the intensities of exciting light transmitted through the sample in the presence and absence, respectively, of the absorbing species, and T is called the transmittance. Although in analytical practice it is customary to measure A and ε at some nominal analytical wavelength, Beer's law is, strictly speaking, valid only for A and ε integrated over the entire spectral band. For Gaussian bands, ε is proportional to ε_m, a fact that justifies using absorbances and absorptivities measured at the band maximum provided that they are measured using light that is highly monochromatic. However, for $\varepsilon \neq \varepsilon_m$ and for non-Gaussian spectral bands, the absorbance measured at a single nominal wavelength will not necessarily be truly proportional to the concentration of absorbers and will give less than ideal analytical results.

The magnitude of the transition moment integral depends in large part on the molecular electronic distributions in the ground state and the excited state participating in the absorptive transition. In particular, the spatial properties of the orbitals accommodating the optical electron in the ground and excited states are important in determining the magnitude of R. The $n \rightarrow \pi^*$ transitions of N-heteroaromatics, azaaromatics, and carbonyl compounds involve promotion of an electron from an sp^2 orbital to a π^*-orbital directed at right angles to the former. There is little or no overlap between the n-sp^2-orbital and the π^*-orbital and, as a result, the probability of promoting the n-electron to the π^*-orbital through a region of space quantum mechanically forbidden to it (this amounts to ionization and recapture of the n-electron) is low. This results in a very small value of R and therefore of ε_m. Typically, $n \rightarrow \pi^*$ transitions have $\varepsilon_m < 1000$ and are said to be symmetry-forbidden transitions. Although $n \rightarrow \pi^*$ transitions play a major role in the practicability of fluorescence or phosphorescence analysis, they are usually of little or no value in quantitative trace analysis by absorptiometry. The $n \rightarrow \sigma^*$ transitions are also symmetry-forbidden.

The $\pi \rightarrow \pi^*$ transitions of unsaturated, conjugated, and aromatic molecules and their derivatives are usually allowed transitions corresponding to intense absorption bands of $\varepsilon = 1000\text{--}100{,}000$. However, in some of the long-wavelength transitions of the highly symmetrical unsubstituted aromatic molecules (e.g., benzene and naphthalene), the ground and excited states are highly symmetrical in their electron distributions. As a result, these transitions entail no change in electronic dipole moment and theoretically $R = 0$. These transitions which include the longest-wavelength absorption bands of the aromatic hydrocarbons are weak (e.g., for benzene, $\varepsilon_m = 200$). Transitions having low intensity because of highly symmetrical charge distributions in ground and excited states are also said to be symmetry-forbidden.

If heteroatomic or exocyclic substituents are introduced into a highly symmetrical aromatic ring the substituents, by virtue of breaking down the electronic symmetries of highly symmetrical molecules, can intensify ordinarily forbidden transitions. Thus, the longest-wavelength absorption band of benzoic acid is about five times more intense than that of benzene, and those of quinoline and isoquinoline are about 10 times more intense than that of naphthalene. The greater the extent to which a substituent interacts with the excited states of the parent hydrocarbon, the greater will be the breakdown in the symmetry of the electronic distribution of the excited states and the greater will be the transition dipole moment R. This means that the strongly interacting substituents will not only produce greater red-shifting of the absorption bands but will also give rise to more intense absorption bands than will weakly interacting substituents.

E.   The Shape and Structure of Absorption Bands

The electronic transition from the ground state to an electronically excited state will often entail changes in vibrational as well as electronic energy. This results from excitation of the molecule to one of the various vibronic levels of the excited electronic state. This process can be illustrated by a diagram of potential energy versus internuclear separation. Such a diagram is very difficult to represent for a polyatomic molecule, but it can be instructive for a diatomic molecule (Fig. 5).

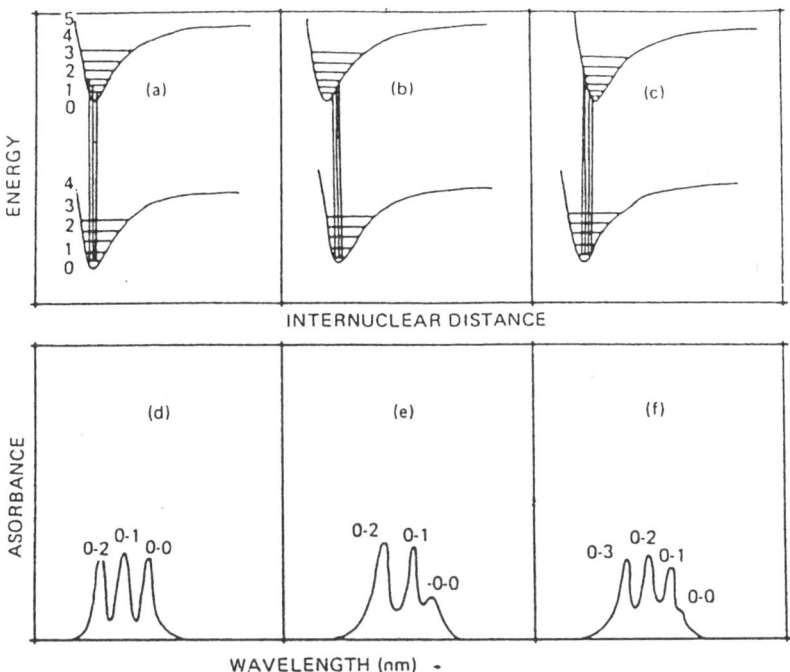

Fig. 5   The influence of the Franck-Condon principle on the appearance of the absorption band in a diatomic molecule where the equilibrium internuclear separations are (a) identical in the ground and excited states, (b) smaller in the excited state, and (c) greater in the excited state. The spectra representations (d), (e), and (f) correspond to the situations depicted in (a), (b), and (c), respectively. The numbers in (a) represent the vibrational quantum numbers in the ground and excited states and in (d), (e), and (f) the transitions between these sublevels.

The horizontal distances between the minima in Fig. 5 represent the differences in equilibrium internuclear separation between ground and excited states. The vertical distances between the curves represent the differences in energy between the ground and excited states for any given internuclear separation. Figure 5a shows the potential energy curves for the case where the ground and excited states have the same internuclear separation. In Figs. 5b and 5c are shown the circumstances occurring when the excited-state equilibrium internuclear separations are more and less compressed, respectively, than the ground-state equilibrium geometry.

Absorptive electronic transitions in molecules occur about 10 times faster than vibrational transitions. This is the basis of the Franck-Condon principle, which states that electronic transitions occur without change of position of the nuclei. Therefore, in Fig. 5 an electronic transition can be represented as a vertical line between a vibrational level in the ground state and one in the excited state to which transition takes place. At room temperature most absorbing molecules in the ground electronic state will occupy the lowest vibrational level. Transitions from this vibronic state to the various vibronic levels of the excited state will occur with intensities that depend σ/n vibrational as well as electronic factors. The multitude of vibronic transitions that accompany each electronic transition give the electronic absorption band its breadth—i.e., its bandlike appearance in contrast to the linelike appearance of an atomic absorptive transition. Among the factors that govern the vibrational components of the electronic transition is spatial overlap of the vibrational states, i.e., the probability that a given internuclear separation will be the same for both vibrational states involved in the transition.

In Fig. 5a, where the equilibrium internuclear separations of ground and excited states are identical, the lowest vibrational level of the ground electronic state, to a first approximation, coincides with all of the vibrational levels of the excited state and all of the possible vibronic transitions are likely to appear in the corresponding absorption band (Fig. 5d). However, in Figs. 5a and 5b the lowest vibrational level of the ground state coincides poorly in space (or not at all) with the lower vibrational levels of the excited state. In these cases the lowest-energy (longest-wavelength) vibronic states will appear weakly or be absent from the corresponding absorption bands. The maxima of the absorption bands will then occur near the centers or toward the short-wavelength sides of the absorption bands (Figs. 5e and 5f) rather than on the long-wavelength side as in Fig. 5d. Figure 5d is typical of the absorption spectra of unsubstituted aromatic hydrocarbons. Figures 5e and 5f are typical of the absorption spectra of functionally substituted aromatic hydrocarbons. In substituted hydrocarbons, low-energy torsional vibrations due to the substituents are often imposed on the vibrational structure, which, for the most part, corresponds to breathing vibrations of the ring. In this case the vibrational structure of the absorption bands may be blurred or obliterated, giving rise to structureless bands with the same general intensity distribution as shown in Figs. 5e and 5f.

F. The Influence of the Solvent on the Absorption Spectrum

The solvents in which absorption spectra are observed play a substantial role in determining the spectral positions and shapes and affect to a somewhat lesser degree the intensities of absorption spectral bands. The effects of the solvent on the absorption spectra are determined by the nature and relative strengths of the interactions of the solvent molecules with the ground and electronically excited singlet states of the absorbing solute molecules.

Solvent interactions with solute molecules are predominantly electrostatic in nature and may be classified as induced dipole–induced dipole, dipole–induced dipole, dipole–dipole, or hydrogen bonding. The position of an absorption-band maximum in one solvent relative to that in another depends on the relative separations between ground- and excited-state energies in either solvent and therefore on the relative strengths of ground- and excited-state solvent stabilization.

Induced dipole–induced dipole and dipole–induced dipole interactions (van der Waals forces or dispersion forces) are the weakest of solvent–solute interactions and are predominant in solutions of nonpolar solutes in nonpolar or polar solvents and/or polar molecules in nonpolar solvents. These interactions, which are closely related to the polarizabilities of the solute and solvent, account for the small shifts to lower frequency of the absorption spectra of molecules upon going from the gas phase to solutions in nonpolar media. Presumably, the π,π^* excited states of the nonpolar aromatic hydrocarbons and other conjugated hydrocarbons are more polarizable than the ground state, leading to stabilization of the excited states relative to the ground state and shifts of the absorption bands to the red.

If the π,π^* or intramolecular charge-transfer excited state of a polar (substituted) molecule has a higher dipole moment than its ground state (most molecules fall into this class), the excited state will be more stabilized by interaction with a polar solvent than the ground state. As a result, upon going from a less polar to a more polar solvent, the absorption spectrum will shift to longer wavelengths. Although this generalization is often made, it is, strictly speaking, applicable only to dipolar aprotic solvents or, in general, to solvents in which hydroyen bonding between solvent and solute is either weak or nonexistent. In a few cases, the ground state of the solute is more polar than the excited state. In this circumstance, going to a more polar solvent stabilizes the ground state more than the excited state, causing a shift to higher frequency with increasing solvent polarity (dielectric strength)—a feature seen with 9-aminoacridine.

Dipole–dipole interactions decrease with the third power of the distance between interacting molecules. As a result, dilution of a solution of a polar solute, in a polar solvent, with a solvent of lower polarity results in an essentially continuous decrease in the dipole–dipole interaction with increasing mole fraction of the solvent of lower polarity. In molecules whose excited states are more polar than their ground states, this means that the spectrum will blue-shift continuously as the mole fraction of nonpolar solvent increases.

Hydrogen-bonding solvents having positively polarized hydrogen atoms are said to be hydrogen-bond donor or protic solvents. They interact with the nonbonded and lone electron pairs of solute molecules. Hydrogen-bonding solvents having atoms with lone or nonbonded electron pairs are said to be hydrogen-bond acceptor or basic solvents. They interact with positively polarized hydrogen atoms on electronegative atoms belonging to the solute molecules (e.g., in $-COOH$, $-NH_2$, $-OH$, $-SH$).

Hydrogen bonding is a shorter-range and therefore stronger interaction than nonspecific dipole–dipole interaction. It involves some degree of covalency and is manifested only in the primary solvent cage of the solute. Dilution of a solution of a solute in a hydrogen-bonding solvent by a nonpolar, nonhydrogen-bonding solvent will not normally disrupt the hydrogen bonding in the primary solvent cage and will therefore cause very small spectral shifts. Because most hydrogen-bonding solvents are also polar, hydrogen bonding and nonspecific dipolar interaction are usually both present as modes of solvation of functional molecules. Accordingly, the spectral shifts actually observed upon going from one solvent to another are a composite of dipolar and hydrogen-bonding effects which may be constructively or destructively additive. This makes it rather difficult to interpret solvent-induced spectral shifts unambiguously.

Due to the involvement of nonbonded and lone electron pairs in $n \rightarrow \pi^*$ intramolecular charge-transfer transitions, hydrogen-bonding effects play a major role in the appearances of these spectra. Dipolar effects are most pronounced in intramolecular charge-transfer spectra because of the large dipole-moment changes accompanying the associated transitions.

Nonbonded electron pairs interact strongly with protic solvents. Because the orbital accommodating the n-electrons has two electrons in the ground state and only one in the n,π^* excited state, the ground state is more stabilized by hydrogen bonding with a protic solvent than the excited state. This means that the absorption bands arising from $n \rightarrow \pi^*$ transitions should move to higher energies or shorter wavelengths when an aprotic solvent is replaced by a protic solvent. This is indeed the case: the $n \rightarrow \pi^*$ shoulder at 270 nm observed on the longest-wavelength π,π^* band of pyridine dissolved in hexane disappears when the spectrum is taken in ethanol or water. Presumably the $n \rightarrow \pi^*$ band of pyridine blue-shifts and is submerged in the $\pi \rightarrow \pi^*$ band at shorter wavelengths. The $n \rightarrow \pi^*$ bands of benzaldehyde and acetophenone at ~330 nm move to distinctly shorter wavelengths upon going from aprotic to protic solvents. It has been suggested that the blue-shifting of the $n \rightarrow \pi^*$ bands and the red-shifting of π,π^* bands (or intramolecular charge-transfer bands) upon going to protic (and invariably more polar) solvents be used to distinguish between these types of transitions. This, however, is not advisable. Although $n \rightarrow \pi^*$ absorption bands invariably move to shorter wavelengths upon going to more protic solvents, many $\pi \rightarrow \pi^*$-type transitions (especially those involving l electrons) also blue-shift in highly protic solvents.

Protic solvents interacting with lone pairs on functional groups which are electron-withdrawing in the excited state (e.g., $=N$, $C=O$) enhance charge transfer by introducing a partial positive charge into the electron-acceptor group. This interaction stabilizes the charge-transfer excited state relative to the ground state, so the absorption spectra are expected to shift to longer wavelengths with increasing hydrogen-bond donor capacity of the solvent. Increasing hydrogen-bond donor capacity of the solvent produces shifts to shorter wavelengths when the solvent interacts with lone pairs on functional groups which are electron donors in the excited state (e.g., $-OH$, $-NH_2$). Hydrogen-bond acceptor solvents produce shifts to lower frequency when solvating hydrogen atoms on functional groups which are electron donors in the excited state (e.g., $-OH$, $-NH_2$). This is effected by the partial withdrawal of the positively charged proton from the functional groups, thereby facilitating transfer of electronic charge away from the functional group. Solvation of hydrogen atoms on functional groups which are charge-transfer acceptors in the excited state (e.g., $-COOH$) inhibits charge transfer by leaving a residual negative charge on the functional group. Thus, the latter interaction results in shifting of the spectrum to shorter wavelengths.

G. The Influence of Coordination by Metal Ions

The coordination of absorbing molecules by metal ions is actually an acid–base reaction, with the metal ion acting as a Lewis acid (electron-pair acceptor) and the ligand acting as a Lewis base (electron-pair donor). In this regard, the coordination of ligands by non-transition metal ions is analogous to the protonation of the ligand. In the latter case, the hydrogen ion functions as the Lewis acid. As a result, many of the changes of the electronic spectra of the ligands, produced by metal ions coordination, are similar to the changes caused by protonation of substituents described earlier (e.g., both protonated and zinc-coordinated 8-quinolinol absorb at ~ 350 nm). However, the analogies between electronic spectral changes in the ligand produced by protonation and those produced by coordination with transition metal ions are weak, and often there are spectral phenomena observed with transition metal ion coordination that have no equivalent in coordination with hydrogen ion.

The spectra of transition metal complexes contain absorption bands in the visible, showing that they possess closely spaced electronic energy levels. In these species the central metal ion is surrounded by a certain number of ligands, which are generally either negative ions, as in $[Fe(CN)_6]^{3-}$, or dipolar molecules whose negative ends are oriented toward the metal ion as in $[Co(NH_3)_6]^{3+}$. The transition metal ions all possess incompleted shells. The d-orbitals all have the same energy in the isolated metal ions. However, when surrounded by ligands, some d-orbitals are raised in energy more than others. Electronic transitions between the lower-energy d-orbitals (occupied) and higher-energy d-orbitals (unoccupied) give rise to the ligand field spectra which give transition metal ions their characteristic colors (e.g., pink for manganese and green for nickel). A detailed description of the ligand field theory, which treats ligand field spectra quantitatively or semiquantitatively, is beyond the scope of this chapter. Because they are symmetry-forbidden, the ligand-field absorption spectra are very weak ($\varepsilon = 10$) and therefore not useful for trace analysis. For example, $Cu(H_2O)_4^{2+}$, which is blue in concentrated solutions, appears colorless in a 1×10^{-4} M solution. However, transition metal complexes often demonstrate very intense absorption bands ($\varepsilon = 10^3-10^5$) in the visible and near-ultraviolet, which are analytically useful at trace concentrations. These bands originate from transitions involving the promotion of the d-electrons of the metal ion to vacant π^*-orbitals of the ligand or from π-electrons of the ligand to the d-orbitals of the metal ion. Since the $d \rightarrow \pi^*$ and $\pi \rightarrow d$ transitions involve upper and lower states with very different electronic distributions, they are strongly allowed. This accounts for their high intensities. The intense colors of the Fe(II)–phenanthroline complex, the Fe(III)–phenol complexes, and the Cu(I)–biquinolyl complex are all examples of $d \rightarrow \pi^*$ transitions which have been used in spectrophotometric pharmaceutical analysis. In addition, the colors of oxyhemoglobin, the cytochromes, vitamin B_{12}, and the oxidizing enzyme P-450 are all due to $d \rightarrow \pi^*$ transitions.

IV. INSTRUMENTATION

The basic spectrophotometer generally consists of a light source from which a given wavelength or range of wavelengths is selected by a wavelength selection device. The radiation selected is directed through the analytical sample and the transmitted light monitored by a detector. The light intensity measured by the detector is subsequently compared to that transmitted by a reference substance, the ratio being displayed usually as an absorbance but less commonly as a percent transmittance on a readout device.

A. Light Sources

There are two commonly used sources of light in UV–visible absorption spectrophotometry, hydrogen or deuterium discharge lamps and incandescent filament lamps. The discharge lamps emit most of their light output in the range 200–360 nm. They consist of a quartz envelope filled with hydrogen or deuterium at low pressure (approx 5 torr), into which project the closely spaced ends of an anode and a cathode. The passage of an electric current across the tips of the anode and cathode leads to excitation of the intervening gas molecules, which subsequently dissociate into photoexcited atoms that emit energy in the form of ultraviolet radiation. The light output with deuterium lamps is about 3–5 times brighter than that in hydrogen discharge lamps. In some instruments the light emitted is collected and focused into the wavelength selection device by use of a concave mirror.

Incandescent lamps commonly find use as sources of visible light. They consist basically of a metal wire filament, usually of tungsten, which is sealed inside an evacuated glass envelope. The filament is heated by the passage of an electric current and then emits a broad band of energy with a maximum dependent on the temperature of the filament. For example, a tungsten filament heated to 2860 K has a radiation maximum in the near-infrared at about 1000 nm. Only about 15% of the light output from the filament occurs in the visible region, with most occurring in the infrared. Generally, higher temperatures will shift the maximum to shorter wavelengths, but this does compromise the lifetime of the filament. Most of the current instruments have both kinds of lamps, with a switching device permitting scanning over the whole UV–visible range (200–800 nm) in one sweep. Improvements in lamp technology have made it possible to employ deuterium lamps over the whole range.

In some of the more sophisticated instruments, lasers have been employed as light sources. These have the advantage of generating a high-intensity monochromatic light output which is useful for high-absorbance samples; a consequence of this is improved sensitivity and resolution. Laser beams are also more easily focused than beams from conventional light sources. This is useful when the sample cell is small in size, as in flowing streams. Tunable diode lasers are most commonly used for spectrophotometric analyses. Unfortunately, due to their high costs, lasers are not commonly used as light sources for routine analysis by UV–visible spectrophotometry. Their availability has led, however, to the development of newer spectrophotometric techniques such as photoacoustic and thermal lens spectrometry.

High-intensity line sources still occasionally provide a simple source of high-intensity monochromatic light and have found use in portable instruments, designed for specific analyses.

B. Wavelength Selection Devices

The wavelength selection device in most instruments is a monochromator. The monochromator consists of an entrance slit, a dispersive device, a collimator, and an exit slit. The slits are narrow planar apertures that are used to isolate a narrow band of light. The entrance slit is placed between the source and the collimator and the exit slit is placed between the focusing device and the sample. The widths of the slits are frequently variable and significantly influence the quality and accuracy of the analysis. Collimators are either lenses or concave mirrors which serve to align the incident light beam into parallel light rays which then impinge on the dispersive element.

There are two types of dispersive elements, prisms and gratings. The prisms are constructed of transparent material of known refractive index. In prisms the incident light is refracted to varying extents, depending on its wavelength, generating a range of wavelengths that impinge upon the exit slit. The different materials give rise to prisms of different angular dispersion. Prisms are generally good for the separation of light of shorter wavelengths, but they are less efficient for the separation of light of longer wavelengths. Furthermore, the resolving power (R) of prisms as defined by

$$R = b \, d\eta/d\lambda \tag{8}$$

where b is the base width of the prism and $d\eta/d\lambda$ is the dispersive power of the prism, which is characteristic of the prism material. From Eq. (8) it can be seen that it would require large prisms to efficiently separate long wavelengths of light. Such large prisms would not only be difficult to construct, but would result in unduly expensive instruments. As a result, grating monochromators are more commonly used nowadays. These may be either reflection or transmission gratings. A transmission grating consists of a series of finely spaced parallel lines on a transparent material, whereas a reflection grating consists of a series of equally spaced parallel grooves cut into a reflecting surface. Reflection gratings are more commonly used in UV–visible spectrophotometers, because the entire optical system can then be contained within a smaller volume.

In a grating the incident beam is diffracted at each of the surfaces generating an interference pattern. The different wavelengths of light undergo constructive interference at different angles, permitting wavelength selection by pivoting the grating through different angles, thereby focusing different wavelengths of light onto the exit slit. The higher-order interference patterns can be removed by the use of appropriate filters. Older instruments employed filters as a means of wavelength selection. These were either interference or absorption filters. Interference filters consist of a thin layer of transparent dielectric medium between two thin reflective metal films, whose thicknesses are carefully controlled. A portion of the light in the incident beam is transmitted through the first metal film and undergoes a series of reflections through the dielectric medium between the films, generating an interferometric pattern. Those constructive interferences of second or higher order emerge from the second metal surface. The filters typically allow through a small band of wavelengths of spectral band width 10–15 nm with a maximum percent transmission of about 40%. The higher orders of constructive interferometric light that emerge from the filter are less intense with increasing order and can be selected by the use of additional filters. The interference filters permit through more light than monochromators with the same bandpass, but lack the versatility of monochromators. Interference filters are usually available for the ultraviolet, visible, and infrared regions of the electromagnetic spectrum. Absorption filters, on the other hand, select a desired range of wavelengths by absorbing undesired wavelengths. They typically consist of a colored plate of glass. The filters may be bandpass or cutoff filters. Bandpass filters transmit a band of light of about 30–250 nm in width and generally transmit 5–20% of the incident light. Cutoff filters transmit essentially 100% of the incident light beyond a certain wavelength, cutting off light of shorter wavelengths in short-wavelength cutoff filters and longer wavelengths in long-wavelength cutoff filters. The cutoff wavelength is generally defined as the wavelength where the absorbance of the filter is unity (10% transmittance). Absorption filters lack the selectivity possible with monochromators and are available only for the visible region. They are used, however, for eliminating unwanted orders of light emerging from interference filters.

C. Sample Cells

Sample cells are commonly made of quartz or fused silica and readily transmit light of 200- to 800-nm wavelength. However, less expensive cells, constructed of high-quality glass and, more recently, plastic are available but are useful largely in the visible region of the spectrum, though some plastics do permit measurements at wavelengths as short as ~300 nm. The cells are available in various shapes and sizes, many of which are constructed for specialized applications. The two most common shapes are cylindrical and rectangular. The cylindrical cells are often used in low-cost filter photometers, whereas the rectangular cells are used in high-precision instruments. Cell path length is largely dependent on the application and ranges from 5 μm to 5 m. Smaller sample volumes with extremely long path lengths have also been employed. In these cases the increased path length at constant cell volume provides additional analytical sensitivity. Flow-through cells have found use in flow injection analysis and liquid chromatography as well as in process analysis. In flowing streams the cell volumes employed can vary considerably, ranging from 5 to 50 μL depending on the analytical sensitivity required. However, the path length employed is commonly 1 cm. The choice of cell shape or size is largely dependent on the instrument employed and the application.

Autosampling devices have also been developed for automated instruments. These are often microprocessor-controlled and have found use in the analysis of large samples, eliminating the operator error often associated with long analysis periods.

D. Detectors

1. Photoemissive and Photomultiplier Tubes

The photoemissive tube consists of an anode and a cathode enclosed in a transparent envelope. The cathode consists of a semicircular metal sheet plated with a thin layer of photoemissive alkali metal (cesium or rubidium). A negative potential of about 90 V is applied across the gap between the cathode and anode, which facilitates electron migration to the anode following electron release by the impact of light photons on the cathode. The magnitude of the current is proportional to the intensity of the incident light. Photomultiplier tubes have the additional feature of possessing multiple anodes, referred to as dynodes, which are coated with an electron-rich material of low ionization potential such as BeO, GaP, or CsSb. Across each subsequent dynode an increasing potential is applied, which serves to both accelerate the electrons and enhance electron yield. As a result, high electronic gains can be achieved. A stable and regulated power supply is important for consistency of yield for a given light intensity. Both photoemissive and photomultiplier tubes generate amplifiable currents. However, they have limited ranges of optimum spectral response and are both limited by thermal and shot noise.

2. Photovoltaic Cells

The photovoltaic cell consists of a metal plate enclosed in a plastic case. Sandwiched between the plate and a glass window is a thin layer of semiconducting material such as selenium, plated with a thin layer of gold or silver. When light impinges upon the silver layer, it emits electrons by the photoelectric effect. These electrons pass through the selenium layer and onto the metal plate. The current generated by this effect is proportional to the intensity of incident light and can be measured. Despite the advantage of not requiring an external power source, the signal is not easily amplified due to the low internal impedance. As a result, photovoltaic cells find limited use in portable instruments. Other disadvantages include limited sensitivity, limitation of their usefulness to dilute samples, and the fact that they have slow response times and exhibit fatigue with time.

3. Photodiode Array Detectors

Photodiode array detectors are an offshoot of semiconductor technology. In semiconductors, impurities have been added to pure silicon to create two classes of materials. The addition of arsenic, bismuth, phosphorous, or antimony creates a pentavalent material (n-type) that is able to function as a donor of electrons. The addition of trivalent elements such as aluminium, boron, gallium, indium, etc., to silicon gives rise to the p-type material, in which the trivalent material is able to accept electrons to make up for its electron deficiency. The trivalent metals have only three electrons in their outer shells. Application of a potential difference to n-type material causes electron movement from negative to positive potential (cathodic). With p-type material the applied potential difference is manifested as apparent movement of positive charge from positive to negative potential (anodic).

Photodiode array detectors consist of an array of p-n type semiconductor diodes mounted onto a semicircular plate facing the light source. In these instruments the transmitted light beam from the sample is split into its component wavelengths by a dispersive element, and the various wavelengths of light fall onto different photodiodes on the semicircular plate. Interaction of a photon with the semiconducting layer of the diode generates a flow of current, the magnitude of which is proportional to the intensity of transmitted light. The advantage here is that temporal scanning is no longer necessary and the whole spectrum can be obtained almost instantaneously. The quality of spectra obtainable with diode array instruments is, however, limited by the spectral wavelength range as well as the number of diodes covering the entire spectral range. Diode array instruments, despite having a slower response time, have found application as detectors in chromatographic separations, where the spectrum of the emerging solutes may be used for semiquantitative characterization of the solutes. The diode array detectors are significantly useful in multiple signal detection, peak identification, and peak purity determinations.

E. Readout Devices

Readout devices convert the electrical signal emerging from the detectors to an analog or digital signal that is more understandable to the operator. The electrical signal is either converted to a plot of voltage versus time or digitized as numerical values of voltage versus time. For scanning instruments the time can be read as wavelength for a consistent known scan rate. In the diode array instruments all measurements are recorded simultaneously, and can be issued as a single spectrum or incorporated into a three-dimensional profile with time. The use of microprocessors in data acquisition and handling has facilitated the development of simultaneous multicomponent determinations over a range of wavelengths as opposed to single-wavelength measurements. They have also facilitated extensive manipulation of the data obtained from the instruments, such as smoothing, overlaying, derivatization, etc.

F. Instrument Configuration

The simplest instrument configuration is the single-beam configuration, in which a single light beam is transmitted from the source through the described modules

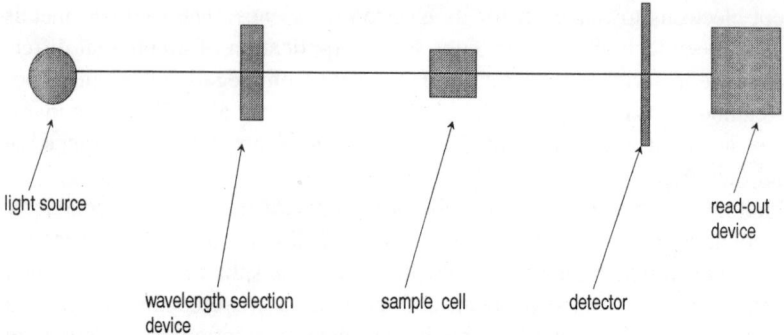

Fig. 6 Schematic diagram for the layout of a single-beam UV–visible spectrophotometer.

to the detector (Fig. 6). In double-beam instruments the light beam emerging from the source is split into two separate beams for the sample and the reference paths, respectively. This modification is associated with increased instrumental cost and lower light energy throughput as a result of the splitting of the source beam (Fig. 7). The double-beam instruments have largely superceded the single-beam instruments because in the double-beam instruments it is possible to eliminate the instability and drift arising from temporal differences in scanning the reference and sample cells.

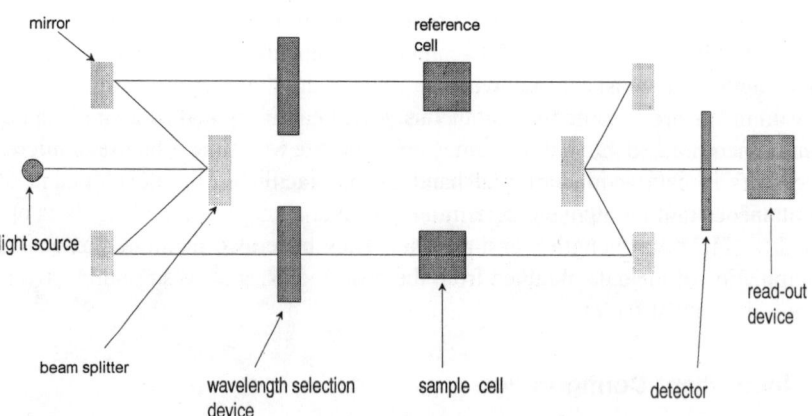

Fig. 7 Schematic diagram for layout of a double-beam UV–visible spectrophotometer.

Forward and reverse optical designs have also been tried. In the reverse optical arrangement the wavelength selection device is placed between the sample cell and the detector, as opposed to being between the source and the sample cell in the forward optical arrangement. The reverse optical design is reported to improve detection limits as a result of greater light throughput, but does not eliminate stray light as well as does the forward optical arrangement.

V. QUALITATIVE AND QUANTITATIVE ANALYSES

A. Qualitative Analysis

The ultraviolet–visible spectra of most compounds are of limited value for qualitative analysis and have been largely superseded by the more definitive infrared and mass spectroscopies. Qualitative analytical use of ultraviolet–visible spectra has largely involved describing compounds in terms of the positions and molar absorptivities of their absorption maxima, occasionally including their absorption minima. Indeed, some organic compounds are still characterized in terms of the number of peaks in the UV–visible spectrum and their absorbance ratios. This is usually the case in phytochemistry and photodiode array chromatography and when the analyst has a limited range of compounds to work with whose spectra are known to differ. In the pharmacopeias, however, absorbance ratios have found use in identity tests, and are referred to as Q-values in the U.S. Pharmacopia (USP).

Ultraviolet–visible spectrophotometry has also been applied to titrimetry. In this case the variation in the absorbance of the analyte with addition of titrant is used to obtain a spectrophotometric profile from which titration end points and/or equilibrium constants, etc., can be determined. This has been applied to the whole range of titrations in which a chromophore is generated. These include acid–base, redox, and complexometric titrations.

B. Quantitative Analysis

Ultraviolet–visible spectrophotometry is perhaps the most widely used spectrophotometric technique for the quantitative analysis of chemical substances as pure materials and as components of dosage forms. It has found increasing usefulness as a means of assaying pharmaceutical substances described in the pharmacopeias.

Pharmaceuticals are usually marketed as formulations containing more than just the active ingredient(s). The other components, referred to as excipients, are added to the formulation to enhance efficacy, improve the appearance of the product, or facilitate certain stages in production. All of these other materials present in the dosage form can, and often do, complicate the analyses of the active ingredient(s). Hence, in most spectrophotometric analyses of pharmaceuticals, a separation stage is often included in the analytical procedure. Physical isolation of the active ingredient from all other dosage form ingredients is not always possible, and when these other excipients are known to interfere with the spectrum of the analyte they are referred to as interferents. Interferents are broadly considered to be those substances that modify the shape of the absorption spectrum, usually contributing to the absorption measured. The presence of interferents in an analytical sample can be inferred from the failure of the absorption curve to return to zero absorption in regions where the analyte is known not to absorb. For purposes of this discussion we can divide spectrophotometric analyses into single-component and multicomponent analyses. Single-component analysis is more commonly applied, and it involves the isolation of the analyte of interest from other dosage form excipients. Multicomponent analysis, on the other hand, does not involve isolation of the analyte of interest from the all the other possible interferents present in the dosage form, but rather involves the retention of one or more interferents whose spectrophotometric profiles in the analytical solution are known. The use of UV-visible spectrophotometry in quantitative analysis is centered around the Beer-Lambert law, which relates the absorbance of an analyte to its concentration:

$$A = \varepsilon\, Cd \tag{9}$$

where A is the absorbance, ε is the molar absorptivity (L mole/cm), c is the concentration (mole/L), and d is the path length (cm). The Beer-Lambert law applies rigorously only to integrated absorption bands from absorption-versus-energy plots. In these cases the area under the curve is directly proportional to the concentration of the absorbing species. It is possible, however, to apply the law to nominal single-wavelength analyses, especially when the analytical wavelength is at the absorption maximum. The Beer Lambert law, however, does not necessarily hold at single wavelengths that are significantly different from the wavelength of maximum absorption, especially when the absorbance is rapidly changing.

1. Monocomponent Analysis

In a simple analysis, the analyte can be quantified by measuring its absorbance at a given wavelength, then substituting for absorbance (A) molar absorptivity (ε), and pathlength (d) in Eq. (9) and solving for concentration (c). If the molar absorptivity (ε) is not known and a pure sample of analyte is available, a calibration curve (of absorbance versus concentration) can be drawn. The slope of the calibration curve is given by the product $\varepsilon\, d$. When the path length (d) is known, the molar absorptivity (ε) can be calculated. Occasionally, a single sample may be used, but this is less reliable than drawing a calibration curve. Linear interpolation on the calibration curve can also be used to read the concentration of an analyte whose absorbance is known. This approach is particularly useful when apparent deviations from the Beer-Lambert law are observed. In applying the Beer-Lambert law to analyses it is important that the solvent does not interfere with the analysis. The solvent must not only dissolve the analyte, but must also have a low volatility and be transparent in the spectral region of interest. Thus volatile solvents necessitate that the sample cell be sealed during the analysis or that the measurements be carried out as quickly as possible. Similarly, solvents with a UV cutoff in the near-UV do not permit analytical measurement at wavelengths shorter than the cutoff wavelength.

Ionizable compounds may have to be dissolved in buffered solutions, to ensure that only one form of the analyte exists in solution. The pH employed is commonly at least two pH units above or below the pK_a, depending on which pH yields the optimal chromophore. Just as pH equilibria can be employed, chemical complexation equilibria can also be employed to improve analytical selectivity.

A number of organic compounds have found use as chemical derivatization reagents. Usually these have served as chromogenic reagents that generate a chromophore upon interaction with the analyte. Organic compounds used for this purpose include the crown ethers, diazotizing reagents, and the porphyrines. Their uses have been largely for metal analysis, as many metal complexes are chromogenic. In some cases, however, the complexing reagents have served to enhance selectivity by facilitating extraction of the analyte—usually by ion-pair extraction.

2. Multicomponent Analysis

Spectrophotometric multicomponent analyses are based on mathematically processing a composite absorption spectrum made up of the spectra several components that contribute additively to the overall spectrum. In all cases, some idea of the nature of the contributing spectra is required for the mathematical processing of the composite sample spectrum. A host of techniques of varying complexity have been reported in the literature.

a. Experimental Techniques Applied to Multicomponent Analysis

Difference spectrophotometry involves the exploitation of the ability to chemically modify the spectrophotometric profile of the analyte alone in the presence of other possible interferents. The analyte may be modified by alteration of pH or through chemical reaction in either the reference or sample cell. pH-induced difference spectrophotometry is most commonly employed, due to its simplicity, though reagents to covalently modify the analyte have also found use. The selective modification of the analyte alone in the presence of interferents permits quantitation on the basis of spectral differences between the otherwise identical reference and test solutions. A useful test for indicating whether selective modification of the analyte alone has been achieved involves confirming that there is zero absorbance at the isosbestic points of the two spectral species of the analyte.

Derivative spectra are literally the derivatives of the normal spectra. Their analytical advantage stems from the fact that the slopes of the spectra of substances of narrow spectral bandwidth are usually higher in magnitude than those of substances with broad spectral bandwidth. As a result, the analyses of substances with narrow spectral bandwidth can often be performed in the presence of substances with broad spectral bandwidth, where the spectrum of the analyte appears as a shoulder on the spectrum of the broad-bandwidth interferant. The differentiation procedure was initially carried out manually. However, electronic differentiation techniques developed more recently have simplified the application of the techniques to pharmaceutical analyses. It is worth noting that though higher derivatives appear to improve resolution, the spectra obtained are also significantly distorted by noise. A trade-off is therefore required.

b. Mathematical Correction Techniques

The simultaneous-equations method is a simple case of multicomponent analysis that is applicable to the simultaneous determination of two absorbing species present in a solution. When two or more absorbing species are present in the cell and the Beer-Lambert law is obeyed, the absorbance at a given wavelength is the sum of the absorbances of the two species at that wavelength. That is,

$$A_{tot} + A_1 + A_2 \tag{10}$$

where A_{tot} is total absorbance and A_1 and A_2 are the absorbance of species 1 and 2, respectively, at the wavelength of measurement.

From Eq. (10) we can therefore say that

$$A_{tot} = \varepsilon_1 C_1 l_1 + \varepsilon_2 C l_2 \tag{11}$$

for each wavelength of measurement. Hence, if the molar absorptivities ε_1 and ε_2 are determined from standard solutions, two simultaneous equations can be written whose solution would afford the concentrations of the absorbing species. For n species, n analogous equations can be drawn, permitting the calculation of the concentrations of all n species. It is worth noting that significant errors may be introduced into the calculation if the individual spectra overlap considerably. A number of modern instruments incorporate microprocessors that are preprogrammed to solve simultaneous equations of this sort. In these cases, the operator specifies the appropriate wavelength values and the predetermined molar absorptivities (ε) of the species to be determined. The microprocessor then calculates the concentrations. The absorbance ratio method is a modification of the simultaneous-equations method which confers on the analysis the advantage of requiring less stringent experimental technique. In the technique, two wavelengths are employed for measurement, one of which is at an isosbestic point for the two spectra. An equation can be drawn relating the fraction of the absorbing components to the absorption at one wavelength and the absorption at the isosbestic point used to eliminate one concentration variable in the equation drawn for the other wavelength.

Thus it has been shown that in a two-component system, the fraction of a component f_p is given by the relation

$$f_p = \frac{Q_m - Q_q}{Q_p - Q_q} \tag{12}$$

where $Q_m = A_2/A_1$, $Q_x = a_{p2}/a_{p1}$, $Q_q = a_{q2}/a_{q1}$, a_{pi} and a_{qi} are the absorbances of pure p and q at wavelength i, and A is the absorbance of the mixture. Hence the concentration of component p is given by

$$c = \frac{Q_m - Q_q}{Q_p - Q_q} \cdot \frac{A_1}{a_{p1}} \tag{13}$$

because at the isosbestic point the molar absorptivities for the two components p and q are identical.

In most analyses it may be possible to separate the analyte(s) from the other dosage form ingredients, but this is not always the case. When this is not the case, the dosage form excipients may be able to interfere with the analyses. To deal with this problem, a number of specific techniques have been developed for the analysis of a variety of analytes in the presence of interferants.

The absorption spectrum of interferants is commonly linear, but nonlinear interferant absorption has been reported. A number of mathematical techniques have been developed to correct for nonlinear interfering absorption. Most of the correction techniques are based on assuming that the interferents have an absorption profile that can be represented by some mathematical function. The simpler correction techniques, such as the geometric correction technique(s), assume a linear interferant absorption profile. A basic approach to the technique can be seen from the three-point geometric correction technique, a modification of which has found applicability to the analysis of vitamin A in fish oils. Higher-order functions have also been used to describe the interfering absorption, and in these cases more involved formulas have been developed.

More involved mathematical correction techniques have also been developed; these include, among others, the use of orthogonal polynomial techniques to correct for the distortion of spectra induced by interferants.

The orthogonal function method has been used for the correction of irrelevant absorption in multicomponent spectrophotometric analysis. Each component makes a fundamental contribution shape to the overall shape of the spectrum, the spectrum being considered as a composite of these contributing spectra. The contribution of each component is represented by a coefficient whose magnitude is in part linearly concentration-dependent. In applying the method a part of the spectrum is selected in which the analyte and interferent show significant variation in their contributions to the overall shape of the spectrum. Since being suggested in the 1960s, the orthogonal polynomial method has been applied to the analyses of a number of compound preparations. Algorithms have also been proposed for computerization of the calculations.

Multivariate analytical methods have also been applied to the analysis of drug substances. The methods have a significant component of matrix analysis, and the Beer-Lambert law is basically rewritten in matrix form, permitting matrix analysis of absorbance data. A number of other mathematical algorithms have also been developed for the quantitation of analytes in multicomponent mixtures. These have either been iterative methods or methods based on multiple least-squares regression. The multiple least-squares regression methods require a knowledge of all the components of the multicomponent mixture, whereas the iterative methods such as the Kalman or the simplex method are less restrictive in the sense that interferents whose spectra are not known need not be included in the database.

The general protocol is transformation of the absorption data to reduce noise. Then, for an n-dimensional data matrix a square covariance matrix is obtained by a series of transformations. If the data matrix is not square, its square is obtained by multiplying the matrix by its transpose. The covariance matrix is then analyzed relative to the calibration matrix to determine concentrations in the unknown mixture. The eigenanalysis performed on the covariance matrix generates eigenvectors, the number of which should correspond to the number of components in the mixture. When the number of eigenvalues is greater than the number of components, this suggests interaction.

Analyte–analyte and analyte–matrix interactions can and often do complicate the analysis, usually showing up as additional significant eigenvalues. A heavy reliance on appropriate choice of working wavelength is a drawback of these techniques. The wavelength range should include 10–20% of the baseline, but application to quantitative UV–visible analysis has been reported with varying degrees of success.

3. Errors in Spectrophotometric Measurements

The errors that arise in spectrophotometric measurements arise from either of two sources. They may arise from instrumental factors or from chemical factors.

Instrumental errors can arise from several sources. Electronic noise in the detector, referred to as Johnson or shot noise, is a primary source of error. A less important source of error is flicker in the light source.

The ideal absorbance range for most measurements is in the region of 0.5 to 1.5 absorbance units for most modern instruments, as the concentration-versus-absorbance curve is relatively linear between these absorbance values. Other instrumental factors affecting analytical accuracy include spectral slit width. As slit width is increased,the fine structure of absorption bands is lost as the incident light is no longer monochromatic, a requirement for the Beer-lambert law to hold. The area under the absorption band is less affected by the monochromaticity of light than the intensity at a particular wavelength, and for this reason, more accurate measurements tend to make use of integrated areas under the absorption band rather than intensities at the peak maxima. Therefore it is useful to indicate the slit width employed in calculating molar absorptivities. The scan rate is another instrumental factor that can introduce error in measurement. This is an important consideration when entire spectra are employed. Generally, fast scan rates tend to distort spectra in the direction of the scan, altering the positions of both maxima and minima as well as diminishing peak intensities. This introduces both qualitative and quantitative errors into the measurement. The distortions arise from the relatively slower response time of the recorder as compared to the rate of signal change. The distortion is best countered by slowing down the scan rate, to give the instrument time to average out instrumental noise. This provides a more accurate measurement.

The combination of instrumental factors that gives rise to measurement errors forms the basis for the greater tolerances seen in pharmacopeial limits set for compounds determined by instrumental techniques when compared to those determined by, say, titrimetry.

A number of chemical factors may also contribute to errors in measurement. These factors generally lead to deviations from the Beer-Lambert law, and can largely be controlled once they are recognized as potential sources of error. Solute–solute interaction, for example, whether it leads to aggregation or precipitation of the aggregate, diminishes the apparent concentration of the analyte of interest. Aggregation of hydrophobic polycyclic aromatics at high concentrations in aqueous media is one such example where deviation from the Beer-Lambert law would be seen. Similarly, dimerization of molecules, for example, carboxylic acids, or even polymerization of analyte in solution may also lead to apparent deviation from the Beer-Lambert law. Both can be controlled to some extent by use of appropriately diluted solutions.

Ionization or even complexation of the analyte in solution can also lead to apparent deviation from the Beer-Lambert law. Again, by appropriate control of pH or complexation conditions it is possible to ensure that only one form predominates in solution, permitting quantitation of the absorbing species. Measurement of absorbance at the isosbestic point has been used to counter this problem. However, this approach is limited by the fact that the absorbance of the analyte at this point is frequently not high enough.

Fluorescence from absorbing species in solution may also contribute to interference. However, this kind of interference is rare and minimal because, first, the lower source intensities employed in UV–visible spectrophotometry imply that fewer fluorescent species become excited, and second, the fluorescence emitted, if any, will be too weak to significantly influence the accuracy of measurement.

It is perhaps for all these possible contributions to analytical error that it is important to carry out a quantitative analysis by employing a calibration curve obtained from a calibration series, as this will not only confirm adherence to the Beer-Lambert law, but also correct for errors introduced by the various factos described.

BIBLIOGRAPHY

1. Beckett, JB Stenlake. Practical Pharmaceutical Chemistry. 4th ed. Athione Press, 1988.
2. J Griffiths. Colour and Constitution of.Organic Molecules. New York: Academic Press, 1976.
3. HH Jaffe, M Orchin. Theory and Applications of Ultraviolet Spectroscopy. New York: Wiley, 1962.
4. JN Murrell. Theory of Electronic Spectra of Organic Molecules. New York: Wiley, 1963.
5. JW Rabalais. Principles of Ultraviolet Photoelectron Spectroscopy. New York: Wiley, 1977.
6. L. Sommer. Analytical Absorption Spectrophotometry in the Visible and Ultraviolet. Amsterdam, New York: Elsevier, 1989.

6

Immunoassay Techniques

Jean W. Lee
MDS Pharma Services, Lincoln, Nebraska

Wayne A. Colburn
MDS Pharma Services, Phoenix, Arizona

I. INTRODUCTION

The introduction of radioimmunoassay (RIA) in 1959 by Rosalyn S. Yalow and Solomon A. Berson for the measurement of insulin in human plasma (1) revolutionized endocrinology and the clinical chemistry laboratory. RIA combines the specificity of an antigen–antibody reaction with the sensitivity of radioactivity measurement. Other nonisotopic labels have been developed with similar sensitivity but without the safety and regulatory complexities in handling radioisotopes (2). Immunoassay (IA) is an indirect method which measures the effect of varying concentrations of a compound/analyte in the test fluid on an in-vitro reaction of the specific antibody and the antigen. The label may reside on the antigen or the antibody. Depending on whether limited or excess reagent assay protocol is used, analyte concentration is inversely or directly proportional to the formed antibody–antigen complex, respectively, and can be determined through mathematical calculation.

A wide variety of compounds can be quantified by IA. These range from large polymeric proteins, nucleic acids, receptors, and structural proteins, to small-molecular-weight haptens of drugs or their metabolites (3). IA can also be designed to measure the amount of antibody in a test system, where the antibody becomes the analyte. IA belongs to the class of binding assays which also includes specific binding proteins, receptors, and nucleic acid probes.

In 1968, G. C. Oliver et al. were the first to publish an RIA procedure for the measurement of a drug (digitoxin) in biological fluid (4). Since then, IA has proven useful in the quantification of many drugs and biologics (5,6). In general, however, the pharmaceutical chemist has been reluctant to use IA unless all other methods of analysis have proven fruitless. This is due largely to a misunderstanding that the immunoanalytical method, being a biological assay, is a "black box." The lack of appropriate resources for method development in many bioanalytical laboratories results in a lack of understanding of the method's strengths, weaknesses, and inappropriate methods of application.

With proper planning and trained personnel, IA can be utilized as a specific, accurate, and precise method with sensitivity below the femtomole (10^{-15} mole) level. Using specific antibodies and automation, rapid assay throughput for samples of limited volumes can be achieved. The purpose of this chapter is to provide a perspective of the present state of IA technology and to give insight into the theory, techniques, and applicability of this method to pharmaceutical substances. The role of IA and related binding assays in the future of pharmaceutical analysis is also discussed.

II. THEORY AND PRINCIPLE

A. The Antibody-Antigen Reaction

The basic IA measures the reaction of an analyte with its specific antiserum. A labeled antigen or antibody is used as the tracer to quantitate the extent of the reaction. The label can be radioisotopic or nonisotopic. In enzymatic IA, the antigen–antibody reaction is linked to an enzyme reaction which releases a chromophore detected by colorimetric, UV, or fluorimetric devices.

When the antigen–antibody complex (AgAb) formation is at equilibrium with the free antigen (fAg) and free antibody (fAb),

$$\text{Ag} + \text{Ab} \rightleftharpoons \text{AgAb}$$

the equilibrium constant (K) is defined as

$$K = \frac{\text{AgAb}}{[\text{fAg}] \, [\text{fAb}]}$$

or

$$\frac{[\text{AgAb}]}{[\text{fAb}]} = K[\text{fAg}]$$

The fraction of Ab occupied can be expressed by

$$\frac{[\text{AgAb}]}{[\text{AgAb}] + [\text{fAb}]}$$

or

$$\frac{1}{1 + \dfrac{1}{K[\text{fAg}]}}$$

IAs are classified by several groupings. If the antigen–antibody complex is separated from the unbound reactants before quantitation, it is referred to as a heterogeneous assay. For homogeneous assays, the detection of the bound complexes can be differentiated from the unbound reactants, and no separation process is needed. Another classification is competitive and noncompetitive IA. The measure of occupied or unoccupied Ab determines whether an IA is competitive or noncompetitive, not whether the label is on the Ag or Ab. A broader classification is limited-reagent assay and excess-reagent assay (7,8). They will be discussed in more detail.

Limited-reagent methods measure unoccupied antibodies; therefore, optimal sensitivity is achieved as antibody approaches zero. Excess-reagent methods measure occupied antibodies; therefore, optimal sensitivity is achieved as antibody approaches infinity. Table 1 compares the general features of these two types of assays.

B. Limited-Reagent Assay

Many conventional RIAs follow limited-reagent assay protocols. The following scheme depicts the AgAb reaction:

Table 1 Limited Versus Excess-Reagent Assays

Limited reagent	Excess reagent
Measures the unoccupied Ab [fAb]	Measures the occupied Ab [AgAb]
Analyte observed	Reagent observed
Indirect measurement, relies on subtraction from total	Direct measurement on the occupied [AgAb]
Signal is maximum at the minimum of [AgAb]	Signal is maximum at maximum of [AgAb]
Maximum sensitivity as [Ab] → 0	Maximum sensitivity as [Ab] → ∞
[Analyte] ∝ 1/[Ab]	[Analyte] ∝ [Ab]
Assay time usually longer, dependent on time to reach equilibrium	Assay time usually shorter
Sensitivity dependent on K	Sensitivity independent on K
Usually less cross-reactivity against other compounds	Usually more cross-reactivity against other compounds
Generally competitive assays	Generally noncompetitive assays

$$Ag \quad + \quad Ab \quad \begin{array}{c} \xrightarrow{\quad k \quad} \\ \xleftarrow{\quad} \end{array} \quad AgAb$$
$$Ag^* \qquad \qquad \begin{array}{c} \xrightarrow{\quad} \\ \xleftarrow{\quad k \quad} \end{array} \quad Ag^*Ab$$

With limited amount of Ab, the unlabeled antigen (analyte) competes with the labeled antigen Ag^* for limited binding sites. Bound fraction (AgAb) is separated from free (Ab), and the signal [Ag^*Ab] complex (the Ab fraction not occupied by the analyte) is measured. The amount of analyte is inversely proportional to the bound [Ag^*Ab] complex in a hyperbolic function as in Fig. 1. Methods for transforming or linearizing these functions are presented in the section on data reduction (Sec. V).

C. Excess-Reagent Assay

Wide (9) and Miles and Hales (10) developed the first immunoradiometric (IRMA) assays where excess Ab was labeled. Later, "two-site" or "sandwich" assays using an excess amount of first Ab to capture the analyte from the sample matrix, and a labeled second Ab provided the signal for quantitation (11).

IRMA: $Ag + Ab^* \rightleftharpoons AgAb^*$

Sandwich assay:

$Ag + Ab1 \rightleftharpoons Ag{-}Ab1 + Ab2^* \rightleftharpoons Ab1{-}Ag{-}Ab2^*$

Bound fraction is separated from free; the signal [$AgAb^*$] or [$Ab1{-}Ag{-}Ab2^*$] complex (the Ab fraction occupied by the analyte) is measured. The amount of analyte is proportional to the bound complex in a hyperbolic function. Methods similar to those used for transforming or linearizing limited-reagent assays can be applied to these excess-reagent functions.

In homogeneous assays, bound complex releases a differentiated signal from the unbound reactants, so no separation process is needed. Whether to measure the occupied or unoccupied Ab can be determined in a similar fashion as in the case of heterogeneous assay as described above.

D. Precision and Sensitivity

Sensitivity and precision are interrelated in an assay system. Precision is defined as the reproducibility of replicate analyses at different levels of the analyte, within assay and between assays. The variability of the measurement is dependent on the concentration of the analyte being measured. Sensitivity is defined as the level of measure for a nonzero quantity that can be measured with a predetermined precision. The term "low limit of quantitation" (LLOQ) is often used as

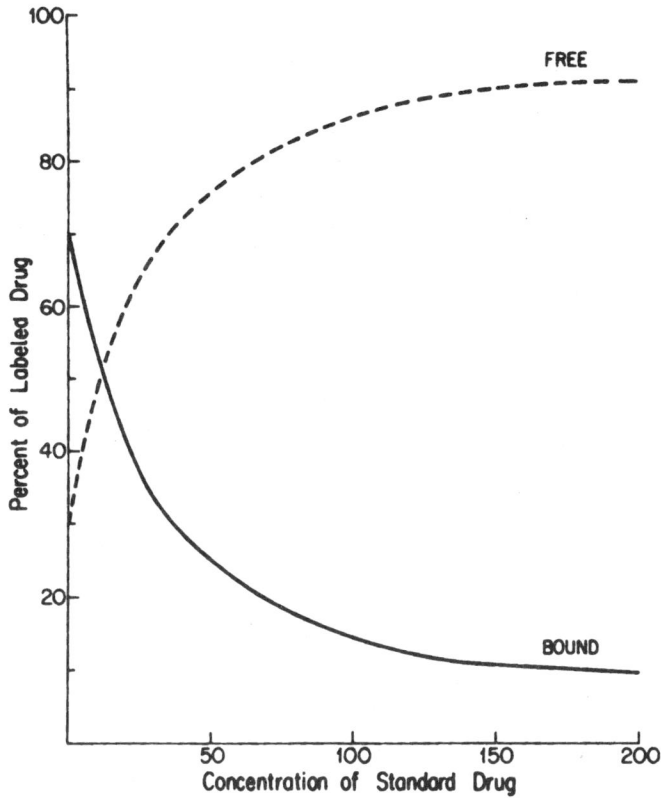

Fig. 1 The dose–response relationship of the unlabeled drug and the antibody-bound labeled drug in a limited-reagent assay. The fraction of antibodies unoccupied by the unlabeled drug (i.e., bound by the labeled drug) is measured as shown in the BOUND curve. As the concentration of the unlabeled drug increases, the percent labeled drug bound by the antibody decreases. On the other hand, in an excess-reagent assay, the fraction of antibodies occupied by the drug is measured. As the drug concentration increases, the signal increases similarly to the FREE curve shown for the limited-reagent assay.

an empirical assay parameter for the lowest concentration that can be reliably quantified with acceptable precision both within and between assays. Acceptance criteria for LLOQ should be defined for each assay system and should adhere to industry standards for bioanalytical sample analysis (12). Usually the LLOQ is the lowest acceptable standard with precision and accuracy of no more than 20% C.V. and relative error at five repetitive determinations.

E. Accuracy

Accuracy is a measure of the ability of the assay to quantify the true value of the standard substance. This definition incorporates both precision and specificity. It implies that, to be accurate, the procedure must be reproducible as well as unbiased by potentially interfering compounds. However, for the purpose of characterizing the IA, it is important to know if the assay can measure the "true" value of a standard substance in the absence of potentially interfering compounds. This can be accomplished by repetitive measurements of known concentrations of the standard drug. From these data, it is possible to determine a mean and standard deviation as well as any possible bias that might be characteristic of the assay system.

Because sensitivity is the principal advantage of IA, optimization of assay sensitivity is paramount. Fundamental requirements to optimize assay sensitivity are assay protocol design and reagent concentrations. Computer optimization techniques to improve sensitivity have been reported (13–16). Because of the complexity and variability of assay protocols, empirical approaches are more widely practiced. Ekins compared the sensitivities achieved by different method designs (17,18). Advances in chemiluminescent and time-resolved fluorescent labels enables IA sensitivity to reach attamole levels (18–20).

F. Specificity

Specificity is the freedom from interference caused by substances other than the intended compound. Interference can be caused by (a) heterologous antibody populations, (b) cross-reactivity with structurally related compounds, and (c) nonspecific interference due to low-molecular-weight compounds that alter the reaction conditions.

Heterology can be the result of immunizations with an impure antigen or a generalized antibody response to a specific antigen. Causes of nonspecificity can be minimized by (a) purifying immunogens prior to immunization of test animals, (b) choosing the appropriate animal species, and (c) immunizing large numbers of animals to increase the chance of obtaining an "ideal" antiserum. Monoclonal antibodies can be produced without a purified immunogen as long as vigorous screening is performed to assure identification of a true monoclone with the desired specificity. Two-site assays using two antibodies, each directed against different distinctive determinants of the antigen, provide additional assay specificity.

Cross-reactivity due to structural identity for certain immunoreactive functional groups is the most common cause of IA interference. Structural similarity can occur between analogs or metabolites and the parent drug. Cross-reactivity also occurs with endogenous substances that are immunochemically similar to certain drugs, e.g., thyroxine and diazepam (21). Specificity of an antibody can be directed to certain functional groups by choosing sites for conjugation to protein which are remote from the groups that impart specificity (22). Specific IAs had been developed for the androgens, progestins, estrogens, theophylline, and digoxin using this approach (23–27).

Precision, accuracy, and specificity have meaning only for the concentration tested. Therefore, it is imperative to test the range of concentrations of drug and potentially interfering substances that will be encountered during sample analysis. For example, a cross-reactivity of 1% with cortisone concentrations of 1–50 ng/mL is acceptable for a prednisone RIA because after a standard 20-mg dose of prednisone, the prednisone concentrations in plasma will exceed or equal the cortisone concentrations. In contrast, similar cross-reactivity with cholesterol at these concentrations would render the assay useless without prior separation of prednisone and cholesterol, because cholesterol is present in concentrations of 150–250 mg/mL in plasma ($\sim$1000 $\times$ prednisone). It is apparent that cross-reactivity per se is not the problem. Cross-reactivity in combination with the anticipated concentrations of the cross-reacting substance determines the resultant assay interference.

Nonspecific interference can be encountered as a result of changes in temperature, ionic strength, and pH, or as a result of the presence of hemolysis or excessive quantities of bilirubin, heparin, and urea. Any of these factors can alter the composition of the incubation medium and affect the kinetics or equilibrium of the antigen–antibody reaction. Nonspecific interference contributes to assay variability and results in a decrease in sensitivity. This is particularly prevalent in early enzyme IA applications. Assay sensitivity can be greatly improved with increased assay specificity.

When the source of interference is attributed to the sample matrix, the term "matrix effect" is used. This may be caused by specific or nonspecific interference, or both. It will be discussed later, in Sec. III.F.

Besides choosing the appropriate Ab, the assay method can be designed or manipulated to improve assay specificity using (a) protein precipitation, (b) liquid/liquid or solid-phase extraction, (c) HPLC separation of the analyte from the interfering compounds, (d) sample dilution with buffer or control matrix, or (e) an affinity solid phase (e.g., antibody-coated microtiter plate or polystyrene beads) to capture the analyte followed by wash steps. Affinity-purified antibodies and protein blockers are used in EIA to decrease nonspecific binding in plate assays. Increasing incubation time to reach equilibrium also improves binding specificity.

Precision, accuracy, and specificity of the IA procedure can be verified by comparison of results from actual samples using both the IA and an alternative analytical technique if one is available. If an alternative technique is not available, the IA results should be compared with the without prior chromatographic separation of potentially interfering substances. If discrepancies are observed, the chromatographic separation should be included as an integral part of the IA technique.

III. METHODS IN IMMUNOASSAY

Development of an IA technique for a specific drug is dependent on four processes: (a) preparation and purification of an antigenic form of the drug—this may be the drug itself, but for conventional drugs it requires a drug–protein conjugate; (b) production and characterization of specific antibodies; (c) production and purification of a nonexchangeable label on either the drug compound or antibody; (d) development of a suitable assay design, which often includes steps to separate interfering compounds and free drug from bound antibody–drug complexes. These four development processes are not totally independent of each other and usually proceed from initiation to completion with some overlap. The fundamental procedures necessary to fulfill these requirements are discussed in this section.

A. Conjugate Preparation

Most drugs are low-molecular-weight and/or low-immunogenicity (hapten), because drugs that are immunogenic in humans are generally not extremely useful as therapeutic agents. To make them immunogenic, these compounds must be coupled to a suitable protein carrier. In some instances a suitable functional group, such as a carboxyl, thiol, amino group, or an active hydrogen, is present in the drug molecule. In other cases the drug requires derivatization to include an appropriate functional group so that the compound can be linked to the carrier protein.

Drugs of high molecular weight (polypeptides and polysaccharides > 2000–5000 Da, nucleic acids > 5000 Da) from human sources are not immunogenic to humans but are usually immunogenic in animals. Many humanized/ chimeric peptides, proteins, and antibodies are somewhat immunogenic in humans and should be potent immunogens in selected animal species. Conjugation may not be required for these compounds to produce antibodies in animals.

Another use for drug conjugation is for signal labeling. For example, the drug will be conjugated to an enzyme in EIA; to biotin, avidin, or streptavidin in biotin amplification; to a protein for adsorption onto a solid-phase support; or to polystyrene beads by direct covalent linkage.

Several reactive groups on proteins can be used for conjugation with the drug moiety. These groups include the terminal amino and carboxyl groups, the ϵ-amino group of lysine, the carboxyl groups of aspartic and glutamic acid, the phenolic groups of tyrosine, and sufhydryl group of cysteine (28–35). Reactive groups of tyrosine, histidine, and tryptophane had been used to react with standard drugs by diazotization. A few popular reagents are listed in Table 2.

If the drug lacks a suitable reactive group, sometimes one can be introduced through reduction, oxidation, or disulfide exchange. Photoreactive cross-linkers are also used for compounds that are difficult to conjugate (36).

Bovine serum albumin (BSA), gamma globulins, egg albumin, hemocyanin, fibrinogen, and thyroglobulin have been used in preparation of drug–protein conjugates. The use of synthetic polymers and polypeptides to increase the number of reactive sites on the carrier has not met with much success. Although the number of drug molecules linked to the carrier molecule is substantially increased, the titer and specificity of these antisera show no improvement over those obtained using conventional conjugates. It is important to design the conjugating position and linkage bridge so that antigenic determinants will be remote from the site of conjugation (37). Some authors have improved specificity by introducing reactive functional groups into the steroid nucleus for conjugation that are far removed from the sites desired for specificity determination (38). It is useful to

Table 2 Conjugating Reagents Used for Various Functional Groups

Functional group	Reagent	Conjugate	References
R_1—COOH[a]	Carbodiimide	R_1—CO—O—$\overset{\displaystyle NH-R'}{\underset{\displaystyle \mid}{C}}$=N—R″	30–34
	R′—N=C=N—R″		
R_2—COOH[a]	NHS esters[b]	R_1—CO—R_2	
R—NH$_2$		R—NH—CO—R′	28, 29
	R'-COO–N (succinimide)		
R—SH	Maleimides	R'–N (maleimide) S-R	35
R—SH Nonselective	Alkyl/aryl halides Arylazides —photoactivation	R′—Phenyl—N—R	36

[a] Sequential reactions conjugating R_1 with R_2 using carbodiimide.
[b] NHS esters = N-hydroxysuccinimyl esters.

prepare several hapten–protein conjugates with two different carriers and at several hapten–carrier coupling ratios. For example, a highly immunogenic carrier such as keyhole limpet hemocyanin at high coupling ratio is used to prepare the conjugate for an immunogen, while the soluble BSA at a low coupling ratio results in a conjugate for hapten immobilization on the solid-phase support. BSA can be modified to increase the immunogenicity by substituting the anionic carboxyl groups with cationic aminoethylamide groups. Higher and prolonged antibody responses were reported from immunogens conjugated to cationic BSA (39).

Proteins from the same biosynthetic origin share common subunits. Antibodies were raised selectively against the subunit carrying the specific determinants. For example, antibodies produced from the differentiating β-subunit of the pituitary hormones as immunogen have better selectivity than those from the native hormone, which contains both α- and β-subunits (40).

Once a suitable reactive group is present in the drug molecule, the method for conjugating drug to protein must be selected. General methods for preparing and characterizing steroid–protein conjugates have been described by Erlanger et al. (41,42). Similar methods for preparing drug–protein conjugates have been described by Erlanger (43) and by Landon and Moffat (5). The most commonly used methods for conjugating drug to protein include the carbodiimide (44), carbonyl–diimidazole (45), mixed anhydride (46), Schotten-Baumann (47), Mannish (48), glutaraldehyde (49), and diazotization (50) reactions.

Bifunctional reagents are often used for cross-linking carrier protein to hapten or protein to protein. Reagents that react with two or more identical functional groups are homo-bifunctional cross-linkers. Mixed products result from inter- and intramolecular conjugation forming dimers and polymers. Reagents that react with two or more different functional groups are hetero-bifunctional cross-linkers. To minimize self-conjugation and polymerization and to decrease side-reaction products, sequential and controlled reaction of one functional group at a time should be conducted. The other unreacted functional groups can be protected (blocked), or the reaction conditions controlled to favor one over the other functional group. It is also important to conduct reactions under mild conditions, including appropriate pH and temperature, with an optimal hapten-to-protein ratio.

Various lengths of spacer arm can be used for optimal steric effects. Several authors have investigated the effect of the number of hapten molecules per protein (51), the bridge used to couple hapten to carrier (52–54), the protein carrier used (55), and the site of conjugation (56–58) on the production of antisera. From these and other studies, it has been ascertained that several aspects of immunogen production are relevant to the final antibody titer and specificity. About 5–20 hapten molecules per carrier molecule yields the highest titered antisera. A bridge containing four carbon atoms between hapten and carrier yields more specific and higher titered antisera. The site of conjugation is the most important controllable factor in determining antibody specificity. A higher hapten/protein ratio of the conjugate will increase immunogenicity; however, it may decrease the immunoactivity and the enzyme tracer activity. Therefore, one hapten/protein ratio should be used to prepare a conjugate for tracer/solid-phase support and another one to prepare the immunogen.

After the reaction, the conjugate should be separated from the reagents and side products after using size-exclusion, ion-exchange, or affinity liquid chromatography. The purified fraction is collected and the hapten–protein ratio is characterized. For monoclonal antibody production, it is not as important to purify and characterize the conjugate because of later clone selection. If the conjugate is to be used for enzyme-tracer or solid-phase support coating, the conjugate must be tested for the activity of enzyme or immunocomplexation.

B. Labeling Methods

1. Radioisotopic

The two most commonly used radiolabels in RIA are ^{3}H and ^{125}I, although ^{14}C, ^{131}I, ^{57}Co, ^{75}Se, and ^{32}P have also been used. ^{3}H and ^{125}I are commonly used because they provide adequate activity (1 Ci/mM or greater) and have long enough half-lives (12 years and 60 days, respectively). Tritium has the advantage of direct incorporation into the molecular structure of the drug, whereas ^{125}I has the advantage of high activity and ease of counting.

a. Tritium Labels

Drugs that are used extensively for research are generally available with tritium labels. However, custom ^{3}H labeling is required for many substances. Unless a radiochemical laboratory is available in-house, this is usually accomplished more efficiently, if not more economically, using a reliable commercial source (59). Specific tritium labels are generally obtained by reducing an appropriate precursor in the presence of ^{3}H. The resulting labeled compounds usually possess sufficient activity for RIA use and are less susceptible to tritium exchange in aqueous solution.

b. Iodination

Labeling drugs with ^{125}I for the use in RIA procedures was introduced with the first drug RIA (digitoxin) by Oliver et al. in 1968 (4). The use of a tyrosine methyl ester (TME) conjugate of digitoxin was introduced for ^{125}I labeling in the same publication. Along with the TME derivative, histidine (60), hydrazone (61), tyrosine, tyramine, and histidine residues (5) have been coupled with biological substances and drugs to serve as sites for iodination. Techniques have been developed to iodinate carrier prior to conjugation using acylating (62) and chelating (63) agents as well as polycationic compounds (64). This process decreases iodination damage to the hapten, which has been an inherent disadvantage of iodine labels.

Iodination can be performed using one of many procedures, including the chloramine-T (65), monochloride (66), exchange (67), and enzymatic (68) iodination methods. The most suitable iodination procedure depends on the stability of the drug and the specific activity that is sufficient to meet the sensitivity requirements of the assay. Harsh radioiodination methods tend to damage the antigen, whereas the milder iodination processes may not yield high specific activity. The goal is to obtain a labeled form of the drug with sufficient label to yield the desired sensitivity but at the same time maintain the structural integrity of the molecule.

It also must be cautioned that conjugating the carrier for iodination at the same position as the drug–protein conjugate for antibody production may result in antisera which have higher affinity for the carrier or bridge than for the drug (69). The ^{125}I-tyramine radiolabel was introduced into the 6-position through a carboxymethyloxime bridge to estradiol to allow exposure of the discriminating epitopes on the A and D rings of the steroid.

c. Quality Control and Comparison of ^3H and ^{125}I Labels

Quality control of ^3H- and ^{125}I-radiolabeled compounds is imperative. The labeled compounds should be tested for identity, radiochemical purity, and specific activity. The two most common quality-control problems for iodinated drugs are (a) damage during the reaction and (b) radiation damage following synthesis. Chromatographic separation of labeled antigen from the excess reagents and multiple-iodinated products should be conducted immediately after iodination. Hunter (70) reviewed several iodination methods and discussed the causes of iodination damage. Increases in nonspecific binding counts, change in slope, and decreases in sensitivity are signs of radiation damage.

^{125}I labels usually have higher radioactivity than ^3H labels. The counting time for a large number of samples are much shorter for ^{125}I than for ^3H labels because ^{125}I can be counted in a multiwell counter at a much shorter time (generally 1 min for ^{125}I, versus 10 min counting time for ^3H). On the other hand, most compounds can be tritiated, but not all can be successfully radioiodinated. Some compounds can form adducts of one or more iodine groups which have different binding reactions with the antibody. ^{125}I labels have shorter shelf-lives than ^3H labels because of the shorter half-life and higher energy causing radiation damage. Multiple ^{125}I labels per molecule of high radioactivity and a need for derivatization before iodination may change the nature of the ligand, causing low affinity. ^{3}H labels usually have higher affinity but lower sensitivity.

2. Nonisotopic Labels

Radioisotope requires radiation safety processes to monitor their use, expensive equipment, training of personnel in safe handling of the materials, and waste management. In addition to safety issues, nonisotopic labels are more efficient signals than radioisotopes. For ^{125}I, only one signal per 7.5×10^6 molecules is being detected, while at least one signal per molecule will be detected for nonisotopic labels. A chemiluminescent label gives one detectable event per labeled molecule, while a fluorescent label gives many detectable events per labeled molecule. The signal is amplified for enzyme labels by every reaction generated. In the last decade, nonisotopic labels have surpassed the popularity of radioisotopic labels (71).

a. Enzyme Labels

Theoretically, enzyme immunoassays (EIA) can offer greater versatility in assay designs and better sensitivity than RIA (17,72–74). The drug or the antibody is labeled with an enzyme. The assay protocol depends on whether occupied Ab (e.g., enzyme captured in the Ag–Ab–Enz complex) or unoccupied Ab (e.g., Ag–Enz) is measured. For detection, a chromogenic substrate is added. The enzyme reaction releases the chromophore product which is then quantified by colorimetric, fluorimetric, or luminometric measurement. Drugs can be conjugated to an enzyme using similar conjugation methods as those used in the preparation of antibodies. Extra care must be used to preserve the enzyme activity. The antibody can be labeled with an enzyme using protein-to-protein linkers. Either homo- or hetero-bifunctional cross-linking can be used for conjugation. The enzymes that are widely used and their chromogenic substrates are listed in Table 3.

b. Fluorescence Labels

Fluorimetric detection usually provides greater sensitivity than colorimetric detection (75). A fluorescent signal can be obtained by using fluorescent labels or enzyme substrates. Fluoresceins, rhodamines, and umbeliferones are the more commonly used labels. However, the results have not met the theoretical expectation because of interferences from light scattering, high background fluorescence, and quenching (76,77). Polarized fluorescence detection results in less background noise. Fluorescence polarization immunoassay (FPIA) has been used in the commercial TDx™ immunoassays for many drug-monitoring programs in clinical chemistry laboratories (78).

Table 3 Some Enzymes and Substrates Commonly Used for EIA

Enzyme	Substrates	λ_m (nm)
Horseradish peroxidase (HPR)	H_2O_2/ABTS OPD or TMB	415, 492, 450
Alkaline phosphatase (AP)	PNP	405
β-Galactosidase	ONPG	420
Urease	Urea/bromcresol yellow	588
Urease peroxidase	Glucose + HPR chromogen	
Glucose oxidase/peroxidase	Couple enzyme reaction	
	Glucose + HPR chromogen	
Acetylcholinesterase	Acetylthiocholine, DTNB	412

ABTS = 2,2'-azino-di(3-ethylbenzthiazoline sulfonic acid-6)
OPD = o-phenylenediamine
TMB = 3,3',5,5'-tetramethylbenzidine
PNP = p-nitrophenyl phosphate
ONGP = o-nitrophenyl-β-D-galactopyranoside
DTNB = 5,5'-dithio-bis-(2-nitrobenzoic acid)

Time-resolved luminescent labels using lanthanides with decay times of micro- to milliseconds have been developed. Because the decay time is distinctly shorter than that of the background fluorescence from the biological matrix, background noise can be totally eliminated by taking the reading after the rapid decay of background fluorescence (79). The fluorescence signal of lanthanides are enhanced by chelate formations with β-diketones (80–82).

The common fluorogenic substrates used for three widely used enzyme tracers are p-hydroxyphenylacetic acid for HRP, 4-methylumbelliferyl phosphate for alkaline phosphatase, and 4-methylumbelliferyl-β-D-galacto-pyranoside for β-galactosidase.

c. Luminescence Labels

Luminescence can be triggered by chemical reactions (chemiluminescence) or biochemical reactions (bioluminescence) (83). Light is produced by the decay of the molecule from electronically excited to ground state. Chemiluminescent labels provide high sensitivity because a much greater signal can be observed (84). One type of label is the conjugates of luminols, acridinium esters, or their derivatives (85). Many of these are patented for commercial assays. A second type of luminescence label uses conventional enzyme labels, such as peroxidase and alkaline phosphatase (AP), which catalyze reactions that generate luminescence (86,87). Analogs of substituted phenols and naphthols, 6-hydroxylbenzo-thiazol, tetramethylbenzidine, OH radicals, steroids, azide, and many others have been found to enhance the luminescent signal (88–90). For example, horseradish peroxidase catalyzes the oxidation of luminol to the excited state in the presence of hydrogen peroxide. A fluorescence signal is produced when the excited molecule decays to aminophthalate. The intensity of the signal is enhanced by the addition of phenols.

Oxidant (H_2O_2) $\searrow$ Redox enzyme $\longrightarrow$ N_2 Enhancer
Luminol $\nearrow$ Excited State $\longrightarrow$ Aminophthalate

Bioluminescence reactions of NADP or ATP catalyzed by luciferase enzyme labels with electrochemical detection have been investigated (91–93).

d. Signal Amplification

Signals can be amplified by two types of mechanisms. The first uses a high accumulation of the enzyme labels, while the second uses a high yield of the enzyme product from a multiple-enzyme cascade (94,95).

The biotin system is an example of the first type. Avidin is a glycoprotein in egg white which has four active sites with high binding affinity for biotin ($K_d = 10^{-15}$M) (96). A high number of the small-molecular-weight biotin can be conjugated to an antibody or enzyme without affecting the biological activity. The biotinylation factor is one amplification; binding of four biotins to each avidin represents another amplification; and each avidin can be conjugated to multiple analytes for an additional amplification factor. Two- to 100-fold increases in sensitivity over the conventional method were reported. Streptavidin, a neutral protein, has replaced the positive-charged avidin in many applications, resulting in less nonspecific binding (97,98).

Enzyme product signals can be amplified by enzymatic cycling systems. For example, as shown in Fig. 2, when alcohol dehydrogenase/diaphorase and

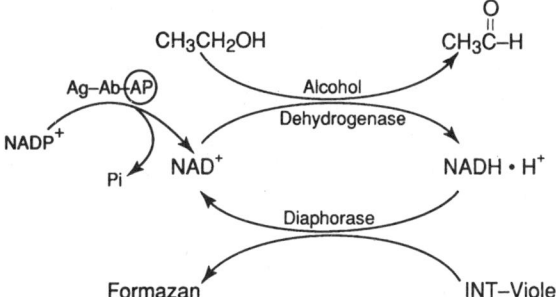

Fig. 2 Enzyme amplification for alkaline phosphatase. The primary substrate is NADP⁺. The product NAD⁺ is recycled through a coupled-enzyme redox system of alcohol dehydrogenase and diaphorase for continuous generation of the color product Formazan.

excess alcohol and NADP$^+$ were added to the alkaline phosphatase system, the redox cycle amplified the product yield 40-fold, and to a final visible product in red color (99–101). However, extreme care must be used during the assay to prevent nonspecific reactions.

C. Antibody Production

1. Polyclonal Antibodies

Polyclonal or monoclonal antibody production can be chosen, depending on the resources available and the method design. For polyclonal antibodies, a suitable animal species can be chosen based on the animal facility. It is necessary to immunize multiple animals to assure reasonable success because of the variability in immunogenic responses from one animal to another. There are many immunization schemes and various adjuvants are used by different laboratories. A common procedure involves mixing the purified immunogen with emulsified Freund's adjuvant in a concentration of about 10–50 mg/mL. One milliliter of the emulsion is injected intradermally, subcutaneously, and/or intramuscularly at weekly or monthly intervals into multiple sites along the back and flank of a suitable animal species. Other adjuvants, such as the muranmyl peptide analogs and carbohydrate-based adjuvants, have been used to enhance immunogenicity with less harmful effect to the animal (102). Antibody titer is determined about 3 months after the initial immunization. Booster injections, using small quantities of immunogen (103,104), are continued at 1- to 3-month intervals until satisfactory titers are obtained. Sufficient blood is then collected and antiserum is harvested for characterization. Antiserum from different animals, and even from the same animal at different times, will have different characteristics. Each antiserum lot from a single animal must be characterized with respect to affinity (105,106) and specificity. If antisera from different blood samplings are pooled, the pooled antiserum should be recharacterized to establish the precision, accuracy, and specificity of the pool. Superior antisera can be aliquoted into small ampoules and stored at about −70°C or lyophilized for storage. After thawing the antiserum, it should be stored in the refrigerator. Repeated freezing and thawing should be avoided.

If the antisera are partially nonspecific, it may be possible to improve the specificity using fractionation (107), immunoadsorption (108–110), or immunosaturation (111) techniques.

2. Monoclonal Antibodies

A single clone of lymphocytes is formed by a hybridoma through fusion of a sensitized lymphocyte and a myeloma cell, producing an antibody with a unique structure (112). Usually, immunized mouse spleen cells are fused with myeloma mutant cells which are deficient in hypoxanthine guanine phosphoridyl transferase (HGPRT) in the presence of polyethyleneglycol (113). A pure preparation of the immunogen is not required for immunization. Antibodies produced by the hybridomas are screened against the antigen, selecting reactions with specific epitopes or immuno-complexation that parallel certain bioactivities. Assays used for clone screening must be efficient to provide fast results over a huge number of possible cell lines. Automated microtiter plate assays are generally used for the screening process. If possible, the immunoassay method intended to be used for the drug should also be used for monoclonal antibody screening.

Molecular biology can be used to identify the peptide sequences which can be synthesized as immunogens for monoclonal antibody production. After a few hybridoma clones are selected, they are diluted and allowed to grow again to assure that true monoclones are selected. Large-scale production takes place in cell culture or in-vivo ascites fluid (113–115). The original cell lines are kept in cryopreserved aliquots in liquid nitrogen. The antibodies are often purified with protein A, protein G, or HPLC column before use (116,117).

Generally, polyclonal antibodies are easier to produce, and high-affinity polyclonal antibodies can be obtained. Monoclonal antibodies are more specific to a certain epitope. They provide continuous production of exactly the same defined reagent and are more preferable for excess-reagent assays. The double sandwich technique has used two antibodies from monoclonals or combinations of mono- and polyclonals, with specificity against two different epitopes of the analyte. One antibody functions as a capturing antibody for the analyte and the other as the label carrier (118).

3. Fab Fragments and Chimeric Antibodies

Mass or steric configuration of large immunoglobulins may interfere with binding between analyte and the Fab-binding region of the antibody molecule. Steric hindrance is one of the causes of a high-dose hook effect, where the response decreases with increases in analyte concentrations (119,120). Papain can be used to digest the antibody into three components: two Fab fragments and one Fc fragment. Each Fab fragment contains one binding site; the Fc fragment contains no binding sites, but it is responsible for most of the mass and potential antigenicity of the antibody. The Fc component of the immunoglobulin molecules can cause interferences because of binding of complement proteins and rheumatoid factors to Fc (121). After enzyme hydrolysis, the Fab fragments can be purified by liquid chromatography. Papain immobilized on agarose gel provides easy control of digestion and separation of the crude digest from the enzyme (122,123). A protein A column can also be used for purification (124–126). It binds the Fc and intact IgG molecules, allowing the Fab fragment to be eluted. Chimeric (hybrid) antibodies can be constructed to distinguish between one site on an antigen and another on an enzyme (127,128). Both specificity and sensitivity can be increased using chimeric antibodies (129–131). Antibody engineering through excision and replacement of the C-region has produced mouse/human chimeras of murine V-region coupled with human C-region for therapeutic antibodies, invivo diagnosis, and radioimaging (132,133).

D. Antibody Characterization

An antiserum is characterized by its affinity and specificity of binding to the desired analyte relative to other substances in the sample matrix. It is not necessary to determine titers for monoclonal antibodies, because their affinity and specificity are unique.

Affinity is often defined by antibody titer, i.e., dilution required to reach a specified percent binding with the analyte. For polyclonal antibodies, titer should be determined for each blood-sample collection. Various dilutions of the antisera are allowed to bind the labeled antigen. Antibody dilution required to obtain 30% or 50% binding is defined as the titer. High titer is necessary for efficient use of antiserum. At a high dilution, only antibodies of high avidity from a few clones will be bound to the antigen; antibodies with low affinity will not be active. Therefore, polyclonal antibodies with high titer used at a high dilution may have better specificity than those of low titer. Yalow and Berson (134) observed that extensive dilution of a heterologous antibody population would ultimately result in essentially a single class of the highest-affinity sites remaining at a significant concentration. Dilution of both antiserum and sample can minimize the influence of potentially interfering substances (both specific and nonspecific). This is restricted only by the activity of the labeled hapten and the equilibrium constant K of the antiserum. Therefore, higher-affinity antiserum is more amenable to this type of specificity improvement.

Scatchard plots can be constructed to estimate the affinity constant: r/[antigen] is plotted versus r, where r is the molar ratio of bound antigen to antibody (135,136). For monoclonal antibodies, a straight line is obtained, with the slope equal to K. For polyclonal antibodies, the average K of multiple binding clones can be calculated. However, it will be difficult to assess K if a complex curvilinear plot is obtained. The same IA method used for sample analysis should be used to test binding avidity.

It is useful to run preliminary standard curves of the few selected antisera to have an initial look at the curve range, sensitivity, and linearity.

A specific antibody is desirable because it enables an IA to be performed with limited or no sample clean-up. For reagent-excess methods, it is very important to choose specific antibodies. Cross-reactivity of polyclonal antiserum from each blood collection or each clone of monoclonal antibodies is tested against known metabolites, drug degradants, concomitant drugs, and the protein carrier used in immunogen conjugates. The B/B_0 ratio is plotted versus the concentrations of the test antigens. The percent cross-reactivity at ED_{50} is calculated by the ratio of the drug concentration over that of the test compound's concentration at 50% displacement.

E. Separation Techniques and Immunoassay Protocol Designs

1. Precipitation of Antigen–Antibody Complex

For most IA, it is necessary to separate free from bound labeled antigen. Several methods that employ physicochemical and immunological separation have been devised. Physical methods include filtration, chromatography, electrophoresis, charcoal–dextran adsorption, and adsorption on ion-exchange resins. Chemical methods include organic solvents, such as ethanol, dioxane, and polyethylene glycol (PEG), or salts, such as sodium, zinc, and ammonium sulfate, to precipitate antibody-bound hapten. The most widely used precipitation procedure for bound antigen is the physiological second-antibody method. This method employs an antibody against the gamma-globulin of the animal species used to produce antibody to the drug substance to precipitate the drug–antibody complex. For example, if the anti-drug serum is prepared in rabbit, an anti-rabbit gamma-globulin serum is used to precipitate the drug–antibody complex. Pre-precipitation of primary and secondary antibody (137) or pre adsorption of the secondary antibody on solid-phase support (138) can be used to decrease the incubation time for the second antibody reaction. Many physicochemical methods tend to be lengthy, harsh, and nonspecific. PEG and charcoal–dextran are used more often than others. PEG is used to facilitate precipitation of the secondary–primary antibody and drug-binding complex. Charcoal–dextran is often used for tritium-labeled RIA because of the convenience of decanting the drug–antibody complex solution into the counting vials, leaving behind the unbound drug, which is adsorbed on the charcoal particles.

2. Protein Immobilization on Solid Phase

Protein immobilization on solid phase has provided versatility in separation techniques and IA formats. Most proteins can be adsorbed onto plastic or glass solid-phase support surfaces (139). Cellulose acetate, cellulose nitrate, polymethylmethacrylate disk, filter paper, and other synthetic solid-phase supports have been used (140,141). Noncovalent hydrophobic, electrostatic, and hydrophilic bonds form between proteins and the surface of the solid phase (142–144). Covalent linkages between proteins and solid-phase supports can also be developed (145). Some solid-phase supports with activated surface groups are commercially available. Surfaces can be test tubes (coated tubes) (146,147), polystyrene or glass beads (148,149), magnetized particles, dipsticks, or microtiter plates (150). Dependent on the IA design, adsorbed protein can be drug–protein conjugates (e.g., drug–BSA and drug–thyroglobulin), secondary or primary antibodies, or enzyme conjugates. Generally, overnight incubation of the protein solution with the solid-phase support in the refrigerator or at room temperature is used. Coated solid-phase support is rinsed with buffer; the residual uncovered sites of the solid-phase surfaces are blocked with another protein such as casein (from nonfat dry milk) or BSA to decrease nonspecific binding (151). After being washed, the solid-phase support can be stored over a period of time. The physical chemistry of protein immobilization has been studied on the varies types of solid-phase support. These include protein interactions under various conditions of protein concentration versus surface area, surface type, buffer pH, ionic strength, incubation time, and temperature (152).

Immobilization may alter the binding activity of the protein (153–156). It is important to establish the parameters of protein immobilization steps to assure consistence in the amount of protein adsorbed and the biological activities of the immobilized protein. Second, it is important to monitor nonspecific binding among lots of solid-phase support. Many ELISA techniques are unable to achieve theoretical sensitivity because of the background noise from nonspecific binding. Storage stability of the immobilized protein solid-phase support must be established on various storage conditions. Bleeding or leaching of protein from the solid phase during storage and assay process should also be tested.

Protein immobilization has offered tremendous ease in separation techniques and flexibility in assay design protocols. For example, coated-tube assays are designed with simple procedures of adding sample and radiotracer to coated tubes, incubating, decanting, and counting. Polystyrene beads have increased surface area as an advantage over coated tubes. Application of multichannel pipettes, 96-well microtiter plates, 96-probe plate washers, and plate readers have increased assay throughput as well as flexibility in assay designs.

One widely used format is the double-antibody sandwich technique (145,157) shown in Fig. 3. This technique requires that two antibodies against the analyte's specific epitopes be prepared. The first antibody (Ab1) is immobilized onto a solid-phase support (SP) by noncovalent adsorption or covalent binding. The residual sites of the solid-phase surfaces are blocked by protein. This SP–Ab1 can be prepared ahead and stored for a period of time. The drug analyte in the sample is captured by the immobilized Ab1. After washing, a second antibody labeled (Ab2*) with the signaling agent (radioisotopic, enzyme, or other nonisotopic label) is introduced and the double-antibody immunocomplex SP–Ab1–Ag–Ab2* is formed. This assay protocol provides high specificity based on two antibodies binding to two separate epitopes of the analyte. Background noise is decreased as a result of specific binding and washing steps.

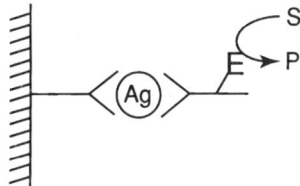

Fig. 3 Double-antibody sandwich technique. The capturing antibody (Ab1) is immobilized onto the solid phase. The analyte from the sample is captured by forming Ab1–Ag immunocomplex. After washing off the extraneous materials from the sample, the reporting antibody (Ab2, which has an enzyme label in this illustration) is introduced. The double-antibody sandwich is formed: Ab1–Ag–Ab2E. Compounds not recognized by both Ab1 and Ab2 will be washed away. A chromogenic substrate is added to produce the product for detection. The occupied Ab2 by the Ag is measured; this is an excess-reagent assay. As the sample analyte concentration increases, the signal responses increases proportionally.

The double-antibody sandwich technique is applicable to large molecules. Small analytes have difficulty forming the double-antibody sandwich immunocomplex. The smallest analytes reported that have been used in a double sandwich assay are peptides of around 10 amino acid residues (158). The double-antibody sandwich technique measures the occupied antibodies using an excess-reagent assay protocol. A limited-reagent assay protocol can also be designed, as shown in Fig. 4. In this format, the antibodies are immobilized onto the solid phase,

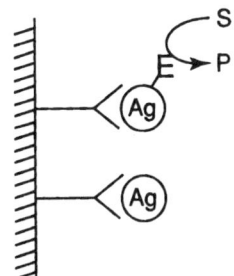

Fig. 4 Antibody immobilization in a limited-reagent assay format. The antibody is immobilized onto the solid-phase support. Labeled antigen and sample are introduced, and they compete with one another to form immunocomplex (Ab–Ag or Ab–AgE) with the limited antibody sites on the solid-phase support. After washing, the substrate is added to produce the detecting product. The antibody unoccupied by the sample analyte is measured; as the concentration increases, the signal responses decreases.

similar to Fig. 3. However, instead of using a labeled second antibody, the antigen is labeled. The labeled antigen competes against the unlabeled analyte in the sample to form immunocomplex, SP–Ab–Ag or SP–Ab–Ag*, with the limited antibody sites on the solid phase. The unoccupied antibodies are measured. This assay protocol is applicable to both large and small molecules. For small molecules, except for tritium labels, an appropriate spacer arm may be placed between the label (e.g., ^{125}I or nonisotopic label) and the antigen to avoid different steric configuration problems of the labeled antigen (159).

A small antigen can be immobilized onto a solid phase by covalent conjugation to a carrier protein which is easily adsorbed on the solid phase. Figure 5 illustrates such an assay design. It is convenient to use secondary antibodies against the animal host species which is used to produce the primary antibodies. Labeled secondary antibodies can be a common, indirect signal carrier for the primary antibodies. Figures 6a and 6b show design expansions of Figs. 3 and 5, respectively, using labeled secondary antibodies. Antiserum against mouse, rabbit, sheep, goat, and guinea pig are commercially available in various labels as well as in biotinylated derivatives.

Protein-immobilization techniques have gained popularity in many research laboratories. It is paramount to take all precautions to decrease background noise in the assay design. Therefore, it is better to: (a) use affinity-purified antibodies, (b) use an antibody–enzyme conjugate that has high specific activity, and (c) use high-quality reagents to prepare assay buffers and blocking solutions (e.g., water purified from organic matter and electrolytes, peroxide-free Tween-20 and Triton-X 100, protease- and lipid-free BSA). For each batch of new antibodies

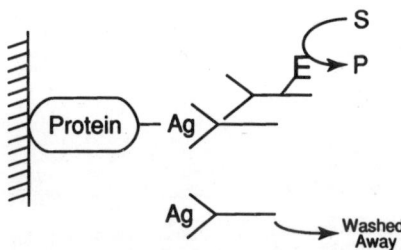

Fig. 5 Immobilization of analyte–protein conjugate. A small molecule can be conjugated to a carrier protein such as BSA or thyroglobulin and be immobilized onto a solid phase: antigen–protein–solid phase (Ag–P–SP). Labeled primary or secondary antibody (AbE) and sample (Ag) are introduced. The analyte in the sample competes with the immobilized Ag to form immunocomplex (Ab–Ag or AbE–Ab–Ag–P–SP) with the limited amount of antibody. Ab–Ag is washed away, and the AbE that is not occupied by the sample analyte is detected by substrate signal development.

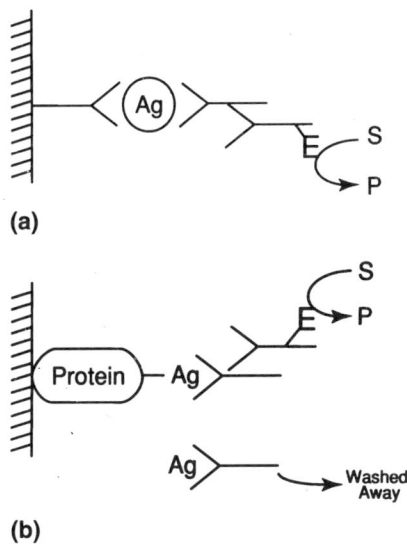

(a)

(b)

Fig. 6   Assay design using a general secondary antibody label. It is more cost-effective to label the secondary antibody against the animal host species than to label the primary antibody. Assay protocols of Figs. 3 (Fig. 6a) and 5 (Fig. 6b) can be modified to use this strategy. Variations of enzyme amplification or biotinylation can also be introduced.

and conjugates, titers should be determined. A two-component checkerboard titer for microtiter plates can be set up to test the interaction of the two components to optimize their concentrations. Serial dilutions are made in each direction for each component on the plate. Incubation time, temperature, shaker speed, and plate washing speed should be tested, optimized, and followed consistently throughout the study.

3.   Disequilibrium Assays

In a limited-reagent assay, sensitivity can be increased using disequilibrium principles (160–162). Drug is allowed to preincubate with the antibody to reach an equilibrium before the tracer is added and then incubated for a short time (disequilibrium). Under such conditions, the number of antibody-binding sites available for the labeled antigen are minimized and sensitivity is maximized. However, this method does not always improve sensitivity (163) and may result in decreased specificity (164). An alternative, more productive method to improve sensitivity is to obtain high-titer antisera and high-activity antigen so that the concentration of both label and antibody can be decreased, resulting in increased sensitivity (134). Another approach is to use an excess-reagent method (72).

4. Homogeneous Assays

When binding of analyte to antibody changes the kinetics of light emission, it allows discrimination of free versus bound label. Separation of the bound from the free label is not required for quantitation, and direct measurement can be made in the reaction solution. Homogeneous methods are simple, rapid, and can be readily automated. Examples include widely used commercial devices and kits for drug monitoring: EMIT™ (Enzyme Multiplied Immunoassay), which applies the fluorescence-excitation energy transfer differences (165), and TDx™, which uses the differences in the fluorescence polarization between the bound and free labels (166). Large numbers of hapten-type molecules for drugs of abuse, therapeutic drugs, and small-molecular-weight hormones are available in kits for automated assays in clinical laboratories.

An enzyme-channeling IA was first developed by Ullman utilizing two consecutive enzymatic reactions (167–169). Two enzyme labels are brought in close proximity via an antibody–antigen binding complex. The first reaction product is present in high local concentrations and results in a large signal by being the substrate of the second reaction. Enzyme-channeling IA on dipsticks has been applied for therapeutic drugs based on the principles of enzyme immunochromatography (170).

Recombinant DNA technology was applied to IA in the development of the homogeneous cloned enzyme donor IA (CEDIA™) (171–173). The enzyme β-galactosidase from *Escherichia coli* is a tetramer formed spontaneously from inactive monomers. Mutation of the Z gene in the lac operon of *E. coli* and recombinant techniques yield inactive enzyme fragments that can be used as an enzyme donor (ED), while the other fragments becomes an enzyme acceptor (EA). The haptenic analyte is conjugated to ED. The reaction of ED and EA is affected by the analyte in the sample competing with the ED–analyte conjugate for antibody binding. Antibodies are immobilized onto the solid-phase support. Sample, enzyme-donor, and substrate reagents are introduced. As the sample analyte concentration increases, more antibody-binding sites are occupied by the analyte and become unavailable to bind the hapten-conjugated ED which then forms the active enzyme with EA, resulting in an increased response. This assay design works well for small molecules. Commercial kits have been developed for thyroxine, cortisol, vitamin B_{12}, folate, phenytoin, and digoxin, with a growing menu for other haptens.

F. Assay Specificity

1. Matrix Effect

Matrix effects can be a problem for immunoassays, especially for a method without any prior sample clean-up. It can be caused by either nonspecific or specific interferences from the sample matrix and reagents. Possible matrix effects can be assessed by comparing a standard curve with calibrators prepared in assay buffer versus those in the intended sample matrix. The amount of nonspecific binding (NSB), maximum percent of binding in the absence of analyte (B_{max}), and the slope will provide an early indication of possible matrix interferences. High NSB, low percent binding to the antibody, and a shallow slope often indicate problems that could be caused by matrix effect.

Selectivity against the endogenous compounds in the sample matrix is tested on separate matrix samples from at least six undosed individuals (or various lots of control matrix from commercial sources). If there are structural similarities between the drug and endogenous compounds, a larger number of control lots (e.g., ≥ 20) should be tested. Control matrix is tested for NSB, B_{max}, and recovery of a known amount of analyte added at or near the limit of quantitation. If the NSB and B_{max} from various control matrix lots are similar, the standard deviation of B_0 (B_{max} − NSB) is an indication of the variability of the noise level, which can be used for the estimation of the limit of detection. If a sample from an undosed individual exhibits a response deviating more than one standard deviation from mean response, interfering material is most likely present. To determine whether interference is specific or nonspecific, a parallelism test should be conducted.

A useful tool to test matrix effect is parallelism. Test samples from a clinical trial and/or samples from various control batches, with known amounts of analytes added, are diluted with control samples containing no analyte, and these are used as standard calibrator preparations. Various dilutions (e.g., 2-, 4-, 6-, 8-, 10-, and 20-fold) are prepared and analyzed against the standard calibrators. The dose–response curves of the diluted samples are compared to those of the standard calibrators. A parallel line of the test (or spiked) sample shows that the compound present in the sample has the same antigen–antibody binding response as the analyte and, therefore, is very probably the analyte itself. If the line is not parallel to the standard curve line, and the concentrations at higher dilutions agree with one another, the matrix effect is nonspecific and could be overcome by dilution.

Sometimes, matrix effects can be corrected by normalization of the NSB/B_0. This must be tested by showing that the interference is consistent in multiple samples taken at different times from the untreated individual. In other instances, predose samples from each patient (or test animal) are used to construct the calibrator standards for quantitation. In such cases, it may be better to do sample clean-up to eliminate matrix interferences, because it is not always possible to obtain adequate predose sample volumes from each subject.

2. Sample Clean-up

The simplest processing step to overcome matrix effect is to dilute the sample threefold or more in buffers containing chaotropic or chelating agents such as

Tween-20, Triton-X 100, and EDTA. In other cases, matrix effect and potentially interfering substances of drug metabolites and analogs should be removed by sample clean-up preceding the IA. Sample extraction as applied to other analytical methods can be applied to IA. Therefore, these methods will not be elaborated here. Briefly, liquid/liquid and solid-phase extractions are most commonly used. HPLC-IA has the combined advantages of the selectivity of HPLC and the sensitivity of IA (IA becomes a sensitive detector for the HPLC method). However, sample clean-up by HPLC is labor-intensive and has not been popular for large numbers of samples. New techniques in perfusion chromatography using porous immunoaffinity columns coupled with enzyme chemiluminescence detectors may be a novel way to address this problem (174).

For every sample clean-up method, it is important to investigate the recovery of the parent drug as well as the potentially interfering compounds. Any organic solvent must be evaporated completely and the residue reconstituted in the assay buffer to avoid altering the antibody-binding activity.

The sample extraction step can also serve as a concentration step to increase the assay sensitivity. Direct IAs generally use a sample volume of 0.1 mL or less, while it is common to extract 1.0 mL of human biological fluid.

3. Cross-validation with Another Method

Because IA methods are generally more economical and/or faster, a strategy in drug development is to use IA method as a main method to analyze large numbers of samples and to use another method to analyze a subset of these samples. Correlation plots of one method against another over various concentrations are constructed for method comparison. For example, cyclosporine IA methods using polyclonal and monoclonal antibodies had been compared with HPLC methods on samples from various types of patients (175–179). Several authors have concluded that the correlations were excellent for samples from normal volunteers and some patient types. An investigator in a drug-development and monitoring program should consider using comparative methods; a method that is fast and easy to handle can be chosen for high-volume analysis, while a second, more elaborate method can be used to validate the first.

4. Exploiting Nonspecific Antibodies

During the course of drug development, several analogs may be considered as candidates for investigation. Immunogens designed not to discriminate among epitope differences of the analogs can be used to produce nonspecific antibodies. The antibodies will recognize the common structure, and one method can be used for several drug analyses. This will save a lot of time and cost. If certain analogs are possible precursors/metabolites of one another, separation methods prior to IA could be used to provide the required selectivity.

Broughton et al. (180) have demonstrated that the use of nonspecific antisera to gentamicin could also be applied to the analysis of sisomicin if gentamicin were not present in the samples, because cross-reactivity with other substances was minimal. We have successfully used anti-prednisone antiserum to measure cortisone in serum in the absence of prednisone. This was possible because cross-reactivity with other compounds was not significant (181).

Drug-monitoring programs in clinical chemistry also use nonspecific antibodies to detect drugs of abuse, such as cannabinoids, opiates, and bezodiazepines. References are listed in Table 4 in Sec. VIII. Investigators deliberately pooled and mixed antisera against several analytes to develop a common IA method for multiple analytes. For example, antiserum against testosterone (T) and antiserum against 5α-dihydrotestosterone (DHT) were used to analyze samples for both T and DHT simultaneously (182). A multivariable (three-dimensional) standard curve was created which allowed the independent estimation of T and DHT concentrations when both T and DHT were present in samples. The method is valid as long as both assays are precise, and the procedure avoided the need for tedious, time-consuming chromatographic separation.

IV. COMMERCIAL KITS

A. Application to Pharmacokinetic and Pharmacodynamic Measurements

Many RIA and EIA kits are readily available for drugs of abuse and for drugs whose plasma concentrations have been correlated with therapeutic and toxic effects. The individual components of the kit, such as control standards, labeled hapten, and antisera, are also available from many commercial sources. The ease of using such kits is apparent, but the buyer must be aware of the potential pitfalls as well. Often the kits are made for purposes other than drug quantitation in clinical studies, drug pharmacokinetics for bioavailability assessment, or pharmacodynamics of biochemical markers (183). Instead, their intended uses are for the detection of drug abuse (toxic concentrations), therapeutic doses, and diagnosis of diseases or certain physiological conditions. The lowest limit of the calibrator range is often greater than that of the desired limit of quantitation for a pharmacokinetic or pharmacodynamic study. To save reagent and labor costs, the number of standards in a calibrator set is limited (usually no more than five standards), and they are not run every time a sample set is run. Calibration curves are run weekly or monthly and the curve parameters are stored until quality-control samples (QCs) do not perform according to predetermined standards. Calibrator standards and QCs may or may not be in the same intended matrix of the patient samples. Calibrators are usually prepared in protein-based buffer or stripped serum. If the stripped serum is prepared by adsorption to activated charcoal (184), the matrix will be very different from the intended matrix. It is better to prepare the stripped serum using adsorption to specific antibodies linked covalently to a solid-phase support (such as an immunoaffinity column), as long as antibody leaching from the support is monitored (185). Commercial kit QC values are generally determined by the mean values from multiple laboratories and not by the theoretical values of the QC preparation.

B. Adhering to Industry Guidelines During Kit Use

The lists of commercial IA kits are growing, especially for therapeutic drug monitoring and biochemical markers (186). Automation is available for some of these kits. The pharmaceutical bioanalytical laboratory should be able to take advantage of the convenience of commercial kits if the following six cautions are exercised.

1. Purity or potency of the reference-standard compound must be documented. A certificate of analysis should be obtained from the supplier. Whenever possible, a universal concentration or activity unit should be used to provide a constant basis for comparison among studies.
2. Standards and QCs should be prepared from separate weighings. If kit standards are used, it is prudent to check their accuracy against an in-house preparation or standards from another lot or source.
3. Standards and QCs should be prepared in the same matrix as the intended samples.
4. For analytes that have endogenous levels, standards can be prepared in "treated" or surrogate matrix. Comparison of responses of spiked analyte in the standard matrix versus the sample matrix should be performed to demonstrate the lack of matrix effects.
5. Sufficient standards concentrations are required to establish a complete standard curve. The concentrations of standard curve ranges should cover the expected sample concentrations.
6. Three levels of QCs are prepared over the standard curve range. If dilutions are to be made, QCs of similarly high concentrations should be prepared to mimic the samples.

It is the responsibility of the individual laboratory to rigorously evaluate each new lot of IA kits for sensitivity, precision, accuracy, and specificity to assure assay reliability. Lot-to-lot variation in kit reagents has been verified by several authors with respect to digoxin RIA kits (187–189). It is necessary to assure the availability of a large number of kits with the same antibody lot from the manufacturer before the evaluation. Once the evaluation is found to be satisfactory, reservation of these kits should be made to cover the entire study. If a tracer lot has to be changed during a study (such as ^{125}I label, which has a short shelf life), QC performance should be checked to ensure that assay performance is similar to that of the previous tracer lot. For a new lot of enzyme marker, conditions of the enzyme reaction (e.g., various enzyme dilutions and incubation times) should be checked by running a validation curve.

C. Improving Assay Specificity and Sensitivity

Many commercial kits for drug monitoring are designed for a higher concentration range than the range that is suitable for pharmacokinetic or pharmacodynamic studies. Often the commercial kits may not be specific for the parent compound of interest. For example, many kits developed for drug monitoring are intended for urine samples, where the concentrations are high and the method measures both parent and polar conjugates excreted in the urine. Sample extraction can separate the parent compound from the polar metabolites as well as serve as a sample concentration procedure to increase the assay sensitivity (190).

D. Using Commercial Reagents

Besides commercial kits, many immunochemical reagents are available (191,192): anti-drug antibodies, anti-enzyme antibodies, radioisotopic labels, nonisotopic labels, secondary antibodies, and biotinylated conjugates. Because of the flexibility in IA designs, investigators can consider several options from the availability of the commercial materials in addition to in-house resources in preparing the IA reagent components.

V. DATA REDUCTION AND QUALITY CONTROL

A. Data Reduction

Based on the law of mass action, the concentration–response curve in a limiting-reagent reaction can be expressed mathematically according to the following equation:

$$\frac{B}{T} = \frac{q}{[S + S^*]}$$

where q is the binding capacity of the antibody, T is the total activity in the system, B is the bound activity, and S and S^* are the unlabeled and labeled antigen, respectively. The curve has the form of a hyperbola (Fig. 1). However, the most common graphic display method is to plot log concentration versus response (Fig. 7). The resulting sigmoidal curve is commonly displayed throughout IA literature. Variance in data points along the curve is reasonably uniform. Many equations have been derived and empirical plotting procedures have been devised

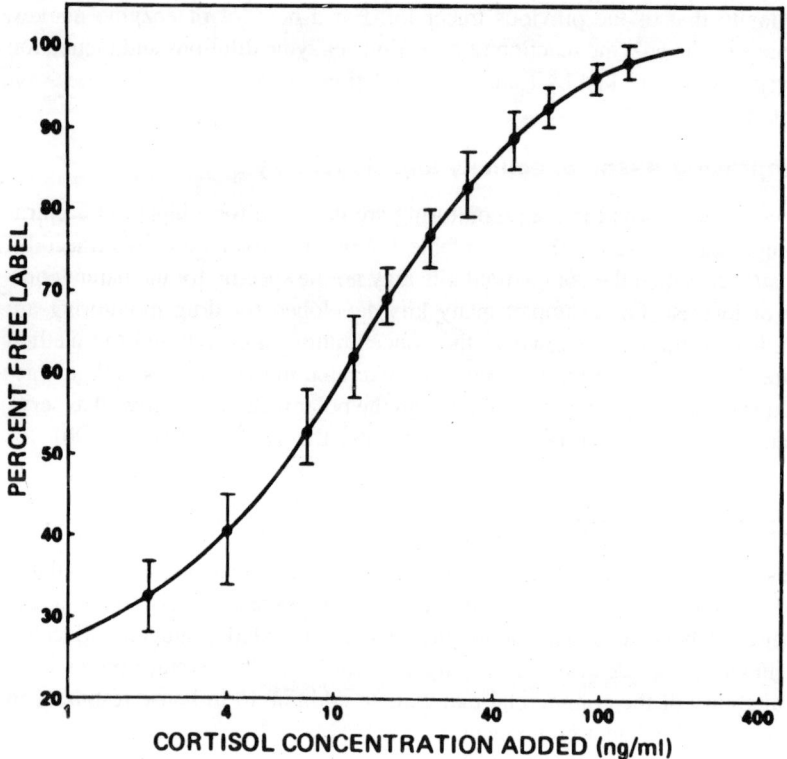

Fig. 7 Typical RIA log concentration–response curve. (From Ref. 647.)

in an attempt to linearize data processing. Several standard calibration plots are presented in Fig. 8. The most common linearizing method used in IA data reduction is the following log-logit transformation:

$$\text{Logit } y = \ln\left[\frac{y}{1 - y}\right]$$

where $y = B/B_0$, B is the bound-labeled analyte, and B_0 is the maximum binding of labeled analyte when no unlabeled analyte is present. Both B and B_0 are subtracted for the background blank. This method of transforming IA data has been studied extensively by Rodbard and co-workers (193–195), as well as by other authors (196–198). However, because this transform, like any other, introduces nonuniform variance in the linear calibration curve (199), weighted regression is required to obtain a calibration curve with normally distributed error (200). Computer programs have been written to fit these complex curves (197,198,201).

Several other linearizing methods have been published (199,202,203), but nonuniform variance is inherent in these methods as well. No single linearized plotting method will be appropriate for all IA data (200,204). It may be necessary to investigate several plotting methods before choosing the most appropriate one. This problem can be circumvented if the sigmoidal log concentration–response curve is retained.

Another widely used curve-fitting algorithm is the four-parameter logistic (4PL), which is a different version of the log-logit method (195,205).

$$y = \frac{a - d}{1 + (X/c)^b} + d$$

where y = response, X = dose, $a = B_0$ (response at zero dose), d = response at infinite dose, c = dose resulting in response halfway between a and d (ED_{50}), and b = exponent, usually near $+1$. If b is negative, the roles of a and d are reversed, and the equation can be applied to excess-reagent IA methods.

When the curve is asymmetric around the two sides of the ED_{50}, another term is introduced to provide an estimate for this asymmetry and a better fit. This becomes the five-parameter logistic (5PL) model:

$$y = \frac{a - d}{[1 + (X/c)^b]^m} + d$$

If $m = 1$, the equation is the same as the 4PL.

Data regressions based on the law of mass action are generally adequate for most situations. However, this model only retains validity in liquid-phase reactions at equilibrium without cooperativity. Reactions that involve solid-phase, multiple cooperative binding, and not reaching equilibrium, deviate from the model. Therefore, empirical equations that are not based on the law of mass action have been used for curve fitting also. Among these, polynomial (205) and spline functions are often used (206–209). Polynomial regression can be a second-order (parabolic) or third-order (cubic) function:

$$Y(X) = a_0 + a_1 X + a_2 X^2$$

or

$$Y(X) = a_0 + a_1 X + a_2 X^2 + a_3 X^3$$

Spline functions using piecewise third-order polynomials can be used to fit each interval between standards. The entire function and its first and second derivatives are continuous:

$$Y(X) = a_{0j} + a_{1j} X + a_{2j} X^2 + a_{3j} X^3$$

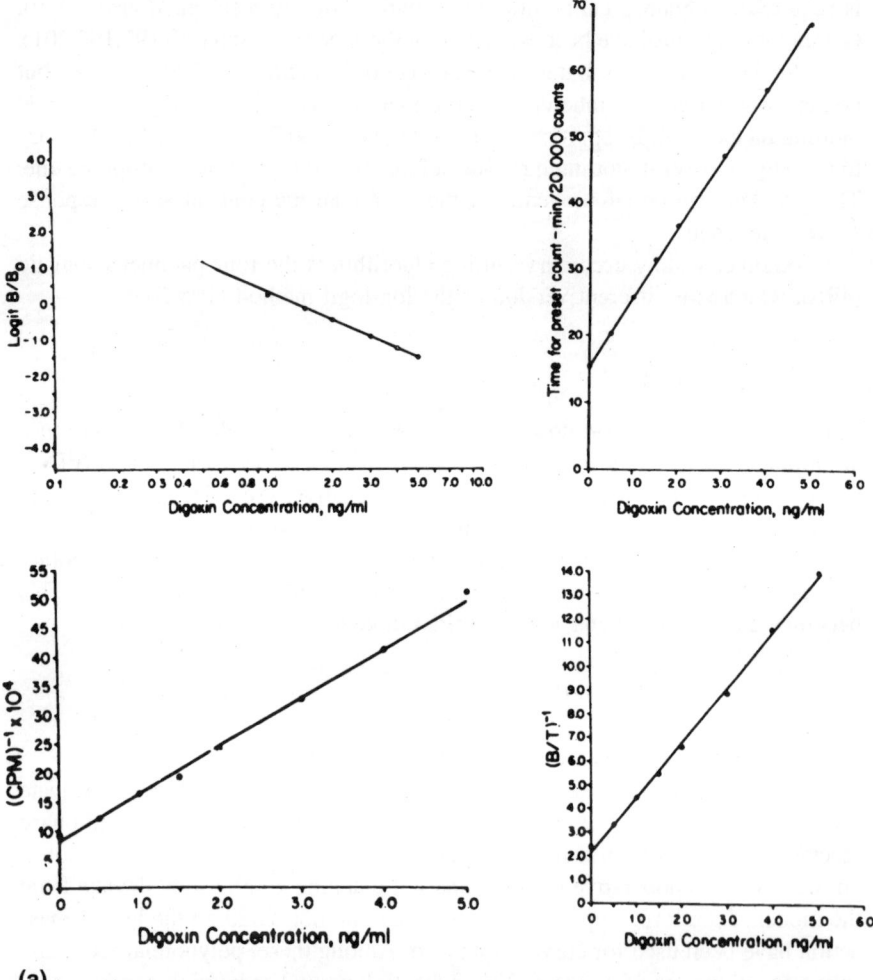

Fig. 8 Data reduction methods to linearize immunoassay concentration–response curves. (From Ref. 647.)

where the constants a_{ij} are between standard j and standard $(j + 1)$. Iterative curve-fitting programs have been written for polynomial equations with constraints that result in no maxima, no more than one minima, and only one turning point (210).

More standard points are needed to adequately define a curvilinear function than to define a linear function. Weighted least-square regression methods are

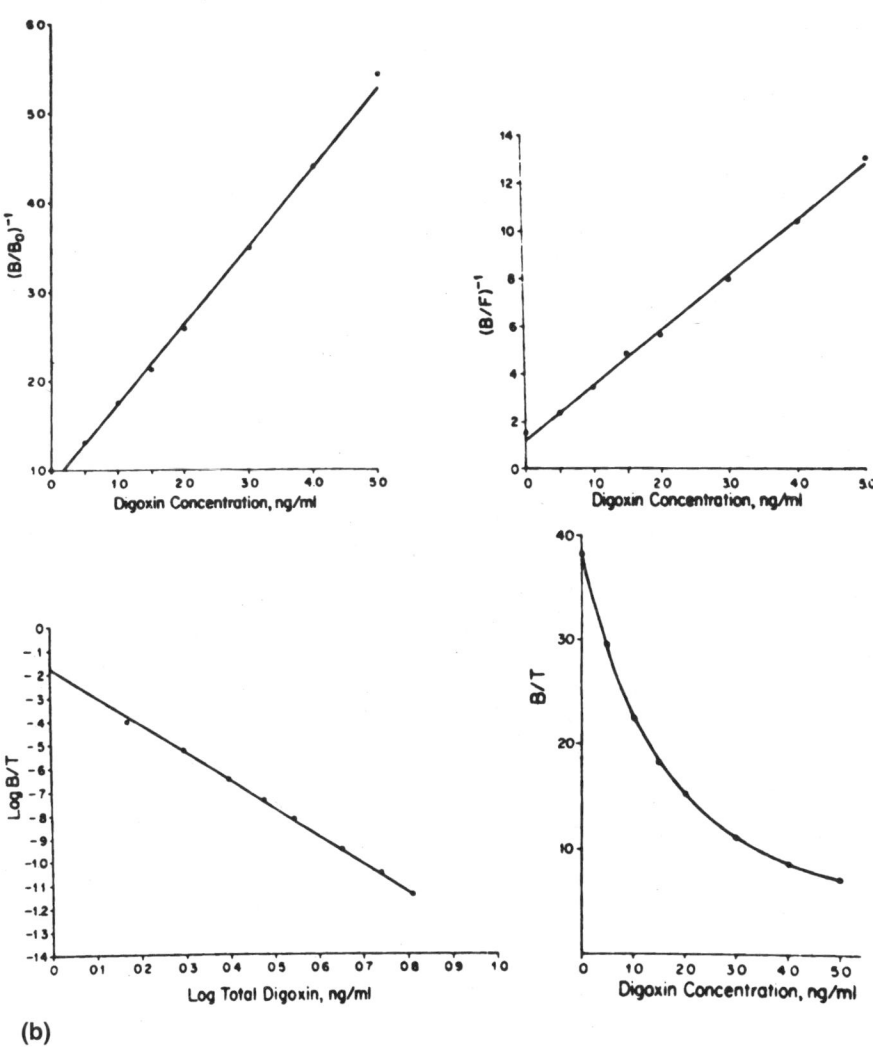

(b)

generally used, where the sum of the weighted residual squares of Y is minimized to improve the fit. The linear portion on a log-log, logit-log, or 4- and 5-PL regression should have a slope that approaches unity. A slope of less than unity indicates that the assay will have great variability even with a very small change in response.

Deviations from the linearized logit-log plot in the low- and high-concentration regions may indicate that improper background corrections have been applied due to nonspecific binding (210), hindered binding (211), antibody heterogeneity, and/or high-dose hook effect (212).

Acceptance criteria for accuracy and precision of standards and QCs must be determined during method validation, and are analogous to acceptance criteria for chromatographic methods. IAs may not be as inherently precise as chemical methods, because IAs measure a reaction rather than a physicochemical property of the analyte. In cases where internal standards are not used for recovery correction, two to three replicate assays may be conducted on a single sample to improve precision. Despite all of the available mathematical transformations, it is important to remember that this is not a linear system and caution must be used as the concentrations approach either the upper or lower end of the standard curve. For example, variability becomes too large to be acceptable as the B/B_0 value goes beyond <0.1 or >0.9 for most limited reagent assays.

B. Quality Control Samples

1. Preparation of QC Samples

Because IA methods tend to be subjected to matrix effects, it is imperative to prepare QCs in the same matrix as the intended sample. Guidelines for pharmaceutical assays recommend that at least three QC concentrations (low, intermediate, and high on the standard curve) be batched with the unknown samples for analysis. This means that QCs are prepared at the same time samples are generated and stored with samples under identical conditions. In some cases, the standard curve range for an IA is not able to cover the entire range of concentrations observed in the unknown samples. When this occurs, QCs concentrations in the range of the observed unknown sample concentrations should be prepared and diluted along with the unknown samples before analysis to evaluate the influence of the dilution process. Parallelism of the high-QC sample should be evaluated.

Sometimes a clinical site may deviate from the study protocol-defined sample-collection procedures. For example, instead of collecting blood into heparinized tubes, the sample may be collected into a tube containing another anticoagulant, such as citrate or EDTA. Under these conditions, it is necessary to cross-validate each of the plasma/serum matrices: QC samples in the deviant matrix must be prepared and processed along with the deviant samples to ensure that the method is validated in the new matrix.

2. Interlaboratory and Commercial QC Samples

In addition to QCs prepared in-house, QCs prepared in another laboratory at the same location or at another location can be used to assess interlaboratory accuracy and precision. For example, as part of the CLIA 88 proficiency test program,

QCs are prepared at a central laboratory and sent to various clinical chemistry laboratories that participate in the program. The mean observed values among the laboratories, not the theoretical values of the QCs, are used for laboratory proficiency evaluation. Criteria are set differently for different tests by CLIA 88 regulations. Also, bioanalytical laboratories involved in a drug-development program may exchange QCs and test samples to ensure lab-to-lab reproducibility in the analyses.

QCs for analytes used in certain high-volume analyses are available from commercial sources. However, unless they are prepared in the exact matrix that is of interest to the analyst, they are of no practical use.

VI. AUTOMATION

In the last decade, automation for IA has developed to meet the challenging needs in hospital clinical chemistry laboratories to provide fast throughput and random-access sample analysis. Technologies in solid-phase support and nonisotopic labels enable high-throughput automation of homogeneous to heterogeneous IA. In this section, we briefly discuss automation available in clinical chemistry and how this can be utilized together with other semiautomation advances in analytical chemistry for pharmaceutical IA.

A. Batch Process

1. Homogeneous Assays

The early automated instruments were designed for batch-mode analysis. The first complete automated assays were homogeneous assays such as the Syva EMIT™ (165,213,214) and Abbott TDx™ (215). General-purpose open systems such as chemical analyzers with automatic pipetting, incubation controls (on time and temperature), and data processing can be applied to homogeneous assays. Systems such as the Hitachi 911™, which provides multiple reagent-addition steps, offers the flexibility for the researcher to optimize the IA design protocol. Enzyme inhibition analysis correlated with drug concentrations can also be processed in this type of system.

Closed automated systems designed for particular tests in an instrument have been developed. They are closed systems because the reagents and IA protocols were controlled by the manufacturer. Examples are the Roche Cobas FARA II™, DuPont ACA™, and Miles Immuno 1™ analyzers. The reagents are expensive, and in many cases the applications are fixed and may not be adaptable for PK and PD study purposes.

2. Heterogeneous Assays

Heterogeneous assays are more versatile, and sensitivity can be improved with assay protocol changes. Advances in nonisotopic labels, specific antibodies, and protein immobilization on solid phase enable the development of automation for heterogeneous assays. Using the excess-reagent assay design, incubation time is usually less than 30 min, instead of hours in the limited-reagent assays. Stringent control of incubation time and temperature, and accurate, automated liquid-delivery systems allow rapid, nonequilibrium reactions to be performed with satisfactory precision.

Chemiluminescent and fluorescent labels provide increased sensitivity with less background, for a much higher signal-to-noise ratio. Some examples of automated analyzers are Abbott IMx™ and Commander™, Baxter Stratus™, Kodak Amerlite™, and Hybritech Photon QA™. With the increased sensitivity and expanded menu for drug monitoring, researchers in pharmaceutical bioanalytical laboratories should be able to utilize such automation, provided that cautions in Secs. IV.B are taken into consideration.

3. Semiautomation

Multiple modules for automation such as pipetting, extraction for sample cleanup, incubation of the antigen–antibody reaction, separation of the bound/free, signal detection, and data processing can be applied separately or can be physically or computationally linked together. Many of these devices are used in other bioanalytical methods. For example, various models of robotics, such as Zymate™, Biomak™, and Tecan™ are used in many laboratories to increase throughput and precision. They are especially helpful in microtiter-plate assay designs.

B. Continuous Random Access

Emergency testing required in hospitals drives the technology for continuous, fast, random-access IA automation. Throughput varies from 30 to 150 tests per hour in most automated systems. Many of these systems consolidate previous small models of homogeneous and heterogeneous assays. Examples are Abbott AxSYM™, Miles Immuno 1™, DuPont ACA+™, Ciba ACS-180+™, Tosoh AIA-1200DX™, BMC-300™, Sanofi Access™, DPC Imulite™, Behring Magnum™, and Pharmacia-Wallach autoDELFIA™ (216,217). Recently, electrochemiluminescence (ECL) IA has been automated by Boehringer Mannheim in the ELECYS™ system, with a major increase in sensitivity (218,219).

C. Method Development

In a pharmaceutical laboratory, batch processing may be more appropriate. Total automation with continuous access may be used for a very large drug-development program. In a bioanalytical laboratory where method development needs are many, semiautomation of discrete blocks is more cost-effective. Generally, modules of discrete procedures can be designed and linked together according to specific method designs. Some considerations are use of barcode systems, solid-phase supports (microtiter plates, beads, latex particles, etc.), washing devices, pipetting—accuracy, precision, carry-over problems, handling viscous samples and samples with clots or particles, and validation of the automated processes step by step against manual processes.

VII. RELATED METHODS

A. Receptors, Binding Proteins, and Nucleic Acids

IA measures the reaction of antigen–antibody binding. Other binding reactions can be utilized for the quantitation of pharmaceutical compounds. One of the oldest binding assays for folate uses the folate-binding protein found in bovine milk. The major metabolite of leucovorin, N^5-methyltetrahydrofolic acid, is analyzed by a protein-binding method. Other examples are porcine intrinsic factor for vitamin B_{12} (220) and corticosteroid-binding globulin for cortisol (221) assays.

Receptor assays have been developed for opioid peptides (222), benzodiazepines (223), oxitropium bromide (224), atenolol (225,226), atrial natriuretic peptide (227), benidipine (228), several calcium channel blockers (229–232), CNS drugs (233–237), and estrogens (238). Many of these receptor assays reported have been compared with IA methods.

Since the binding of a drug to the receptor triggers subsequent physiological events that lead to efficacy /toxicity, receptor assays can provide a link between pharmacokinetic and pharmacodynamic measures. Receptor assays are generally not specific for one analyte. Pharmacologically active metabolites bind to the receptor, although binding constants may be different. When more than one active analyte is present in the sample, apparent activity (the sum of analytes' activities as a result of the concentration and binding constant) will be measured and reported as "parent drug-equivalent concentration." Such measurement may be conceptually difficult for a purist analytical chemist to accept. However, it could be more relevant to measure the apparent bioactivity than the pure analyte when active metabolites (known and unknown) are present—a potential advantage for this approach.

The drug–antibody binding reaction sometimes can be found to parallel that of drug–receptor binding reaction. When that happens, the ability to correlate immunoactivity with bioactivity provides a vital tool for efficient drug monitoring. Miller et al. illustrated this principle by testing four IA methods against the Na^+-K^+ ATPase-based receptor assay on digoxin and four of the metabolites (239). The major differences between the IA methods was found in digoxigenin, where only one of the four methods tested showed a correlation between immunoactivity and bioactivity. Wahyono et al. also correlated binding activities of digoxin and metabolites with RIA (240,241).

The complementary reaction of nucleic acids has one of the strongest binding affinities found in nature. Nucleic acid probe assays have been developed extensively in research on genetics, forensics, microbiology, and oncology (242,243). Applications of nucleic acid probe assays for pharmaceutical analysis are still limited at this time and are beyond the scope of this chapter.

B. Hyphenated Immunoassays

1. HPLC-IA

HPLC methods have been used to separate analytes prior to IA. In this case, IA becomes a sensitive detector for HPLC. Some of the HPLC-IA methods are listed in Sec. VIII, Table 4. The method is cumbersome and tedious, because fraction collections, dry-down, and reconstitution steps are labor-intensive and time-consuming. Perfusion chromatography technique has been developed using antibodies immobilized on porous flow-through beads of large surface area packed in a cartridge (174). Automation is designed using an excess-reagent protocol to allow fast reaction time. The cartridge can be regenerated and reused several hundred times. Enzyme and fluorescent labels can be incorporated into the assay system for sensitivity and flexibility in design.

2. Immunoaffinity Columns Coupled to CE or MS

The specificity of IA and antibody immobilization on solid-phase support techniques can be utilized for sample clean-up and concentration before capillary electrophoresis and mass spectrophotometry. For example, analysis of prostacyclins requires sensitivity in the pg/mL range and selectivity from multiple endogenous prostaglandin compounds. Immunoaffinity sample clean-up before GCMS has been used for thromboxane B_2 analogs (244–246), iloprost (247), and nocloprost (248). Immunoaffinity sample clean-up before HPLC determinations was reported for clenbuterol (249) and pravastatin (250).

VIII. APPLICATION OF IA TECHNIQUES

A. General Comments

This section includes a general review of the literature in a table format. It is not intended to be comprehensive, but should permit the reader to envision the diverse application of IA to many classes of drugs. The reader can draw analogies from this information which may be applicable to a compound of interest. Drug compounds are listed under several classifications (broadly adapted from Goodman and Gilman (251)) for the convenience of the reader; many of them could be included in a different category because of other pharmacological uses and effects.

Because of drug-monitoring programs in clinical chemistry, many IA methods have been developed for opiates, steroid hormones, peptide hormones, CNS drugs, drugs of abuse, and drugs with a narrow therapeutic window. Many IA methods have been developed recently for endogenous biochemical markers. Only a few of these are included in the table as illustration. It would be impossible to cover so many biochemical markers in this space.

RIA was the predominant method used in earlier literature; many nonisotopic methods for drug compounds are now more widely used than RIA. Comparisons to HPLC, GCMS, and radioreceptor methods are listed in the table, as well as HPLC-IA methods.

B. Table of Applications

Table 4 lists applications of IA in pharmaceutical analysis.

IX. FUTURE TRENDS

Future trends in IA reside in several areas, including (a) new labels and improved sensitivity, (b) automation, (c) simultaneous multianalyte assays, and (d) genetic engineering. The trend must be to exploit the current advantages of IAs, which include sensitivity, simplicity, limited cost, and speed. Our goal must be to make the assays faster, better, and more accessible. However, it must be remembered that what is appropriate and acceptable in the clinical chemistry laboratory may not be acceptable or may be much less appropriate in the biopharmaceutical laboratory. Only trends that may apply to the pharmaceutical analyst will be discussed here.

A. New Labels

Improved sensitivity, and with it specificity of antibody, will continue to be a need for improved IAs. The most convenient and direct way to achieve this is

Table 4 Application of IA in Pharmaceutical Analysis

Analyte	IA[a]	Method comparison and comment	References
1. Drugs that act at synaptic and neuroeffector junctional sites			
Albuterol	RIA		252
	RIA		253
Amphetamine	RIA	Horse urine	254
	FIA		255
Atenolol	RRP	With GC	225, 226
Atropine	RIA		256–258
	RRP		259
Benzoylecgonine (abuse)	RIA	Blood spot	260, 261
Benzphetamine	EIA	With GC	262
Clenbuterol	EIA	Bovine tissue, urine	263
	EIA	Immunoaffinity column	264
Clomipramine	RIA		265, 266
Clonidine	RIA		267
Cocaine (abuse)	RIA		268, 269
Cotinine	RIA		270
Desipramine	EIA		271
	RIA		272
Detomidine	RIA		273
Ergotamine	RIA		274
Fenoterol	RIA		275
L-Hyoscyamine	RIA	With GC	276
Imipramine	RIA	With GC-FID GCMS	277
	EIA		271
Labetalol	RRP		278
Mabuterol	EIA	Homogeneous	279

Drug	Method	Note	Reference
Maprotiline	RIA	With HPLC	280
Methamphetamine	RIA		281, 282
Nicergoline	RIA		283
Nicotine	RIA		270, 284, 285
Normetanephrine	FIA		286
Oxitropium bromide	RRP		224
Paraquat	RIA		287, 288
Pirenzepine	RIA		289
	RIA	Automation	290
Procaterol	RIA		291
Propranolol	RIA	Stereospecific	292, 293
Pseudoephedrine	RIA	Stereospecific	294
Reserpine	RIA		295
Ritodrine	RIA		296
RP-42068	RIA		297
Scopolamine	RRP		298–301
	EIA		302
	RIA		303
Tubocurarine			
2. Drugs that act on the CNS			
Alfentanil	RIA	With GC	304
Amitriptyline	RIA		305
	EIA		271
Amylobarbitone	EMIT	With RIA	306
Antipyrine	RIA		307, 308
Barbiturates	RIA		37, 309
	EIA		310
Benzodiazepines	EMIT		311
	RIA	Blood, urine	312, 313
	RIA	Hair sample	314
	RRP		223
	IA	Blood	315

Table 4 Continued

Analyte	IA[a]	Method comparison and comment	References
Benztropine	RIA		316
Bromocriptine	RIA		274, 317
Buprenorphrine	RIA		318, 319
	RIA	With HPLC, GCMS, horse urine	320
	EIA		321
Bupropion	RIA		322
Butorphanol	RIA		323
Caffeine	RIA		324
	EIA		325
Cannabinoids (abuse)	RIA		326, 327
	RIA	Δ9THC & metabolites	328
	RIA	With GCMS, blood, urine	329
	EIA	Urine	330
Carbamazepine	EIA	With chromatography, whole blood	331
Chlordiazepoxide	RIA		332
Chlorpromazine	RRP		234
	RIA		333
	RIA	N-oxide metabolite	334, 335
Clonazepam	RIA		336
CNS in general	FPIA	With EIA	337
Codeine	RIA		338–340
Diazepam	RIA		341, 342
Dimethylamide lysergic acid (abuse)	RIA		343
Doxepin	RIA	And metabolite	344
Etorphine	RIA		345

Compound	Technique	Note	Page
Fentanyl	RIA		346–348
	RRP	With GCMS, RIA	235
	RRP	Fentanyl analogs	236, 237
Flunitrazepam	RIA	Solid phase	349
Fluphenazine	RIA		350
	HPLC-RIA		351
	RIA		352–354
	RIA	Sulfoxide metabolite	355
	RIA	N-oxide metabolite	356
	RIA	With GCMS	357, 358
Flupenthixol	RIA		359, 360
Flurazepam	RIA		361
Galanthamine	EIA		362
Haloperidol	RIA		363–366
Hydrocodone	RIA		367
Lacidipine	HPLC-RIA		368
Lamotrigine	FIA		369
Levorphanol	RIA		370
Lidocaine	FPIA	With GC, EIA	371
Lomotil	RIA	Diphenoxylic acid	372
Loperamide	RIA		373
Lormetazepam	RIA		374
Meperidine	RIA		375, 376
Mesoridazine	RIA	With HPLC	377, 430
Methadone	RIA		378, 379
	RIA	Stereospecific	380
Methaqualone	RIA	With TLC	381
	RIA	Metabolite	382
Metoclopramide	RIA		36, 383

Table 4 Continued

Analyte	• IA[a]	Method comparison and comment	References
Morphine	RIA		37, 189, 384–389
	EIA		390
	RIA	With GCMS, blood	391
	RIA	With GCMS, vitreous humor	392
Nalmefene	RIA		393
Naloxone	RIA		394
Nitrazepam	RIA		395
Nomifensine	RIA	With GC	396
Nordiazepam	RIA		397
Nortriptyline	RIA		398, 399
	RIA	With HPLC, GC	400, 401
Nuvenzepine	RRP		233
Opiod peptides	RRP		222
	HPLC-RIA	With RIA, radioreceptor, HPLC, MS	402, 403
Oxidiazole-benzo	RRP	With HPLC	404
Pentazocine	RIA		405, 406
Pentobarbital	RIA	Enantioselective	407
	RIA		408, 409
Phenobarbitone	FPIA	With HPLC	410, 411
Phencyclidine (abuse)	RIA		412, 413
Phenothiazine	RIA		414
Phenytoin (DPH)	RIA		415–417
	RIA	Saliva	418, 419
	RIA	Non-protein bound	420
	RIA	With GC	421
	EIA	Automation	422
	EIA	CEDIA	423

Pimozide	RIA	HPLC, acid metabolite	424
Progabide	RRP	Metabolites	425
Rilmazafone	HPLC-EIA	With HPLC	426
Rolipram	RIA	With GCMS	427
Sufentanil	RIA	With HPLC	428
Sulforidazine	RIA		429–431
Sulpiride	RIA	With HPLC	432
Thioridizine	RIA		429, 430
Triazolam	RIA	Nordiazepam, lorazepam, alprazolam	433
Tricyclic	EIA	With GC antidepressant	434
Trifluoperazine	RIA		435
Valproic acid	EIA	Non-protein bound, with HPLC	436
Zonisamide	EIA		437
3. Anti-inflammatory drugs			
Anti-inflammatory protease	EIA	With HPLC	438
Chlorpheniramine	RIA		439
Colchicine	RIA		440, 441
Iloprost	RIA		247, 442
Indomethacin	RIA		443
SCH40120	EIA		444
Sulindac	RIA	Metabolites	445
Theophylline	FIA	Homogeneous	446
	RIA		447
	HPLC-EIA	Whole blood	448, 449
	FPIA	With RIA, EIA, HPLC	450
	EIA	Liposome	451
	EIA	Electrochemical detection	452
	FPIA	Bedside detection	453

Table 4 Continued

Analyte	IA[a]	Method comparison and comment	References
4. Drugs that affect renal function and electrolyte metabolism			
Arginine (8) vasopressin	RIA		454
Atrial natriuretic peptide	RIA		455, 456
	EIA		457
	RRP	With RIA	227
Bumetanide	RIA		458
Vasopressin	RIA		459
5. Drugs that act on the cardiovascular system			
Acetylstrophanthidin	RIA		460
Altiopril Ca	RIA		461
Benidipine HCl	RIA		462
	RRP		228
Captopril	RIA		463
Cardiac glycosides	RIA		464, 465, 474
CI-906	RIA	Metabolite	466
Ceruletide	HPLC-RIA		467
Cilazapril	EIA	Metabolite	468
Digitoxin	RIA		4, 469
	RIA-HPLC	& Metabolites	470, 480
Digoxin	RIA		471, 472
	FIA		473
	RIA		474–476
	EIA	Automation	477, 478
	RIA	With HPLC	479, 480
	HPLC-RIA		481, 482
	RIA	Saliva	483

Enalaprilat	RIA		484
Flecainide	EIA		485
Guanethidine	RIA		486
ISF-2405	RIA	Metabolite of cadralazine	487
Isradipine	RRP		231
Lanatoside C	RIA		488
Lisinopril	RIA		484
Medigoxin	RIA		489
Mepiradipine	RRP		490
Minoxidil	RIA		491
Naftopidil	RRP		229
Nicardipine	RRP		230, 231
Nifedipine	RRP		232
Ouabain	RIA		492
Perindopril	RIA		493, 494
SQ29, 852	RIA		495
SQ27, 519	EIA	Metabolite of fosinopril	496
Saralasin	RIA		497
Vinpocetine	EIA		498
Zabicipril	RIA	Metabolite	499
	EIA	Metabolite	500
6. Anti-infective agents			
Acyclovir	RIA	Scintillation proximity	501
Amikacin	RIA		502
	FIA		503
Berlopentin	EIA		504
Chloramphenicol	RIA		505
	RIA	Microtiter plate	506
Chloroquine	EIA, RIA		507
CI-937	RIA	Metabolites	508

Table 4 Continued

Analyte	IA[a]	Method comparison and comment	References
Cinobufagin	EIA		509
Clindamycin	RIA		510
Dibekacin	FPIA	With HPLC, FIA, RIA	511
Desiclovir	RIA	Prodrug of acyclovir	512
Erythromycin	RIA	Emycin derivatives	513
Gentamicin	RIA		514
	EIA	With RIA	515, 516
5-Hydroxymethyl			
Deoxyuridine	RIA		517
Isoniazid	RIA		518
Mithramycin	EIA		519
Monensin	EIA		520
Navelbine	RIA		521, 522
Netilmicin	RIA		523
Penicillin	RIA		524
Penicilloyl groups	RIA		525
Retrovir	RIA		526
Ribavirin	RIA	With HPLC	527
Sisomicin	RIA		180
Teicoplanin	FPIA		528
Tetracycline	RIA		529
Tobramycin	RIA		530
Zalcitabine	RIA		531
Zidovudine	EIA	With FIA	532
Zidovudine	RIA		533
	HPLC-RIA	Metabolites	534
	RIA	With HPLC, phosphorylated-AZT	535

	Method	Notes	Reference
7. Chemotherapeutic agents for neoplastic diseases			
Adriamycin	RIA		536
Bleomycin	RIA		537
Daunomycin	RIA		536
Etoposide	RIA		538
Methotrexate	RIA		539, 540
	FPIA	With HPLC	541
	HPLC-RIA		542
Taxanes & taxol	EIA		543, 544
Vinblastine	RIA		545
Vincristine	RIA	Microbiological	545, 546
8. Immunosuppressive drugs			
Cyclosporine	RIA, FPIA	With HPLC	175, 176
Cyclosporine	RIA	With HPLC	177, 178
	EIA	With HPLC, RIA	179
	RIA	Metabolites	547
	EIA		548
Tacrolimus	EIA	FK506, whole blood	549
9. Hormones and hormone antagonists			
Cortisol	FPIA	Free, in ultrafiltrate paper as solid phase	550
D-Norgestrel	EIA		551
Detirelix	RIA	D-Ala 10 LHRH	552, 553
Danazol	RIA		554
Dexamethasone	RIA		555
	RIA		556–559
	EIA		560
Dienogest	RIA		561, 562
Diethylstilbestrol	RIA		563

Table 4 Continued

Analyte	IA[a]	Method comparison and comment	References
Estradiol	EIA	With RIA	564
Estriol	FIA		565
Estrogens	RRP	With EIA	238
Ethinyl estradiol	RIA	Sephadex, LH-20 sample clean-up	566
	HLPC-RIA		567
	RIA		568, 576
Fluoxymesterone	RIA	With GCMS	569
Fluticasone-17-propionate	RIA		570
Human growth hormone	RIA	Solid-phase Extraction	571
4-Hydroxyandrostenedione	RIA		11, 572
25-Hydroxyvitamin D₃	EIA		573
Insulin	RIA	Immunoaffinity column	574
Mestranol	RIA		138, 575
Methylprednisolone	RIA		576
Mometasone furoate			577
Norethindrone	RIA	SCH-32088	578
	HPLC-RIA	Sephadex LH20 sample clean-up	566, 579
	RIA		567
Normegesterol acetate	EIA		580
Nortestosterone	EIA		581
Prednisolone	RIA		579
Prednisone	RIA		582, 583
Progesterone	EIA	Saliva	584
Progesterone	EIA	Paper as solid phase	585
Propylthiouracil	RIA		551
Somatomedin-C	RIA		586
			587

Somatostat in analogs	RIA		588
Testosterone	RIA		38, 181
	EIA	Plasma, salvia	589
Thyroliberin	RIA	With HPLC	590
Triamcinolone	RIA		591
TSH	EIA		19
10. Drugs that affect the blood and blood-forming organs			
Erythropoietin	RIA	Recombinant erythroproietin	592–599, 607
CSF-1 (colony-stimulating factor)	RIA		600
Tissue plasminogen activator	EIA		601
11. Antiulcers, antidiarrheals, and GI therapeutics			
BW942C	RIA		602
Carbenoxolone	RIA		603
General	EIA		604
12. Miscellaneous			
Antipyrine	RIA	Saliva	605
Desmosine	RIA		606
Glipizide (antidiabetic)	RIA		607
Glisoxepide	RIA		608
Glycyrrhetic acid	RIA		609
	EIA		610
4-Hydroxy-2-(4-methylphenyl) benzothiazole	EIA		611
ICI-200, 800	RIA	Elastase inhibitor	612
Idazoxan	RRP		613
Indacrinone	RIA	Stereospecific	614
Levomethadyl acetate	RIA	Stereospecific	615
MK-852	RIA	Fibrinogen receptor antagonist	616

Table 4 Continued

Analyte	IA[a]	Method comparison and comment	References
13. PD biochemical markers			
Adenosine	RIA		617
Angiotensin peptides	RIA		618
Angiotensin II	HPLC-RIA		619
cAMP	EIA		620
	RIA		621
CD4, CD8	EIA	T-lymphocytes	622
cGMP	EIA		620
Hirudin	RIA		623–625
Histamine	FIA	With RIA; glass fiber as solid–phase support	626
IGFBP-2	RIA		627
IGFBP-3	RIA		628
Interleukin-6	RIA		629, 630
Laminin	EIA		631
Leukotrienes	EIA		632
Melatonin	RIA		633
Osteocalcin	EIA		634
Protein kinase C isoenzymes	EIA		635
Renin	RIA		636
Thromboxane B_2	EIA	With immunoaffinity column GCMS	637
TNF_α	RIA		638
	EIA	Binding protein I	639
14. Drug antibodies			
Anti-growth hormone	RIA		640
Anti-insulin	EIA		641
Anti-streptokinase	EIA		642
General	EIA		643

a FIA = fluoroimmunoassay, RRR = radioreceptorassay, FPIA = fluorimetric polarization immunoassay to improve the labels that are available for detection. For example, biosynthetic enzymes that do not exist today may be the labels of tomorrow. Bi- or multifunctional enzymes with more ruggedness and specificity can be produced repeatedly with the same quality (644). Alternative detection methods such as electroluminescence, bioluminescence and phosphorescence, or detection methods that are only now being conceptualized, may be applied to provide increased sensitivity and specificity in the future.

B. Automation

Pharmaceutical and biotechnology companies have not yet taken full advantage of the proliferation in clinical chemistry automation technology. Flexible assay protocols can be developed to take advantage of existing technology so that a single system can be applied to a variety of drug assay methods. Similarly, manufacturers of automation equipment should consider the pharmaceutical and biotechnology industries' needs and collaborate with the bioanalytical laboratories for future development.

Computers should take the drudgery out of method optimization. Expert systems should be able to compile all previous IA experience, in much the same way as QSAR programs design drug molecules to fit receptors, and then apply it to evaluate a multitude of assay parameters in a factorial design. Potential applications of computer technology to the IA laboratory holds much promise. Remember that 10 to 15 years ago, most IAs depended on manual curve-fitting or number crunching in a centralized mainframe computer. There is no reason to think that similar advances will not be achieved during the next decade.

Detection methods are critical for future automation, since detection can be a rate-limiting step in an automated process. In addition to common detection methods used in manual IA techniques, other detection methods have been applied to automation. For example, ECL technology has been used for nucleic acids in the IGEN™ system. Cell surface markers are detected by the Copalis™ system using a fluorescent conjugate and flow cytometry. New detection methods for drug IAs as well as PD measurements for efficacy and safety should prove useful in the future.

C. Simultaneous Multianalyte IAs

In the realm of biochemical pharmacodynamic markers and endpoints, simultaneous assay methods have the same advantages as they do in the clinical laboratory. However, the analyst must remember that the objectives are different. The assay methods must satisfy the same GLP requirements as a drug assay. The assay results must have the specificity, precision, and accuracy as results that are being analyzed to fulfill BA/BE requirements.

D. Genetic Engineering

The greatest asset that IA has to offer is its inherent sensitivity and its potential selectivity. Recent developments have focused new labels and improving antibody binding through the use of genetic engineering to produce monoclonals. Since conjugation methods have lacked reproducibility, it has not been possible to ensure that the same antigen conjugates or labeling conjugates would result each time they were prepared. Genetic engineering efforts need to continue to focus on highly reproducible preparations of homogeneous monosubstituted peptide–enzyme conjugates that can be applied to IAs (645). By combining genetically engineered antibodies (646) and conjugates, reproducible sources for IA components should become more readily available over the next few years.

X. CONCLUSIONS

Immunoassays and related techniques have been successfully applied to a wide variety of pharmacological agents including drugs, biologicals and pharmacodynamic markers. The principal advantages of IA techniques remain sensitivity and high throughout. Principal disadvantages remain concerns about specificity, need for specially trained individuals, and a general lack of understanding about how these techniques work. The simplicity of IA methods, once an assay protocol is established, negate these concerns. Appropriate use of assay optimization techniques make IA methods sensitive, specific, precise, accurate, and ready for automation.

IA has changed dramatically since the original chapter was written between 1975 and 1978 and published in 1982 (647). Progress has been made, but there is still room for improvement. With the quantum leaps in computer and analytical technology that have taken place in the last ten years, it can only be anticipated that IA will assimilate these along with advances in recombinant technology to revolutionize bioanalytical chemistry. For now, IA is an integral part of the comprehensive pharmaceutical analysis laboratory.

REFERENCES

1. SA Berson, RS Yalow. Quantitative aspects of the reaction between insulin and insulin-binding antibody. J Clin Invest 38:1996, 1959.
2. BD Albertson, FP Haseltine. Non-radiometric Assays, Technology and Application in Polypeptide and Steriod Hormone Detection. New York: Alan R. Liss, 1988.
3. DS Skelley, LP Brown, PK Besch. Radioimmunoassay. Clin Chem 19:146, 1973.
4. GC Oliver, BM Parker, DL Brasfield, CW Parker. The measurement of digitoxin in human serum by radioimmunoassay. J Clin Invest 47:1035, 1968.
5. J Landon, AC Moffat. The radioimmunoassay of drugs. Analyst 101:225, 1976.

6. I Jonsdottir, HP Ekre, P Perlmann. Comparative study of pituitary and bacteria-derived human growth hormone by monoclonal antibodies. Mol Immunol 20:871, 1983.

7. R Ekins. In: R Ekins, A Vollers, A Bartlett, D Bidwell, eds. Merits and disadvantages of different labels and methods of immunoassay. *Immunoassays for the 80s*. Baltimore, MD: University Park Press, 1981, p 5.

8. R Ekins. In: A Albertini, R Ekins, eds. Toward Immunoassays of Greater Sensitivity, Specificity and Speed: An Overview. Monoclonal Antibodies and Developments in Immunoassays. Amsterdam: Elsevier, 1981, p 3.

9. L Wide, H Bennich, SGO Johansson. Diagnosis of allergy by an in-vitro test for allergen antibodies. Lancet 2:1105, 1967.

10. LEM Miles, CN Hales. Labelled antibodies and immunological assay systems. Nature 219:186, 1968.

11. GM Addison, CN Hales. Two site assay of human growth hormone. Horm Metab Res 3:59, 1971.

12. VP Shah, et al. Analytical methods validation—Bioavailability, bioequivalence, and pharmacokinetic studies. J Pharm Sci 81:309, 1992.

13. D Rodbard, JE Lewald. In: E Diczfalusy, A Diczfalusy, eds. Computer analysis of radioligand assay and radioimmunoassay data. Steroid Assay by Protein Binding. Karolinska Symposia on Research Methods in Reproductive Endocrinology, WHO/Karolinska Institut, Stockholm, 1970, p 79.

14. D Rodbard, PJ Munson, A De Lean. Improved curve-fitting, parallelism testing, characterisation of sensitivity and specificity, validation, and optimization for radioligand assays. In: Radioimmunoassay and Related Procedures in Medicine 1, Vienna: International Atomic Energy Agency, 1978, p 469.

15. PJ Goadsby, et al. An interactive, readily transportable program using a log-logit transformation for the analysis of radioimmunoassay data. Comput Meth Programs Biomed 23:263, 1986.

16. PJ Geiger. Radioimmunoassay data handling and calculations with a graphics statistics comptuer program. Biochem Med Metab Biol 48:74, 1992.

17. TM Jackson, RP Ekins. Theoretical limitations on immunoassay sensitivity. J Immunol Meth 87:13, 1986.

18. RP Ekins, S Dakubu. The development of highly sensitive pulsed light, time-resolved fluoroimmunoassays. Pure Appl Chem 57:473, 1985.

19. I Weeks, M Sturgess, K Siddle, et al. A high sensitivity immunochemiluminometric assay for human thyrotrophin. Clin Endocrinol 20:489, 1984.

20. LJ Kricka. Chemiluminescent and bioluminescent techniques. Clin Chem 37:1472, 1991.

21. GC Schussler. Diazepam competes for thyroxine binding sites. J Pharmacol Exp Ther 178:204, 1971.

22. SG Hillier, BG Brownsey, EH Cameron. Some observations on the determination of testosterone in human plasma by radioimmunoassay using antisera raised against testosterone-3-BSA and testosterone-11α-BSA. Steroids 21:735, 1973.

23. S Bauminger, HR Lindner, A Weinstein. Properties of antisera to progesterone and to 17-hydroxyprogesterone elicited by immunization with the steroids attached to protein through position 7. Steroids 21:947, 1973.

24. T Nambara, M Numazawa, Y Tsuchida, T Tanaka, Specificity of antiserum raised

against estradiol using 6-O-carboxymethyloxime-bovine serum albumin conjugate. Chem Pharm Bull 24:1510, 1976.

25. O Chappey, P Sandouk, JM Scherrmann. Modulation of the specificity of anti-hapten antibodies. J Clin Immunoassay 15:51, 1992.

26. AL Neese, LF Soyka. Mis au point d'une methode dosage radioimmunologique pour la theophylline; Application chez l'enfant premature. Clin Pharmacol Ther 5: 633, 1977.

27. VP Butler, JP Chen. Digoxin specific antibodies. Proc Natl Acad Sci (USA) 57: 71, 1977.

28. AJ Lomant, G Fairbanks. Chemical probes of extended biological structures: Synthesis and properties of the cleavable protein crosslinking reagent [^{35}S] dithiobis (succinimidyl propionate). J Mol Biol 104:243, 1976.

29. S Yoshitake, Y Yomada, E Ishikawa, R Masseyeff. Conjugation of glucose oxidase from *aspergillus niger* and rabbit antibodies using N-hydroxysuccinimide ester of N-(4-carboxycyclohexylmethyl)-maleimide. Eur J Biochem 101:395, 1979.

30. FRN Gurd. Carboxymethylation. Meth Enzymol 11:532, 1967.

31. AM Crestfield, S Moore, WH Stein. The preparation and enzymatic hydrolysis of reduced and S-carboxymethylated proteins. J Biol Chem 238:622, 1963.

32. Z Grabarek, J Gergely. Zero-length crosslinking procedure with the use of active esters. Anal Biochem 185:131, 1990.

33. MA Gilles, AQ Hudson, CL Borders Jr. Stability of water-soluble carbodiimides in aqueous solution. Anal Biochem 184:244, 1990.

34. JV Staros, RW Wright, DM Swingle. Enhancement by N-hydroxysulfosuccinimide of water-soluble carbodiimide-mediated coupling reactions. Anal Biochem 156: 220, 1986.

35. DG Smyth, OO Blumenfeld, W Konigsberg. Reaction of N-ethylmaleimide with peptides and amino acids. Biochem J 91:589, 1964.

36. M De Villiers, D Parkin, P Van Jaarsveld, B Vander Walt. A radioimmunoassay for metoclopramide. J Immunol Meth 103:33, 1987.

37. S Spector, B Berkowitz, EJ Flynn, B Peskar. Antibodies to morphine, barbiturates, and serotonin. Pharmacol Rev 25:281, 1973.

38. JPP Tyler, JF Hennam, JR Newton, WP Collins. Radioimmunoassay of plasma testosterone without chromatography: A comparison of four antisera, and the evaluation of a novel approach to liquid scintillation counting. Steroids 22:871, 1973.

39. A Muckerheide, RJ Apple, AJ Pesce, JG Michael. Cationization of protein antigens. J Immunol 138:833, 1987.

40. S Schwarz, P Berger, G Wick. Epitope-selective, monoclonal-antibody-based immunoradiometric assays of predictable specificity for different measurement of choriogonadotropin and its subunits. Clin Chem 31:1322, 1985.

41. BF Erlanger, R Borek, SM Beiser, S Lieberman. Steroid protein conjugate, I. Preparation and characterization of conjugates of bovine serum albumin with testosterone and with cortisone. J Biol Chem 228:713, 1957.

42. BF Erlanger, R Borek, SM Beiser, S Lieberman. Steroid protein conjugates, II. Preparation and characterization of conjugates of bovine serum albumin with progesterone, deoxycorticosterone, and estrone. J Biol Chem 234:1090, 1959.

43. BF Erlanger. Principles and methods for the preparation of drug protein conjugates for immunological studies. Pharmacol Rev 25:271, 1973.
44. GE Abraham, PK Grover. Covalent linkage of hormonal haptens to protein carriers for use in radioimmunoassay. In: WD Odell, WH Daughaday, eds. Principle of Competitive Protein-Binding Assays. Philadelphia: Lippincott, 1974, p 134.
45. UF Axen. N,N'-carbonyldiimidzole as coupling reagent for the preparation of bovine serum albumin conjugates. Prostaglandins 5:45, 1974.
46. JR Vaughan Jr, RL Osato. Preparation of peptides using mixed carbonic-carboxylic acid anhydrides. J Am Chem Soc 74:676, 1952.
47. JF Vaughan Jr. Acylalkylcarbonates as acylating agents for the synthesis of peptides. J Am Chem Soc 73:3547, 1951.
48. NS Ranadive, AH Sehon. Antibodies to serotonin. Can J Biochem. 45:1701, 1967.
49. S Aurameas, T Termynck. The cross-linking of proteins with gluteraldehyde and its use for the preparation of immunoabsorbents. Immunochemistry 6:53, 1969.
50. DH Catlin, JC Schaeffer, MB Liewen. 2-Diazomorphine directed antiserum: Determination of morphine in brain after naloxone challenge in morphine pellet implanted mice. Life Sci 20:123, 1977.
51. B Rubin. Studies on the induction of antibody synthesis against sulfanilic acid in rabbits, I. Effect of the number of hapten molecules introduced in homogeneous protein on antibody synthesis against the hapten and the new antigenic determinants. Eur J Immunol 2:5, 1972.
52. K Wicher, M Schwartz, CE Arbesman, F Milgrom. Immunologic studies of aspirin, I. Antibodies to aspiryl-protein conjugates. J Immunol 101:342, 1968.
53. N Hanna, E Jarosch, S Leskowitz. Altered immunogenicity produced by change in mode of linkage of hapten to carrier. Proc Soc Exp Biol Med 140:89, 1972.
54. K Hoffman, P Samarazeewa, EK Smith, AE Kellie. Radioimmunoassay of steroids, the role of the "bridge" linking the steroid hapten to the protein carrier. J Steroid Biochem 6:91, 1975.
55. CS Walker, SJ Clark, HH Wotiz. Factors involved in the production of specific antibodies to estriol and estradiol. Steroids 21:259, 1973.
56. GD Niswender. Influence of the site of conjugation on the specificity of antibodies to progesterone. Steroids 22:413, 1973.
57. M Koida, M Takahashi, S Muraoka, H Kaneto. Antibodies to BSA conjugates of morphine derivatives: Strict dependency of the immunological specificity on the hapten structure. Jpn J Pharmacol 24:165, 1974.
58. S Spector. Radioimmunoassay of drugs. In Radioimmunoassay and Related Procedures in Medicine, Vol. II. International Atomic Energy Association, 1974, p 233.
59. S Rothchild. Tritium labeling by other methods. In S Rothchild, ed. Advances in Tracer Methodology, Vol. 1. New York: Plenum Press, 1963, p 50.
60. JG Spenney, BJ Johnson, BI Hirschowitz, AA Mihas, R Gibson. An ^{125}I radioimmunoassay for primary conjugated bile salts. Gastroenterology 72:305, 1977.
61. C Gomez-Sanchez, L Milewich, OB Holland. Radioiodinated derivatives for steroid radioimmunossay. Application to the radioimmunoassay of cortisol. J Lab Clin Med 89:902, 1977.
62. AE Bolton, WM Hunter. The labelling of proteins to high specific radioactivities by conjugation to a ^{125}I-containing acylating agent. Biochem J 133:529, 1973.

63. WJ Rzeszotarski, C Paik, WC Eckelman, RC Reba. The synthesis of chelating agents for the preparation of drug derivatives. J Labeled Comp Radiopharm 13: 171, 1977.

64. JR Little, TJ Blanke, KD Little. A new method of radioactive labelling of polyanionic macromolecules for radioimmunoassay. Eur J Immunol 5:373, 1975.

65. FC Greenwood, WM Hunter, GS Glover. The preparation of.[131]I labeled human growth hormone of high specific radioactivity. Biochem J 89:114, 1963.

66. E Samols, HS Wilteams. Trace-labelling of insulin with iodine. Nature 190:1211, 1961.

67. WR Butt. The iodination of follicle-stimulating and other hormones for radioimmunoassay. J Endocrinol 55:453, 1972.

68. JI Thorell, BG Johansson. Enzymatic iodination of polypeptides with ^{125}I to high specific activity. Biochim Biophys Acta 251:363, 1971.

69. EHD Cameron, JJ Scarisbrick, SE Morris, SG Hillier, G Read. Some aspects of the use of ^{125}I-labeled ligands for steroid radioimmunoassay. J Steroid Biochem 5: 749, 1974.

70. WM Hunter. Preparation and assessment of radioactive tracers. Br Med Bull 30: 18, 1974.

71. JP Gosling. A decade of development in immunoassay methodology. Clin Chem 36:1408, 1990.

72. TM Jackson, NJ Marshall, RP Ekins. Optimization of immunoradiometric assays. In: WM Hunter, JET Corrie, eds. Immunoassays for Clinical Chemistry. Edinburgh: Churchill Livingstone, 1983, p 557.

73. T Porstmann, ST Kiessig. Enzyme immunoassay techniques, an overview. J Immunol Meth 150:5, 1992.

74. K Helgert, et al. Enzyme immunoassay—Principle and application. Pharmazie 44: 745, 1989.

75. E Ishikawa. Development and clinical application of sensitive enzyme immunoassay for macromelecular antigens. A review. Clin Biochem 20:375, 1987.

76. E Soini, J Hemmila. Fluoroimmunoassay: Present status and key problems. Clin Chem 25:353, 1979.

77. I Wieder. Background rejection in fluorescence immunoassay. In: Proc 6th Int Conf Immunofluorescence and Related Staining Techniques. Vienna: Elsevier, 1978, p 67.

78. WB Dandliker, GA Feigen. Quantification of the antigen-antibody reaction by the polarization of fluorescence. Biochem Biophys Res Commun 5:299, 1961.

79. IM Roberts, et al. A comparison of the sensitivity and specificity of enzyme immunoassays and time-resolved fluoroimmunoassay. J Immunol Meth 143:49, 1991.

80. I Hemmila, S Dakubu, VM Mukkala, H Siitari, T Lovgren. Europium as a label in time-resolved immunofluorometric assays. Anal Biochem 137:335, 1984.

81. T Lovegren, I Hemmila, K Pettersson, P Halonen. Time-resolved fluorometry in immunoassay. In: WP Collins, ed. Alternative Immunoassays. Chichester, UK: Wiley, 1985, p 203.

82. N Filipescu, WF Sager, FA Serafin. Substituent effects on intramolecular energy transfer. II. Fluorescence spectra of europium and terbium beta-diketone chelates. J Phys Chem 68:3324, 1964.

83. M DeLuca, W McElroy. Bioluminescence and Chemiluminescence. New York: Academic Press, 1981.

84. LJ Kricka. Selected strategies for improving sensitivity and reliability of immunoassays. Clin Chem 40:347, 1994.

85. F Kohen, M Pazzagli, M Serio, J deBoever, D Vanderkerckhove. Chemiluminescence and bioluminescence immunoassay. In: WP Collins, ed. Alternative Immunoassays. New York: Wiley, 1985, p 103.

86. GHG Thorpe, LJ Kricka. Enhanced chemiluminescent reactions catalysed by horseradish peroxidase. Meth Enzymol 133:331, 1986.

87. AP Schaap, H Akhavan, LJ Romano. Chemiluminescent substrates for alkaline phosphatase application to ultrasensitive enzyme-linked immunoassays and DNA probes. Clin Chem 35:1863, 1989.

88. LJ Kricka. Ligand Binder Assays. New York: Marcel Dekker, 1985, p 199.

89. TJN Carter, CJ Groucutt, RAW Stott, GHG Thorpe. Enhanced luminescent and luminometric assay. European Patent 0087959, 1985.

90. LJ Kricka, M O'Toole, GHG Thorpe, TP Whitehead. Enhanced luminescent for luminometric assay. US Patent 4729950, 1988.

91. E Ishikawa, K Kato. Ultrasensitive enzyme immunoassay. Scand J Immunol 8:43, 1978.

92. J Wannlund, M Deluce. A sensitive bioluminescent immunoassay for dinitrophenol and trinitro toluene. Anal Biochem 22:385, 1982.

93. CJ Stanley, RB Cox, MF Cardosi, et al. Amperometric enzyme amplified immunoassay. J Immunol Meth 112:153, 1988.

94. PMS Clark, CP Price. Enzyme amplified immunoassays: A new sensitive assay of thyrotropin evaluated. Clin Chem 32:88, 1986.

95. S Avrameas. Amplification systems in immunoenzymatic techniques. J Immunol Meth 150:23, 1992.

96. G Gitlin, EA Bayer, M Wilchek. Studies of the biotin-binding site of avidin. Biochem J 242:923, 1987.

97. I Chaiet, FJ Wolf. The properties of streptavidin, a biotin-binding protein produced by Streptomycetes. Arch Biochem Biophys 106:1, 1964.

98. RC Duhamel, JS Whitehead. Prevention of nonspecific binding of avidin. Meth Enzymol 184:201, 1990.

99. A Johannsson, DH Ellis, DL Bates. Enzyme amplification for IAs detection limit of 1/100 of an attomole. J Immunol Meth 87:7, 1986.

100. CH Self. Enzyme amplification—A general method applied to provide an immunoassisted assay for placental alkaline phosphatase. J Immunol Meth 76:389, 1985.

101. H Arakawa, M Maeda, A Tsuji. Chemiluminescence enzyme immunoassay for thyroxin with the use of glucose oxidase and a bis(2,4,6-trichlorophenyl) oxalate-fluorescent dye system. Clin Chem 31:430, 1985.

102. L Chedid, E Lederer. Past, present and future of the synthetic immunoadjuvant MDP and its analogs. Biochem Pharmacol 27:2183, 1978.

103. GW Boyd, WS Peart. The production of high-titre antibody against free angiotensin II. Lancet 2 560:129, 1968.

104. J Vaitukaitis, JB Robbins, E Nieschlag, GT Ross. A method for producing specific antisera with small doses of immunogen. J Clin Endocrinol 33:988, 1971.

105. HN Eisen. Equilibrium dialysis for measurement of antibody-hapten affinities. In: HN Eisen, ed. Methods in Medical Research, Vol 10. Chicago: Yearbook Medical Publishing, 1964, p 106.

106. PM Keane, WHC Walker, J Gauldie, GE Abraham. Thermodynamic aspects of some radioassays. Clin Chem 22:70, 1976.

107. DL DiPietro, RD Brown, CA Strott. A pregnenolone radioimmunoassay utilizing a new fractionation technique for sheep antiserum. J Clin Endocrinol Metab 35: 729, 1972.

108. JB Robbins, J Haimovich, M Sela. Purification of antibodies with immunoadsorbents prepared using bromoacetyl cellulose. Immunochemistry 4:11, 1967.

109. RJ Hill. Elution of antidbodies from immunoadsorbents: Effect of dioxane in promoting release of antibody. J Immunol Meth 1:231, 1972.

110. LG Hoffmann, C-Y Kuo. Immunoadsorbents for the isolation of high-affinity antihapten antibodies in high yield. J Immunol Meth 15:101, 1977.

111. TM Murray, HT Keutmann. The immunochemical specificity of antisera to bovine parathyroid hormone: An approach to region-specific radioimmunoassay. J Endocrinol 56:493, 1973.

112. G Galfre, C Milstein. Preparation of monoclonal antibodies: Strategies and procedures. In: S Colowick, N Kaplan, eds. Methods in Enzymology 73. New York: Academic Press, 1981, p 1.

113. JW Littlefield. Selection of hybrids from matings of fibroblasts in vitro and their presumed recombinants. Science 145:709, 1964.

114. CA Caulcott, R Boraston, C Hill, et al. Production and purification of monoclonal antibodies. In: WP Collins, ed. Complementary Immunoassays. Chichester, UK: Wiley, 1988, p 27.

115. G Kohler, C Milstein. Continuous culture of fused cells secreting specific antibody of predefined specificity. Nature 256:495, 1975.

116. RR Hardy. Purification and characterization of monoclonal antibodies. In: DM Weir, LA Herzenberg, C Blackwell, LA Herzenberg, eds. Handbook of Experimental Immunology, Immunochemistry, 1. Oxford: Blackwell Scientific, 1986, p 13.1.

117. JR Deschamps, EK Hidreth, D Derr, JT August. A high-performance liquid chromatography procedure for the purification of mouse monoclonal antibodies. Anal Biochem 147:451, 1985.

118. LM Boscato, GM Egan, MC Stuart. Specificity of two-site immunoassays. J Immunol Meth 117:221, 1989.

119. LM Boscato, MC Stuart. Incidence and specificity of interference in two-site immunoassays. Clin Chem 32:1491, 1986.

120. JC Cresto, RS Yalow. Anomalously ascending standard curves in the radioimmunoassay of insulin, human growth hormone and ACTH. Horm Metab Res 5:1, 1974.

121. TH Weber, KI Kapyaho, P Tanner. Endogenous interference in immunoassays in clinical chemistry. A review. Scand J Clin Lab Invest 50(suppl 201): 77, 1990.

122. DW Rea, ME Ultee. A novel method for controlling the pepsin digestion of antibodies. J Immunol Meth 157:165, 1993.

123. SJ Boguslawski, et al. Improved procedure for preparation of F (ab')$_2$ fragments of mouse IgGs by papain digestion. J Immunol Meth 120:51, 1989.

124. A Coulter, R Harris. Simplified preparation of rabbit fab fragments. J Immunol Meth 59:199, 1983.
125. JW Goding. Use of staphylococcal protein A as an immunological reagent. J Immunol Meth 20:241, 1978.
126. J Rousseaux, et al. Optimal conditions for the preparation of Fab and F (ab') fragments from monoclonal IgG of different rat IgG subclasses. J Immunol Meth 64: 141, 1983.
127. JL Guesdon, F Naquira Velarde, S Avrameas. Solid phase immunoassays using chimera antibodies prepared with monoclonal or polyclonal anti-enzyme and anti-erythrocyte antibodies. Ann Immunol 1396:265, 1983.
128. B Porstmann, S Avrameas, T Ternynck, T Porstmann, B Micheel, JL Guesdon. An antibody chimera technique applied to enzyme immunoassay for human alpha-1-feto-protein with monoclonal and polyclonal antibodies. J Immunol Meth 66:179, 1984.
129. C Milstein, AC Cuello. Hybrid hybridoma production of bi-specific monoclonal antibodies. Immunol Today 5:299, 1984.
130. G Spira, A Bargellesi, JL Teillaud, MD Scharff. The identification of monoclonal class switch variants by Sib selection and an ELISA assay. J Immunol Meth 74: 307, 1984.
131. MR Suresh, AC Cuello, C Milstein. Bi-specific monoclonal antibodies from hybridomas. Meth Enzymol 121:210, 1986.
132. LK Sun, P Curtis, E Rekowicz-Szulcyznska, et al. Chimeric antibody with human constant regions and mouse variable region directed against carcinoma-associated antigen 17-1A. Proc Natl Acad Sci (USA) 84:214, 1987.
133. L Riechmann, M Clark, H Waldmann, G Winter. Reshaping human antibodies for therapy. Nature 332:323, 1988.
134. RS Yalow, SA Berson. Special problems in the radioimmunoassay of small poly-peptides. In: M Margoulies, ed. Protein and Poly-peptide Hormones. Excerpta Medica Foundation, Amsterdam, 1969, p 71.
135. G Scatchard. The attractions of proteins for small molecules and ions. Ann NY Acad Sci 51:660, 1949.
136. MW Steward. Introduction to methods used to study antibody-antigen reactions. In: DM Weir, ed. Handbook of Immunological Methods. 3rd ed. Oxford: Blackwell Scientific, 1981, p 16.1.
137. CN Hales, PJ Pandle. Immunoassay of insulin-antibody precipitate. Biochem J 88: 137, 1963.
138. CA Velasco, HS Cole, RA Camerini-Davalos. Radioimmunoassay of insulin, with use of an immunosorbent. Clin Chem 20:700, 1974.
139. K Catt, GW Tregear. Solid-phase radioimmunoassay in antibody-coated tubes. Science 158:1570, 1967.
140. BA Baldo, ER Tovey. Protein Blotting: Methodology, Research and Diagnostic Applications. Basel: S Karger, 1989.
141. LA Blankstein, L Dohrman. An advanced affinity membrane for immunodiagnostic tests, Am Clin Products Rev 33, 1985.
142. AJ Pesce, DJ Ford, M Gaizutis, VW Pollak. Binding of protein to polystyrene in solid-phase immunoassay. Biochem Biophys Acta 492:399, 1977.
143. LA Cantarero, JE Butler, JW Osborne. The adsorptive characteristics of proteins

for polystyrene and their significance in solid-phase immunoassays. Anal Biochem 105:375, 1980.

144. JL Brash, DJ Lyman. Adsorption of plasma proteins in solution to uncharge surfaces. J Biomed Mater Res 3:175, 1969.

145. JH Peterman, PJ Tarcha, VP Chu, JE Butler. The immunochemistry of sandwich ELISAs. IV. A comparison of the antigen capture capacity of antibodies adsorbed or covalently attached to polystyrene. J Immunol Meth 111:271, 1988.

146. E Engvall, K Jonsson, P Perlmann. Enzyme-linked immunosorbent assay. II. Quantitative assay of protein antigen, immunoglobulin G, by means of enzyme labelled antigen and antibody coated tubes. Biochem Biophys Acta 251:427, 1971.

147. EE Howell, J Nasser, KJ Schray. Coated tube immunoassay: Factors affecting sensitivity and effects of reversible protein binding to polystyrene. J Immunoassay 2: 205, 1981.

148. HB Bull. Adsorption of bovine serum albumin on glass. Biochem Biophys Acta 19:464, 1956.

149. E Ishikawa, Y Hamaguchi, M Imagawa, et al. An improved preparation of antibody-coated polystyrene beads for sandwich enzyme immunoassays. J Immunoassay 1: 385, 1980.

150. M Kemeny, SJ Challacombe. Microtiter plates and other solid-phase supports, In: DM Kemeny, SJ Challacombe, eds. ELISA and Other Solid-Phase Immunoassay. London: Wiley, 1987, p 31.

151. RP Pratt, P Roser. Comparisons of blocking reagents for ELISA. Nunc Bulletin 181:1, 1984.

152. PJ Tarcha. The chemical properties of solid-phases and their interaction with proteins. In: LE Butler, ed. Immunochemistry of Solid-Phase Immunoassay. Boca Raton, FL: CRC Press, 1991, p 27.

153. SJ Kennel. Binding of monoclonal antibody to protein in fluid phase and bound to solid supports. J Immunol Meth 55:1, 1982.

154. ME Soderquist, AG Walton. Structural changes in proteins absorbed on polymer surfaces. J Colloid Interface Sci 75:386, 1980.

155. SE Dierks, JE Butler, HB Richerson. Altered recognition of surface antigen-bound antibodies in the ELISA. Mol Immunol 23:403, 1986.

156. EW Voss Jr, RW Watt. Steric orientation of hapten groups and the effects. Immunochemistry 14:741, 1977.

157. KS Joshi, JE Butler. The immunochemistry of sandwich ELISAs. V. The capture antibody performance of polyclonal antibody-enriched fractions prepared by various methods. Mol Immunol 29:971, 1992.

158. C Creminon, O Dery, Y Frobert, J-Y Couraud, P Pradelles, J Grassi. Two-site immunometric assay for substance P with increased sensitivity and specificity. Anal Chem 67:1617, 1995..

159. LX Tiefenauer, RY Andres. Biotinyl-estradiol derivatives in enzyme immunoassays: Structural requirements for optimal antibody binding. J Steroid Biochem 35: 633, 1990.

160. WM Hunter. Optimization of RIA: Some simple guidelines, In: WM Hunter JET Corrie, eds. Immunoassays for Clinical Chemistry. London: Churchill Livingstone, 1983, p 69.

161. HR Lindner, E Perel, A Friedlander, A Zeitlen. Specificity of antibodies to ovarian hormones in relation to the site of attachment of the steroid hapten to the peptide carrier. Steroids 19:357, 1972.

162. A Zettner, PE Duly. Principles of competitive binding assays (saturation analysis) II. Sequential saturation. Clin Chem 20:5, 1974.

163. R Malvano, E Rolleri, U Rosa. Standardization and control of steroid radioimmunoassays. In: Radioimmunoassay and Related Procedures in Medicine, Vol II. Vienna: International Atomic Energy Agency, 1974, p 112.

164. JJ Pratt, MG Woldring. Radioimmunoassay specificity and the "first-come, first-served effect." Clin Chim Acta 68:87, 1976.

165. A Jacklitsch. Separation-free enzyme immunoassay for haptens. In: TT Ngo and HM Lenhoff, eds. Enzyme-mediated Immunoassay. New York: Plenum Press, 1985, p 33.

166. WB Dandliker, RJ Kelly, J Dandliker, J Farquhar, J Levin. Flourescence polarization immunoassay theory and experimental method. Immunochemistry 10:219, 1973.

167. EF Ullman, PL Khanna. Fluorescence excitation transfer immunoassay. Meth Enzymol 74:28, 1981.

168. DJ Litman, TM Hanlon, EF Ullman. Enzyme channeling immunoassay. A new homogeneous enzyme immunoassay technique. Anal Biochem 106:223, 1980.

169. I Gibbons, R Armenta, RK DiNello, et al. Nonseparation enzyme channeling immunometric assays. Meth Enzymol 136:93, 1987.

170. RF Zuk, VK Ginsberg, T Houts, et al. Enzyme immunochromatography—A quantitative immunoassay requiring no instrumentation. Clin Chem 31:1144, 1985.

171. DR Henderson, SB Friedman, JB Harris, et al. "CEDIA," a new homogeneous immunoassay system. Clin Chem 32:1637, 1986.

172. PL Khanna, TE Worthy. A recombinant protein-based homogeneous immunoassay. Am Clin Lab, Oct 14, 1989.

173. PL Khanna, RT Dworschack, WB Manning, et al. A new homogeneous enzyme immunoassay using recombinant enzyme fragments. Clin Chim Acta 185:231, 1989.

174. M Evans, DA Palmer, JN Miller, MT French. Flow injection fluorescence immunoassay for serum phenytoin using perfusion chromatography. Anal Proc 31:7, 1994.

175. KT Kivisto, et al. Therapeutic cyclosporine monitoring. Comparison of radioimmunoassay and high performance liquid chromatography methods in organ transplant recipients. Ther Drug Monit 12:353, 1990.

176. A Lindholm, et al. Comparative analyses of cyclosporine in whole blood and plasma by radioimmunoassay, fluorescence polarization immunoassay, and high performance of liquid chromatography. Ther Drug Monit 12:344, 1990.

177. DW Holt, et al. HPLC and Cyclo-Trac SP radioimmunoassay compared for monitoring cyclosporine. Clin Chem 38:442, 1992.

178. RG Buice et al. Analytical methodologies for cyclosporine pharmacokinetics—A comparison of radioimmunoassay with high performance liquid chromatography. J Liq Chromatogr 10:421, 1987.

179. G Schumann, D Petersen, PF Hoyer, K Wonigeit. Monitoring cyclosporine A (Ciclosporin, INN) concentrations in whole blood: Evaluation of the EMIT™ assay

in comparison with HPLC and RIA. Eur J Clin Chem Clin Biochem 31:381, 1993.

180. A Broughton, JE Strong, GP Bodey. Radioimmunoassay of sisomicin. Antimicrob Agents Chemother 9:247, 1976.

181. WA Colburn, AR DiSanto, SS Stubbs, RE Monovich, KA DeSante. Pharmacokinetic interpretation of plasma cortisol and cortisone concentrations following a single oral administration of cortisone acetate to human subjects. J Clin Pharmacol 20:428, 1980.

182. DEH Llewelyn, SG Hillier, GF Read. The use of multivariable standard curves in the radioimmunoassay of testosterone and 5α-dihydrotestosterone. Steroids 28:339, 1976.

183. JW Lee, JD Hulse, WA Colburn. Surrogate biochemical markers: Precise measurement for strategic drug and biologics development. J Clin Pharmacol 35:464, 1995.

184. P Carter. Preparation of ligand-free human serum for radioimmunoassay by adsorption on activated charcoal. Clin Chem 24:362, 1978.

185. L Peng, GJ Calton, J Burnett. Stability of antibody attachment in immunosorbent chromatography. Enzyme Microb Technol 8:681, 1986.

186. RH Ng. Product Information, Laboratorian Desk Reference. 2nd ed. Wayne, MI: Clinical Ligand Assay Society, 1993.

187. NP Kubasik, NS Norkus, HE Sim. Comparison of commercial kits for radioimmunoassay: II. The radioimmunoassay of serum digoxin using iodinated tracer. Clin Biochem 7:307, 1974.

188. AA MacKinney Jr, GH Burnett, RL Conklin, GW Wasson. Comparison of five radioimmunoassays and enzyme bioassay for measurement of digoxin in blood. Clin Chem 21:857, 1975.

189. H Muller, EH Graul, H Brauer. Different results produced by five radioimmunoassays for determination of digitalis in serum. Eur J Clin Pharmacol 10:227, 1976.

190. JW Lee, JE Pedersen, TL Moravetz, et al. Sensitive and specific radioimmunoassays for opiates using commercially available materials. I: Methods for the determinations of morphine and hydromorphone. J Pharm Sci 80:284, 1991.

191. RV Weimer. MRS Catalog Primary Antibodies. 3rd ed. Brimingham, MI: MRS, 1995.

192. W Linscott's Directory. 7th ed. Santa Rosa, CA, 1993.

193. D Rodbard, W Bridson, PL Rayford. Rapid calculation of radioimmunoassay results. J Lab Clin Med 74:770, 1969.

194. D Rodbard, JE Lewald. Computer analysis of radioligand assay and radioimmunoassay data. Acta Endocrinol suppl 64(147):79, 1970.

195. D Rodbard, DM Hutt. Statistical analysis of radioimmunoassays and immunoradiometric (labelled antibody) assays: A generalized weighted, iterative, least-squares method for logistic curve fitting. In: Radioimmunoassay and Related Procedures in Medicine, Vol I. Vienna: International Atomic Energy Agency, 1974, p 165.

196. WG Duddleson, AR Midgley Jr, GD Nswinder. Computer program sequence for analysis and summary of radioimmunoassay data. Comput Biomed Res 5:205, 1975.

197. KF Hatch, E Coles, H Busey, SC Goldman. End-point parameter adjustment on a

small desk-top programmable calculator for logit-log analysis of radioimmunoassay data. Clin Chem 22:1383, 1976.

198. HE Grotjan Jr, E Steinberger. Radioimmunoassay and bioassay data processing using a logistic curve fitting routine adapted to a desk top computer. Comput Biol Med 7:159, 1977.

199. DS Riggs. The Mathematical Approach to Physiological Problems. Cambridge, MA: MIT Press, 1963, p 58.

200. D Rodbard, RH Lenox, HL Wray, D Ramseth. Statistical characterization of the random errors in the radioimmunoassay dose-response variable. Clin Chem 22:350, 1976.

201. D Rodbard, SW McClean. Automated computer analysis for enzyme-multiplied immunological techniques. Clin Chem 23:112, 1977.

202. AA Fernandez, HG Loeb. Practical applications of radioimmunoassay theory. A simple procedure yielding linear calibration curves. Clin Chem 21:1113, 1975.

203. R Ekins, G Newman, J Riordan. Theoretical aspects of 'saturation' and radioimmunoassay. In: R Hayes, F Goswitz, B Murphy, eds. Radioisotopes in medicine: In Vitro Studies. US Atomic Energy Commission, Oak Ridge, Tennessee, 1968, p 61.

204. AV Tembo, MA Schork, JG Wagner. Statistical survey of "saturation analysis" calibration curve data for prednisolone, prednisone and digoxin. Steroids 28:387, 1977.

205. D Rodbard, Y Feldman. Kinetics of two-site immunoradiometric ("sandwich") assays. I. Mathematical models for simulation, optimization, and curve fitting. Immunochemistry 15:71, 1978.

206. M Kraupp, et al. Evaluation of radioimmunoassays—Comparison of dose interpolation calculations by 4 parameter logistic and spline functions. J Clin Chem 24: 1023, 1986.

207. SL Ng et al. Cubic regression analysis for radioimmunoassay data processing. Comput Meth Prog Biomed. 34:273, 1991.

208. M Gripenberg, G Gripenberg. Expression of antibody activity measured by ELISA. Anti-ssDNA antibody activity characterized by the shape of the dose-response curve. J Immunol Meth 62:315, 1983.

209. V Guardabasso, D Rodbard, PJ Munson. A model-free approach to estimation of relative potency in dose-response curve analysis. Am J Physiol 252:E357, 1987.

210. JE Butler, JH Peterman, TE Koertge. The amplified enzyme-linked immunosorbent assay (a-ELISA). In: TT Ngo, HM Lenhoff, eds. Enzyme-Mediated Immunoassay. New York: Plenum Press, 1985, p 241.

211. TE Koertge, JE Butler. The relationship between the binding of primary antibody to solid-phase antigen in microtiter plates and its detection by the ELISA. J Immunol. Meth 83:283, 1985.

212. D Rodbard, Y Feldman, ML Jaffe, LEM Miles. Kinetics of two-site immunoradiometric ("sandwich") assays. II. Studies on the nature of the "high-dose" hook effect. Immunochemistry 15:77, 1978.

213. KE Rubenstein, RS Schneider, EF Ullman. "Homogeneous" enzyme immunoassay. A new immunochemical technique. Biochem Biophys Res Commun 47:846, 1972.

214. GL Rowley, KE Rubenstein, J Huisjen, EF Ullman. Mechanism by which antibodies inhibit hapten-malate dehydrogenase conjugates. J Biol Chem 250:3759, 1975.

215. ME Jolley, SD Stroupe, KS Schwenzer. Fluorescence polarization immunoassay. Monitoring aminoglycoside antibiotics in serum and plasma. Clin Chem 27:1190, 1981.

216. RH Ng. Immunoassay automation. J Clin Immunoassay 14:59, 1991.

217. RH Ng. Automated immunoassay instruments: The user's perspective. J Clin Immunoassay 15:194, 1992.

218. S Krais, H Lenz, A Rotter, M Simmeth, N Franken. Electro-chemiluminescent 3rd generation TSH assay using the random-access analyzer ELECSYS. Clin Chem 41:S52, 1995.

219. P Bailk, R Vogel, D Mayr, S Richter, N Franken. Electro-chemiluminescent immunoassay for troponin T using the random-access analyzer ELECSYS. Clin Chem 41:S60, 1995.

220. PL Khanna, RT Dworschack, WB Manning, JD Harris. A new homogeneous enzyme immunoassay using recombinant enzyme fragments. Clin Chem 185:231, 1989.

221. BEP Murphy, A Barta. One-tube radiotransinassay for determination of cortisol at ambient temperature. Clin Chem 33:1137, 1987.

222. KI Shestak, et al. Analysis of new opioid-like peptides by the radioreceptor assay. Ukr Biokhim Zh 62:23, 1990.

223. ML Belyaeva, et al. Radioreceptor assay of benzodiazepines and its use in clinical practice. Zh. Nevropatol Psikhiatr. 90:147, 1990.

224. K Ensing, et al. Application of a radioreceptor assay in a pharmacokinetic study of oxitropium bromide in healthy volunteers after single IV, oral, and inhalation doses. Eur J Clin Pharmacol 37:507, 1989.

225. B Heibel, et al. Bioequivalence of 2 atenolol formulations in healthy volunteers— Evaluation by ergometric exercise, gas liquid chromatography, and radioreceptor assay. Eur J Pharmacol 183:2394, 1990.

226. G Sitzler, et al. Bioequivalence of 2 atenolol formulations in healthy volunteers— Evaluation and prediction of effect kinetics at beta-adrenoceptors in vivo by means of a radioreceptor assay. Meth Find Exp Clin Pharmacol 13:129, 1991.

227. SJ Capper, et al. Specificities compared for a radioreceptor assay and a radioimmunoassay of atrial natriuretic peptide. Clin Chem 36:656, 1990.

228. A Ishii, et al. Sensitive radioreceptor assay of the calcium antagonist benidipine hydrochloride in plasma and urine. Arzneim-Forsch. 38:1733, 1988.

229. HO Borbe, et al. Radioreceptor assay for the determination of alpha-1-adrenoceptor binding material in rat plasma following single oral administration of naftopidil. Arzneim-Forsch 40:253, 1990.

230. C Gerbeau, et al. A radioreceptor assay for determination of nicardipine in human serum using [^{3}H]-PN 200-110 as radioligand. J Immunoassay 11:271, 1990.

231. JF Lebigot et al. Development of a test radioreceptor for nicardipine and isradipine a pharmacokinetic application. Arch Mal Coeur Vaiss 80:716, 1987.

232. K Nguyen, et al. A novel radioreceptor assay for measuring nifedipine levels in biological fluids. Clin Res 35:A379, 1987.

233. G Caselli, et al. Determination of nuvenzepine in human plasma by a sensitive ^{3}H-pirenzepine radioreceptor binding assay. J Pharm Sci 80:173, 1991.

234. J Krska, et al. Determination of chlorpromazine in serum by radioreceptor assay and HPLC. Ann Clin Biochem, 23:340, 1986.

235. ME Alburges, et al. Evaluation of a radioreceptor assay to measure fentanyl and fentanyl-like drugs using gas chromatography/mass spectrometry and radioimmunoassay techniques. FASEB J 2:1807, 1988.

236. DE Rollins, et al. A radioreceptor assay to detect fentanyl-like designer drugs. Clin Pharm 45:186, 1989.

237. ME Alburges, et al. Fentanyl receptor assay—Development of a radioreceptor assay for analysis of fentanyl and fentanyl analogs in urine. J Anal Toxicol 15: 311, 1991.

238. TA Aasmundstad, et al. Estrogen receptor analysis—Correlation between enzyme immunoassay and immunohistochemical methods. J Clin Pathol 45:125, 1992.

239. JJ Miller, RW Straub Jr, R Valdes Jr. Digoxin immunoassay with cross-reactivity of digoxin metabolites proportional to their biological activity. Clin Chem 40:1898, 1994.

240. D Wahyono, M Piechaczyk, C Mourton, JM Bastide, B Pau. Novel antidigoxin monoclonal antibodies with different binding specificities for digoxin metabolites and other glycosides. Hybridoma 9:619, 1990.

241. D Wahyono, M Piehaczyk, JM Scherrmann, C Girard, J Grenier, JC Mani, et al. Highly specific radioimmunoassay for digoxin using a monoclonal antibody selected for lack of interference by digoxin-like immunoreactive substances in cord blood sera. Ther Drug Monit 13:113, 1991.

242. JD Watson, M Gilman, J Witkowski, M Zoller. Recombinant DNA. New York: Freeman, 1992.

243. MA Innes, DH Gelfand, JJ Sninsky, and TJ White. PCR Protocols: A Guide to Methods and Applications. San Diego, CA: Academic Press, 1990.

244. HL Hubbard, TD Eller, DE Mais, et al. Extraction of thromboxane B_2 from urine using an immobilized antibody column for subsequent analysis by gas chromatography-mass spectrometry. Prostaglandins 33:149, 1987.

245. C Chiabrando, A Benigni, A Piccinell, et al. Antibody-mediated extraction negative-ion chemical ionization mass spectrometric measurement of thromboxane-B_2 and 2,3-dinorthromboxane B_2 in human and rat urine. Anal Biochem 163:255, 1987.

246. M Ishibashi, K Watanabe, Y Ohyama, et al. Novel derivatization and immunoextraction to improve microanalysis of 11-dehydrothromboxane B_2 in human urine. J Chromatogr 562:613, 1991.

247. W Krause, U Jakobs, PE Schulze, et al. Development of antibody-mediated extraction followed by GC/MS (antibody/GC/MS) and its application to iloprost determination in plasma. Prostaglandins Leukot Med 17:167, 1985.

248. W Krause, U Jakobs, R Buckenauer, et al. Synthesis of a deuterated analogue and development of antibody/GC/MS for the determination of nocloprost in plasma. Prostaglandins 40:283, 1990.

249. W Haasnoot, ME Ploum, RJA Paulussen, et al. Rapid determination of clenbuterol residues in urine by high-performance liquid chromatography with on-line automated sample processing using immunoaffinity chromatography. J Chromatogr. 519:323, 1990.

250. C Dumouseaux, S Muramatsu, W Takasaki et al. Highly sensitive and specific determination of pravastatin sodium in plasma by high-performance liquid chromatography and laser-induced fluorescence detection after immobilized antibody extraction. J Pharm Sci 83:1630, 1994.

251. A Goodman Gilman, TW Rall, AS Nies, P Taylor, eds. Goodman and Gilman's The Pharmalogical Basis of Therapeutics. 8th ed. New York: Pergamon Press, 1990.

252. JCK Loo, et al. A specific radioimmunoassay (RIA) for salbutamol (albuterol) in human plasma. Res Commun Chem Pathol Pharmacol 55:283, 1987.

253. A Adam, et al. Radioimmunoassay for albuterol using a monoclonal antibody— Application for direct quantification in horse urine. J Immunoassay 11:329, 1990.

254. SJ Mule, E Whitlock, D Jukofsky. Radioimmunoassay of drugs subject to abuse: Critical evaluation of urinary morphine-barbiturate morphine, barbiturate and amphetamine assays. Clin Chem 21:81, 1975.

255. G Gallacher, et al. An improved label for amphetamine fluoroimmunoassay. Ther Drug Monit 11:607, 1989.

256. A Fasth, J Sollenberg, B Sorbo. Production and characterization of antibodies to atropine. Acta Pharm 12:311, 1975.

257. RJ Wurzberger, RL Miller, HG Boxenbaum, S Spector. Radioimmunoassay for atropine. J Pharmacol Exp Ther 203:435, 1977.

258. R Virtanen, et al. Radioimmunoassay for atropine and L-hyoscyamine. Acta Pharmac Toxicol (Copenh) 47:208, 1980.

259. K Enging, et al. Determination of atropine in plasma by a direct radioreceptor assay. Pharm Week 9:321, 1987.

260. LO Henderson, et al. Radioimmunoassay screening of dried blood spot materials for benzoylecgonine. J Anal Toxicol 17:42, 1993.

261. SJ Mule, D Jukofsky, M Kogan, A De Pace, K Verebey. Evaluation of the radioimmunoassay for benzoylecgonine (a cocaine metabolite) in human urine. Clin Chem 23:796, 1977.

262. RD Budd, NC Jain. Benzphetamine Zey IA-GC, Un. J Anal Toxicol 2:241, 1978.

263. G Degand, et al. Determination of clenbuterol in bovine tissues and urine by enzyme immunoassay. J Agric Food Chem 40:70, 1992.

264. RJH Picket, MJ Sauer. Determination of clenbuterol in bovine urine by enzyme immunoassay following concentration and clean-up by immunoaffinity chromatography. Anal Chim Acta 275:268, 1993.

265. GF Read, D Riad-Fahmey. Determination of a tricyclic antidepressant clomipramine anafranil in plasma by a specific radioimmunoassay procedure. Clin Chem 24: 36, 1978.

266. B Jarrott, S Spector. Disposition of clonidine in rats as determined by radioimmunoassay. J Pharmacol Exp Ther 207:195, 1978.

267. D Arndts, H Stahle, CJ Struck. A newly developed precise and sensitive radioimmunoassay for clonidine. Arzneim-Forsch 29:532, 1979.

268. K Robinson, RN Smith. Radioimmunoassay of benzoylecgonine in samples of forensic interest. J Pharm Pharmacol 36:157, 1984.

269. B Kaul, SJ Millian, B Davidow. The development of radioimmunoassay for detection of cocaine metabolites. J Pharmacol Exp Ther 199:171, 1976.

270. JJ Langone, HB Gjika, H VanVunakis. Nicotine and its metabolites. Radioimmuno-assays for nicotine and cotinine. Biochemistry 12:5025, 1973.

271. S Pankey, et al. Quantitive homogeneous enzyme immunoassays for amitriptyline, nortriptyline, imipramine, and desipramine. Clin Chem 32:768, 1986.

272. DJ Brunswick, B Needelman, J Mendels. Radioimmunoassay of imipramine and desmethylimipramine. Life Sci. 22:137, 1978.

273. O Vakkuri, et al. Radioimmunoassay of detomidine, a new benzylimidazole drug with analgesic sedation properties. Life Sci 40:1357, 1987.

274. HF Schran, HJ Schwarz, KC Talbot, LJ Loeffler. Specific radioimmunoassay of ergot peptide alkaloids in plasma. Clin Chem 25:1928, 1979.

275. KL Rominger, et al. Radioimmunological determination of fenoterol. Part II: antise-rum and tracer for the determination of fenoterol, Arzneim-Forsch 40:887, 1990.

276. A Martinsen, et al. Comparison of radioimmunoassay and capillary gas chromatog-raphy in the analysis of L-hyoscyamine from plant material. Phytochem Anal 2: 163, 1991.

277. KK Midha, C Charette, JK Cooper, IJ McGilveray. Comparison of a new GLC-AFID method with a GLC-MS selected ion monitoring technique and a radioimmu-noassay for the determination of plasma concentrations of imipramine and desipra-mine. J Anal Toxicol 4:237, 1980.

278. JG Kelly, K McGarry, K O'Malley. Radioreceptor assay for labetalol. J Clin Phar-macol 12:258, 1981.

279. I Yamamoto, et al. Enzyme immunoassay for mabuterol, a selective beta-2-adrener-gic stimulant in the trachea. J Immunoassay 6:261, 1985.

280. C Goyot, et al. Maprotiline measurement (tetracyclic antidepressant) by high per-formance liquid chromatography and by radioimmunoassay—Clinical use. Pathol Biol 35:1132, 1987.

281. V Spiehler, et al. Elimination of interferences from ephedrine and related com-pounds in Coat-A-Count methamphetamine radioimmunoassay. Clin Chem 39:172, 1993.

282. S Inayama, Y Tokunaga, E Hosoya, T Nakadate, T Niwaguchi, K Aoki, S Saito. A rapid and simple screening method for methamphetamine in urine by radioimmu-noassay using a 1251-labeled methamphetamine derivative. Chem Pharm Bull (Tokyo) 28:2779, 1980.

283. CA Bizollon, JP Rocher, P Chevalier. Radioimmunoassay of nicergoline in biologi-cal material. Eur J Nuclear Med 7:318, 1982.

284. S Matsukara, N Sakamoto, H Imura, et al. Radioimmunoassay of nicotine. Biochem Biophys Res Commun 64:574, 1975.

285. JJ Langone, J Franke, H VanVanakis. Nicotine and its metabolites. Radioimmuno-assay for gamma-(3-pyridyl)-gamma-oxo-N-methylbutyramide. Arch Biochem Bi-ophys 164:536, 1974.

286. G Gallacher, et al. Development and validation of a fluoroimmunoassay for urinary normetanephrine. Ann Clin Biochem 29:492, 1991.

287. T Levitt. Determination of paraquat in clinical practice using radioimmunoassay. Proc Anal Div Chem Soc 16:72, 1979.

288. D Fatori, WM Hunter. Radioimmunoassay for serum paraquat. Clin Chim Acta 100:81, 1980.

289. CA Homon, et al. A selective radioimmunoassay for the determination of pirenze-pine in plasma and urine. Ther Drug Monit 9:236, 1987.

290. P Tanswell, et al. Automated monoclonal radioimmunoassays for pirenzepine, a selective muscarinic receptor antagonist, in plasma and urine. J Immunol Meth 93: 247, 1986.

291. GC Nordblom, et al. Development of radioimmunoassays for measurement of the specific bronchodilator procaterol in human urine and plasma. J Clin Immunol 15: 258, 1992.

292. K Kawashima, A Levy, S. Spector. Stereospecific radioimmunoassay for propran-olol isomers. J Pharmacol Exp Ther 196:517, 1976.

293. GP Mould, J Clough, BA Morris, G Stout, V Marks. A propranolol radioimmunoas-say and its use in the study of its pharmacokinetics following low doses. Biopharm Drug Dispos 2:49, 1981.

294. JWA Findlay, et al. Stereospecific radioimmunoassays for dextro pseudoephedrine in human plasma and their application to bioequivalency studies. J Pharm Sci 70: 624, 1981.

295. A Levy, K Kawaskima, S Spector. Radioimmunoassay for reserpine. Life Sci 19: 1421, 1976.

296. M Van Lierde, JP Desager, C Harvengt, K Thomas. Ritodrine serum levels: Influ-ence of dose and route of administration. Int J Clin Pharmacol Ther Toxicol 22: 382, 1984.

297. M Surtees, et al. Radioimmunoassay of a novel anti-allergic compound (RP-42068) and its application to the analysis of RP-42068 in plasma. Br J Clin Pharmacol 27: P111, 1989.

298. NM Cintron, et al. A sensitive radioreceptor assay for determining scopalamine concentration in plasma and urine. J Pharm Sci 76:328, 1987.

299. YJ Markkanen, et al. Serum antimuscarinic activity after a single dose of oral sco-polamine hydrobromide solution measured by radioreceptor assay. Oral Surg Oral Med Oral Pathol 63:534, 1987.

300. L Stoll, et al. A simple but highly sensitive radioreceptor assay for the determination of scopolamine and biperiden in human plasma. Res Commun Chem Path Pharma-col 64:59, 1989.

301. J Schnabel, et al. Applications of scopolamine radioreceptor assay to pharmacoki-netic studies in rats and humans. Clin Chem 36:1045, 1990.

302. K Hagemann, et al. Monoclonal antibody based enzyme immunoassay for the quan-titative determination of the tropane alkaloid. scopolamine. Planta Med 58:68, 1992.

303. PE Horowitz, S Spector. Determination of serum tubocurarine concentration by radioimmunoassay. J Pharmacol Exp Ther 185:94, 1973.

304. S Bjorkman, et al. Determination of alfentanil in serum by radioimmunoassay or capillary column gas liquid chromatography—A comparison of the assays. Acta Pharm Nord 1:211, 1989.

305. GP Mould, G Stout, GW Aherne, V Marks. Radioimmunoassay of amitriptyline and nortriptyline in body fluids. Ann Clin Biochem 15:221, 1978.

306. B Law, AC Moffat. The evaluation of an homogeneous enzyme immunoassay (Emit) and radioimmunoassay for barbiturates. J Forensic Sci Soc 21:55, 1981.

307. RL Chang, AW Wood, WR Dixon, AH Conney, KE Anderson, J Eiseman, AP Alvares. Antipyrine: Radioimmunoassay in plasma and saliva following administration of a high dose and a low dose. Clin Pharmacol Ther 20:219, 1976.

308. RL Chang, AW Wood, WR Dixon. ^{125}I-radioimmunossay for antipyrine. Life Sci 22:855, 1978.

309. EJ Flynn, S Spector. Determination of barbiturate derivatives by radioimmunoassay. J Pharmacol Exp Ther 181:547, 1972.

310. AT Watson, JE Manno, BR Manno. Quantitation of barbiturates by a modification of the EMIT®-tox™ serum barbiturate assay. J Anal Toxicol 7:257, 1983.

311. W Roos, et al. A homogeneous enzyme immunoassay for benzodiazepine alkaloids (EMIT). Pharmazie 42:213, 1987.

312. CP Goddard, et al. An I-125 radioimmunoassay for the direct detection of benzodiazepines in blood and urine. Analyst 111:525, 1986.

313. H Schutz, et al. Improved enzyme immunological screening procedure for benzodiazepines in urine after extrelut-enrichment. Aerztl. Lab 34:130, 1988.

314. JJ Sramek, et al. Detection of benzodiazepines in human hair by radioimmunoassay. Ann Pharmacol 26:469, 1992.

315. W Huang, et al. Immunoassay detection of nordiazepam, triazolam, lorazepam, and alprazolam in blood. J Anal Toxicol 17:365, 1993.

316. H He, et al. Development of a sensitive and specific radioimmunoassay for benztropine. J Pharm Sci 82:1027, 1993.

317. IS Bevan. Sensitive and specific bromocriptine radioimmunoassay with iodine label—Measurement of bromocriptine in human plasma. Ann Clin Biochem 23:686, 1986.

318. AJ Bartlett, JG Lloyd-Jones, MJ Rance, IR Flockhart, GJ Dockray, MR Bennett, RA Moore. The radioimmunoassay of buprenorphine. Eur J Clin Pharmacol 18: 339, 1980.

319. L Debrabandere, et al. Development of a radioimmunoassay for the determination of buprenorphine in biological samples. Analyst 118:137, 1993.

320. P Daenens, et al. Routine detection of buprenorphine in horse urine—Possibilities and limitations of the combined use of radioimmunoassay, liquid chromatography and gas chromatography/mass spectrometry. Anal Chim Acta 275:295, 1993.

321. GK Tiong, et al. Enzyme immunoassay of buprenorphine. N-S Arch Pharmacol 338:202, 1988.

322. NB Mehta. Design and synthesis of a hapten for the radioimmunoassay of bupropion. J Pharm Sci 75:410, 1986.

323. KA Pitman, et al. Serum levels of butorphanol by radioimmunoassay. J Pharm Sci 69:160, 1980.

324. CE Cook, CR Tallent, EW Amerson, MW Myers, JA Kepler, GF Taylor, HD Christensen. Caffeine in plasma and saliva by a radioimmunoassay procedure. J Pharmacol Exp Ther 199:679, 1976.

325. JV Aranda, et al. Caffeine enzyme immunoassay in neonatal and pediatric drug monitoring. Ther Drug Monit 9:97, 1987.

326. JR Soares, SJ Gross. Separate radioimmune measurements of body fluid-Δ^9-THC. Life Sci 19:1711, 1976.

327. JD Teale, EJ Forman, LJ King, EM Piall, V Marks. The development of a radioim-

munoassay of cannabinoids in blood and urine. J Pharm Pharmacol 27:465, 1975.

328. B Law, PA Mason, AC Moffat, LJ King, A novel ^{125}I radioimmunoassay for the analysis of delta 9-tetrahydrocannabinol and its metabolites in human body fluids. J Anal Toxicol 8:14, 1984.

329. AJ Clatworthy, et al. Gas chromatographic mass spectrometric confirmation of radioimmunoassay results for cannabinoids in blood and urine. Forensic Sci Int 46: 219, 1990.

330. AJ McNally, et al. A convenient enzyme immunoassay system for detection of drugs of abuse—Cannabinoids in urine. Clin Chem 33:972, 1987.

331. M McCombs, et al. A noninstrumented enzyme immunochromatographic assay for the quantification of carbamazepine in whole blood. Clin Chem 33:1019, 1987.

332. WR Dixon, R Lucek, J Earley, C Perry. Chlordiazepoxide, a new more sensitive and specific radioimmunoassay. J Pharm Sci 68:261, 1979.

333. K Kawashima, R Dixon, S Spector. Development of radioimmunoassay for chlorpromazine. Eur J Pharmacol 32:195, 1975.

334. JW Hubbard, KK Midha, IJ McGilveray, JK Cooper. Radioimmunoassay for psychotropic drugs I: Synthesis and properties of haptens for chlorpromazine. J Pharm Sci 67:1563, 1978.

335. PK Yeung, et al. Radioimmunoassay for the N-oxide metabolite of chlorpromazine in human plasma and its application to a pharmacokinetic study in healthy humans. J Pharm Sci 76:803, 1987.

336. WR Dixon, RL Young, R Ning, A Liebman. Radioimmunoassay of the anticonvulsant agent clonazepam. J Pharm Sci 66:235, 1977.

337. N Ratnaraj, et al. Correlation between fluorescent polarization immunoassay and enzyme immunoassay of anticonvulsant drugs, and stability of calibration graphs. Analyst 111:517, 1986.

338. AR Gintzler, E Mohacsi, S Spector. Radioimmunoassay for the simultaneous determination of morphine and codeine. Eur J Pharmacol 38:149, 1976.

339. BH Wainer, FW Fitch, J Fried, RM Rothberg. Immunochemical studies of opioids: Specificitics of antibodies against codeine and hydromorphone. Clin Immunol Immunopathol 3:155, 1974.

340. JW Findlay, RF Butz, FM Welch. A codeine radioimmunoassay exhibiting insignificant cross-reactivity with morphine. Life Sci 19:389, 1976.

341. B Peskar, and S Spector. Quantitative determination of diazepam in blood by radioimmunoassay. J Pharmacol Exp Ther 186:167, 1973.

342. R Dixon, T Crews. Diazepam: Determination in micro samples of blood, plasma, and saliva by radioimmunoassay. J Anal Toxicol 2:210, 1978.

343. WA Ratcliffe, SM Fletcher, AC Moffat, JG Ratcliffe, WA Harland, TE Levitt. Radioimmunoassay of lysergic acid diethylamide (LSD) in serum and urine by using antisera of different specificities. Clin Chem 23:169, 1977.

344. R Virtanen, JS Salonen, M Scheinin, E Iisalo, V Mattila. Radioimmunoassay for doxepin and desmethyldoxepin. Acta Pharmacol Toxicol (Copenh) 47:274, 1980.

345. JD Robinson, BA Morris, V Marks. Development of a radioimmunoassay for etorphine. Res Commun Chem Pathol Pharmacol 10:1, 1975.

346. GL Henderson, J Frincke, CY Leung, M Torten, E Benjamini. Antibodies to fentanyl. J Pharmacol Exp Ther 192:489, 1975.

347. M Michiels, R Hendriks, J Heykants. A sensitive radioimmunoassay for fentanyl. Plasma level in dogs and man. Eur J Clin Pharmacol 12:153, 1977.

348. RJH Woestenborghs, et al. Assay methods for fentanyl in serum—Gas liquid chromatography versus radioimmunoassay. Anesthesiology 67:85, 1987.

349. GL Henderson, et al. Rapid screening of fentanyl (china white) powder samples by solid phase radioimmunoassay. J Anal Toxicol 14:172, 1990.

350. R Dixon, W Glover, J Earley. Specific radioimmunoassay for flunitrazepam. J Pharm Sci 70:230, 1981.

351. SA Goldstein, H Van Vunakis. Determination of fluphenazine, related phenothiazine drugs and metabolites by combined high-performance liquid chromatography and radioimmunoassay. J Pharmacol Exp Ther 217:36, 1981.

352. DH Wiles, M Franklin. Radioimmunoassay for fluphenazine in human plasma. Br J Clin Pharmacol 5:265, 1978.

353. G McKay, et al. Radioimmunoassay of plasma fluphenazine using monoclonal antibodies, J Pharm Sci 76:20, 1987.

354. ES Lo, et al. A highly sensitive and specific radioimmunoassay for quantitation of plasma fluphenazine. J Pharm Sci 77:255, 1988.

355. KK Midha, et al. Radioimmunoassay for fluphenazine sulfoxide in human plasma. J Pharmacol Methods 19:63, 1988.

356. M Aravagiri, et al. Therapeutic monitoring of steady state plasma levels of the N4' oxide metabolite of fluphenazine in chronically treated schizophrenic patients determined by a specific and sensitive radioimmunoassay. Ther Drug Monit 12: 268, 1990.

357. G McKay, et al. Development and application of a radioimmunoassay for fluphenazine based on monoclonal antibodies and its comparison with alternate assay methods. J Pharm Sci 79:240, 1990.

358. M Aravagiri, et al. Radioimmunoassay for 7-hydroxy metabolite of fluphenazine and its application to plasma level monitoring in schizophrenic patients treated long term with oral and depot fluphenazine. Ther Drug Monit 16:21, 1994.

359. JD Robinson, D Risby. Radioimmunoassay for flupenthixol in plasma. Clin Chem 23:2085, 1977.

360. A Jorgensen. A sensitive and specific radioimmunoassay for cis (z)-flupenthixol in human serum. Life Sci 23:1533, 1978.

361. W Glover, J Earley, M Delaney, R Dixon. Radioimmunoassay of flurazepam in human plasma. J Pharm Sci 69:601, 1980.

362. A Pouler, B Deus-Neumann, MH Zenk. Enzyme immunoassay for the quantitative determination of galanthamine. Planta Med 59:44, 1993.

363. BR Clark, BB Tower, RT Rubin. Radioimmunoassay of halperidol in human serum. Life Sci 20:319, 1977.

364. S Yamazumi, S Miura. Haloperidol concentrations in saliva and serum: Determination by the radioimmunoassay method. Int Pharmacopsychiatry 16:174, 1981.

365. H Suzuki, et al. Determination of haloperidol in human serum by radioimmunoassay. J Pharmacobiodyn. 3:250, 1980.

366. Y Terauchi, et al. Direct radioimmunoassay for haloperidol in human serum. J Pharm Sci 79:432, 1990.
367. IL Honigberg, JT Stewart. Radioimmunoassay of hydromorphone and hydrocodone in human plasma. J Pharm Sci 69:1171, 1980.
368. M Pellegatti, et al. Validation of a high performance liquid chromatographic radio-immunoassay method for the determination of lacidipine in plasma. J Chromatogr-Bio 573:105, 1992.
369. JM Sailstad, JWA Findlay. Immunofluorometric assay for lamotrigine (lamictal) in human plasma. Ther Drug Monit 13:433, 1991.
370. R Dixon, T Crews, E Mohacsi, C Inturrisi, K Foley. Levorphanol: A simplified radioimmunoassay for clinical use. Res Commun Chem Pathol Pharmacol 32:545, 1981.
371. E Bertol, et al. Comparison of lidocaine by fluorescence polarization immunoassay, enzyme immunoassay, and high resolution gas chromatography. J Anal Toxicol 11:112, 1987.
372. LS Jackson, et al. The evaluation and application of a radioimmunoassay for the measurement of diphenoxylic acid, the major metabolite of diphenoxylate hydro-chloride (Lomotil), in human plasma. J Pharmacol Meth 18:189, 1987.
373. M Michiels, R Hendriks, J Heykants. Radioimmunoassay of the antidiarrhoeal lop-eramide. Life Sci 21:451, 1977.
374. M Humpel, B Nieuweboer, W Milius, H Hanke, H Wendt. Kinetics and biotransfor-mation of lormetazepam. II. Radioimmunologic determinations in plasma and urine of young and elderly subjects: First-pass effect. Clin Pharmacol Ther 28:673, 1980.
375. DH Catlin, JC Schaeffer, JF Fischer. Production and characterization of antibodies to meperidine. Res Commun Chem Pathol Pharmacol 11:245, 1975.
376. BH Wainer, WE Wung, JH Hill, FW Fitch, J Fried, RM Rothberg. The production and characterization of antibodies reactive with meperidine. J Pharmacol Exp Ther 197:734, 1976.
377. KK Midha, et al. Development of a radioimmunoassay procedure for mesoridazine and its comparison with a high performance liquid chromatographic method. Ther Drug Monit 9:464, 1987.
378. DL Roerig, RIH Wang, MM Mueller, DL Lewand, SM Adams. Radioimmunoassay compared to thin-layer and gas-liquid chromatography for detecting methadone in human urine. Clin Chem 22:1915, 1976.
379. CT Liu, FI Adler. Immunologic studies on drug addition, I. Antibodies reactive with methadone and their use for detection of the drug. J Immunol 111:472, 1973.
380. F Bartos, GD Olsen, RN Leger, D Bartos. Stereospecific antibodies to methadone. I. Radioimmunoassay of d, l-methadone in human serum. Res Commun Chem Pathol Pharmacol 16:131, 1977.
381. RD Budd, FC Yang, WJ Leung. Mass screening and confirmation of methaqualone and its metabolites in urine by radioimmunoassay-thin-layer chromatography. J Chromatogr 190:129, 1980.
382. AR Berman, JP McGrath, RC Dermisohn, JA Cella. Radioimmunoassay of metha-qualone and monohydroxy metabolites in urine. Clin Chem 21:1878, 1975.
383. B Van der walt, et al. A radioimmunoassay for metoclopramide. J Immunol Meth 130:33, 1987.

384. DR Stanski, L Paalzow, PO Edlund. Morphine pharmacokinetics: GLC assay versus radioimmunoassay. J Pharm Sci 71:314, 1982.

385. H VanVunakis, E Wasserman, L Levine. Specificity of antibodies to morphine. J Pharmacol Exp Ther 180:514, 1972.

386. M Koida, M Takahashi, H Kaneto. The morphine 3-glucuronide directed antibody: Its immunological specificity and possible use for radioimmunoassay of morphine in urine. Jpn J Pharmacol 24:707, 1974.

387. S Spector. Quantitative determination of morphine in serum by radioimmunoassay. J Pharmacol Exp Ther 178:253, 1971.

388. P Leclerc, et al. Radioimmunoassay of serum morphine with commercially available reagents. Clin Biochem 20:297, 1987.

389. K Quinn, et al. Determination of morphine in human plasma by radioimmunoassay utilizing a preliminary liquid solid extraction. J Pharm Biomed Anal 6:15, 1988.

390. D Donohue, et al. A convenient enzyme immunoassay system for detection of drugs of abuse—Morphine in urine. Clin Chem 33:970, 1987.

391. V Spiehler, et al. Unconjugated morphine in blood by radioimmunoassay and gas chromatography mass spectrometry. J Forensic Sci 32:906, 1987.

392. AM Bermejo, et al. Morphine determination by gas chromatography mass spectroscopy in human vitreous humor and comparison with radioimmunoassay. J Anal Toxicol 16:372, 1992.

393. R Dixon, J Hsiao, W Taaffe, E Hahn, R Tuttle. Nalmefene: Radioimmunoassay for a new opioid antagonist. J Pharm Sci 73:1645, 1984.

394. BA Berkowitz, SH Ngai, J Hempstead, S Spector, Disposition of naloxone: Use of a new radioimmunoassay. J Pharmacol Exp Ther 195:499, 1975.

395. R Dixon, R Lucek, R Young, R Ning, A Darragh. Radioimmunoassay for nitrazepam in plasma. Life Sci 25:311, 1979.

396. IM McIntyre, TR Norman, GD Burrows, KP Maguire. Determination of nomifensine plasma concentrations: A comparison of radioimmunoassay and gas chromatography. Br J Clin Pharmacol 12:691, 1981.

397. R Dixon, W Glover, J Earley, M Delaney. Desmethyldiazepam, a specific radioimmunoassay. J Pharm Sci 68:1470, 1979.

398. R Lucek, R Dixon. Specific radioimmunoassay for amitriptyline and nortriptyline in plasma. Res Commun Chem Pathol Pharmacol 18:125, 1977.

399. KP Maguire, GD Burrows, TR Norman, BA Scoggins. A radioimmunoassay for nortriptyline and other tricyclic antidepressants in plasma. Clin Chem 24:549, 1978.

400. JD Robinson, RA Braithwaite, S Dawling. Measurement of plasma nortriptyline concentrations radioimmunoassay and gas chromatography compared. Clin Chem 24:2023, 1978.

401. JF Sayegh. A simplified radioimmunoassay of plasma nortriptyline in depressed patients compared with high pressure liquid chromatography and gas liquid chromatography. Neurochem Res 11:193, 1986.

402. GH Fridland, et al. Measurement of opioid peptides with combinations of reversed phase high performance liquid chromatography, radioimmunoassay, radioreceptorassay, and mass spectrometry. Life Sci 41:809, 1987.

403. DM Desiderio, et al. Opioid peptide measurements with reversed phase HPLC,

radioimmunoassay, radioreceptorassay and mass spectrometry. Pharm Week 10: 34, 1988.

404. L Nordholm, et al. Determination of an oxidiazole substituted 1,4-benzodiazepine in plasma by high performance liquid chromatography with ultraviolet detection and by a radioreceptor assay. J Chromatogr-Bio 494:257, 1989.

405. JE Peterson, M Graham, WF Banks, D Benziger, EA Rowe, S Clemans, J Edelson. Plasma pentazocine radioimmunoassay. J Pharm Sci 68:626, 1979.

406. TA Williams, KA Pittman. Pentazocine radioimmunoassay, Res Commun Chem Pathol Pharmacol 7:119, 1974.

407. CE Cook, et al. Pharmacokinetics of pentobarbital enantiomers as determined by enantioselective radioimmunoassay after administration of racemate to humans and rabbits. J Pharmacol Exp Ther 241:779, 1987.

408. H Satoh, Y, Kuriowa, A Hamada, T Uematsu. Radioimmunoassay for phenobarbital. J Biochem 75:1301, 1974.

409. A Chung, SY Kim, LT Cheng, A Castro. Phenobarbital specific antisera and radioimmunoassay. Experientia 29:820, 1973.

410. AR Ashy, YM El-Sayed, SI Islam. Comparison of fluorescence polarization immunoassay and high performance liquid chromatography for the quantitative determination of phenytoin, phenobarbitone, and carbamazepine in serum. J Pharm Pharmacol 38:572, 1986.

411. AM Sidki, et al. Direct determination of phenobarbital in serum or plasma by polarization FIA. Ther Drug Monit 4:397, 1982.

412. B Levine, et al. Evaluation of the Coat-A-Count radioimmunoassay for phencyclidine. Clin Chem 34:429, 1988.

413. SM Owens, J Woodworth, M Mayersohn. Radioimmunoassay for phencyclidine in serum. Clin Chem 28:1509, 1982.

414. M Adamczyk, et al. Immunoassay reagents for psychoactive drugs. Part 3. Removal of phenothiazine interferences in the quantification of tricyclic antidepressants. Ther Drug Monit 15:436, 1993.

415. CE Cook, JA Kepler, HD Christensen. Antiserum to dephenylhydantoin: Preparation and characterization. Res Commun Chem Pathol Pharmacol 5:767, 1973.

416. JD Robinson, BA Morris, GW Aherne, V Marks. Pharmacokinetics of a single dose of phenytoin in man measured by radioimmunoassay. Br J Clin Pharmacol 2:345, 1975.

417. HD Christensen, E Amerson, MW Myers, CE Cook. Comparison of diphenylhydantoin and phenobarbital serum and blood levels. Pharmacologist 16:228, 1974.

418. CE Cook, E Amerson, WK Poole, P Lesser, L O'Tuama. Phenytoin and phenobarbital concentration in saliva and plasma measured by radioimmunoassay. Clin Pharmacol Ther 18:742, 1975.

419. JW Paxton, B Whiting, KW Stephen. Phenytoin concentrations in mixed, parotid and submandibular saliva and serum measured by radioimmunoassay. Br J Clin Pharmacol 4:185, 1977.

420. JW Paxton, FJ Rowell, JG Ratcliffe, DG Lambie, R Nanda, ID Melville, RH Johnson. Salivary phenytoin radioimmunoassay, a simple method for the assessment of non-protein bound drug concentrations. Eur J Clin Pharm 11:71, 1977.

421. ML Orme, O Borga, CE Cook, F Sjoqvist. Measurement of diphenylhydantoin in

0.1 ml plasma samples: Gas chromatography and radioimmunoassay compared. Clin Chem 22:246, 1976.

422. BC Sallustio, et al. Unbound plasma phenytoin concentrations measured using enzyme immunoassay technique on the Cobas Mira Analyzer—In vivo effect of valproic acid. Ther Drug Monit 14:9, 1992.

423. J van der Weide, et al. Evaluation of the cloned enzyme donor immunoassay for measurement of phenytoin and phenobarbital in serum. Ther Drug Monit 15:344, 1993.

424. LJM Michiels, JJP Heykants, AG Knaeps, PAJ Janssen. Radioimmunoassay of the neuroleptic drug pimozide. Life Sci 16:937, 1975.

425. C Brunet, et al. Acid metabolite of progabide pharmacokinetics following single administration in the rabbit with special references to HPLC and H-muscimol radio-receptor assay. Eur J Drug Metab Pharmacokinet 14:257, 1989.

426. G Kominami, et al. Combined high performance liquid chromatography and enzyme immunoassay for active metabolites of a sleep inducer (rilmazafone), a ring opened derivative of benzodiazepines, in human plasma. J Chromatogr-Bio 417: 216, 1987.

427. W Krause, et al. High performance liquid chromatographic procedure for specificity testing of radioimmunoassay—Rolipram. J Chromatogr-Bio 573:303, 1992.

428. RJ Woestenborghs, et al. Assay methods for sufentanil in plasma. Radioimmunoassay versus gas chromatography-mass spectrometry. Anesthesiology 80:666, 1994.

429. KK Midha, et al. A new radioimmunoassay of thioridazine: its comparison with a high performance liquid chromatographic method and another radioimmunoassay. Ther Drug Monit 9:227, 1987.

430. BS Chakraborty, MS Sardessai, TJ Jaworski, KK Midha, EM Hawes. Synthesis and properties of haptens for the development of radioimmunoassays for thioridazine, mesoridazin, and sulforidazine. Pharm Res 4:207, 1987.

431. KK Midha, et al. Development of radioimmunoassay procedure for sulforidazine and its comparison with a high performance liquid chromatographic method. Ther Drug Monit 10:205, 1988.

432. A Mizuchi, S Saruta, N Kitagawa, Y Miyachi. Development of radioimmunoassay for sultopride and sulpiride. Arch Int Pharmacodyn Ther 254:317, 1981.

433. H Ko, ME Royer, JB Hester, KT Johnston. Radioimmunoassay of triazolam. Anal Lett B 10:1019, 1977.

434. S Vandel, et al. Results compared for tricyclic antidepressants as assayed by gas chromatography and enzyme immunoassay. Therapie 47:41, 1992.

435. KK Midha, JW Hubbard, JK Cooper, EM Hawes, S Fournier, P Yeung. Radioimmunoassay for trifluoperazine in human plasma. Br J Clin Pharmacol 12:189, 1981.

436. H Liu, et al. Determination of free valproic acid—Evaluation of the Centrifree system and comparison between high performance liquid chromatography and enzyme immunoassay. Ther Drug Monit 14:513, 1992.

437. K Kalbe, et al. Competitive binding enzyme immunoassay for zonisamide, a new antiepileptic drug, with selected paired enzyme labeled antigen and antibody. Clin Chem 36:24, 1990.

438. M Sasaki, et al. A highly sensitive enzyme immunoassay for anti-inflammatory protease drug in blood. J Pharm Sci 76:S94, 1987.

439. KK Midha, G Rauw, G McKay, JK Cooper, J McVittie. Subnanogram quantitation of chlorpheniramine in plasma by a new radioimmunoassay and comparison with a liquid chromatographic method. J Pharm Sci 73:1144, 1984.

440. NH Ertel, JC Mittler, S Akgun, SL Wallace. Radioimmunoassay for colchicine in plasma and urine. Science 193:233, 1976.

441. JM Scherrmann, L Boudet, R Pontikis, HN Nguyen, E Fournier. A sensitive radioimmunoassay for colchicine. J Pharm Pharmacol 32:800, 1980.

442. M Hildebrand, et al. Development, validation and practical use of a sensitive and specific radioimmunoassay for the determination of iloprost. Eicosanoids 3:165, 1990.

443. LE Hare, CA Ditzler, DE Duggan. Radioimmunoassay of indomethacin in biological fluids. J Pharm Sci 66:486, 1977.

444. CJ Wang, Z Tian, V Reyes, CC Lin. A specific enzyme immunoassay (EIA) with selective extraction for quantitation of a topical anti-inflammatory agent, SCH40120, in human plasma. J Pharm Biomed Anal 13:121, 1995.

445. LE Hare, CA Ditzler, M Hichens, A Rosegay, DE Duggan. Analysis of sundilac and metabolites by combined isotope dilution-radioimmunoassay. J Pharm Sci 66:414, 1977.

446. TM Li, JL Benovic, RT Buckler, JF Burd. Homogeneous substrate labeled fluorescent immunoassay for theophylline in serum. Clin Chem 27:22, 1981.

447. CE Cook, ME Twine, M Myers, E Amerson, JA Kepler, GF Taylor. Theophylline radioimmunoassay: synthesis of antigen and characterization of antiserum. Res Commun Chem Pathol Pharmacol 13:497, 1976.

448. MP Habib, et al. Evaluation of whole blood theophylline enzyme immunochromatography assay. Chest 92:129, 1987.

449. R Chen, et al. An internal clock reaction used in a one step enzyme immunochromatographic assay of theophylline in whole blood. Clin Chem 33:1521, 1987.

450. A Alonso, et al. Comparative study multicenters of theophyllines titrate by different techniques—Immunology with fluorescent polarization, radioimmunology, immunoenzymology and chromatography in liquid phase. Med Armees 14:499, 1986.

451. TG Wu, et al. Liposome based flow injection enzyme immunoassay for theophylline. Mikrochim Acta 1:187, 1990.

452. DA Palmer, et al. Flow injection electrochemical enzyme immunoassay for theophylline using a protein-A immunoreactor and p-aminophenyl phosphate/p-aminophenol as the detection system. Analyst 117:1679, 1992.

453. U Klotz. Comparison of theophylline blood levels measured by the standard TDx assay and a new patient-side immunoassay cartridge system. Ther. Drug Monit 15:462, 1993.

454. AL Gerbes. A highly sensitive and rapid radioimmunoassay for the determination of arginine (8)-vasopressin. Eur J Clin Chem Clin Biochem 30:229, 1992.

455. FM Rosmalen, et al. A sensitive radioimmunoassay of atrial natriuretic peptide in human plasma—Some guidelines for clinical applications. Z. Kardiol. 77:20, 1988.

456. A Clerio, G Opocher, D Pelizzola, et al. Evaluation of the analytical performance of RIA methods for measurement of atrial natriuretic peptides: A multicenter study. J Clin Immunoassay 14:251, 1991.

457. S Hashida, et al. Novel and sensitive noncompetitive enzyme immunoassay (hetero-

2-site enzyme immunoassay) for alpha human atrial natriuretic peptide in plasma. J Clin Lab Anal 5:324, 1991.

458. WR Dixon, RL Young, A Holazo, ML Jack, RE Weinfeld, K Alexander, A Liebman, SA Kaplan. Bumetanide radioimmunoassay and pharmacokinetic profile in humans. J Pharm Sci 65:701, 1976.

459. JA Tenhaaf, et al. Radioimmunoassay, a goal or a tool—The setup of a reliable, fast and cheap radioimmunoassay for vasopressin in biological samples. J Control Release 21:23, 1992.

460. K Selden, MD Klein, TW Smith. Plasma concentration and urinary excretion kinetics of acetyl strophanthidin. Circulation 47:744, 1973.

461. Y Hinohara, et al. A radioimmunoassay for MC-838 (altiopril calcium), a novel angiotensin converting enzyme inhibitor. J Pharmacobiodyn 11:411, 1988.

462. S Akinaga, et al. Determination of the calcium antagonist benidipine hydrochloride in plasma by sensitive radioimmunoassay. Arzneim.-Forsch. 38:1738, 1988.

463. FM Duncan, VI Martin, BC Willians, EAS Al Dujaili, CRW Edwards. Development and optimization of a radioimmunoassay for plasma captopril. Clin Chim Acta 131:295, 1983.

464. JE Doherty. Digitalis serum levels: Clinical use. Ann Intern Med. 74:787, 1971.

465. GC Oliver Jr, B Parker. Clinical application of a radioimmunoassay for digoxin. Circulation Suppl. 41, 42(III):187, 1970.

466. BM Michniewicz, JC Hodges, BG England, T Chang, CJ Blankley, BD Nordblom. A radioimmunoassay for the diacid metabolite of CI-906, a potent angiotensin-I converting enzyme inhibitor. J Clin Immunoassay 10:111, 1987.

467. G Kominami, H Okabe, K Imoda, et al. Combined high-performance liquid chromatography and radioimmunoassay for ceruletide and its metabolites in dog plasma and urine. J Pharm Biomed Anal 12:413, 1994.

468. H Tanaka, et al. Enzyme immunoassay discrimination of a new angiotensin converting enzyme (ACE) inhibitor, cilazapril, and its active metabolite. J Pharm Sci 76:224, 1987.

469. TW Smith. Radioimmunoassay for serum digitoxin concentration: Methodology and clinical experience. J Pharmacol Exp Ther 175:372, 1970.

470. SRCJ Santos, et al. Simultaneous analysis of digitoxin and its clinically relevant metabolites using high performance liquid chromatography and radioimmunoassay. J Chromatogr -Bio 419:155, 1987.

471. TW Smith, VP Butler, E Haber. Determination of therapeutic and toxic serum digoxin concentrations by radioimmunoassay. N Engl J Med 281:1212, 1969.

472. SL Pippin, FI Marcus. Digoxin immunoassay with use of [3H]digoxin vs. [125]tyrosine-methyl-ester of digoxin. Clin Chem 22:286, 1976.

473. R Stenzel, B Reckmann. Cross-reactivity of anti-digoxin antibodies with digitoxin depends on tracer structure. Clin Chem 38:2228, 1992.

474. SM Fletcher, G Lawson, AC Moffat. Radioimmunoassay of cardiac glycosides in hemolyzed blood derivation of serum levels. J Forensic Sci Soc 19:183, 1979.

475. VP Butler, JP Chen. Digoxin specific antibodies. Proc Natl Acad Sci (USA) 57: 71, 1967.

476. RG Stoll, MS Christensen, E Sakmar, et al. The specificity of the digoxin radioimmunoassay procedure. Res Commun Chem Pathol Pharmacol 4:503, 1972.

477. L Nielsen, et al. A sensitive, automated, non-isotopic enzyme immunoassay for the determination of digoxin. Clin Chem 33:1014, 1987.

478. SJ Danielson, et al. Improved labels for use in digoxin multilayer enzyme immuno-assays. Clin Chem 33:923, 1987.

479. PH Hinderling, et al. Comparative in vivo evaluation of a radioimmunoassay and a chromatographic assay for the measurement of digoxin in biological fluids. J Pharm Sci 75:517, 1986.

480. J Plum, et al. Detection of digoxin, digitoxin, their cardioactive metabolites and derivatives by high performance liquid chromatography and high performance liquid chromatography-radioimmunoassay. J Chromatogr 377:221, 1986.

481. MH Gault, L Longerich, M Dawe, et al. Combined liquid chromatography/radioimmunoassay with improved specificity for serum digoxin. Clin Chem 31:1272, 1985.

482. AJ Oosterkamp, H Irth, M Beth, KK Unger, UR Tjaden, J van de Greef. Bioanalysis of digoxin and its metabolites using direct serum injection combined with liquid chromatography and on-line immunochemical detection. J Chromatogr 653:55, 1994.

483. S Mahmod, et al. Radioimmunoassay of salivary digoxin by simple adaptation of a kit method for serum digoxin—Saliva serum ratio and correlation. Ther Drug Monit 9:91, 1987.

484. PJ Worland, et al. Radioimmunoassay for the quantitation of lisinopril and enalaprilat. J Pharm Sci 75:512, 1986.

485. JP Freche, et al. Enzyme immunoassay for flecainide. Clin Chim Acta 190:139, 1990.

486. LJ Loeffler, AW Pittman. Development of radioimmunoassay for guanethidine. J Pharm Sci 68:1419, 1979.

487. Y Terauchi, et al. Determination of ISF-2405, an active metabolite of cadralazine, in plasma and tissues by radioimmunoassay. J. Pharmacol. 10:758, 1987.

488. AC Moffat. Interpretation of post mortem serum levels of cardiac glycosides after suspected over dosage. Acta Pharmacol Toxicol 35:386, 1974.

489. ER Garrett, PH Hinderling. Pharmacokinetics of beta methyl digoxin in healthy humans part 4 comparisons of radioimmunoassays total radioactivity and specific assays of beta methyl digoxin and digoxin in plasma. J Pharm Sci 66:806, 1977.

490. S Yamada, et al. Determination of a novel calcium channel antagonist, mepiradipine, in plasma by radioreceptor assay using [^{3}H]-PN-200-100. Pharm Res 9:1227, 1992.

491. ME Royer, H Ko, TJ Gilbertson, JM McCall, KT Johnston, R Stryd. Radioimmunoassay of minoxidil in human serum. J Pharm Sci 66:1266, 1977.

492. R Selden, TW Smith. Oubian pharmacokinetics in dog and man, determination by radioimmunoassay. Circulation 45:1176, 1975.

493. L Doucet, et al. Radioimmunoassay of a new angiotensin converting enzyme inhibitor (perindopril) in human plasma and urine—Advantages of coupling anion exchange column chromatography with radioimmunoassay. J Pharm Sci 79:741, 1990.

494. H Vandenberg, et al. A new radioimmunoassay for the determination of the angiotensin converting enzyme inhibitor perindopril and its active metabolite in plasma

and urine—Advantages of a lysine derivative as immunogen to improve the assay specificity. J Pharm Biomed Anal 9:517, 1991.

495. B Stouffer, et al. A specific radioimmunoassay for the measurement of a new ACE inhibitor, SQ29852 in plasma. Clin Chem 34:1259, 1988.

496. S Mantha, et al. Development of an enzyme immunoassay (EIA) for SQ27, 519, the active phosphinic-carboxylic diacid of the prodrug fosinopril, in human serum— A comparison with the SQ27, 519 RIA, Clin Chem 34:1153, 1988.

497. WA Pettinger, K Keeton, K Tanaka. Radioimmunoassay and pharmacokinetics of saralasin in the rat and hypertensive patients. Clin Pharmacol Ther 17:146, 1975.

498. B Reck, et al. Development of a sensitive enzyme immunoassay for the determination of vinpocetine in human plasma. Arzneim.-Forsch. 42(2):1171, 1992.

499. E Lelievre, et al. Radioimmunoassays for a new angiotensin converting enzyme inhibitor, zabicipril, and its active metabolite, zabiciprilat, in human plasma. J Pharm Sci 81:1065, 1992.

500. E Ezan et al. Enzyme immunoassays for a new angiotensin converting enzyme inhibitor, zabicipril, and its active metabolite in human plasma—Application to pharmacokinetic studies. Ther Drug Monit 15:448, 1993.

501. SM Tadepalli, et al. Scintillation proximity radioimmunoassay for the measurement of acyclovir. Clin Chem 36:1044, 1990.

502. CD Ashby, JE Lewis, JC Nelson. Measurement of 3 amino glycoside antibiotics with a single radioimmunoassay system. Clin Chem 24:1734, 1978.

503. SG Thompson, JF Burd. Substrate labeled fluorescent immunoassay for amikacin in human serum. Antimicrob. Agents Chemother 18:264, 1980.

504. HU Simon, et al. Determination of berlopentin in human plasma by enzyme immunoassay. Pharmazie 46:139, 1991.

505. RN Hamberger. Chloramphenicol-specific antibody. Science 152:203, 1966.

506. J Pohlschmidt, et al. Modified radioimmunoassay (RIA) for chloramphenicol (CAP) using antibody coated microtiter plates. Arch Lebensmettelhyg 43:3, 1992.

507. C Escande, et al. Sensitive radioimmunoassay and enzyme linked immunosorbent assay for the simultaneous determination of chloroquine and its metabolites in biological fluids. J Pharm Sci 79:23, 1990.

508. GD Nordblom, et al. Development of a radioimmunoassay for the anthrapyrazole chemotherapy agent CI-937 and the pharmacokinetics of CI-937 in rats. Cancer Res 49:5345, 1989.

509. K Aoki, et al. Enzyme immunoassay for cinobufagin. Chem Pharm Bull 34:1184, 1986.

510. TJ Gilbertson, RP Stryd. Radioimmunoassay for clindamycin. Clin Chem 22:828, 1976.

511. B Rollman, et al. Dibekacin assay in serum by automated fluorescence polarization immunoassay (Abbot TDX)—Comparison with high performance liquid chromatography, substrate labeled fluorescent immunoassay and radioimmunoassay. J Pharm Biomed Anal 4:53, 1986.

512. RP Quinn, et al. Radioimmunoassay for desciclovir, 2-(2-amino-9H-purin-9-yl) methoxy ethanol, a prodrug for the antiviral acyclovir. J Immunoassay 8:247, 1987.

513. Y Tanaka, et al. Radioimmunoassay for erythromycin derivatives. J Antibiot 41:258, 1988.

514. P Longmore, RC Atkins, D Casley, CI Johnston. Radioimmunoassay as an improved method for measurement of serum levels of gentamicin. Med J Austal 1: 738, 1976.

515. PJ Wills, R Wise. Rapid simple enzyme immunoassay for gentamicin. Antimicrob. Agents Chemother. 16:40, 1979.

516. JC Rotschafer, C Morlock, L Strand, K Crossley. Comparison of radioimmunoassay and enzyme immunoassay methods in determining gentamicin pharmacokinetic parameters and dosages. Antimicrob. Agents Chemother 22:648, 1982.

517. JA Vilpo, et al. Radioimmunoassay of 5-hydroxymethyl-2′-deoxyuridine. J Immunol Meth 103:41, 1987.

518. R Schwenk, K Kelly, KS Tse, AH Sehon. A radioimmunoassay for isoniazid. Clin Chem 21:1059, 1975.

519. K Fujiwara, et al. Enzyme immunoassay for the quantification of mithramycin using beta-deuterium-galactosidase as a label. Cancer Res 46:1084, 1986.

520. ME Mount, et al. Production of antibodies and development of enzyme immunoassay for determination of monensin in biological samples, J Assoc Offc Anal Chem 70:201, 1987.

521. R Rahmani, et al. Oral administration of [H³] navelbine in patients—Comparative pharmacokinetics using radioactive and radioimmunologic determination methods. Anti-Cancer Drug 2:405, 1991.

522. P Bore, et al. Pharmacokinetics of a new anticancer drug, navelbine, in patients— Comparative study of radioimmunologic and radioactive determination methods. Cancer Chemother Pharmacol 23:247, 1989.

523. A Broughton, JE Strong, LK Pickering, J Knight, GP Bodey. Radioimmunoassay and radioenzymatic assay of a new amino glycoside antibiotic netilmicin. Clin Chem 24:717, 1978.

524. TH Elsasser, et al. Methodoligical considerations for penicillin radioimmunoassay. J. Immunoassay 8:73, 1987.

525. JM Wal, G Bories. Radioimmunoassay of penicilloyl groups in biological fluids. FEBS Lett 57:9, 1975.

526. RP Quinn. Radioimmunoassay for retrovir, an antihuman immunodeficiency virus drug. J. Immunoassay 10:177, 1989.

527. GG Granich, et al. High performance liquid chromatography (HPLC) assay for ribavirin and comparison of the HPLC assay with radioimmunoassay. Antimicrob. Agents Chemother 33:311, 1989.

528. H Cox, et al. Evaluation of a novel fluorescence polarization immunoassay for teicoplanin. Antimicrob. Agents Chemother 37:1924, 1993.

529. BA Faraj, FM Ali. Development and application of a radioimmunoassay for tetracycline. J Pharmacol Exp Ther 217:10, 1981.

530. A Broughton, JE Strong, LK Pickering, GP Bodey. Radioimmunoassay of iodinated tobramycin. Antimicrob Agents Chemother 10:632, 1976.

531. WL Roberts, et al. Solid-phase extraction combined with radioimmunoassay for measurement of zalcitabine (2′, 3′-dideoxycytidine) in plasma and serum. Clin Chem 40:211, 1994.

532. SM Tadepalli, et al. Determination of zidovudine concentration in serum by enzyme

linked immunosorbent-assay and by time resolved fluoroimmunoassay. J Acq Immune Defic Syndr 3:19, 1990.

533. S Cox, et al. Serum levels and catabolism of 3'-azido-3'-deoxythymidine in vivo measured using a specific radioimmunoassay. J Virol Meth 30:89, 1990.

534. JT Slusher, et al. Intracellular zidovudine (ZDV) and ZDV phosphates as measured by a validated combined high pressure liquid chromatography radioimmunoassay procedure. Antimicrob Agents Chemother 36:2473, 1992.

535. B Stretcher, et al. Measurement of phosphorylated zidovudine in peripheral blood leukocytes by radioimmunoassay. Clin Chem 36:1104, 1990.

536. H VanVunakis, JJ Langone, LJ Riceberg, L Levine. Radioimmunoassays for adriamycin and daunomycin. Cancer Res 34:2547, 1974.

537. A Broughton, JE Strong. Radioimmunoassay of bleomycin. Cancer Res 36:1418, 196.

538. H Miyazaki, et al. Measurement of plasma etoposide by radioimmunoassay. J Pharm Biomed Anal 5:11, 1987.

539. V Raso, R Schrieber. A rapid and specific radioimmunoassay for methotrexate. Cancer Res. 35:1407, 1975.

540. GW Aherne, EM Piall, V Marks. Development and application of a radioimmunoassay for methotrexate. Br J Cancer 36:608, 1977.

541. W Cosolo, et al. Comparison of high performance liquid chromatography and the Abbott fluorescent polarization radioimmunoassay in the measurement of methotrexate. J Chromatogr-Bio 494:201, 1989.

542. T Anzai, et al. Separation and identification of methotrexate and its metabolites, 7-hydromethotrexate and polyglutamates, in human tissues by reversed phase high performance liquid chromatography coupled with radioimmunoassay. J Chromatogr-Bio 415:445, 1987.

543. PG Grothaus, et al. Taxane-specific monoclonal antibodies: Measurement of taxol, baccatin III, and "total taxanes" in *Taxus brevifolia* extracts by enzyme immunoassay. J Natural Products, 1995; 58:1003.

544. PG Grothaus, et al. An enzyme immunoassay for the determination of taxol and taxanes in taxus sp tissues and human plasma. J Immunol Meth 158:5, 1993.

545. JD Teale, JM Clough, V Marks. Radioimmunoassay of vinblastine and vincristine. Br J Clin Pharmacol 4:169, 1977.

546. A Huhtikangas, et al. Specific radioimmunoassay for vincristine. Planta Med 53:85, 1987.

547. PE Wallemaco, et al. Cross-reactivity of cyclosporine metabolites in 2 different radioimmunoassays in which the same specific monoclonal antibody is used (technical note). Clin Chem 36:385, 1990.

548. RG Morris, et al. Experiences with the enzyme-multiplied immunoassay cyclosporine specific assay in a therapeutic drug monitoring laboratory. Ther Drug Monit 15:410, 1993.

549. R D'Ambrosio, et al. Improved procedures for enzyme immunoassay of tacrolimus, (FK506) in whole blood. Clin Chem 40:159, 1994.

550. EG Lentjes, et al. Free cortisol in serum assayed by temperature-controlled ultrafiltration before fluorescence polarization immunoassay. Clin Chem 39:2518, 1993.

551. AN Eremin, et al. Enzyme immunoassay of cortisol and progesterone with polymer modified paper. J Anal Chem 48:100, 1993.

552. RP Blye, et al. Development and use of a radioimmunoassay for D-(−)-norgestrel 17-beta-cyclo-pentanecarboxylate. Steroids 48:27, 1986.

553. TG Watson, et al. A sensitive direct radioimmunoassay for assessing D-norgestrel levels in human plasma. Ann Clin Biochem 25:280, 1988.

554. C Nerenberg, et al. Radioimmunoassay of detirelix (N-Ac-D-Nal(2)1, D-p-Cl-Phe2, D-Trp3, D-hArg (ET) 26, D-Ala10-luteinizing hormone releasing hormone) in plasma or serum, J Immunoassay 9:245, 1988.

555. TA Williams, J Edelson, RW Ross, Jr. A radioimmunoassay for danazol 17-alpha pregna-2,4-dien-20-yno 2 3-d-isoxazol-17-ol. Steroids 31:205, 1978.

556. M Hichens, AF Hogans. Radioimmunoassay for dexamethasone in plasma. Clin Chem 20:266, 1974.

557. AW Meikle, LG Lagerquist, FH Tyler. A plasma dexamethasone radioimmunoassay. Steroids 22:193, 1973.

558. J English, J Chakraborty, V Marks, A Parke. A radioimmunoassay procedure for dexamethasone plasma and urine levels in man. Eur J Clin Pharmacol 9:239, 1975.

559. ES Lo, et al. Direct radioimmunoassay procedure for plasma dexamethasone with a sensitivity at the picogram level. J Pharm Sci 78:1040, 1989.

560. G Watanabe et al. Enzyme immunoassay for serum dexamethasone using 4-(carboxymethylthio)-dexamethasone as a new hapten. Steroids 57:1781, 1992.

561. DK Hieu, et al. Radioimmunoassay of the progestagen dienogest using different methods of plasma sample preparation. Pharmazie 41:711, 1986.

562. G Hobe, et al. Determination of the progestagen dienogest in plasma and saliva by radioimmunoassay. Pharmazie 41:772, 1986.

563. GE Abraham, EM Reifman, JE Buster, J DiStephano, JR Marshal. Production of specific antibodies against diethylstilbestrol. Anal Lett 5:479, 1972.

564. J Grenier. Plasma estradiol-17-beta assay, using an enzyme immunoassay kit—Comparison with a specific radioimmunoassay. J Steroid Biochem 33:833, 1989.

565. A Ius, et al. Direct time-resolved fluoroimmunoassay of estriol in serum. J Steroid Biochem 39:189, 1991.

566. HA Zacur, et al. Ethinyl estradiol and norethindrone radioimmunoassay following sephadex LH-20 column chromatography. Clin Chim Acta 204:209, 1991.

567. GJL Lee, et al. Determination of ethynylestradiol and norethindrone in a single specimen of plasma by automated high performance liquid chromatography and subsequent radioimmunoassay. J Liq Chromatogr 10:2305, 1987.

568. PN Rao, et al. Generation of specific antibodies for radioimmunoassay for ethynylestradiol using ethynylestradiol-3-O-carboxymethyl ether-bovine serum albumin conjugate. Clin Chem 36:1099, 1990.

569. W Kuhnz, et al. Radioimmunological analysis of ethinyl-estradiol in human serum—Validation of the method and comparison with a gas chromatographic mass spectrometric assay. Arzneim-Forsch 43-1:16, 1993.

570. WA Colburn. Radioimmunoassay for fluoxymesterone (Halotestin(R)). Steroids 25:43, 1975.

571. BM Bain, et al. A sensitive radioimmunoassay, incorporating solid phase extraction, for fluticasone 17-propionate in plasma. J Pharm Biomed Anal 11:557, 1993.

572. J Girard, et al. Human growth hormone in urine—Development of an ultrasensitive radioimmunometric assay and its application. J Endocrinol 112:240, 1987.

573. J Khubieh, et al. Radioimmunoassay of the anticancer agent 4-hydroxyandrostenedione in body fluids. J Steroid Biochem 35:377, 1990.

574. K Shimada, et al. Enzyme immunoassay for plasma 25-hydroxyvitamin D-3 employing immunoaffinity chromatography as a pretreatment method. J Steroid Biochem 44:93, 1993.

575. WD Gennaro, JD Van Norman. Quantitation of free, total and antibody bound insulin in insulin treated diabetics. Clin Chem 21:873, 1975.

576. PN Rao, ADI Pena, JW Goldzieher. Antisera for radioimmunoassay of 17α-ethynylestradiol and mestranol. Steroids 24:803, 1974.

577. WA Colburn, RH Buller. Radioimmunoassay for methylprednisolone (Medrol®). Steroids 22:687, 1973.

578. CJ Wang, et al. A competitive enzyme immunoassay for the direct determination of momentasone furoate (SCH-32088) in human plasma. J Pharm Biomed Anal 10:473, 1992.

579. EHJM Jansen, et al. Comparison between spectrophotometric and chemiluminescence detection in enzyme immunoassays for nortestosterone. Anal Chim 227:109, 1989.

580. KG Nygren, P Lindberg, K Martinsson, WTK Boser, EDB Johansson. Radioimmunoassay of norethindrone: Peripheral plasma levels after oral administration to humans and rhesus monkeys. Contraception 9:265, 1974.

581. E Ezan, et al. Enzyme immunoassay for normegestrol acetate in human plasma. J Steroid Biochem 46:507, 1993.

582. WA Colburn, RH Buller. Radioimmunoassay for prednisolone. Steroids 21:833, 1973.

583. A Olivesi. Specific micro-radioimmunoassay for prednisolone in serum. Clin Chem 29:1358, 1983.

584. WA Colburn. Radioimmunoassay for prednisone. Steroids 24:95, 1974.

585. II Bolaji, et al. Assessment of bioavailability of oral micronized progesterone using a salivary progesterone enzyme immunoassy. Gynecol Endocrinol 7:101, 1993.

586. DS Cooper, et al. Studies of propylthiouracil using a newly developed radioimmunoassay. J Clin Endocrinol Metab 52:204, 1981.

587. PC Kao, et al. Assay of somatomedin-C by cartridge extraction prior to radioimmunoassay with antiserum developed against synthetic somatomedin-C. Ann Clin Lab Sci 18:120, 1988.

588. AV Schally, et al. Radioimmunoassay for octapeptide analogs of somatostatin—Measurement of serum levels after administration of long acting microcapsule formulations. Proc Natl Acad Sci (USA) 85:5688, 1988.

589. A Turkes, AO Turkes, BG Joyce, GF Read, D Riad-Fahmy. A sensitive solid phase enzymeimmunoassay for testosterone in plasma and saliva. Steroids 33:347, 1979.

590. L Duntas, et al. Pharmacokinetic profiles of intravenously, nasally and orally applied thyroliberin in humans—Determination by radioimmunoassay and fast protein liquid chromatography. Acta Endocrinol 114:81, 1987.

591. JCK Loo, N Jordan. A radioimmunological assay for triamcinolone in plasma. Res Commun Chem Pathol Pharmacol 23:493, 1979.

592. JC Ergrie, et al. Development of radioimmunoassays for human erythropoietin using recombinant erythropoietin as tracer and immunogen. J Immunol Meth 99:235, 1987.

593. M Mason-Garcia, JW Brokins, BS Beckman, JW Fisher. Comparison of recombinant and human erythropoietin as antigen in the radioimmunoassay. J Clin Immunoassay 11:135, 1988.

594. K Matsubara, et al. Radioimmunoassay for erythropoietin using anti-recombinant erythropoietin antibody with high affinity. Clin Chim Acta 185:177, 1989.

595. JW Fisher, et al. Development of a new radioimmunoassay for erythropoietin using recombinant erythropoietin. Kidney Int 38:969, 1990.

596. RG Kendall, et al. Storage and preparation of samples for erythropoietin radioimmunoassay. Clin Lab Haematol 13:189, 1991.

597. AF Serre, et al. Comparison of 2 radioimmunoassays for the measurement of serum erythropoietin. Nephrol Dial Transplant 7:367, 1992.

598. R Deacon, et al. Invalidity from nonparellelism in a radioimmunoassay for erythropoietin accounted for by human serum antibodies to rabbit IgG. Exp Hematol 21: 1680, 1993.

599. MH Schlageter, et al. Radioimmunoassay of erythropoietin—Analytical performance and clinical use in hematology. Clin Chem 36:1731, 1990.

600. RK Shadduck, et al. Development of a radioimmunoassay for human macrophage colony stimulating factor (CSF-1). Ann NY Acad Sci 554:156, 1989.

601. M Ranby, N Bergsdorf, T Nilsson, G Mellbring, B Winblad, G Bucht. Age dependence of tissue plasminogen activator concentrations in plasma, as studied by an improved enzyme-linked immunosorbent assay. Clin Chem 32:2160, 1986.

602. JM Sailstad, et al. Pharmacokinetics of the antidiarrheal pentapeptide, BW942C, in dog, monkey and man determined by a new radioimmunoassay. Fed Proc 46: 867, 1987.

603. BM Peskar, BA Beskar, JC Turner. Radioimmunoassay for carbenoxolone. J Pharm Pharmacol 28:720, 1976.

604. T Okumura, et al. Enzyme immunoassay for the drug of antiulcer using avidin biotin system. Chem Pharm Bull 39:1779, 1991.

605. RL Chang, AW Wood, WR Dixon, AH Conney, KE Anderson, J Eiseman, AP Alvares. Antipyrine radioimmunoassay in plasma and saliva following administration of a high dose and a low dose. Clin Pharmacol Ther 20:219, 1976.

606. B Starcher, et al. Fractionation of urine to allow demosine analysis by radioimmunoassay. Ann Clin Biochem 29:72, 1992.

607. E Maggi, E Pianezzola, G Valzelli. Radioimmunoassay of glipizide in human plasma. Eur J Clin Pharmacol 21:251, 1981.

608. B Nieuweboer, D Gabriel, K Lubke. A method for radioimmunological assay for glisoxepide. Arzneim-Forsch 26:1633, 1976.

609. M Kanaoka, et al. Radioimmunoassay of glycyrrhetic acid. J Pharm Sci 76:S202, 1987.

610. M Kanaoka, et al. Studies on the enzyme immunoassay of bioactive constituents contained in oriental medicinal drugs—Enzyme immunoassay of glycyrrhetic acid. Chem Pharm Bull 36:8, 1988.

611. S Miyairi, et al. Enzyme immunoassay for 4-hydroxy-2-(4-methyphenyl)benzothiazole. J Pharm Biomed Anal 11:469, 1993.

612. PC Davis, et al. A specific radioimmunoassay for the measurement of ICI-200, 880, an elastase inhibitor, in human serum. J Pharm Biomed Anal 11:549, 1993.

613. AC Lane, et al. Determination of idazoxan in plasma by radioreceptor assay. J Pharm Biomed Anal 6:787, 1988.

614. KP Vyas, et al. Radioimmunoassays for the enantiomeric components of indacrinone and their phenolic metabolites. J Immunoassay 8:179, 1987.

615. KL McGilliard, GD Olsen. Sterospecific radioimmunoassay of alpha-1 acetylmethadol (LAAM). J Pharmcol Exp Ther 215:205, 1980.

616. AS Yuan, et al. Determination of MK-852, a new fibrinogen receptor antagonist, in plasma and urine, by radioimmunoassay. J Pharm Biomed Anal 11:427, 1993.

617. J Linden, et al. The precise radioimmunoassay of adenosine—Minimization of sample collection artifacts and immunocrossreactivity. Anal Biochem 201:246, 1992.

618. AC Lawrence, et al. An alternative strategy for the radioimmunoassay of angiotensin peptides using amino-terminal directed antisera—Measurement of 8 angiotensin peptides in human plasma. J Hypertens 8:715, 1990.

619. JR Voelker, SL Cobb, RR Bowsher. Improved HPLC-radioimmunoassay for quantifying angiotensin in plasma. Clin Chem 40:1537, 1994.

620. JK Horton, et al. Enzyme immunoassays for the estimation of adenosine 3′,5′-cyclic monophosphate and guanosine 3′,5′-cyclic monophosphate in biological fluids. J Immunol Meth 155:31, 1992.

621. CK Daniels, et al. A solid-phase radioimmunoassay for cyclic AMP. J Pharmacol Toxicol Meth 31:41, 1994.

622. D Carriere, et al. Two-site enzyme immunoassay of CD4 and CD8 molecules on the surface of T Lymphocytes from healthy subjects and HIV-1-infected patients. Clin Chem 40:30, 1994.

623. M Lackmann, et al. Radioimmunoassay for the detection of active site specific thrombin inhibitors in biological fluids—Heparin affects the binding of hirudin to alpha-thrombin. Thromb Res 63:609, 1991.

624. M Lackmann, et al., Radioimmunoassay for the detection of active site specific thrombin inhibitors in biological fluids—Assay characteristics and quantitation of recombinant hirudin. Thromb Res 63:595, 1991.

625. B Mille, et al. Two-site immunoassay of recombinant hirudin based on two monoclonal antibodies. Clin Chem 40:734, 1994.

626. M Andersson, et al. Measurement of histamine in nasal lavage fluid—Comparison of a glass fiber based fluorometric method with 2 radioimmunoassays. J Allergy Clin Immunol 86:815, 1990.

627. WF Blum, et al. Clinical studies of IGFBP-2 by radioimmunoassay. Growth Regulat 3:100, 1993.

628. SCC Hughes, et al. Radioimmunoassays for IGFBP-3 using different site specific antisera—Divergence of which reflect the presence of circulating protease. J Endocrinol 135:P73, 1992.

629. JV Peppard, et al. A simple and rapid radioimmunoassay for human interleukin-6. J Immunol Meth 148:23, 1992.

630. H Brailly, et al. Total interleukin-6 in plasma measured by immunoassay. Clin Chem 40:116, 1994.

631. K Iwata. One step sandwich enzyme immunoassay for human laminin using monoclonal antibodies. Clin Chim Acta 191:211, 1990.

632. C Antoine, et al. Development of enzyme immunoassay for leukotrienes using acetylcholinesterase. Biochim Biophys Acta 1075:162, 1991.

633. DK Lahiri, et al. Factors that influence radioimmunoassay of human plasma melatonin—A modified column procedure to eliminate interference. Biochem Med Metab Biol 49:36, 1993.

634. DA Monaghan, et al. Sandwich enzyme immunoassay of osteocalcin in serum with use of an antibody against human osteocalcin. Clin Chem 39:942, 1993.

635. S Shimohama, et al. Assessment of protein kinase-C isozymes by 2-site enzyme immunoassay in human brains and changes in alzheimers disease. Neurology 43: 1407, 1993.

636. H Shionoiri, et al. Measurement of active renin by solid-phase radioimmunoassay using monoclonal antibodies. Am J Med Sci 300:138, 1990.

637. R Djurup, et al. Rapid, direct enzyme immunoassay of 11-keto-thromboxane B2 urine, validated by immunoaffinity/gas chromatography-mass spectrometry. Clin Chem 39:2470, 1993.

638. A Reuter, et al. Radioimmunoassay of tumor necrosis factor TNF-alpha and its clinical application by direct determination in serum. J Leukoc Biol 42:604, 1987.

639. GR Adolf, et al. A monoclonal antibody based enzyme immunoassay for quantitation of human tumor necrosis factor binding protein-I, a soluble fragment of the 60 KDa TNF receptor, in biological fluids. J Immunol Meth 143:127, 1991.

640. L Disilvio, et al. A radioimmunoprecipitation assay for antibodies to growth hormone. J Endocrinol 112:239, 1987.

641. T Kohno, et al. A highly sensitive enzyme immunoassay for anti-insulin antibodies in human serum. J Clin Lab Anal 1:170, 1987.

642. VJJ Bom. Solid phase enzyme immunoassay of antistreptokinase antibodies in human plasma. Thromb Haemost 65:1268, 1991.

643. E Ishikawa, et al. Principle and applications of ultrasensitive enzyme immunoassay (immune-complex transfer enzyme immunoassay) for antibodies in body fluids. J Clin Lab Anal 7:376, 1993.

644. E Kopetzki, K Lehnert, P Buckel. Enzymes in diagnostics: Achievements and possibilities of recombinant DNA technology. Clin Chem, 40:688, 1994.

645. A Witkowski, et al. Preparation of β-Galactosidase conjugates for competitive binding assays by posttranslational modification of recombinant proteins. Anal Chem 67:1301, 1995.

646. SO Leung, et al. Engineering a unique glycosylation site for site-specific conjugation of haptens to antibody fragments. J Immunol 154:5919, 1995.

647. WA Colburn. Radioimmunoassay and related immunoassay techniques, In: JW Munson, ed. Pharmaceutical Analysis Modern Methods, Part A. New York: Marcel Dekker, 1982, p. 381.

7

Applications of Capillary Electrophoresis Technology in the Pharmaceutical Industry

Charles J. Shaw and Norberto A. Guzman
The R. W. Johnson Pharmaceutical Research Institute, Raritan, New Jersey

I. INTRODUCTION

Pharmaceutical companies are facing new challenges in the starting millennium. Drug discovery is a high-stakes game; therefore, innovative strategies are being planned for the research and development of new and novel breakthrough products (1–7). Genomics, high-throughput screening, robotics, combinatorial chemistry, proteomics, proteinomimetics, informatics, and miniaturization are some of the strategies that are revolutionizing drug discovery efforts (8–18). In addition, the genome age will change biology forever, providing sequence blueprints for numerous bacteria, fungi, plants, and animals, and thus facilitating the understanding of human behavior, disease, and other health issues. Nevertheless, as the Human Genome Initiative nears completion, it is becoming increasingly clear that the behavior of gene products is difficult or impossible to predict from the gene sequence (19–21).

Genomics, or the analysis of gene sequences, allows identification of novel proteins which are potential "drug candidate molecules" targeted as therapeutic agents (22,23). The ability to analyze entire genomes is accelerating gene discovery (8) and presents unique scientific opportunities for understanding the structure and function of the proteins coded for those genes (24–27). The exponential growth of knowledge obtained about genes, from the early experiments of Mendel to the most recent high-throughput sequencing of DNA by capillary array electro-

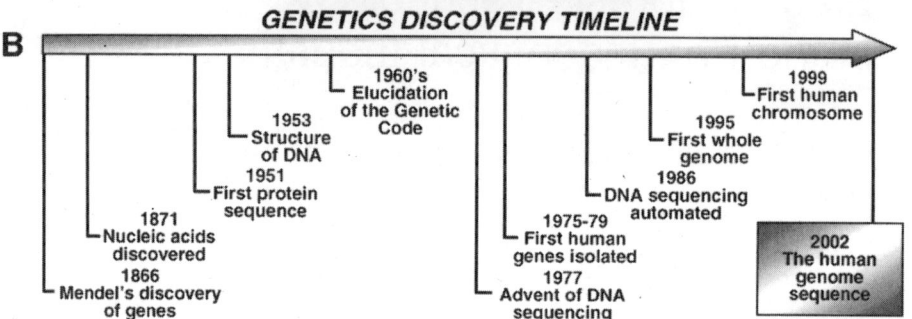

Fig. 1 Genetic discovery timeline. This schematic is a brief chronological outline and essence of the various events that took place for more than a century regarding gene studies. The information covers the period from 1866 to 1871, which describes the pioneering work of Mendel and Miesch in the discovery of genes and DNA, respectively, through the beginning of the massive DNA sequencing that started in the early 1980s. A powerful tool that accelerated the sequencing of the human genome was the use of high-throughput DNA sequencers based on multicapillary-array electrophoresis. (A) Representation of a model system of coupling conventional slow-pace techniques with modern fast-pace and high-throughput techniques for the advancement of science. (B) Representation of some of the key steps that were needed in the genomic era to reach the massive undertaking of sequencing the human genome. (Adapted from Ref. 28.)

phoresis, is illustrated in the genetics discovery timeline (Ref. 28 and Fig. 1). A complete understanding of the DNA code, which underlies human physiology, is crucial to unlocking the mystery of normal functioning and disease at the molecular level (29–32). Comprehending how that happens is, literally, having the keys to life. Furthermore, it is the gateway to developing better drugs more rapidly. The entire history of medical research is that once a disease is well understood, it is only a matter of time until that disease or its effects are ameliorated.

Proteomics is a field that promises to bridge the gap between sequence and cellular behavior. It aims to study the dynamic protein products of the genome and their interactions, rather than focusing on the simple static DNA blueprint of a cell. Understanding the proteome poses an even greater challenge than sequencing the genome (33–42).

High-throughput screening techniques and combinatorial chemistry have greatly expanded the compound universe from which to pick our development candidates (43). Combinatorial chemistry produces extensive chemical libraries and screens them for potential pharmacological activity by searching among a large source of chemical entities for promising lead compounds (10–13). Once a promising lead compound is identified in a preliminary screen, it is turned over to chemists for complete characterization, modification, and eventually for synthesis in bulk amounts. Combinatorial chemistry in its various manifestations enables the generation of new molecules with unprecedented speed. As a result, the supply of lead compounds has never been greater. However, the evolution from promising lead to useful drug requires much optimization, ranging from the reorganization of key structural features on the molecule to the reconfiguration of asymmetric centers. Biocatalysis is particularly well suited to these tasks. In fact, researchers must do more work after finding hits from screening combinatorial libraries of pure compounds. A million-compound library may not be any more valuable than a ten-thousand-compound library. What is really more important is the types of compounds in the library, their diversity and physical properties, molecular weight, substructure, and ease of incorporation (44,45).

It is noteworthy that a compound with a great receptor affinity and selectivity, but with poor biopharmaceutical properties for formulation or delivery, is rarely regarded as ineligible to enter development. Desirable characteristics of the "well-designed compound" can include increase potency, higher selectivity, reduced toxicity, good solubility, excellent stability, effective permeability, and greater bioavailability. The overall aim of combinatorial chemistry is to make new molecules faster, more cheaply, and in numbers large enough for high-throughput screening (HTS). High-throughput screening refers to the integration of technologies to rapidly assay thousands of chemical compounds simultaneously in search of biological activity and a potential drug (13,43,46). The application of HTS techniques represents a major increase in automation of the drug discovery laboratory.

The success of drug development in the pharmaceutical industry, which uses genomics and combinatorial chemistry strategies as well as proteomics, is contingent upon the availability of a battery of tools (13,38,39,47,48). Capillary electrophoresis (CE), or separations on micromachined planar substrates such as microchips (MC), are outstanding state-of-the-art modern analytical tools to assist these strategies, as well as to provide additional information through the development of numerous and diverse kind of applications aimed at the diagnostic and pharmaceutical industry (49–68).

Once a potential useful drug exits the phase of drug discovery, it enters the phase of drug development. From the first part of drug development until the drug is approved, a series of analytical, chemical, toxicological, and ADME (absorption, distribution, metabolism, and excretion) studies are carried out. The timeline for drug development from preclinical research to postmarketing surveillance is illustrated in Ref. 69 and Fig. 2. The entire process from discovery to lead candidate selection, through development, and to registration is highly integrated, with overlap and sharing of the informational content across the various phases (the preclinical phase or Phase 0, the three clinical phases 1, 2, and 3, and even the postmarketing surveillance phase or pharmacovigilance phase, which is commonly referred as Phase 4). Currently, pharmaceutical companies are trying to optimize the science of drug development, from production to evaluation, in order to manufacture a successful drug product (70). Physicochemical characterization studies of the drug start in the early part of the discovery process and continue until completion of the finished product. Stringent requirements by regulatory agencies (71–73) require fine characterization of drug substances, isomeric forms, degradation products, excipients, impurities, contaminants, and possible micromolecular interaction of components in finished pharmaceutical products.

TIMELINE FOR DRUG DEVELOPMENT

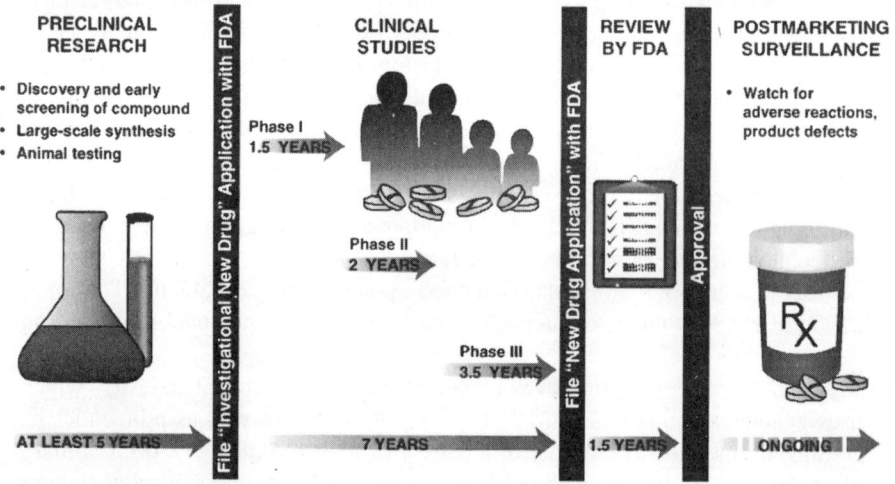

Fig. 2 Timeline for drug development. The schematic representation of the time and steps needed in the development of a successful drug to reach the market. The diagram represents the process of manufacturing a drug, starting at the early part of the discovery and screening of a target compound, continuing until reaching the market. The entire process must comply with a series of stringent requirements by regulatory agencies. (Modified from Ref. 69.)

Furthermore, the requirements for stability tests, which are usually performed on hundreds of samples, require reliable, rugged, and fast analytical tests that are crucial in the control and research laboratories in order to meet these demands. Sample preparation may not be a glamorous part of chemical analysis, but it is where most of the errors occur and where most of the analytical time is spent. One way of improving sample preparation is to automate these procedures to shorten analysis time and increase throughput by parallel formats. Advances in micro-electro-mechanical systems (MEMS) has afforded accurate and reproducible injection volumes by means of microelectronic and micromechanical pumping, even at volumes of less than one microliter. New technology has recently emerged that has the potential to take automation and miniaturization to a level unimaginable just a few years ago, drastically reducing costs and sample analysis time. A comparison of analysis time and sample costs utilizing various analytical techniques is presented in Tables 1 and 2.

Pharmaceutical analysts rely heavily on separation science to perform the physicochemical characterization required from drug development to stability testing of approved drugs (74–80). Many challenges are presented for analytical chemists (see Table 3), since potential pharmaceutical drugs have a wide range of physicochemical properties, and a characterization study may take from one week to several months. For years, gas chromatography (GC) and high-

Table 1 Comparative Analysis Time and Cost for the Assessment of a Pharmaceutical Drug (Various Formulations)[a]

Cost and analysis time	Methodology	HPLC	CE
Cost per Analysis (US $)	Column	150–450	10–100
	Number of uses	100–500	40–200
	Solvents (and reagents)	40–200[b]	4.0–50[b]
	Solvent disposal	30–400	0.5–10
	Total cost (per sample)	2.0–5.0	0.5–1.0
Time for analysis (min)	Equipment setup	30–60	15–45
	Sample preparation	10–30	10–30
	Run	15–30	10–60
	Total	55–120	30–120

[a] Multiple formulations are routinely analyzed in the pharmaceutical laboratory. Samples range from a simple formulation (having just buffer as an excipient) to very complicated formulations (containing numerous excipients). The finished drug product could be in a simple solution, or in a complex cream. Thus the matrix, separation medium, and experimental conditions are the limiting factors for the column and capillary life. The numbers in the table are based on the worse-case scenario. Furthermore, improvements in HPLC column packings and in the capillary electrophoresis column chemistries require certain optimal conditions, which are sensitive to drastic changes. These latter optimal conditions may also influence the numbers described in the table
[b] Some reagents may include cyclodextrins and/or other expensive buffer additives.

Table 2 Comparative Analysis Time and Cost for a Typical Analyte
Characterization Using Various Analytical Separation Techniques

Technique	Total analysis time (min)	Throughput	Processing of assay points per day	Cost per sample (US $)
GC	30–60	Medium	Up to 50	>2.0
HPLC	30–60	Medium	Up to 50	>2.0
CE	15–45	High	>1000[a]	<0.025
Microchip	0.2–2.0	High	>5000[a]	<0.005

[a] When utilizing an array system.

performance liquid chromatography (HPLC) have been the methods of choice
for analytical tests performed routinely in the pharmaceutical laboratory. A major
requirement of these quantitative techniques is a high degree of accuracy and
precision. These techniques include the determination of physicochemical characterization studies on the drug substance and drug product. Although HPLC has
been accepted for the last three decades, a fairly recent technology, capillary
electrophoresis (capillary-based and microchip-based) is gaining popularity in
the laboratory. CE and MC are versatile techniques which have evolved in recent
years into the forefront of analytical methodology. Capillary electrophoresis and
CE microchip are powerful modern techniques for separating components in simple and complex matrices. It is a family of separation methods that includes free-
zone electrophoresis, capillary gel electrophoresis, capillary isotachophoresis,
capillary affinity electrophoresis, capillary electrochromatography, and micellar
electrokinetic chromatography (81–86). These methods are capable of handling
the identification and characterization of a diverse group of substances over a
wide concentration range. Using capillary electrophoresis, substances can be sep-

Table 3 Main Challenges for
Analytical Chemists

Resolution
Sensitivity
Speed
Labor
Throughput
Cost saving
Miniaturization
Understanding chemical reac-
  tions at the molecular level

arated in a variety of background electrolytes, including aqueous buffers, partially aqueous buffers, and completely nonaqueous buffers, when dealing with insoluble substances (87–92). The sample volumes required for analysis are very small, and a variety of specialized detectors can be employed (93–101).

The main advantage of modern CE over traditional or conventional chromatographic techniques is the ability to produce a higher number of theoretical plates in a liquid-phase separation, relative simplicity, full automation, and cost effectiveness. Efficient separations result from:

> The high electric fields which can be used effectively because of the high surface-area-to-volume ratio offered by the small-diameter capillary tube, permitting an efficient heat dissipation
> Flat flow generated by the electroosmotic flow

Recently, CE has benefited from several major innovations. The instrumentation has been improved by the concept of parallel processing or multiplexing (see Refs. 102–108 and Sec. VIII). Characterizing so many compounds creates significant bottlenecks. Analysts who test many samples can now use parallel sample introduction into an array of capillaries or microchip channels. This design is different from single-probe autosamplers, which transfer samples one at a time. Multiplexing is the process by which multiple samples are analyzed in a single assay. Here, the single-capillary apparatus is extended to a multiple-capillary array instrument. This implementation allows the throughput of the system to be increased by the number of capillaries in the array. In addition, these new systems offer improved detection schemes, such as laser-induced fluorescence and mass spectrometry.

Another recent innovation deals with the chemistry of the capillary wall. For example, capillaries have been improved by coating their inner surfaces with novel chemistries, or by adding a replaceable dynamic coating solution, allowing better resolution of analytes (109–114). Carrier solutions or background electrolytes can be enriched with a variety of additives, permitting the separation of closely related substances at baseline resolution (see Refs. 115–118 and Sec. II). These improvements have lead to better precision, selectivity, and accuracy of separated analytes. Precision and accuracy is improved for both peak migration and peak area counts for the simple and complex analytes.

Although CE technology has many advantages as an analytical tool, it has some disadvantages which are currently under investigation for improvements. For example, two major CE drawbacks frequently discussed by investigators are (a) the low concentration limits of detection (CLOD) and (b) the lack of detectors which provide structural information. Recent advances have addressed both of these issues. Enhancements in CLOD have been achieved by the advent of a number of preconcentration methods, in particular the use of on-line solid-phase microextractors or analyte concentrator-microreactors (see Refs. 50,51,58,68,

119,120 and Sec. VII). Chemically selective detectors (mass spectrometers, laser-induced fluorescence detectors, etc.) have been available for more than a decade. This solves some of the problems in CE detection expressed since the inception of CE technology.

Capillary electrophoresis has a number of advantages compared to HPLC which are immediately apparent. In the pharmaceutical industry, these include:

Rapid method development and optimization
Faster analysis time
A choice of separation modes
The use of low volumes for expensive reagents
A reduction in the generation of toxic organic waste
Relatively lower operating cost

Capillary electrophoresis has become extremely important in several application areas: (a) enantiomeric separations can be achieved by the simple addition of a chiral additive to the run buffer; (b) rapid and distinctive profiles for different isoform groups can be obtained by reversed-charge CE separations of heterogeneous mixtures; and (c) inorganic and organic ion determination has improved significantly with new detection systems.

In this overview, we will discuss some of the most common applications in the pharmaceutical industry environment that utilize capillary electrophoresis and microchip technology. Additional information can be found elsewhere (50–69,121–128). The pie chart, which is presented in Fig. 3, summarizes the advantages of capillary electrophoresis. These results are based on a recent survey of pharmaceutical scientists (129).

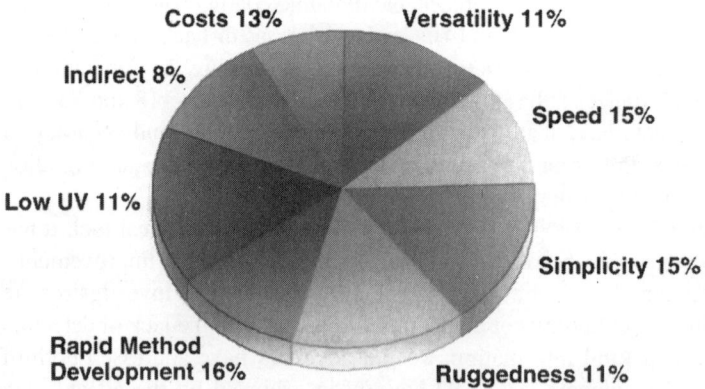

Fig. 3 Graphical representation of the advantages of capillary electrophoresis. (Modified from Ref. 129.)

II. ENANTIOMERIC SEPARATIONS

Life is heavily dependent on chirality. For example, the human body metabolizes only the D-enantiomer of glucose and produces only L-amino acids. Chiral discrimination is frequently encountered in biological systems. A key question in the chemical development and pharmacological testing of new pharmaceutical compounds is the purity of the stereochemically drug substance in use (130–135). The disaster with thalidomide (N-alpha-phthal-imidoglutarimide) (136), prescribed as a sedative to pregnant women in the late 1950s and early 1960s, is a classic example that has been used to illustrate the significance of drug stereochemistry (137–140). The R isomer has the desired sedation effect, while the S isomer causes the fetal deformity phocomelia. The accurate determination of enantiomeric purity is a major concern for the analytical chemist. The stereochemistry of the molecule is provided when documentation is submitted to the U.S. Food and Drug Administration (FDA) for the manufacture and control of the drug substance (141,142). The FDA's policy statement for development of the new stereo isomeric drugs can be found on the internet at URL: http://www.fda.gov/cder/guidance/stereo.htm.

During chemical development of the intermediates and final drug, a selective and sensitive technique is needed to measure enantiomeric purity, because the pharmacological and pharmacokinetic differences between drug enantiomers are often significant. By definition, stereoisomers are isomers that are alike one another with respect to the way the atoms are joined, but differ from each other by the position of the atoms in space. This includes enantiomers and diastereomers. An equimolar mixture of a pair of enantiomers is known as a racemate. A molecule having at least one pair of enantiomers is a chiral compound. Enantiomeric molecular entities are nonsuperimposable mirror images, while diastereomers are not mirror images. Diastereomer examples are *cis*, *trans*, *meso*, *Z* and *E* isomers. The Prague-Ingold terminology (143,144) is used when dealing with enantiomers and, due to their importance, this discussion will deal only with enantiomers. The reader is referred to the listed reviews (145–148) for a more complete description of stereochemistry.

Enantiomeric resolution can be obtained by direct and indirect methods. The derivatization of a mixture of enantiomers with an optical active agent (149,150) into diastereomers is the indirect mode. The diastereomers are separated using achiral HPLC or GC. The direct mode depends on the formation of labile diastereomers (formed by hydrogen bonding, dipole–dipole, π–π and/or hydrophobic interactions) between the enantiomers and chiral environment, a chiral stationary phase, with which they interact. HPLC, using chiral stationary phases and derivatization reactions (151–156), has dominated this field, but the increased selectivity and versatility of CE, as well as the ease of incorporating various chiral selectors, make it a viable alternative. The number of chiral selector additives (157–160) available make CE a true alternative to HPLC. CE is also more economical and has a much higher efficiency when compared to HPLC. The separation method in which CE and micellar partition chromatography are combined is known as electrokinetic chromatography (EKC). Micellar EKC (MEKC) uses a pseudo-stationary phase by which the separation of neutral or charged compounds can be obtained. Migration characteristics similar to reversed-phase HPLC separations can be obtained by using SDS micelles or other surfactants. Chiral selectivity is achieved by adding to the MEKC operating buffer an optically active chiral selector in place of micelles as constituents of the buffer (161,162). Preferential interaction of one enantiomer with the chiral selector before the other enantiomer gives the desired resolution.

The chiral buffer additives include antibiotics, cyclodextrins, crown ethers, polysaccharides, proteins, and chiral surfactants. Cyclodextrins are the most versatile and popular chiral selectors (163–168).

A. Cyclodextrins

Cyclodextrins (CD) are naturally occurring cyclic oligosaccharides that have a truncated cone structure. The three different forms of cyclodextrins, α, β, and γ, are based on the size of the cone openings. The hydroxyl groups, which ring the cavity, can be further modified to increase chiral selectivity by chemically changing this functional group with hydroxypropyl or dimethyl, amino, sulfated, or carboxylic groups. These chemical modifications results in ionic chargeable CDs which enhance the separation of ionic or neutral drug enantiomers.

Enantiomeric resolution of solutes that fit within the molecular cavity, which is chiral, results in the formation of an inclusion complex (Ref. 169 and Fig. 4). In general, the enantiomers are separated on the basis of formation constants of the host–guest complexes. The enantiomer that forms the more stable complex has a greater migration time because of this effect. The chiral recognition mechanism for cyclodextrin enantioseparation has been discussed in several works (163–168).

The following applications are provided as examples of the chiral selectors listed below. The purpose is to demonstrate the usefulness of capillary electrophoresis for the resolution and quantitation of enantiomers. Additional applications can be obtained from specific literature searches or from CE reviews on applications to pharmaceutical analysis (163–170).

One interesting application is the separation of the enantiomeric aspartyl dipeptides. Figure 5 shows the electropherogram of Asp-PheOMe and Asp-PheNH$_2$ at pH 3.3. At this pH, the amino groups of the peptides are fully protonated. Small differences of the pK_a of the carboxylic acid groups result in different electrophoretic mobilities of the diastereomers. Some of the β-peptides (β-L/D and β-D/L) are apparently more acidic than the α-isomers (α-L/D and α-D/L) and therefore migrate more slowly (171). Other β-isomers (β-L/L and β-D/D) migrate more quickly than the α-isomers (α-L/L and α-D/D).

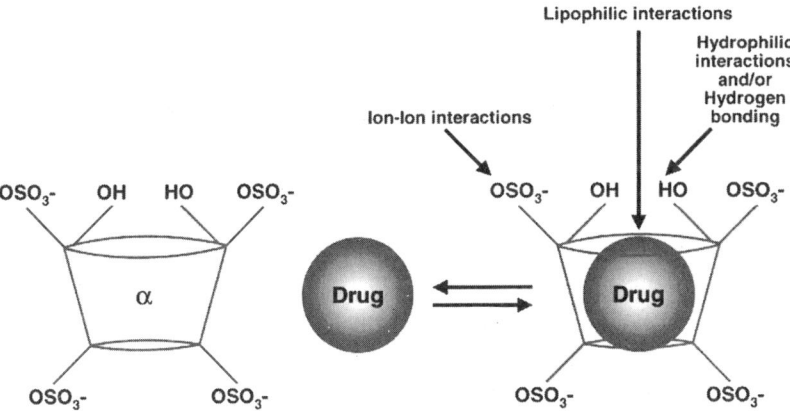

Fig. 4 Schematic representation of the interaction of a representative drug substance with cyclodextrin. A wide range of noncharged and charged (anionic, cationic, and ampho-teric) cyclodextrins have been used as chiral selectors, as well as for the optimized separa-tion of nonchiral compounds using capillary electrophoresis. Cationic and amphoteric cyclodextrins are less commonly used in chiral analysis, and only a few are commercially available. The degree of substitution of a cyclodextrin may vary from one manufacturer to another or even from batch to batch, which may have a detrimental effect on the repro-ducibility and ruggedness of the separation system. (Modified from Ref. 169.)

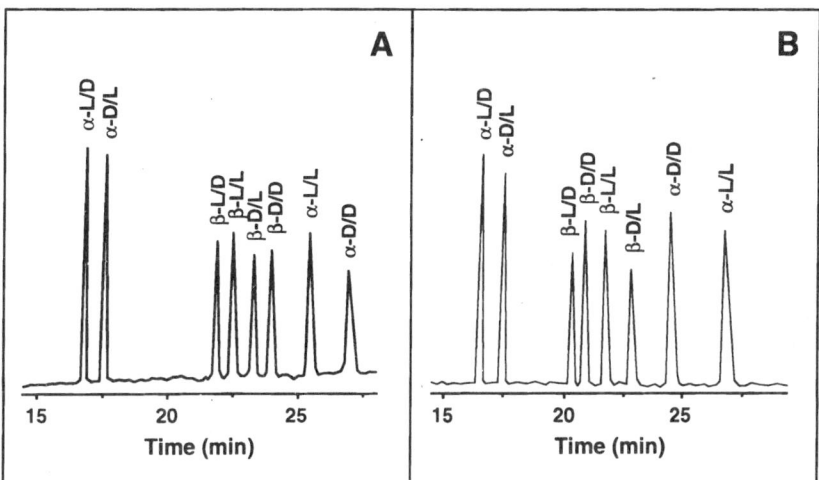

Fig. 5 A typical application of capillary electrophoresis for chiral separations of small peptides. Chiral separation of the isomeric dipeptides (A) of Asp-PheOMe and β-Asp-PheOMe and (B) of Asp-PheNH$_2$ and β-Asp-PheNH$_2$. The separation was carried out at pH 3.3 using 15 mg/mL of carboxymethyl-β-cyclodextrin as the chiral selector. (For de-tails of experimental conditions, see Ref. 171.)

The use of sulfated β-cyclodextrins as chiral resolving agents for the quantitative determination of chiral purity of pharmaceutical compounds has shown great utility (172–178). A systematic method development approach was used by modifying selected parameters such as concentration of the chiral selector, buffer pH, organic modifiers, buffer concentration, temperature, and applied voltage (179–186). This investigation provided an improved understanding of the separation mechanism. The optimized method was validated in terms of the chiral selector, linearity, sensitivity, accuracy, recovery, ruggedness, and precision.

The enantiomeric separation of some racemic antihistamines and antimalarials, namely (+/−)-pheniramine, (+/−)-bromopheniramine, (+/−)-chlorpheniramine, (+/−)-doxylamine, and (+/−)-chloroquine, were investigated by capillary zone electrophoresis (CZE). The enantiomeric separation of these five compounds was obtained by addition of 7 mM or 1% (w/v) of sulfated β-cyclodextrin to the buffer as a chiral selector. It was found that the type of substituent and degree of substitution on the rim of the cyclodextrin structure played a very important part in enhancing chiral recognition (174). The use of sulfated β-cyclodextrin mixtures as chiral additives was evaluated for the chiral resolution of neutral, cyclic, and bicyclic monoterpenes. While there was no resolution of the monoterpene enantiomers with the sulfated β-cyclodextrin, the addition of α-cyclodextrin resulted in mobility differences for the terpenoid enantiomers. Resolution factors of 4–25 were observed. The role of both α-cyclodextrin and sulfated β-cyclodextrin in these separations was discussed (187). The enantiomeric separation of 56 compounds of pharmaceutical interest, including anesthetics, antiarrhythmics, antidepressants, anticonvulsants, antihistamines, antimalarials, relaxants, and broncodilators, was studied. The separations were obtained at pH 3.8 with the anode at the detector end of the capillary. Most of the 40 successfully resolved enantiomers contained a basic functionality and a stereogenic carbon (173).

A method development protocol, based on the charged resolving agent migration model, for the analysis of a minor enantiomer using a single-isomer sulfated cyclodextrin has been used (188). The single isomer heptakis(2,3-diacetyl-6-sulfato)-β-cyclodextrin has been described (188) to resolve a variety of weak base pharmaceuticals in 100% methanol background electrolyte (189). As predicted by the charged resolving agent migration model of enantiomer separations (190), very large separation selectivities were observed for the cationic analytes around the heptakis(2,3-diacyl-6-sulfato)-β-cyclodextrin concentrations where the effective mobilities of the slower migrating enantiomers approached zero. Neutral analytes and acid analytes complexed very weakly with the single-isomer sulfated cyclodextrin in the methanol background electrolyte.

There is a plethora of reports in the literature on CE chiral separations. Approximately 80% use cyclodextrins as the chiral resolving agent. The use of cyclodextrins buffer additives in capillary zone electrophoresis is the most popular means of obtaining chiral resolution (163–167,191–193). In capillary zone electrophoresis, direct separation is obtained by adding the appropriate cyclodextrin, which has previously been dissolved in the buffer electrolyte. The cyclodextrins have been used individually or as a mixture of different cyclodextrin types added to the separation buffer.

Chiral basic drug enantiomers can be separated at low pH using phosphate buffer containing cyclodextrins, while high-pH electrolytes can separate chiral acidic drug enantiomers. A rapid method for developing separation conditions for the chiral resolution of basic or acidic drugs using CD additives has been described (181). Hydroxypropyl-α-cyclodextrin was used to resolve 34 of 86 drugs tested into enantiomeric pairs (194). Using a mixture of anionic charged and neutral cyclodextrins (195), it was possible to resolve a series of acidic nonsteroidal anti-inflammatory drugs, barbiturates, and anticoagulants.

A simple method for the resolution of underivatized aromatic amino acid enantiomers used alpha-cyclodextrins (196). Cyclodextrin concentration, applied voltage, and buffer pH were studied to obtain the best separation conditions. The effects of several cyclodextrins and their concentration on enantiomeric resolution of profen enantiomers were examined. Heptakis-2,3,6-tri-O-methyl-β-cyclodextrin was found to be the only chiral selector for the quantitative enantiomeric resolution of the drug compounds tested (197). Comparison of chiral recognition capabilities of cyclodextrins for the separation of basic drugs has also been investigated (198). The type of substituents and the degree of substitution on the rim of the CD structure played an important role in enhancing chiral recognition. Cyclodextrins with negatively charged substituents and a higher degree of substitution on the rim of the CD structure gave better resolution for the cationic racemic compounds compared with cyclodextrins with neutral substituents. Furthermore, lower concentrations of negatively charged cyclodextrins were necessary to achieve equivalent resolutions as compared to the neutral ones. Three β-cyclodextrin derivatives, carboxymethyl-(CMCD), dimethyl-, and hydroxypropyl-β-cyclodextrin, were tested as chiral selectors for the enantiomeric separation of seven basic drugs using buffers adjusted to pH 3.0 with 100 mM phosphoric acid (199). The best results with respect to chiral resolution were obtained with CMCD. Resolution values up to 23.7 were obtained for several compounds using CMCD in the 5–15 mM concentration range.

Methods development for chiral analyses has been one of the most challenging separation problems for the analytical chemist in the pharmaceutical industry. Racemic drug substances have a variety of chemical structures and several chiral selectors are available for the analyst to choose in order to obtain the enantioselectivity needed for chiral resolution. To alleviate this problem, a fast capillary electrophoresis procedure for the enantiomeric separation of acidic and basic compounds using native and modified cyclodextrins has been described (200). The technique is called cyclodextrin array chiral analysis. A generalized optimi-

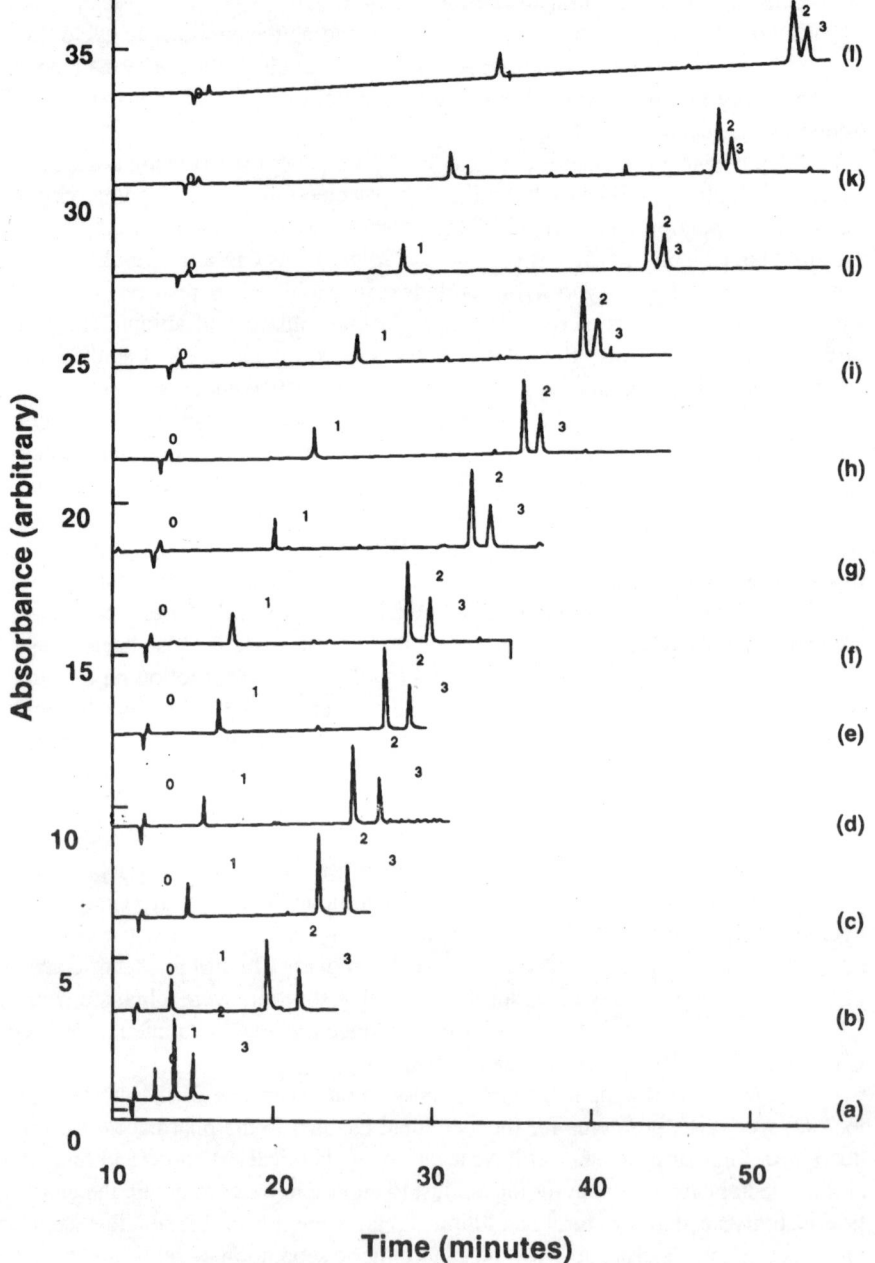

zation scheme eliminates the traditional hit-or-miss approach. The basis for capillary electrophoresis resolution of drug enantiomers and the factors affecting their enantiomeric separation can be found in a review article which describes basic principles and new developments. This review covers the use of CDs and some of their derivatives in the studies of pharmacologically interesting compounds (201).

Although cyclodextrins are used primarily for the enantiomeric separation of racemic compounds, other functions may also be of importance. For example, cyclodextrins have been used to optimize separation conditions of compounds which are not stereoisomers. Figure 6 depicts the separation of phenol, 2-naphthol, and 1-naphthol in the presence of an increasing concentration of sulfobutyl ether β-cyclodextrin (SBE-β-CD). The migration behavior of these analytes changes due to the binding of SBE-β-CD to the analytes, increasing their negative electrophoretic mobility at increasing concentrations (202). Figure 7 shows the separation of eight heterocyclic pharmaceutical compounds that are insoluble in aqueous buffers (169). To achieve optimal resolution, a combination of two conditions permitted successful baseline separations: the presence of a nonaqueous buffer and the addition of sulfated-α-cyclodextrin to the buffer (169).

B. Glycopeptide Antibiotics

Glycopeptide antibiotics have been found to be very effective chiral selectors in the enantiomeric separation of racemic pharmaceutical compounds. Vancomycin, ristocetin A, rifamycins, teicoplanin, kanamycin, streptomycin, and avoparcin have been added to the running buffer to obtain enantioseparation (161,203–207). A few technical modifications, such as coated capillaries and separation conditions in the reverse polarity mode (as opposed to normal polarity mode, where the flow is from anode to cathode) were found to improve sensitivity and increase efficiency (116,208).

The advantages and limitations of vancomycin as a chiral selector in CE has been reported (209,210). This investigation implied that chiral separations

---

Fig. 6 The effect of the addition of a sulfated β-cyclodextrin additive on the nonchiral separation profile of three analytes. The separation was carried out using increasing concentrations of sulfobutyl ether β-cyclodextrin (0–20 mM) in 160 mM borate buffer. The peak numbers correspond to 0 (EOF marker—methanol), 1 (phenol), 2 (2-naphthol), and 3 (1-naphthol). The concentrations of sulfobutyl ether β-cyclodextrin are (a) 0 mM, (b) 1.0 mM, (c) 2.0 mM, (d) 3.0 mM, (e) 4.0 mM, (f) 5.0 mM, (g) 7.5 mM, (h) 10.0 mM, (i) 12.5 mM, (j) 15.0 mM, (k) 17.5 mM, and (l) 20.0 mM. (For details of experimental conditions, see Ref. 202.)

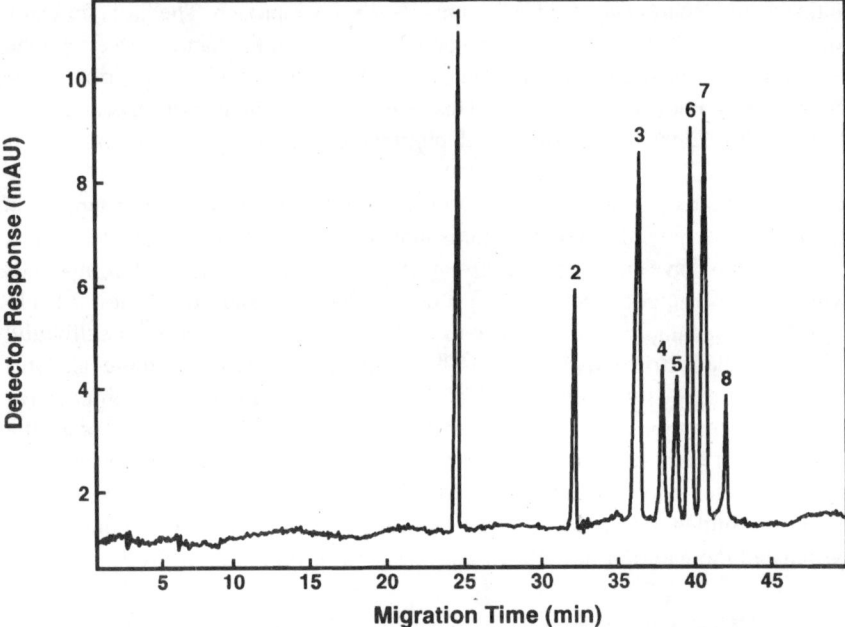

Fig. 7 The effect of the addition of a sulfated α-cyclodextrin additive on the separation profile of eight analytes. The separation was carried out in the presence of a fixed concentration (5 mM) of sulfated α-cyclodextrin in 80 mM phosphate buffer, pH 1.9, and 1:1 (vol/vol) acetonitrile/water. (For details of experimental conditions, see Ref. 169.)

with vancomycin need not be restricted to carboxylic acids. The interactions of vancomycin with analytes are affected by the concentration and chemical composition of the buffer as well as the buffer pH. Experimental and modeling studies were carried out to elucidate the similarities and differences for vancomycin, ristocetin A, and teicoplanin (211). Although these antibiotics have similar structures, they give similar but not identical enantioselectivities. From their studies, the authors invoked the "principle of complementary separations," that is, a partial resolution with one selector can be brought to baseline with one of the other antibiotic chiral selectors. Ristocetin A appeared to have the best antibiotic chiral selector capability for enantioseparations. A list of the most recent publications on the subject of chiral separations achieved by CE where antibiotics were used as chiral selectors can be found in recent review articles (193,204).

Rifamycin B resolved several racemic amino alcohols, such as atenolol, alprenolol, ephedrine, norphenylephrine, salbutanol, terbutaline, and others (212). The charge-charge, hydrogen bonding, and hydrophobic inclusion interactions between the negatively charged rifamycin and the amino alcohol provide chiral recognition. The authors (212) discuss the experimental factors for optimizing the separation. Teicoplanin was used to resolve over 100 anionic racemates. Teicoplanin's enantioselectivity was found to be the opposite of the ansa-type antibiotics (e.g., Rifamycin), which prefer cationic compounds. It was found to form micelles in aqueous solutions which are pH dependent, and in alcohol–water mixtures the teicoplanin precipitated. Additional investigation with vancomycin, ristocetin A, and teicoplanin examined the commonality for chiral recognition (213).

C. Other Chiral Selectors

The separation of primary amine and amino acid enantiomers and/or isomers was successfully accomplished by adding 18-crown-6-tetracarboxylic acid (18C6H4) to buffer constituents. A synergistic effect can be obtained for the resolution of the analytes, which were not resolved by either crown ether or the cyclodextrin, by using a combination of both chiral selectors. The buffer components, pH, and percentage of organic modifier affect resolution (214). The relative advantages and disadvantages of enantiomeric resolution of primary amines by CE and HPLC using chiral crown ethers have been reported (215). The separation techniques were found to be complementary. The separation of dipeptides containing two stereogenic centers were resolved into the four optical isomers with baseline separation (216).

A review of enantiomeric separations by CE using polysaccharides as chiral selectors has been reported (217). Ionic and neutral polysaccharides (e.g., heparin, chondroitin sulfate, dextrin, and maltodextrins) have been used to resolved enantiomers. Racemic acidic drugs were resolved with maltodextrins (218), while heparins and cyclodextrins were used to resolve oxamigue (219). A large number of drugs have been found to bind enantioselectively to proteins.

Alpha-1-acid glycoproteins (220,221), bovine serum albumin (222–224), pepsin, and cellobiohydrolase I (223) have been used as chiral selectors. In general, enantiomeric separation increases with increased protein concentration; however, this results in a sensitivity loss due to high background signal. The use of a partial filling technique for the capillary injection to overcome this limitation has been reported (220,221).

Chiral surfactants have been used for the enantiomeric separation of ionic and hydrophobic compounds. Poly-(sodium-N-undecylenyl-L-valinate) and poly-(L-SUV) can be used to resolve neutral, acidic, and basic compounds (225,226).

III. SMALL-MOLECULE SEPARATIONS

Numerous applications in the literature describe the analysis of small-molecule pharmaceuticals (see Refs. 56,57,59,121–129 and Table 4). A summary of the

Table 4 Determination of Small-Molecular-Weight
Pharmaceutical Drugs by Capillary Electrophoresis[a]

Drug	Mode of separation	References
Alendronate	CZE	227
Amiloride	CZE	228
Amitryptyline	NACE	229
Amoxicillin	CZE	230
Amphetamine	CZE, NACE/MS	231–235
Atracurium besylate	CZE	236
Bambuterol	CZE	237
Barbiturates	CZE/MEKC	238
Bendroflumethiazide	CZE	228
Benzodiazepines	CZE/MEKC	238
Beta blockers	CZE	239
Betametasone	CZE	240
Bile acids	EKC	241
Bumetanide	CZE	228
Caffeine	CEC	242
Calcium antagonist	EKC	243
Captopril	CZE	244
Cephalosphorin	CZE	245
Chlorothiazide	MEKC	246
Cimetidine	CZE	247
Clenbuterol	ITP-CZE	248,249
Clomiphene	EKC	250
Cocaine	CZE	234
Codeine	CZE	251–253
Dexamethasone	MEKC	254
Enalapril	MEKC	255
Ephedrine	CZE	233,256
Ergotamine	CZE	240
Fenoterol	CZE	257
Flufenamic	ACE	258
Flumethasone	MEKC	254
Flurbiprofen	ACE	258
Heroin	CZE	259
Hydrochlorothiazide	MEKC	246
Ibuprofen	CZE	251,252
Imipramine	NACE	260,261
Imipramine maprotiline	ITP	262
β-Lactam antibiotics	CZE, MEKC, ITP	263–265
Leucinostatins	CZE	266
Lidocaine	CZE	267,268
Methamphetamine	CZE	232,256,269

Table 4 Continued

Drug	Mode of separation	References
Methadone	CZE	232
6-Mercaptopurine	CZE	270,271
Methylthiouracyl	CZE	272
Morphine	CZE/MS	234,273
Paracetamol	MEKC	274,275
Penciclovir	CZE	276
Phenothiazines	EKC	277
Pilocarpine	CZE, MEKC	278,279
Polymyxins	CZE	280
Procaine	CZE	267,268
Propylthiouracyl	CZE	272
Ranitidine	CZE	281
Salbutamol	CZE	257,282
Selegiline	CZE	231,256
Sulfadiazine	CZE	283
Sulfamethoxazole	CZE	283
Sulfonamides	CZE, MEKC	284–287
Suramin	MEKC	288–290
Taxol	MEKC	291
Terbutaline	CZE	257,292
Tetracyclines	CZE	293,294
Theophylline	MEKC	242,295
Tiaprofenic	EKC	296
Triamterene	CZE	228
Tropane	MEKC, CZE/MS	297–299
Venlafaxine	EKC	300

[a] The various capillary electrophoresis applications shown in this table are only a selected group of pharmaceutical drugs analyzed by CE. For a more comprehensive studies of drugs determined by capillary electrophoresis, see Refs. 55–57,77,83,86,92,121–128.

drug classes that have been analyzed by the different modes of capillary electrophoresis has been described earlier (301). This review details applications from within a pharmaceutical analytical laboratory and provides selected literature examples. The applications include quantitative analysis of drug related impurities, drug substance in formulations, and chiral analysis.

Representative applications of small-molecule pharmaceuticals included in this section are antibiotics, barbiturates, flavonoids, vitamins, basic drugs, cold medicines, and corticosteroids. An MEKC method was validated according to

United States Pharmacopeia (USP) guidelines for the determination of a cephalosphorin in solution (302). The validation included specificity and linearity, and the methods were repeated by different analysts on different days. The determination of aminoglycoside antibiotics using direct and indirect ultraviolet detection has been reported (303). The separation of six aminoglycosides was accomplished using a background electrolyte of 0.1 M imidazole acetate at pH 5.0 with a small amount of surfactant. Tetracyclines and their degradation products were separated by CZE using electrolytes which contained ethylenediaminetetraacetic acid (EDTA) (304). These antibiotics degrade in acidic media to give anhydrotetracyclines which can epimerize to form other degradants. It is imperative to separate the degradants from the parent compound because of the differences in biological activity. The CZE method was found to be better than the HPLC method reported in the USP.

A set of 25 barbiturates was analyzed using CZE and MEKC. Buffers consisting of 90 mM borate, pH 8.4 (CZE), and 20 mM phosphate, 50 mM sodium dodecylsulfate (SDS), pH 7.5 (MEKC). The methods were evaluated for their suitability in systematic toxicological analysis (STA), especially when a combination of methods having a low correlation is used (305). A solid-phase microextraction device in combination with CE for the determination of barbiturates was described (see 306 and Sec. VII). The detection limit for 10 barbiturates was 0.1 ppm in urine, while the limit of detection was about 3 times poorer in bovine serum (306). Polyacrylamide-coated columns have been used for barbiturates and benzodiazepines. Seven kinds of barbiturates were sucessfully separated with the coated columns without further additives (307). The benzodiazepines, which are electrically neutral solutes, were separated in the presence of SDS. The CE method offered fast and efficient separations of the more hydrophobic solutes.

A mixture of seven hydroxylated and methoxylated flavonoids was resolved by MEKC using 20 mM borate, 25 mM SDS, pH 8.0, with 20% methanol. The capacity factors of the ionized flavonoids were determined after correction from the influence of buffer pH and their dependence versus the SDS micellar concentration. The correlation was found to be linear, and partitioning coefficients between the aqueous and SDS micellar phase were determined. Their values were interpreted from the structural features of these compounds (308). Both qualitative and quantitative determinations of several isoflavone materials have been reported (309). CE combined with electrospray ionization mass spectrometry in the negative mode was employed. Mass spectrometry overcame the low sample loading drawback, which is inherent in the CE technique. Quantitative analysis of synthetic mixtures were carried out using a calibration curve in the range of 0.2 to 20 fmol. The correlation coefficients obtained for the calibration curves ranged from 0.997 to 0.994 for pseudobaptigenin and daidzein, respectively.

After a bibliographic search of the *Analytical Abstracts* database (310), it was concluded that capillary electrophoresis is the best method for the simultaneous determination of water-soluble vitamins. The methods, HPLC, derivative spectrophotometry, and capillary electrophoresis, were summarized, and the criteria for their selection, performance characteristics, advantages and disadvantages, and practicability were discussed. The review contained 52 references. The quantitative analysis of thiamine, nicotinamide, riboflavin, pyridoxine, ascorbic acid, and panothenic acid was achieved using CZE with uncoated capillaries and UV detection (311). The method was very useful for the separation of more complex samples; a mixture of 10 water soluble vitamins completely resolved in 10 min. However, cyanocobalamine could not be separated from the nicotinamide in this CZE system. This was due to the compounds being in the uncharged form at the pH used. They are easily resolved by MEKC when the anionic surfactant dodecylsulfate is added to the run buffer. Six water-soluble vitamins could be resolved in 25 min using 0.02 M phosphate, pH 9.0, buffer containing 0.025 M SDS (312). This method was found to be accurate and reproducible, as well as simpler than a gradient HPLC method. A multivariate optimization approach for the separation of water-soluble vitamins and related compounds by capillary electrophoresis has been reported (313).

A recently proposed method (314) for the separation of fat-soluble vitamins by electrokinetic chromatography was further developed (315). The separation medium consisted of acetonitrile:water (80:20 v/v) and contained tetradecylammonium bromide as a pseudostationary phase. The high acetonitrile content was necessary to keep the hydrophobic vitamins in solution during electrophoresis. With the cathode placed at the capillary outlet, the fat-soluble vitamins were separated based on different hydrophobic interactions with the pseudo-stationary phase. The vitamins migrated in order of decreasing hydrophobicity prior to the electroosmotic flow.

Nonaqueous capillary electrophoresis has been applied to the separation of basic drugs (316). Efficient, rapid, and versatile conditions were obtained with 20 mM ammonium acetate in acetonitrile–methanol–acetic acid (49:50:1). Baseline separations of 9 morphine analogs, 11 antihistamines, 11 antipsychotics, and 10 stimulants could each be obtained within 6 min. Migration times for individual components had Relative Standard Deviation between 0.8% and 3.5%. Using an internal reference, normalized peak areas were between 2.2% and 9.1%. The precision data was reported to be instrument dependent, since excellent results were obtained only when the instrument had precise evaporation- and temperature-control systems.

The separation of neutral and basic drugs by capillary electrophoresis using lauryl(oxyethylene)sulfate as an additive provided excellent separation of protonated organic bases when the surfactant was added to the running electrolyte. The type and concentration of surfactant, as well as composition of the aqueous organic solution, was varied to obtain separations of both neutral and cationic organic compounds (317). A new CZE method for the determination of tacrine, 7-methoxy tacrine, and its metabolites in pharmaceutical and biological samples has been reported (318). Separations took less than 10 min. The detection limits obtained (s/n = 3) were 3 ppb for tacrine and 4 ppb for 7-methoxy tacrine. The methods were suitable for therapeutic monitoring.

A simple and rapid method for the separation of chlorpheniramine maleate, acetaminophen, and vitamin C in cold-medicine ingredients was obtained by capillary electrophoresis with amperometric detection (319). The effects of detection potential and organic modifier, 1,2-propanediol, were investigated. Sample recovery ranged from 94% to 103%. An HPCE method for the determination of dextromethorphan, chlorpheniramine maleate, pseudoephedrine hydrochloride, sodium benzoate, and acetominophen was found to be faster, convenient, and more economical when the CE method was compared to the HPLC method (320). A MEKC method was developed for the simultaneous determination of active ingredients used in cold medicine. Eleven active ingredients were determined using SDS and ethyl sulfate. When peak area ratios were used for quantitation, the R.S.D. ranged from 0.6% to 2.1% (321).

The role of five different anionic surfactants was investigated for the separation of cold medications by MEKC and their resulting electropherograms were compared with those obtained by CZE (322). The migration order of the 12 ingredients was significantly different among the five surfactants. Buffer pH and surfactant concentrations were investigated. The cold-medicine ingredients were separated within 30 min by MEKC with approximately 200,000 theoretical plate number, and selectivity was improved in comparison with CZE. An internal standard technique was used to quantitate some of the active ingredients combined in the commercial preparations.

The separation of endogenous 17- or 18-hydroxylated corticosteroids of the 21-hydroxylated 4-pregnen series was obtained by capillary electrophoresis of their charged borate chelate complexes (323). Aldosterone, 18-hydroxycorticosterone, 18-hydroxy-deoxycorticosterone, cortisone, cortisol and 11-deoxycortisol are separated and resolved with 400 mM borate buffer at pH 9.0. The corticosteroid/borate chelation complex as indicated by CE data correlated well with 11 B-NMR. The separation of corticosteroids and benzothiazin analogs were studied by MEKC and a comparison with CZE was made (324). Bile salts, which have a similar carbon skeleton to the corticosteroids, were used for the separation of these steroids. A short analysis time, 15 min, and a high number of theoretical plates (150,000–350,000) were obtained. Sodium cholate was found to be very effective. The MEKC method was applied to the determination of the drug substance in tablets and cream formulations. An internal standard method was used for quantitation. The purity of the drug substance was also determined.

A selective CZE microassay was developed for the determination of dexamethasone phosphate and its major metabolite, dexamethasone, in tears (325). An internal standard, indoprofen, was used for quantitation. The limits of detection and quantification were 0.5 and 2.0 µg/mL, respectively. The quantitative method was essential for the in-vivo determination of the dexamethasone concentration–time profiles in tears after the application of the anti-inflammatory drug. Two examples of rapid and simple drug analysis in pharmaceutical formulations using capillary electrophoresis can be found in the methods described for the separation of naphazoline, dexamethasone, and benzalkonium in nose drops (326).

The preceding applications have demonstrated that capillary electrophoresis is a versatile and mature technique in the pharmaceutical analysis laboratory.

IV. CAPILLARY ION ANALYSIS

The salt form provides the chemical development group a means of changing the physicochemical and pharmacokinetic properties of a drug candidate. The importance of choosing the correct salt form from the FDA-approved list (327) continues to be a challenge for the synthetic chemist. A salt selection rationale has been provided for basic drugs (328). The analytical group is concerned with the quantitative determination of the free base and conjugate acid. Capillary ion analysis (CIA) is favored over ion chromatography. CIA has separation efficiencies approaching 1 million theoretical plates with high peak capacity (329). Migration times are very rapid, in many cases less than 10 min, and migration can be predicted from ionic equivalent conductance scales published in the literature (330–338).

Indirect UV detection is a common technique which has been applied to the analysis of cations and anions. A UV-absorbing anion plus an electroosmotic flow modifier is added to the electrolyte. The displacement chromophore permits indirect photometric detection. Optimum separations can be achieved by choosing an electrolyte anion which has a mobility similar to the ions of interest. Some investigators have used conductivity detection, while others have used indirect laser-induced fluorescence for anion detection. Detection of the nonfluorescence analyte is obtained by charge displacement of the fluorophore (339,340).

A CZE method was used to separate EDTA complexes of selected trivalent and divalent transition-metal ions. By adding a surfactant to the separation buffer, an improvement in peak shape and short migration times was obtained (341 and Fig. 8). The ions characterized by CIA consist of 147 ionic species, including inorganic anions, inorganic cations, and organic anions (329).

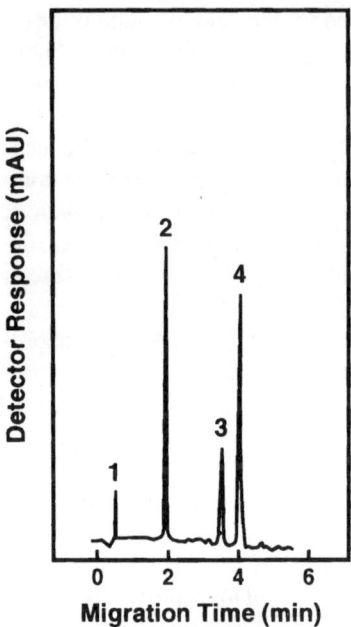

Migration Time (min)

Fig. 8 The effect of the addition of the cationic surfactant tetradecyltrimethylammonium bromide (TTAB) in the separation buffer for the separation of ion complexes with the chelating agent EDTA. TTAB, 0.5 mM, was added to the separation buffer for the separation of NO_3 (1), Cu-EDTA, Pb-EDTA, EDTA (2), Cr-EDTA (3), and Fe-EDTA (4). (For details of experimental conditions, see Ref. 341.)

V. CAPILLARY ELECTROCHROMATOGRAPHY

Capillary electrochromatography (CEC) is a high-efficiency microseparation technique in which mobile-phase transport through a capillary (usually 50- to 100-μm I.D., packed with stationary-phase particles) is achieved by electroosmotic flow instead of a pressure gradient as in HPLC (342–349). The absence of backpressure in electroosmotic flow allows the use of smaller particles and longer columns than in HPLC. In the reversed-phase mode, CEC has the potential to yield efficiencies 5 to 10 times greater than reversed-phase HPLC.

The applications of CEC in the pharmaceutical industry have been discussed by several investigators (350–361). A typical separation for neutral and basic compounds is presented in Fig. 9, illustrating the potential of this technique (358).

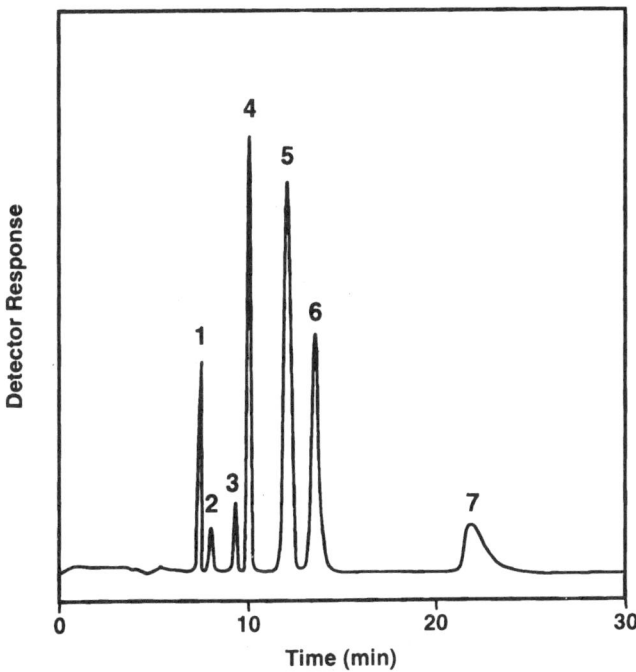

Fig. 9 Separation of neutral and basic.compounds by electrochromatography: thiourea (1), dimethyl phthalate (2), anisole (3), naphthalene (4), nortiptyline (5), amitriptyline (6), clomipramine (7). (For details of experimental conditions, see Ref. 358.)

A commercial CE system and a micropacked capillary was used to separate N—, O—, and S-containing heterocyclic compounds. Migration time reproducibility, linearity, and detector response was found to be comparable to HPLC. A study of the heterocyclic compound's elution order followed that predicted by the octanol–water partition coefficients (354). While chiral CEC provides improved resolution and higher efficiencies, additional work is needed since chiral CEC capillaries are not available commercially. The separation principles and chiral recognition mechanism for the separation of enantiomers have been reviewed (355). Furthermore, a comprehensive collection of drug applications and other compounds of interest has been reported (356). Direct enantiomeric separations by CEC were studied using a capillary packed with alpha-1-acid glycoprotein chiral stationary phase (357). Chiral resolution was achieved for enantiomers of benzoin, hexobarbital, pentobarbital, fosfamide, disopyramide, methoprolol, oxprenolol, and propanolol. The effects of pH, electrolyte concentration, and concentration of organic solvent in the mobile phase on the retention and selectivity were studied. A compilation of enantiomeric CEC separations can be found in Refs. 359 and 360.

Electrochromatography for the most part involves the presence of bead particles, or similar packing materials, to separate analytes, but recently electro-chromatography in the absence of packing materials was reported for the separation of two proteins. The partition of the analytes was accomplished by chemistry that covalently bonded the stationary phase directly to the walls of the capillary instead of bonding it to bead particles. This approach, termed open-tubular capillary electrochromatography, involves etching the inner walls of the capillary surface in order to increase the surface area by a factor of up to 1000, followed by chemical modification to provide the selectivity desired. Figure 10 depicts the separation of turkey and chicken lysozymes using bare and C-18–modified, etched capillary. One major advantage of this system is the absence of a frit structure to hold the packing material. The capillaries used in open-tubular capillary electrochromatography do not have beads or frits, which minimizes bubble formation, eliminates backpressure, and allows for a normal uninterrupted electroosmotic flow (361).

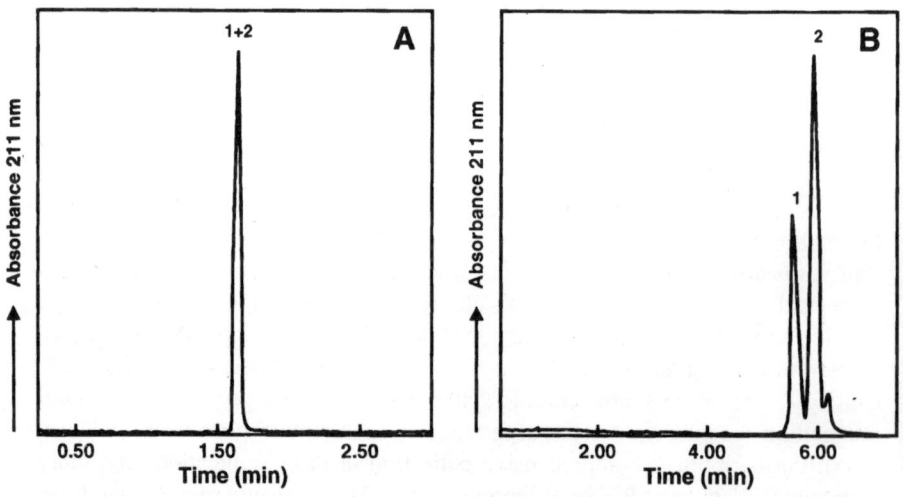

Fig. 10 Separation of turkey (1) and chicken (2) lysozymes on a bare capillary (A) by capillary zone electrophoresis, and C-18 modified etched capillary (B) by open-tubular capillary electrochromatography. (For details of experimental conditions, see Ref. 361.)

VI. BIOMOLECULES

In the late 1980s and 1990s, biopharmaceuticals entered the market and raised the hope that macromolecules, largely proteins, would be a new class of pharmaceutical drugs (362–368). The economic success has been high, reaching many billions of dollars in sales (368). However, the economic arguments as well the failures that some biopharmaceuticals have had are making pharmaceutical companies realign their strategies in the search for new protein drugs. Because some 60 million patients worldwide thus far have benefited from protein pharmaceuticals, we thought it would be useful to the reader to present a summary of the 84 proteins currently approved in the United States or the European Union (see Ref. 368 and Table 5). According to present estimates, around 500 biopharmaceuticals, produced in animals and plants, are currently undergoing clinical trials (368,369).

The majority of protein drugs are formulated as parenteral solutions and they will probably remain as injectables for the near future (370,371). Parenteral protein formulations have a more limited shelf life compared to solid dosage forms (372,373). Additional disadvantages of peptides as drugs have been reported; for example, they often lack validation in animal models, have difficulty in tissue penetration, may be immunogenic, and in general will probably be used only for acute indications (374–377). Progress has been made in various areas to improve effectiveness of the final drug product. There is currently an intense search for new delivery systems and alternative administration routes to produce effective drugs with promising levels of good bioavailability, excellent safety, and a reasonable production cost. The rapid maturity in the field of biopharmaceutical formulation (371) is yielding drug products with enhanced protein stability, as well as maintaining efficacy during storage and transportation.

The field of proteomics is becoming increasingly important as genome sequences are being completed and annotated (33–42,378–383). The discovery of hundreds of thousands of new proteins and peptides will provide the opportunity to map out complex cellular interactions at an unprecedented level of detail. Furthermore, with the advent of the field of proteinomimetics, which utilizes peptide-based molecules to mimic native protein function, many new and novel peptide and nonpeptide compounds will find niches as pharmaceutical drugs (383–386).

Therefore, the ability to characterize thousands of proteins of different physicochemical complexities which are found in nature at various levels of concentrations raises an immediate challenge to protein biochemists. Through the years, a variety of techniques have been used for the identification and characterization of various proteins and peptides in the assay and quality control of drug substance and formulated drug products. Some of these techniques include polyacrylamide gel electrophoresis (PAGE), high-performance liquid chromatogra-

Table 5 Biopharmaceutical Drugs Currently Approved Which are Already in General Medical Use

Product	Cell line production	Company	Therapeutic indication
Recombinate (rh Factor VIII)	Animal cell line	Baxter Healthcare/Genetics Institute	Hemophilia A
Bioclate (rh Factor VIII)	CHO cells	Centeon	Hemophilia A
Kogenate (rh Factor VIII)	BHK cells	Bayer	Hemophilia A
ReFacto (B-domain-deleted rh Factor VIII)	CHO cells	Genetics Institute	Hemophilia A
NovoSeven (rh Factor VIIa)	BHK cells	Novo-Nordisk	Some forms of hemophilia
Benefix (rh Factor IX)	CHO cells	Genetics Institute	Hemophilia B
Revasc (r hirudin)	*S. cerevisiae*	Ciba Novartis/Europharm	Prevention of venous thrombosis
Refludan (r hirudin)	*S. cerevisiae*	Hoechst Marion Roussel Behringwerke	Anticoagulation therapy for heparin-associated thrombocytopenia
Activase (rh tPA)	CHO cells	Genentech	Acute myocardial infarction
Ecokinase (r tPA)	*E. coli*	Gatenus Mannheim	Acute myocardial infarction
Retavase (r tPA)	*E. coli*	Boehringer Mannheim/Centocor	Acute myocardial infarction
Rapilysin (r tPA)	*E. coli*	Boehringer Mannheim	Acute myocardial infarction
Humulin (rh insulin)	*E. coli*	Eli Lilly	Diabetes mellitus
Novolin (rh insulin)	—	Novo Nordisk	Diabetes mellitus
Humalog (rh insulin analog)	*E. coli*	Eli Lilly	Diabetes mellitus
Insuman (rh insulin)	*E. coli*	Hoechst AG	Diabetes mellitus
Liprolog (rh insulin analog)	*E. coli*	Eli Lilly	Diabetes mellitus
NovoRapid (rh insulin analog)	—	Novo Nordisk	Diabetes mellitus
Lantus (rh insulin analog)	*E. coli*	Aventis	Diabetes mellitus

Product	Expression system	Company	Indication
Protropin (r hGH)	E. coli	Genentech	hGH deficiency in children
Humatrope (r hGH)	E. coli	Eli Lilly	hGH deficiency in children
Nutropin (r hGH)	E. coli	Genentech	hGH deficiency in children
BioTropin (r hGH)	—	Biotechnology General	hGH deficiency in children
Genotropin (r hGH)	E. coli	Pharmacia & Upjohn	hGH deficiency in children
Norditropin (r hGH)	—	Novo Nordisk	Treatment of growth failure in children due to inadequate growth hormone secretion
Saizen (r hGH)	—	Serono Laboratories	hGH deficiency in children
Serostim (r hGH)	—	Serono Laboratories	hGH deficiency in children
Glucagen (rh glucagon)	S. cerevisiae	Novo Nordisk	Hypoglycemia
Thyrogen (rh TSH)	CHO cells	Genzyme	Detection/treatment of thyroid cancer
Gonal F (rh FSH)	CHO cells	Ares-Serono	Anovulation and superovulation
Puregon (rh FSH)	CHO cells	N.V. Organon	Anovulation and superovulation
Follistim (rh FSH)	CHO cells	Organon	Infertility
Epogen (rh EPO)	Mammalian cell line	Amgen	Treatment of anemia
Procrit (rh EPO)	Mammalian cell line	Ortho Biotech	Treatment of anemia
Neorecormon (rh EPO)	CHO cells	Boehringer-Mannheim	Treatment of anemia
Leukine (r GM-CSF)	E. coli	Immunex	Autologous bone marrow transplantation
Neupogen (r GM-CSF)	E. coli	Amgen	Chemotherapy-induced nutropenia
Regranex (rh PDGF)	S. cerevisiae	Ortho-McNeil Pharmaceuticals/ Jansen-Cilag	Lower-extremity diabetic neuropathic ulcers
Roferon A (rh IFN-α2a)	E. coli	Hoffmann-La Roche	Hairy-cell leukemia

Table 5 Continued

Product	Cell line production	Company	Therapeutic indication
Infergen (r IFN-α)	*E. coli*	Amgen/Yamanouchi Europe	Chronic hepatitis C
Intron A (r IFN-a2b)	—	Schering Plough	Hairy-cell leukemia, genital warts
Rebetron (combination of ribavirin and rh IFN-α-2b)	*E. coli*	Schering Plough	Chronic hepatitis C
Alfatronol (rh IFN-α2b)	*E. coli*	Schering Plough	Hepatitis B, C, and various cancers
Virtron (rh IFN-α2b)	*E. coli*	Schering Plough	Hepatitis B and C
Betaferon (r IFN-β1b)	*E. coli*	Schering AG	Multiple sclerosis
Betaseron (r IFN-β1b)	*E. coli*	Berlex Laboratories and Chiron	Elapsing/remitting multiple sclerosis
Avonex (rh IFN-β1a)	CHO cells	Biogen	Relapsing multiple sclerosis
Rebif (rh IFN-β1a)	CHO cells	Ares Serono	Relapsing/remitting multiple sclerosis
Actimmune (rh IFN-γ1a)	*E. coli*	Genetech	Chronic granulomatous disease
Proleukin (r IL-2)	*E. coli*	Chiron	Renal cell carcinoma
Neumega (r IL-11)	*E. coli*	Genetic Institute	Prevention of chemotherapy-induced thrombocytopenia
Recombivax (r HBsAg)	*S. cerevisiae*	Merck	Hepatitis B prevention
Comvax (combination vaccine, containing HBsAg)	*S. cerevisiae*	Merck	Vaccination of infants against *H. influenzae* type B and hepatitis B
Tritanrix-HB (combination vaccine, containing r HBsAg)	*S. cerevisiae*	SmithKline Beecham	Vaccination against hepatitis B, diphtheria, tetanus, and perfussis

Twinrix (adult and pediatric forms) (combination vaccine containing r HBsAg)	*S. cerevisiae*	SmithKline Beecham	Immunization against hepatitis A and B
Primavax (combination vaccine, containing r HBsAg)	*S. cerevisiae*	Pasteur Merieux MSD	Immunization against diphtheria, tetanus, and hepatitis B
Procomvax (combination vaccine, containing r HBsAg as one component)	—	Pasteur Merieux MSD	Immunization against *H. influenzae* type B and hepatitis B
Lymerix (r OspA, a protein found on the surface of *B. burgdorferi*)	*E. coli*	SmithKline Beecham	Lyme disease vaccine
Orthoclone OKT3 (Muromomab CD3, murine Mab directed against the T-lymphocyte surface antigen CD3)	—	Ortho Biotech	Reversal of acute kidney transplant rejection
OncoScint CR/OV (Satumomab Pendetide, murine Mab directed against TAG-72, a tumor-associated glycoprotein)	—	Cytogen	Detection/staging/follow-up of colorectal and ovarian cancers
ReoPro (Abciximab, Fab fragments derived from a chimeric Mab, directed against the platelet surface receptor GPIIb/IIIa)	—	Centocor	Prevention of blood clots
Indimacis 125 (Igovomab, murine Mab fragment (Fab2) directed against the tumor-associated antigen CA 125)	—	CIS Bio	Diagnosis of ovarian adenocarcinoma

Table 5 Continued

Product	Cell line production	Company	Therapeutic indication
CEA-scan (Arcitumomab, murine Mab fragment (Fab), directed against human carcino-embryonic antigen, CEA)	—	Immunomedics	Detection of recurrent/metastatic colorectal cancer
MyoScint (Imiciromab-Pentetate, murine Mab fragment directed against human cardiac myosin)	—	Centocor	Myocardial infarction imaging agent
ProstaScint (Capromab Pentetate murine Mab-directed against the tumor surface antigen PSMA)	—	Sorin	Diagnosis of cutaneous melanoma lesions
Tecnemab KI (murine Mab fragments (Fab/Fab$_2$ mix) directed against HMW-MAA)	—	Sorin	Diagnosis of cutaneous melanoma lesions
Verluma (Nofetumomab murine Mab fragments (Fab) directed against carcinoma-associated antigen)	—	Boehringer Ingelheim/NeoRx	Detection of small-cell lung cancer
LeukoScan (Sulesomab, murine Mab fragment (Fab) directed against NCA 90, a surface granulocyte nonspecific cross-reacting antigen)	—	Immunomedics	Diagnostic imaging for infection/inflammation in bone of patients with osteomyelitis

Rituxan (Rituximab chimeric Mab directed against CD 20 antigen found on the surface of B lymphocytes)	—	Genetech/IDEC Pharmaceuticals	Non-Hodgkin's lymphoma
Zenapac (Daclizumab, humanized Mab directed against the α-chain of the IL-2 receptor)	—	Hoffman La-Roche	Prevention of acute kidney transplant rejection
Simulect (Basiliximab, chimeric Mab directed against the α-chain of the IL-2 receptor)	—	Novartis	Prophylaxis of acute organ rejection in allogeneic renal transplantation
Remicade (Infliximab, chimeric Mab directed against TNF-α)	—	Centocor	Treatment of Crohn's disease
Synagis (Palivizumab, humanized Mab directed against an epitope on the surface of respiratory syncytial virus)	—	MedImmune Abbot	Prophylaxis of lower respiratory tract disease caused by respiratory syncytial virus in pediatric patients
Herceptin (Trastuzumab, humanized antibody directed against HER 2, i.e., human epidermal growth factor receptor 2)	—	Genentech	Treatment of metastatic breast cancer if tumor overexpresses HER 2 protein
Humaspect (Votumumab, human Mab directed against cytokeratin tumor-associated antigen)	—	Organon Teknika	Detection of carcinoma of the colon or rectum

Table 5 Continued

Product	Cell line production	Company	Therapeutic indication
Mabthera (Rituximab, chimeric Mab directed against CD 20 surface antigen of B lymphocytes)	—	Hoffman-La Roche	Non-Hodgkin's lymphoma
Pulmozyme (dornase-α, r DNase)	CHO cells	Genentech	Cystic fibrosis
Cerezyme (r B-glucocerebrosidase. Differs from native human enzyme by one amino acid, R495?H, and has modified oligosaccharide component)	E. coli	Genzyme	Treatment of Gaucher's disease
Beromun (rh TNF-α)	E. coli	Boehringer Ingelheim	Adjunct to surgery for subsequent tumor removal, to prevent or delay amputation
Ontak (r IL-2-diphtheria toxin fusion protein that targets cells displaying a surface IL-2 receptor	—	Seragen/Ligand Pharmaceuticals	Cutaneous T-cell lymphoma
Enbrel (r TNFR-IgG fragment fusion protein)	CHO cells	Immunex Wyeth Europa	Rheumatoid arthritis
Vitravene (Fomivirsen, an antisense oligonucleotide)	—	ISIS pharmaceuticals	Treatment of CMV retinitis in AIDS patients

phy (HPLC), mass spectrometry (MS), radioimmunoassays (RIA), immunoblotting procedures, and enzyme-linked immunosorbent assay (ELISA). Although these techniques have been extremely valuable in assessing purity, most of them, with the exception of HPLC and MS, are usually imprecise and semiquantitative. They may have poor concentration detection limits and are usually labor intensive and time consuming. Furthermore, most of these conventional techniques have limited or no automation capability and, in general, method development can be quite complex and expensive (387–392).

Conversely, capillary electrophoresis is becoming the method of choice to characterize proteins and peptides today. The capability of CE for the analysis of proteins and peptides in simple and complex mixtures is well documented (see Refs. 393–403 and Table 6). In the early years of CE, nonspecific adsorption of proteins on the silica capillary wall yielded poor electropherographic results, but numerous efforts were made to decrease these unwanted interactions. Currently, a variety of commercially available capillary columns containing stable chemistries, including polymers which coat the silanol groups on the capillary wall, are generating very effective and reproducible results. Changing the buffer composition with additives and making adjustment in the ionic strength and pH are helpful in the recovery and separation of proteins and peptides. Buffer composition and coated capillaries play a significant role in the separation selectivity of proteins and peptides (401–403). A powerful tool for the characterization of proteins is coating the negative charge of the capillary wall with polybrene or other similar polymeric material and separating the protein by reversed charge. This procedure is extremely useful for glycoproteins or other proteins which contain isoforms. A representative example is presented in Fig. 11. The peptide hormone erythropoeitin can be separated into several isoforms by CE in native conditions (407–409) or in the presence of urea and putrescine modifiers (410,411). A similar

Table 6 Determination and Characterization of Protein Pharmaceutical Drugs by Capillary Electrophoresis

Drug	Mode of separation	References
Albumin	CZE	404,405
r Cytokines	CZE	406,407
r Erythropoietin	CZE	408–412
r Growth hormone	CZE	413
Immunoglobulins	CZE	414–419
r PDGF	CZE	420
r Somatotropin	CZE	421

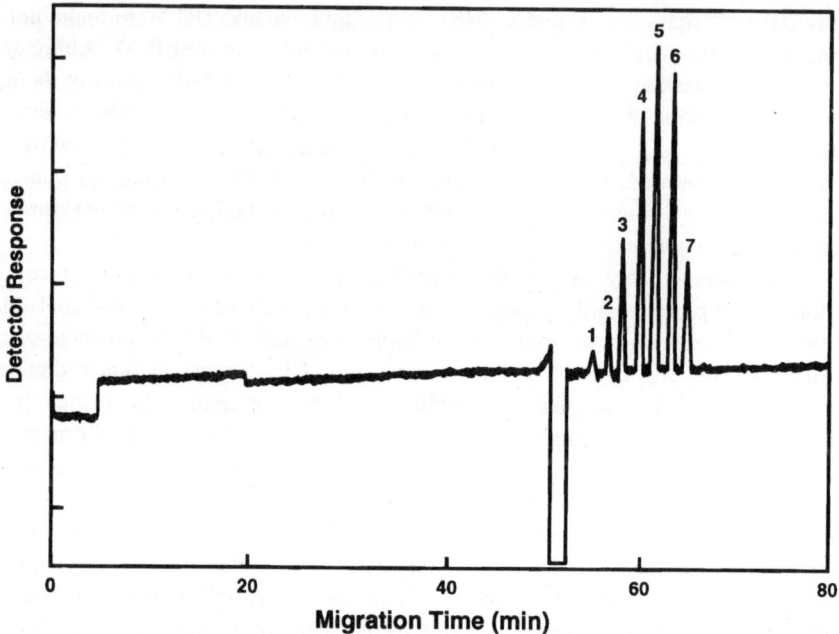

Fig. 11 Capillary zone electrophoresis of recombinant erythropoietin glycoforms under denaturing conditions. The separation buffer consisted of 0.010 M tricine, 0.01 M NaCl, 0.01 M sodium acetate, 7 M urea, and 2.5 mM putrescine, pH 5.5. (For details of experimental conditions, see Refs. 410 and 411.)

procedure has been used to separate the isoforms of platelet-derived growth factor (417).

In addition to open-tubular capillary zone electrophoresis, other modes of CE have been used for the separation of proteins and peptides: capillary gel electrophoresis, capillary isoelectric focusing, and capillary affinity electrophoresis (406–417).

VII. CONCENTRATION SENSITIVITY ENHANCEMENT IN CAPILLARY ELECTROPHORESIS

A full understanding of gene function in health and disease will require a systematic approach that is not inherent in what is known as functional genomics. To determine the role of single proteins in regulatory and metabolic pathways under a broad set of conditions, comprehensive approaches may be required that will enable the prediction of the overall effects of drug–protein interactions on physiological parameters. Isolation and characterization of proteins will be crucial to understanding bioaffinity and molecular recognition of proteins at the molecular level. Approximately 50–70% (depending on the species) of gene products are found as low-abundance proteins. Since many protein and peptide drugs exhibit strong biological potency, dosage forms usually contains a few micrograms of the active drug in the presence of milligram amounts of excipients. In addition, detection of proteins is difficult due to the lack of usable chromophores in easily accessible spectral regions. To address this issue, new procedures have been developed for increasing CE detection sensitivity in the analysis of proteins and peptides as drug substances, or in finished formulated drug products, or for proteomic studies (50,51,58,68,119,120,269,292,393,404–407).

One procedure involves the derivatization of amino-functional groups with fluorescamine and then employing the derivative as a UV chromophore (monitored at 280 nm) to enhance detection sensitivity (404–407). Significant improvements in the detectability of several protein and peptides were reported by fluorescence detection (422–424). To date, several pre- and postcolumn derivatization procedures have been developed for use in capillary electrophoresis employing direct and indirect detection modes and a variety of detection systems, including UV detectors, electrochemical detectors, laser-induced fluorescence detectors, and mass spectrometers. All of these methods are directed at enhancing sensitivity (425–433). The detection method of choice is application specific and dictated by the level of sensitivity necessary. Enhancement of sensitivity is targeted at the identification and characterization of drug substances, isomeric forms, degradation products, excipients, impurities, contaminants, possible molecular interactions of components in finished pharmaceutical products, and metabolites in biological liquids (425–433).

Another procedure for enhancing sensitivity employs a microcartridge device (termed an analyte concentrator–microreactor) inserted in the inlet side of the capillary, which allows large sample volumes to be applied. The microcartridge device is composed primarily of a microchamber, located near the inlet of the capillary, packed with selective or nonselective sorbent material immobilized on a wide range of beads or a porous structure (58,68,119,120,434–445). The packing material is contained by a two frit structures. Alternatively, the affinity ligand can be immobilized directly to a portion of the capillary, to glass wool, or to membranes without the use of frit structures (58,68,119,120,434–440). The microcartridge device is connected to two pieces of separation capillaries, usually glued by an epoxy resin to avoid leaking of circulating fluids, air penetration, and to eliminate bubble formation.

Characterized by high selectivity, the use of immobilized highly specific ligands to a solid support which is traditionally named affinity chromatography. This technique enables high-purity products to be concentrated and separated from a complex feedstock, cell or tissue homogenate, biological fluid, or any other complex mixture, including a pharmaceutical formulation. In affinity chromatography, the molecule to be purified is specifically and reversibly adsorbed by an immobilized ligand. Unbound substances are washed away, and the desired target molecule is recovered in a single desorption step (68,120,407,435,446). Affinity chromatography has relied on a few well-known affinity ligands, such as Protein A, cibacron blue dye, or on animal-derived antibodies. With today's technology, an affinity chromatography ligand can be a selected molecule or a molecule designed to enhance selectivity and to capture the target ligand in the purification or separation processes. Affinity chromatography media that are integrated with ligands designed to optimize individual bioseparation applications are likely to present exciting opportunities for companies engaged in manufacturing biopharmaceuticals, or the technique may be used in the quality control of impurity-free pharmaceutical products (68,120,407,435,446).

Immunoaffinity capillary electrophoresis is rapidly growing in the field of pharmaceutical and diagnostic applications. Several applications have been developed to affinity concentrate proteins and peptides found at low concentrations in simple and complex matrices in order to enhance analyte detectability when separated by CE (68,120,407,435). Figure 12 shows a microphotograph

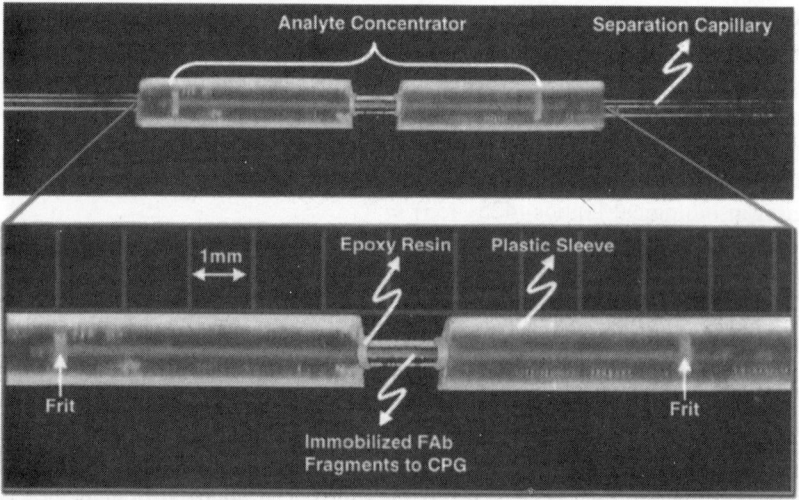

Fig. 12 Microphotograph of an analyte concentrator fabricated with FAb antibody fragments immobilized to controlled-pore glass silica. The irregularly shaped beads were housed between two frit structures. The analyte concentrator device was connected to two separation capillaries by a Teflon sleeve. The plastic connector was glued to the separation capillaries by an epoxy resin. The entire fabrication process was monitored by an stereo microscope. (For details of experimental conditions, see Ref. 120.)

of the "analyte concentrator" device using immobilized antibody fragments to porous glass beads. An application using the immunoaffinity analyte concentrator to enhance sensitivity is illustrated in Fig. 13. In this particular application, the antibodies were immobilized directly to a small portion of the surface of the inner wall of the capillary. Thus, no frit structure was necessary (435).

In recent years, the coupling of capillary electrophoresis to mass spectrometry (CE-MS) has moved at an extraordinary pace, expanding out of the realm of a few and simple analytical chemistry applications into the development

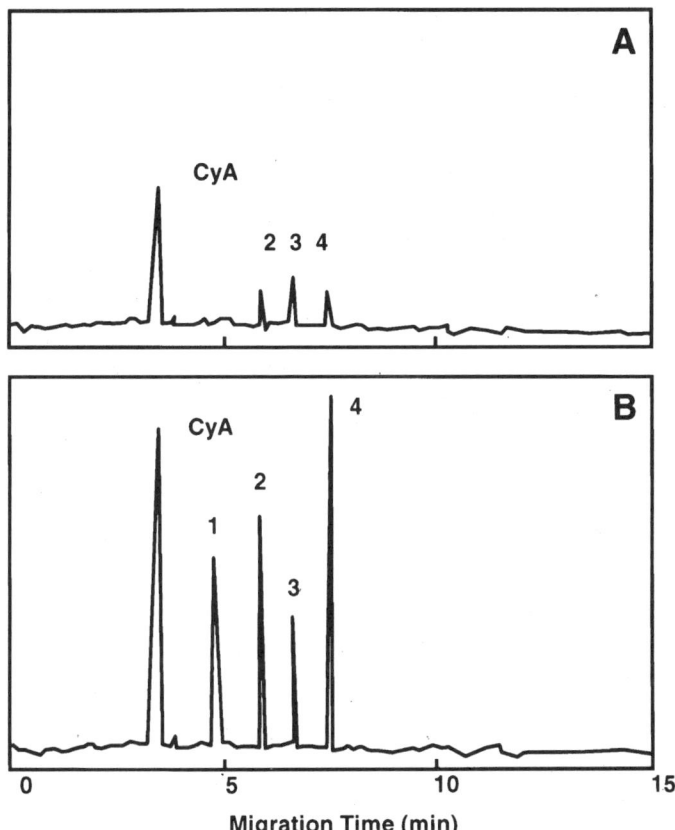

Fig. 13 Immunoaffinity capillary electrophoresis of tear fluid obtained from a patient with no clinical signs of cyclosporin A toxicity (A), and a patient during an episode of systemic toxicity (B). Relevant peaks: CyA, cyclosporin A. Peaks 1–4 represent cyclosporin A metabolites: 1, AM1; 2, AM9; 3, AM1c; 4, AM4N. (Modified from Ref. 435.)

of more complex and useful applications needed in the worlds of biology, medicine, and therapeutics (120,441). Few measurements in biochemistry are as fundamental as the determination of the mass of a chemical substance. In the postgenomic era, determination of mass has become a key tool to facilitate the identity of a protein. Given the specificity of a protease, to generate peptides, and the organism's genome, the identity of a protein is determined by a database search (388,389). It is possible to ionize biomolecules from aqueous buffers at various pHs at atmospheric pressure, but one must consider that the charge state distribution of a protein is sensitive to the solution from which it is ionized (388,389). Proteins ionized from denaturing solution conditions give rise to many more charge states than proteins ionized from native conditions. Deconvolution software allows the transformation of a multiply charged spectrum into a singly charged parent spectrum and calculation of its molecular mass.

Presently, the coupling of capillary electrophoresis to mass spectrometry is generating growing interest in a wide range of applications in the pharmaceutical industry, including the characterization of peptides. Figure 14 depicts a simple model of the principle of CE-MS (441). In this particular case, an analyte concentrator packed with sorbent material is shown near the inlet of the capillary. Figure 15 shows a typical application of CE-MS utilizing a single quadrupole electrospray ionization mass spectrometer. Neurotensin and angiotensin II were separated by CE and the total ion current electropherogram (TICE) was determined. For many interesting pharmaceutical applications of CE-MS and CE-MS/MS, the reader should refers to the references listed (120,427–429,441).

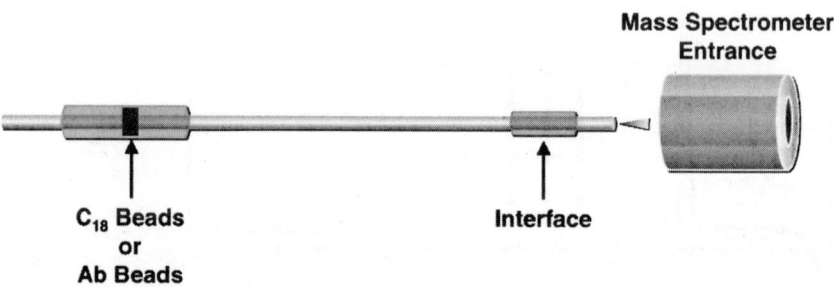

Fig. 14   Schematic representation of a capillary electrophoresis instrumentation coupled in tandem through an interface to an electrospray ionization mass spectrometer. A microchamber affinity device containing an immobilized C-18 or antibody to a porous bead is located near the inlet of the capillary. (Modified from Ref. 441.)

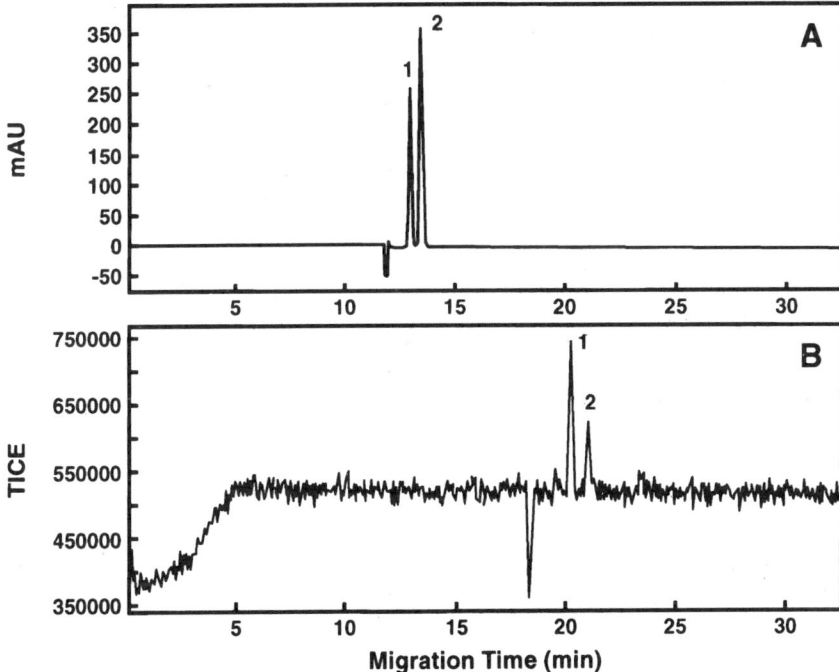

Fig. 15   Capillary electrophoresis-mass spectrometry profile of two peptides preconcentrated on-line prior to separation. (A) Electropherogram of neurotensin (1) and angiotensin (2), concentrated and purified by C-18 immobilized to porous beads and monitored at 195 nm after separation by CE. (B) Total ion-current electropherogram of the separated peptides. The experimental conditions were similar to those described in Ref. 120 for gonadotropin-releasing hormone. The limits of detection for the peptides were approximately 1 to 5 ng/mL, depending primarily on the quality of the analyte concentrator-microreactor.

VIII.   CAPILLARY ARRAY ELECTROPHORESIS

High-throughput CE was the main reason for the advancement in the sequencing of more than 120,000 genes that comprise the human genome. It took more than 15 years to sequence approximately 15–20% of the genome using traditional or classical electrophoretic methods, and less than 2 years to sequence more than 70% with modern capillary electrophoresis instrumentation. The formation of a gene-sequencing factory (Celera Genomics, Rockville, MD) that uses an array of 96 capillary electrophoresis columns simultaneously conducting electrophoresis in every one of the capillaries has been developed for high-throughput analy-

Fig. 16 A partial view of a series of high-throughput DNA sequencers based on multi-capillary-array electrophoresis. (Adapted from Ref. 451.)

sis. Several papers describe a variety of capillary electrophoresis instrumentation for high-throughput analysis (447–450). Figure 16 shows several automated capillary electrophoresis DNA sequencers in operation at Celera Genomics (451).

Why is genomics so fundamental to the success of the biotechnology industry? Genomics companies have the potential to generate information that has immediate and long-term commercial value. A complete understanding of the DNA code that underlies human physiology is key to unlocking secrets to normal functioning and disease, and offers the promise of creating better drugs, faster (452). High-throughput DNA sequencers based on multicapillary-array electrophoresis is one example of the rapid progress occurring in scientific measurement. The next step will be the manufacturing of a proteomic "engine" based on capillary electrophoresis. In most cases, gene sequence reveals little about protein function or disease relevance. Accordingly, the true value of genome sequence information will only be realized after a function has been assigned to all of the encoded proteins. Proteomics seeks to provide functional information for all proteins. Applying proteomics technologies will not only provide validated targets for drug discovery but will also increase the efficiency of the drug discovery process downstream (453).

IX. MICROCHIP CAPILLARY ELECTROPHORESIS

Microchip technology (see Ref. 454 and Fig. 17) is revolutionizing chemical and biochemical testing. The microchip processes fluid rather than electrons. Both electrophoretic and electroosmotic techniques are used to pump the fluid. Pumps, valves, volume-measuring devices, and separation systems are on the microchip's surface. Microchip separation procedures include electrophoresis, chromatography and solid-phase biochemistry. Microchips allow true parallelism, miniaturization, multiplexing, and automation, and these key features provide a set of performance specifications that cannot be achieved with earlier technologies (64–68,454–463).

Several companies are developing analytical instruments and information systems based on the concept of lab-on-a-chip technology (458–463). The chip-

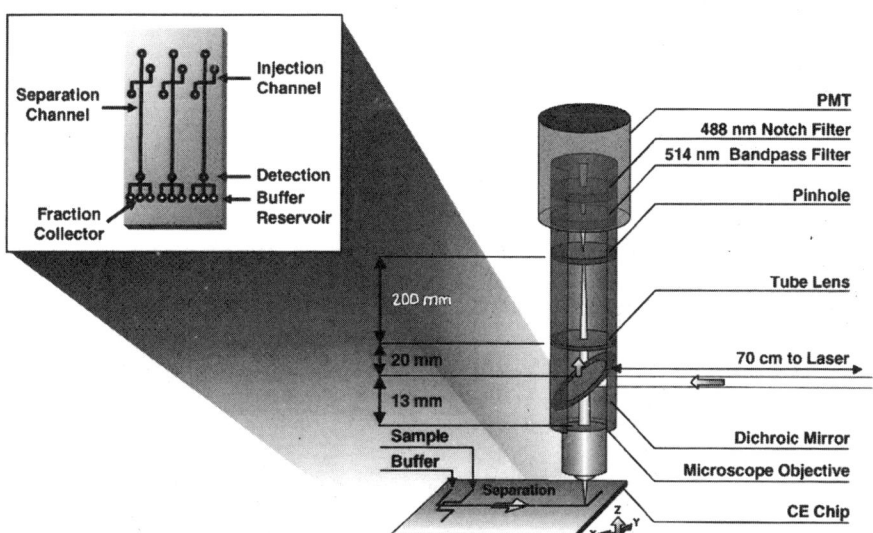

Fig. 17 Schematic representation of a capillary electrophoresis microchip using a confocal epiluminescence microscope as an ultrasensitive detection system to monitor the analytes separated on the separation channel of a CE microchip. (Adapted from Refs. 68 and 454.)

based integrated systems will reduce assay test cost, improve laboratory productivity, and reduce product development cycles for the pharmaceutical industry. At present, gene-expression monitoring appears to be one of the most biologically informative applications of microchip technology. For a number of reasons, microchips are well suited to a wide range of applications including gene-expression analysis, providing an integrated platform for functional genomics. Figure 18 shows a schematic of a microchip platform coupled to a mass spectrometer (463). An application of microchip coupled to mass spectrometry (MC-MS) is depicted in Fig. 19 (463).

Several examples of work in this field have been reported for multiple-sample polymerase chain reaction (PCR) amplifications (464–466) and electrophoretic analysis in a microchip. In one example, the PCR products from four DNA samples were analyzed by microchip gel electrophoresis (466). The ability to analyze a large number of DNA samples in this format was presented. Run-to-run reproducibility in the sizing microchip was very good; percent relative standard deviation $n = 8$ for migration times was better than 0.3%. The accuracy of the 500-base-pair sizing was greater than 98% (466).

Products from 96 PCRs using a single microchip device have been separated (467). This work establishes the feasibility of performing high-through-

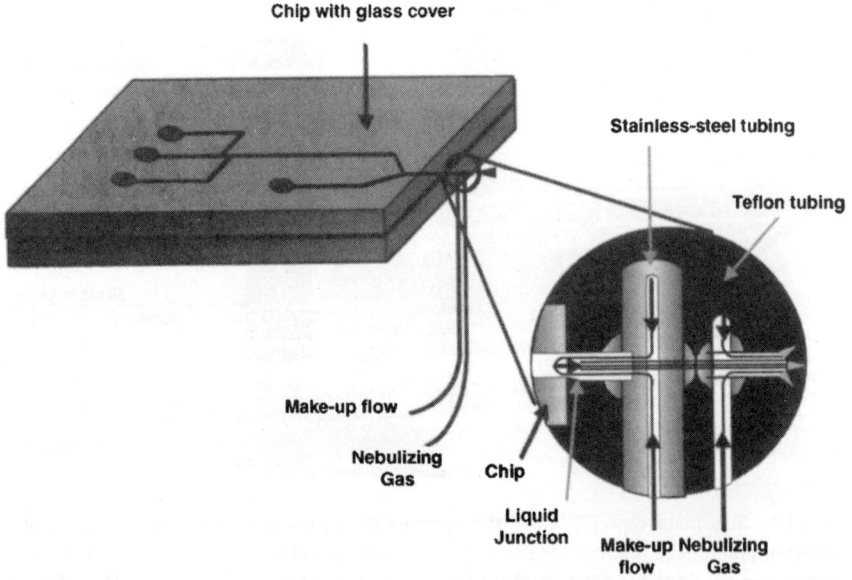

Fig. 18 Schematic representation of the glass chip-based CE/MS apparatus and the expanded view of the coupled microsprayer. (Adapted from Ref. 463.)

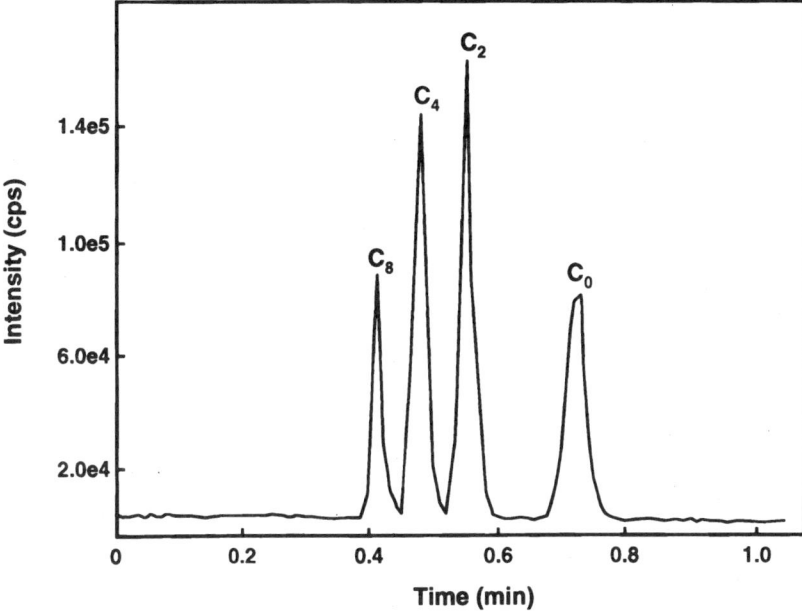

Fig. 19   Quadrupole SIM CE/MS electropherogram for a separation of a synthetic standard mixture containing three acyl carnitine and carnitine with alkyl chain lengths including 8, 4, 2, and 0 carbon atoms, respectively. The separation was carried out using the coupled microsprayer electrospray CE/MS interface shown in Fig. 17. The quaternary cation for each compound (m/z 288, 232, 204, and 162, respectively, according to increasing migration times) was monitored in the SIM mode. (For details of experimental conditions, see Ref. 463.)

put genotyping separations with radial capillary-array electrophoresis microplates.

X.   FUTURE OF CAPILLARY ELECTROPHORESIS IN THE PHARMACEUTICAL INDUSTRY

Drug discovery today is driven largely by biology (molecular genetics, genomic sciences, informatics), but a significant portion is still driven by chemistry. In spite of the biology segment, which will lead the future of drug discovery, there will be a major crossroad of complementation to advance on a road that is sure to be bumpy. In the pursuit of safe and efficacious drugs, pharmaceutical companies must subject possible candidates to a variety of chemical and biological

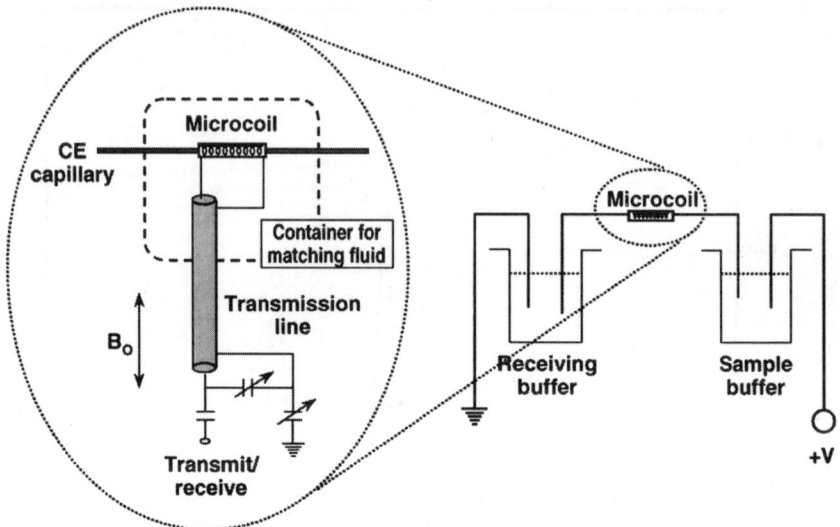

Fig. 20 Schematic representation of the experimental layout for CE/NMR showing the separation capillary and NMR probe. (Adapted from Ref. 471.)

screenings. To comply with stringent regulatory agencies, orthogonal methods will have to be in operation to confirm a complete physicochemical assessment of a drug substance and a drug product.

Capillary electrophoresis, with its inherent separation capabilities, offers an alternative to more established separation techniques for many pharmaceutical analysis applications. Furthermore, the future looks promising, since additional improvements to CE technology will be made to improve characterization of pharmaceutical drugs. For example, one intense area of development is capillary electrophoresis coupled to the mass spectrometer and capillary electrophoresis coupled to nuclear magnetic resonance (468–471). These hyphenated techniques are playing a key role in the characterization of drugs. An example of a CE-NMR design showing the separation capillary and the NMR probe is depicted in Fig. 20. An application of the use of CE-NMR is shown in Fig. 21.

Just as in the computer and Internet revolution, other successful technologies, such as capillary electrophoresis, offer proven and open solutions, deliver robust results, are cost effective and easy to implement, and yield high value. Currently, capillary electrophoresis can quantitate concentration detection limits in the range of 1 ng/mL to 1 pg/mL using on-line immunoaffinity preconcentration with UV or conventional laser-induced fluorescence detection systems (58,

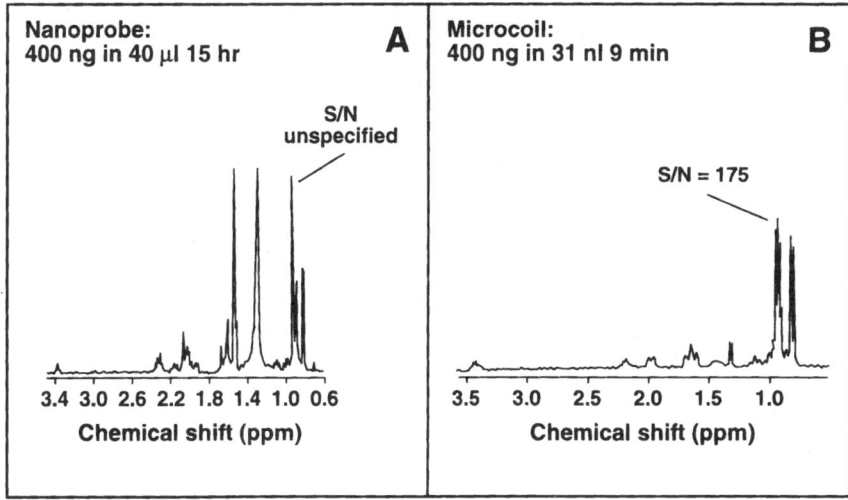

Fig. 21   Proton NMR spectra of 400 ng of menthol using (A) a 40-µL nanoprobe and (B) a microcoil. (Adapted from Ref. 471.)

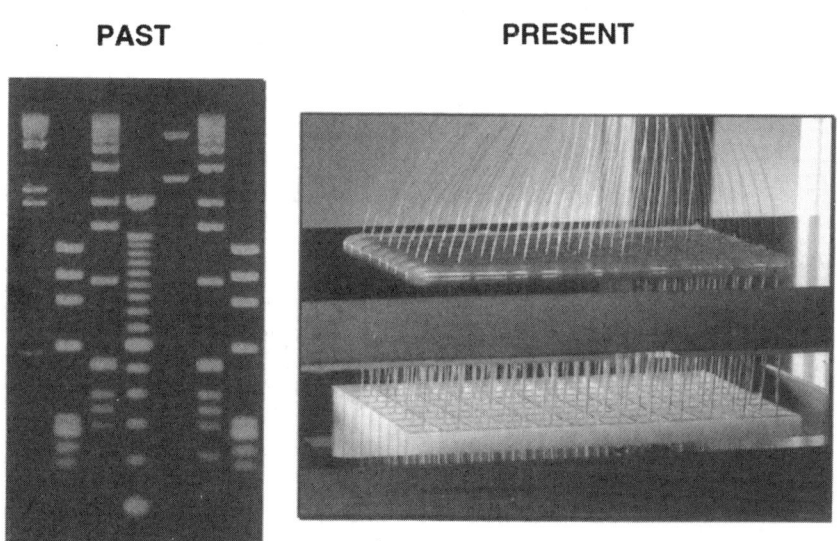

Fig. 22   Representative pictures of biomolecules separated by slab gel electrophoresis (past) and a capillary array instrumentation (present). (Courtesy of SpectruMedix.)

68,120,407,435,472), and it can detect fewer than 1000 molecules with sophisticated laser detectors (473,474).

In summary, capillary electrophoresis is rapidly gaining popularity as a tool with great potential for routine automated assessment of pharmaceutical drugs. Although it is not predicted to replace HPLC technology completely, there are a few examples in which capillary electrophoresis has been chosen as a preferred technology within the pharmaceutical sector for applications such as chiral separation, characterization of biotechnology-derived proteins, and for the quantification of the glycoforms of erythropoeitin (412). With the advent of capillary array electrophoresis, conventional slab gels may become a technique of the past (see Fig. 22). Furthermore, biotechnology drugs are no longer limited to proteins. Cells can be manipulated to produce a high yield of cell metabolites such as antibiotics, vitamins, amino acids, and other small molecules, and capillary electrophoresis has already proven to be of great value for their separation and characterization.

REFERENCES

1. RG Werner. Identification and development of new biopharmaceuticals. Arzneimittelforschung 48:523–530, 1998.
2. J Drews. Drug discovery today—and tomorrow. Drug Discov Today 5:2–4, 2000.
3. S Harris, SM Foord. Transgenic gene knock-outs: Functional genomics and therapeutic target selection. Pharmacogenomics 1:433–443, 2000.
4. ED Zanders. Gene expression analysis as an aid to the identification of drug targets. Pharmacogenonics 1:375–384, 2000.
5. J Drews. Drug discovery: A historical perspective. Science 287:1960–1964, 2000.
6. AD Roses. Pharmacogenetics and the practice of medicine. Nature 405:857–865, 2000.
7. EH Ohlstein, RR Ruffolo Jr, JD Elliott. Drug discovery in the next millennium. Annu Rev Pharmacol Toxicol 40:177–191, 2000.
8. DF Veber, FH Drake, M Gowen. The new partnership of genomics and chemistry for accelerated drug development. Curr Opin Chem Biol 1:151–156, 1997.
9. FS Collins, A Patrinos, E Jordan, A Chakravarti, R Gesteland, L Walters, et al. New goals for the U.S. Human Genome Project: 1998–2003. Science 282:682–689, 1998.
10. MR Spaller, MT Burger, M Fardis, PA Bartlett. Synthetic strategies in combinatorial chemistry. Curr Opin Chem Biol 1:47–53, 1997.
11. D Brown. Future pathways for combinatorial chemistry. Mol Divers 2:217–222, 1997.
12. JF Cargill, M Lebl. New methods in combinatorial chemistry—robotics and parallel synthesis. Curr Opin Chem Biol 1:67–71, 1997.
13. JJ Burbaum, NH Sigal. New technologies for high-throughput screening. Curr Opin Chem Biol 1:72–78, 1997.

14. R Kasher, DA Oren, Y Barba, C Gilon. Miniaturized proteins: The backbone cyclic proteinomimetic approach. J Mol Biol 292:421–429, 1999.
15. N Sleep. Sorting out combinatorial chaos. Modern Drug Discovery 3:37–42, 2000.
16. DJ Ausman. Creating the high-output mind. Modern Drug Discovery 4:19–23, 2001.
17. MJ Felton. Bioinformatics: The child of success. Modern Drug Discovery 4:24–28, 2001.
18. M Rios. Genomics and the future of the pharmaceutical industry. Pharm Technol 25:34–40, 2001.
19. FS Collins. The human genome project and the future of medicine. Ann NY Acad Sci 882:42–55, 1999.
20. FS Collins, KG Jegalian. Deciphering the code of life. Sci Am 281:86–91, 1999.
21. DJ Lockhart, EA Winzeler. Genomics, gene expression and DNA arrays. Nature 405:827–836, 2000.
22. LJ Beeley, DM Duckworth, C Southan. The impact of genomics on drug discovery. Prog Med Chem 37:1–43, 2000.
23. K Pal. The keys to chemical genomics. Modern Drug Discovery 3:46–55, 2000.
24. JM Chen, C Ferec. Genes, cloned cDNAs, and proteins of human trypsinogens and pancreatitis-associated cationic trypsinogen mutations. Pancreas 21:57–62, 2000.
25. JM Thornton, AE Todd, D Milburn, N Borkakoti, CA Orengo. From structure to function: Approaches and limitations. Nat Struct Biol 7(suppl):991–994, 2000.
26. RD Vale, RA Milligan. The way things move: Looking under the hood of molecular motor proteins. Science 288:88–95, 2000.
27. D Eisenberg, EM Marcotte, I Xenarios, TO Yeates. Protein function in the post-genomics era. Nature 405:823–826, 2000.
28. EN Trifonov. Earliest pages of bioinformatics. Bioinformatics 16:5–9, 2000.
29. D Rhodes, JW Schwabe, L Chapman, L Fairall. Towards an understanding of protein-DNA recognition. Phil Trans R Soc Lond B Biol Sci 351:501–509, 1996.
30. MS Clark. Comparative genomics: The key to understanding the Human Genome Project. Bioessays 21:121–130, 1999.
31. C Sander. Genomic medicine and the future of health care. Science 287:1977–1978, 2000.
32. JC Kaplan, C Junien. Genomics and medicine: An anticipation. From Boolean Mendel genetics to multifactorial molecular medicine. CR Acad Sci III 323:1167–1174, 2000.
33. I Humphery-Smith. Replication-induced protein synthesis and its importance to proteomics. Electrophoresis 20:653–659, 1999.
34. PA Haynes, JR Yates 3rd. Proteome profiling-pitfalls and progress. Yeast 17:81–97, 2000.
35. A Pandey, M Mann. Proteomics to study genes and genomes. Nature 405:837–846, 2000.
36. K Gevaert, J Vandekerckhove. Protein identification methods in proteomics. Electrophoresis 21:1145–1154, 2000.
37. MJ Dutt, KH Lee. Proteomic analysis. Curr Opin Biotechnol 11:176–179, 2000.
38. DD Ryu, DH Nam. Recent progress in biomolecular engineering. Biotechnol Prog 16:2–16, 2000.

39. JR Yates 3rd. Mass spectrometry. From genomics to proteomics. Trends Genet 16: 5–9, 2000.

40. S Borman. Proteomics: Taking over where genomics leaves off. Chem Eng News 78:31–37, 2000.

41. G Chambers, L Lawrie, P Cash, GI Murray. Proteomics: A new approach to the study of disease. J Pathol 192:280–288, 2000.

42. RE Banks, MJ Dunn, DF Hochstrasser, JC Sanchez, W Blackstock, DJ Pappin, PJ Selby. Proteomics: New perspective, new biomedical opportunities. Lancet 356: 1749–1756, 2000.

43. B Cox, JC Denyer, A Binnie, MC Donnelly, B Evans, DV Green, JA Lewis, TH Mander, AT Merritt, MJ Valler, SP Watson. Application of high-throughput screening techniques to drug discovery. Prog Med Chem 37:83–133, 2000.

44. L Weber. High-diversity combinatorial chemistry. Curr Opin Chem Biol 4:338–345, 2000.

45. D Gorse, R Lahana. Functional diversity of compound libraries. Curr Opin Chem Biol 4:287–294, 2000.

46. CD Floyd, C Leblanc, M Whittaker. Combinatorial chemistry. Prog Med Chem 36:91–168, 1999.

47. AB Martin, PG Schultz. Opportunities at the interface of chemistry and biology. Trends Cell Biol 9:M24–M28, 1999.

48. MJ Shapiro, JR Wareing. NMR methods in combinatorial chemistry. Curr Opin Chem Biol 2:372–375, 1998.

49. J Cheng, P Fortina, S Surrey, LJ Kricka, P Wilding. Microchip-based devices for molecular diagnosis of genetic diseases. Mol Diagn 1:183–200, 1996.

50. T Blanc, DE Schaufelberger, NA Guzman. Capillary electrophoresis. In: GW Ewing, ed. Analytical Instrumentation Handbook. New York: Marcel Dekker, 1987, pp 1351–1431.

51. NA Guzman, SS Park, D Schaufelberger, L Hernandez, X Paez, P Rada, AJ Tomlinson, S Naylor. New approaches in clinical chemistry: On-line analyte concentration and microreaction capillary electrophoresis for the determination of drugs, metabolic intermediates, and biopolymers in biological fluids. J Chromatogr B 697:37–66, 1997.

52. CL Colyer, T Tang, N Chiem, DJ Harrison. Clinical potential of microchip capillary electrophoresis systems. Electrophoresis 18:1733–1741, 1997.

53. SC Beale. Capillary electrophoresis. Anal Chem 70:279R–300R, 1998.

54. G Kemp. Capillary electrophoresis: A versatile family of analytical techniques. Biotechnol Appl Biochem 27:9–17, 1998.

55. CM Boone, JC Waterval, H Lingeman, K Ensing, WJ Underberg. Capillary electrophoresis as a versatile tool for the bioanalysis of drugs—A review. J Pharm Biomed Anal 5:5–11, 1999.

56. CL Flurer. Analysis of antibiotics by capillary electrophoresis. Electrophoresis 20: 3269–3279, 1999.

57. H Nishi. Capillary electrophoresis of drugs: Current status in the analysis of pharmaceuticals. Electrophoresis 20:3237–3258, 1999.

58. NA Guzman. On-line bioaffinity, molecular recognition, and preconcentration in CE technology. LC-GC 17:16–27, 1999.

59. HJ Issaq. A decade of capillary electrophoresis. Electrophoresis 21:1921–1939, 2000.

60. G Manetto, F Crivellente, F Tagliaro. Capillary electrophoresis: A new analytical tool for forensic toxicologists. Ther Drug Monit 22:84–88, 2000.

61. SN Krylov, NJ Dovichi. Capillary electrophoresis for the analysis of biopolymers. Anal Chem 72:111R–128R, 2000.

62. A Bossi, SA Piletsky, PG Righetti, AP Turner. Capillary electrophoresis coupled to biosensor detection. J Chromatogr A 892:143–153, 2000.

63. LA Larsen, M Christiansen, J Vuust, PS Andersen. High throughput mutation screening by automated capillary electrophoresis. Comb Chem High Throughput Screen 3:393–409, 2000.

64. V Dolnik, S Liu, S Jovanovich. Capillary electrophoresis on microchip. Electrophoresis 21:41–54, 2000.

65. CT Culbertson, SC Jacobson, JM Ramsey. Microchip devices for high-efficiency separations. Anal Chem 72:5814–5819, 2000.

66. G Jiang, S Attiya, G Ocvirk, WE Lee, DJ Harrison. Red diode laser induced fluorescence detection with a confocal microscope on a microchip for capillary electrophoresis. Biosens Bioelectron 14:861–869, 2000.

67. RD Oleschuk, LL Shultz-Lockyear, Y Ning, DJ Harrison. Trapping of bead-based reagents within microfluidic systems: On chip solid-phase extraction and electrochromatography. Anal Chem 72:585–590, 2000.

68. NA Guzman, RE Majors. New directions for concentration sensitivity enhancement in CE and microchip technology. LC-GC 19:14–30, 2001.

69. J Kuhlmann. Alternative strategies in drug development: Clinical pharmacological aspects. Int J Clin Pharmacol Ther 37:575–583, 1999.

70. LJ Lesko, M Rowland, CC Peck, TF Blaschke. Optimizing the science of drug development: Opportunities for better candidate selection and accelerated evaluation in humans. Pharm Res 17:1335–1344, 2000.

71. JP Swann. The 1941 sulfathiazole disaster and the birth of good manufacturing practices. J Pharm Sci Technol 53:148–153, 1999.

72. MA Serabian, AM Pilaro. Safety assessment of biotechnology-derived pharmaceuticals: ICH and beyond. Toxicol Pathol 27:27–31, 1999.

73. MA Friedman, J Woodcock, MM Lumpkin, JE Shuren, AE Hass, LJ Thompson. The safety of newly approved medicines: Do recent market removals mean there is a problem? JAMA 281:1728–1734, 1999.

74. RA Baffi. The role of assay validation in specification development. Dev Biol Stand 91:105–113, 1997.

75. SH Chen, JM Gallo. Use of capillary electrophoresis methods to characterize the pharmacokinetics of antisense drugs. Electrophoresis 19:2861–2869, 1998.

76. C Hartmann, J Smeyers-Verbeke, DL Massart, RD McDowall. Validation of bioanalytical chromatographic methods. J Pharm Biomed Anal 17:193–218, 1998.

77. RH Neubert, MA Schwarz, Y Mrestani, M Platzer, K Raith. Affinity capillary electrophoresis in pharmaceutics. Pharm Res 16:1663–1673, 1999.

78. JM Moore. NMR screening in drug discovery. Curr Opin Biotechnol 10:54–58, 1999.

79. Y Gaillard, G Pepin. Testing hair for pharmaceuticals. J Chromatogr B 733:231–246, 1999.

80. W Muck. Quantitative analysis of pharmacokinetic study samples by liquid chromatography coupled to tandem mass spectrometry (LC-MS/MS). Pharmazie 54: 639–644, 1999.

81. Z El Rassi. Recent developments in capillary electrophoresis and capillary electrochromatography of carbohydrate species. Electrophoresis 20:3145–3155, 1999.

82. A Dermaux, P Sandra. Applications of capillary electrochromatography. Electrophoresis 20:3027–3065, 1999.

83. MR Hadley, P Camilleri, AJ Hutt. Enantiomeric analysis by capillary electrophoresis: Applications in drug metabolism and pharmacokinetics. Electrophoresis 21: 1953–1976, 2000.

84. J Haginaka. Enantiomer separation of drugs by capillary electrophoresis using proteins as chiral selectors. J Chromatogr A 875:235–254, 2000.

85. HH Yarabe, E Billiot, IM Warner. Enantiomeric separations by use of polymeric surfactant electrokinetic chromatography. J Chromatogr A 875:179–206, 2000.

86. G Blaschke, B Chankvetadze. Enantiomeric separation of drugs by capillary electromigration techniques. J Chromatogr A 875:3–25, 2000.

87. M Fillet, I Bechet, V Piette, J Crommen. Separation of nonsteroidal anti-inflammatory drugs by capillary electrophoresis using nonaqueous electrolytes. Electrophoresis 20:1907–1915, 1999.

88. M Tacker, P Glukhovskiy, H Cai, G Vigh. Nonaqueous capillary electrophoresis separation of basic enantiomers using heptakis (2,3-dimethyl-6-sulfato)-beta-cyclodextrin. Electrophoresis 20:2794–2798, 1999.

89. A Karbaum, T Jira. Nonaqueous capillary electrophoresis: Application possibilities and suitability of various solvents for the separation of basic analytes. Electrophoresis 20:3396–3401, 1999.

90. L Geiser, S Cherkaoui, JL Veuthey. Simultaneous analysis of some amphetamine derivatives in urine by nonaqueous capillary electrophoresis coupled to electrospray ionization mass spectrometry. J Chromatogr A 895:111–121, 2000.

91. F Wang, MG Khaledi. Enantiomer separation of drugs by capillary electromigration techniques. J Chromatogr A 875:3–25, 2000.

92. A Prochazkova, M Chouki, R Theurilla, W Thormann. Therapeutic drug monitoring of albendazole: Determination of albendazole sulfoxide, and albendazole sulfone in human plasma using nonaqueous capillary electrophoresis. Electrophoresis 21: 729–736, 2000.

93. JF Banks. Recent advances in capillary electrophoresis/electrospray/mass spectrometry. Electrophoresis 18:2255–2266, 1997.

94. AR Timerbaev. Advances in detection of inorganic ions in capillary electrophoresis. J Capillary Electrophor 5:185–192, 1998.

95. J Ding, P Vouros. Advances in CE/MS. Anal Chem 71:378A–385A, 1999.

96. G Choudhary, A Apftel, H Yin, W Hancock. Use of on-line mass spectrometric detection in capillary electrochromatography. J Chromatogr A 887:85–101, 2000.

97. UB Soetebeer, MO Schierenber, JG Moller, H Schulz, G Grunefeld, P Andresen, G Blasche, G Ahr. Capillary electrophoresis with laser-induced fluorescence in clinical drug development routine application and future aspects. J Chromatogr A 895: 147–155, 2000.

98. WF Nirode, GL Devault, MJ Sepaniak, RO Cole. On-column surface-enhanced

Raman spectroscopy detection in capillary electrophoresis using running buffers containing silver colloidal solutions. Anal Chem 72:1866–1871, 2000.

99. JA Fracassi da Silva, CL do Lago. Conductivity detection of aliphatic alcohols in micellar electrokinetic chromatography using an oscillometric detector. Electrophoresis 21:1405–1408, 2000.

100. K Swinney, DJ Bornhop. Detection in capillary electrophoresis. Electrophoresis 21:1239–1250, 2000.

101. AM García-Campaña, WRG Baeyens, NA Guzman. Chemiluminescence detection in capillary electrophoresis. In: AM García-Campaña and WRG Baeyens, eds. Chemi-luminescence in Analytical Chemistry. New York: Marcel Dekker, 2001, pp 427–472.

102. V Dolnik. DNA sequencing by capillary electrophoresis. J Biochem Biophys Methods 41:103–119, 1999.

103. S Behr, M Matzig, A Levin, H Eickhoff, C Heller. A fully automated multicapillary electrophoresis device for DNA analysis. Electrophoresis 20:1492–1507, 1999.

104. D Schmalzing, L Koutny, O Salas-Solano, A Adourian, P Matsudaria, D Ehrlich. Recent developments in DNA sequencing by capillary and microdevice electrophoresis. Electrophoresis 20:3066–3077, 1999.

105. Y Shi, PC Simpson, JR Scherer, D Wexler, C Skibola, MT Smith, RA Mathies. Radial capillary array electrophoresis microplate and scanner for high-performance nucleic acid analysis. Anal Chem 71:5354–5356, 1999.

106. Z Huang, N Munro, AF Huhmer, JP Landers. Acousto-optical deflection-based laser beam scanning for fluorescence detection on multichannel electrophoresis microchips. Anal Chem 71:5309–5314, 1999.

107. A Hanning, J Westberg, J Roeraade. A liquid core waveguide fluorescence detector for multiple capillary electrophoresis applied to DNA sequencing in a 91-capillary array. Electrophoresis 21:3290–3304, 2000.

108. Y Zhang, X Gong, H Zhang, RC Larock, ES Yeung. Combinatorial screening of homogeneous catalysis and reaction optimization based on multiplexed capillary electrophoresis. J Comb Chem 2:450–452, 2000.

109. Q Gao, ES Yeung. A matrix for DNA separation: Genotyping and sequencing using poly(vinylpyrrolidone) solution in uncoated capillaries. Anal Chem 70:1382–1388, 1998.

110. G Bocaz, AL Revilla, J Krejci, J Havel. Characterization of a sapphire-epoxy coating for capillary electrophoresis. J Capillary Electrophor 5:165–170, 1998.

111. I Miksik, Z Deyl, V Kasicka. Capillary electrophoretic separation of proteins and peptides using pluronic liquid crystals and surface-modified capillaries. J Chromatogr B 741:37–42, 2000.

112. B Verzola, C Gelfi, PG Righetti. Protein adsorption to the bare silica wall in capillary electrophoresis quantitative study on the chemical composition of the background electrolyte for minimizing the phenomenon. J Chromatogr A 868:85–99, 2000.

113. Y Liu, JC Fanguy, JM Bledsoe, CS Henry. Dynamic coating using polyelectrolyte multilayers for chemical control of electroosmotic flow in capillary electrophoresis microchips. Anal Chem 72:5939–5944, 2000.

114. BC Giordano, M Muza, A Trout, JP Landers. Dynamically-coated capillaries allow

for capillary electrophoretic resolution of transferrin sialoforms via direct analysis of human serum. J Chromatogr B 742:79–89, 2000.

115. KA Assi, BJ Clark, KD Altria. Enantiomeric purity determination of propanolol by capillary electrophoresis using dual cyclodextrins and a polyacrylamide-coated capillary. Electrophoresis 20:2723–2725, 1999.

116. TM Oswald, TJ Ward. Enantioseparations with the macrocyclic antibiotic ristocetin A using a countercurrent process in CE. Chirality 11:663–668, 1999.

117. M Chiari, M Cretich, G Crini, L Janus, M Morcellet. Allylamine-beta-cyclodextrin copolymer. A novel chiral selector for capillary electrophoresis. J Chromatogr A 894:95–103, 2000.

118. NA Mohamed, Y Kuroda, A Shibukawa, T Nakagawa, S El Gizawy, HF Askal, ME El Kommos. Enantioselective binding analysis of verapamil to plasma lipoproteins by capillary electrophoresis-frontal analysis. J Chromatogr A 875:447–453, 2000.

119. N Heegaard, S Nilsson, NA Guzman. Affinity capillary electrophoresis: Important application areas and some recent developments. J Chromatogr B 715:29–54, 1998.

120. NA Guzman. Determination of immunoreactive gonadotropin releasing hormone in serum and urine by on-line immunoaffinity capillary electrophoresis coupled to mass spectrometry. J Chromatogr B 749:197–213, 2000.

121. SR Rabel, JF Stobaugh. Applications of capillary electrophoresis in pharmaceutical analysis. Pharm Res 10:171–186, 1993.

122. KD Altria. Application of capillary electrophoresis to pharmaceutical analysis. Methods Mol Biol 52:265–284, 1996.

123. KD Altria, J Bestford. Main component assay of pharmaceuticals by capillary electrophoresis: Considerations regarding precision, accuracy, and linearity data. J Capillary Electrophor 3:13–23, 1996.

124. LA Holland, NP Chetwyn, MD Perkins, SM Lunte. Capillary electrophoresis in pharmaceutical analysis. Pharm Res 14:372–387, 1997.

125. KD Altria, MA Kelly, BJ Clark. Current applications in the analysis of pharmaceuticals by capillary electrophoresis. I. Trends Anal Chem 17:204–214, 1998.

126. KD Altria, MA Kelly, BJ Clark. Current applications in the analysis of pharmaceuticals by capillary electrophoresis. II. Trends Anal Chem 17:214–226, 1998.

127. TD Hatajik, PR Brown. Chiral separations of pharmaceuticals using capillary electrochromatography (CEC). J Capillary Electroph 5:143–151, 1998.

128. TJ O'Shea. Analysis of pharmaceuticals by capillary electrophoresis. In: H Shintani, J Polonsky, eds. Handbook of Capillary Electrophoresis Applications. London: Blackie, 1997, pp 301–323.

129. KD Altria, SM Bryant. Survey into the status of capillary electrophoresis. LC-GC International 10:26–30, 1997.

130. DB Campbell. The development of chiral drugs. Acta Pharm Nord 2:217–226, 1990.

131. IW Wainer. Three-dimensional view of pharmacology. Am J Hosp Pharm 49:S4–S8, 1992.

132. M Gibaldi. Stereoselective and isozyme-selective drug interactions. Chirality 5:407–413, 1993.

133. TS Tracy. Stereochemistry in pharmacotheraphy: When mirror images are not identical. Ann Pharmacother 29:161–165, 1995.

134. AJ Hutt, J O'Grady. Drug chirality: A consideration of the significance of the stereochemistry of antimicrobial agents. J Antimicrob Chemother 37:7–32, 1996.
135. VJ Lee, SJ Hecker. Antibiotic resistance versus small molecules, the chemical evolution. Med Res Rev 19:521–542, 1999.
136. CSW Koehler. How thalidomide was kept out of the U.S. market. Modern Drug Discovery 3:69–72, 2000.
137. MR Juchau. Chemical teratogenesis. Prog Drug Res 41:9–50, 1993.
138. EJ Lien. Chirality and drug targeting: Pros and cons. J Drug Target 2:527–532, 1995.
139. B Vaisman. Mechanism of teratogenic action of thalidomide. Teratology 53:283–284, 1996.
140. JM Larry, KL Daniel, JD Erickson, HE Roberts, CA Moore. The return of thalidomide: Can birth defects be prevented? Drug Safety 21:161–169, 1999.
141. WH De Camp. Chiral drugs: The FDA perspective on manufacturing and control. J Pharm Biomed Anal 11:1167–1172, 1993.
142. ME Donawa. New FDA draft guidance on premarket submissions. Med Device Technol 10:12–14, 1999.
143. RS Cahn, CK Ingold, V Prelog. Specification of molecular chirality. Angew Chem Int Ed 5:385–415, 1966.
144. E Lee, K Williams. Need for precise chiral nomenclature. Clin Pharmacokinet 19:503, 1990.
145. YW Lam. Stereoselectivity: An issue of significant importance in clinical pharmacology. Pharmacotheraphy 8:147–157, 1988.
146. S Topiol. A general criterion for molecular recognition: Implications for chiral interactions. Chirality 1:69–79, 1989.
147. R Bentley. From optical activity in quartz to chiral drugs: Molecular handedness in biology and medicine. Perspect Biol Med 38:188–229, 1995.
148. RC Williams, CM Riley, KW Sigvardson, J Fortunak, P Ma, EC Nicolas, SE Unger, DF Krahn, SL Brenner. Pharmaceutical development and specification of stereoisomers. J Pharm Biomed Anal 17:917–924, 1998.
149. NR Srinivas, LN Igwemezie. Chiral separation by high performance liquid chromatography. I. Review on indirect separation of enantiomers as diastereomeric derivatives using ultraviolet, fluorescence and electrochemical detection. Biomed Chromatogr 6:163–167, 1992.
150. VL Lanchote, PS Bonato, SA Dreossi, PV Goncalves, EJ Cesarino, C Bertucci. High-performance liquid chromatographic determination of mexiletine enantiomers in plasma using direct and indirect enantioselective separations. J Chromatogr B 685:281–289, 1996.
151. S Gorog, M Gazdag. Enantiomeric derivatization for biomedical chromatography. J Chromatogr B 659:51–84, 1994.
152. ME Laethem, FM Belpaire, MG Bogaert. Direct high-performance liquid chromatography determination of diastereomeric oxprenolol glucuronides. J Chromatogr B 675:251–255, 1996.
153. A Berthod, X Chen, JP Kullman, DW Armstrong, F Gasparrini, I D'Acquarica, C Villani, A Carotti. Role of the carbohydrate moieties in chiral recognition on teicoplanin-based LC stationary phases. Anal Chem 72:1767–1780, 2000.

154. OY Al-Dirbashi, M Wada, N Kuroda, M Takahashi, K Nakashima. Achiral and chiral quantification of methamphetamine and amphetamine in human urine by semi-micro column high-performance liquid chromatography and fluorescence detection. J Forensic Sci 45:708–714, 2000.

155. X Chen, H Zou, L Yang, H Wang, O Zhang. Optical resolution of alpha-alkyl phenyl acetonitriles by HPLC on cellulose triacetate chiral stationary phases coated on underivatized silica gel. Chirality 12:599–605, 2000.

156. I D'Acquarica. New synthetic strategies for the preparation of novel chiral stationary phases for high performance liquid chromatography containing natural pool selectors. J Pharm Biomed Anal 23:3–13, 2000.

157. R Vespalec, P Bocek. Chiral separations by capillary electrophoresis: present state of the art. Electrophoresis 15:755–762, 1994.

158. K Verleysen, P Sandra. Separation of chiral compounds by capillary electrophoresis. Electrophoresis 19:2798–2833, 1998.

159. S Fanali, Z Aturki, C Desiderio. Enantioresolution of pharmaceutical compounds by capillary electrophoresis. Use of cyclodextrins and antibiotics. Enantiomer 4: 229–241, 1999.

160. Z El-Rassi. Chiral glycosidic surfactants for enantiomeric separation in capillary electrophoresis. J Chromatogr A 875:207–233, 2000.

161. HH Yarabe, E Billiot, IM Warner. Enantiomeric separations by use of polymeric electrokinetic chromatography. J Chromatogr A 875:179–206, 2000.

162. K Otsuka, S Terabe. Enantiomer separation of drugs by micellar electrokinetic chromatography using chiral surfactants. J Chromatogr A 875:163–178, 2000.

163. DT Burns. Cyclodextrins as versatile chiral recognition reagents for use in a variety of optical and separative analyses. J Pharm Biomed Anal 12:1–3, 1994.

164. E Albers, BW Muller. Cyclodextrin derivatives in pharmaceutics. Crit Rev Ther Drug Carrier Syst 12:311–337, 1995.

165. F Bresolle, M Audran, TN Pham, JJ Vallon. Cyclodextrins and enantiomeric separations of drugs by liquid chromatography and capillary electrophoresis: Basic principles and new developments. J Chromatogr B 687:303–336, 1996.

166. T Loftsson. Increasing the cyclodextrin complexation of drugs and drug bioavailability through addition of water-soluble polymers. Pharmazie 53:733–740, 1998.

167. S Fanali. Enantioselective determination by capillary electrophoresis with cyclodextrins as chiral selectors. J Chromatogr A 875:89–122, 2000.

168. B Koppenhoefer, X Zhu, A Jakob, S Wuerthner, B Lin. Separation of drug enantiomers by capillary electrophoresis in the presence of neutral cyclodextrins. J Chromatogr A 875:135–161, 2000.

169. F Benavente, RA Jackson, JV Weber, NA Guzman. Use of non-aqueous capillary electrophoresis containing sulfated alpha-cyclodextrin for the separation and characterization of a water insoluble drug substance. 24th Int Symp on High Performance Liquid Phase Separations and Related Techniques HPLC 2000, Seattle, WA, abstr P-0732, 2000.

170. M Fillet, P Hubert, J Crommen. Enantiomeric separations of drugs using mixtures of charged and neutral cyclodextrins. J Chromatogr A 875:123–134, 2000.

171. S Sabah, GKE Scriba. Resolution of aspartyl dipeptide and tripeptide stereoisomers by capillary electrophoresis. J Microcol Sep 10:255–258, 1998.

172. K-H Gahm, AM Stalcup. Sulfated cyclodextrins for the chiral separations of cathecolamines and related compounds in the reversed electrophoretic polarity mode. Chirality 8:316–324, 1996.

173. AM Stalcup, K-H Gahm. Application of sulfated cyclodextrins to chiral separations by capillary zone electrophoresis. Anal Chem 68:1360–1368, 1996.

174. LJ Jin, SF Li. Comparison of chiral recognition capabilities of cyclodextrins for the separation of basic drugs in capillary zone electrophoresis. J Chromatogr B 708:257–266, 1998.

175. NJ Munro, J Palmer, AM Stalcup, JP Landers. Charged cyclodextrin-mediated sample stacking in micellar capillary electrophoresis. A simple method for enhancing the detection sensitivity of hydrophobic compounds. J Chromatogr B 731:369–381, 1999.

176. F Wang, MG Khaledi. Non-aqueous capillary electrophoresis chiral separations with sulfated beta-cyclodextrins. J Chromatogr B 731:187–197, 1999.

177. Y Daali, S Cherkaoui, P Christen, JL Veuthey. Experimental design for enantioselective separation of celiprolol by capillary electrophoresis using sulfated beta-cyclodextrin. Electrophoresis 20:3424–3431, 1999.

178. SE Lucangioli, LG Hermida, VP Tripodi, VPT Rodriguez, EE Lopez, PD Rouge, CN Carducci. Analysis of cis-trans isomers and enantiomers of sertraline by cyclodextrin-modified micellar electrokinetic chromatography. J Chromatogr A 871:207–215, 2000.

179. A D'Hulst, N Verbeke. Quantitation in chiral capillary electrophoresis: Theoretical and practical considerations. Electrophoresis 15:854–863, 1994.

180. SG Penn, ET Bergstrom, DM Goodall. Capillary electrophoresis with chiral selectors: Optimization of separation and determination of thermodynamic parameters for binding of tioconazole enantiomers to cyclodextrins. Anal Chem 66:2866–2873, 1994.

181. A Guttman. Novel separation scheme for capillary electrophoresis of enantiomers. Electrophoresis 16:1900–1905, 1995.

182. H Watzig, M Degenhardt, A Kunkel. Strategies for capillary electrophoresis: Method development and validation for pharmaceutical and biological applications. Electrophoresis 19:2695–2752, 1998.

183. M Fillet, P Hubert, J Crommen. Method development strategies for the enantioseparation of drugs by capillary electrophoresis using cyclodextrins as chiral additives. Electrophoresis 19:2834–284, 1998.

184. L Liu, MA Nussbaum. Systematic screening approach for chiral separations of basic compounds by capillary electrophoresis with modified cyclodextrins. J Pharm Biomed Anal 19:679–694, 1999.

185. L Zhou, BD Johnson, C Miller, JM Wyvratt. Chiral capillary analysis of the enantiomeric purity of a pharmaceutical compound using sulfated beta-cyclodextrin. J Chromatogr A 875:389–401, 2000.

186. C Perrin, MG Vargas, YV Heyden, M Maftouh, DL Massart. Fast development of separation methods for the chiral analysis of amino acid derivatives using capillary electrophoresis and experimental designs. J Chromatogr A 883:249–265, 2000.

187. K-H Gahm, LW Chang, DW Armstrong. Chiral separation of monoterpenes using mixtures of sulfated β-cyclodextrins and α-cyclodextrin as chiral additives in the

reversed-polarity capillary electrophoresis mode. J Chromatogr A 759:149–155, 1997.

188. JB Vincent, G Vigh. Systematic approach to methods development for the capillary electrophoretic analysis of a minor enantiomer using a single-isomer sulfated cyclodextrin. A case study of L-carbidopa analysis. J Chromatogr A 817:105–111, 1998.

189. JB Vincent, G Vigh. Nonaqueous capillary electrophoretic separation of enantiomers using the single-isomer heptakis(2,3-di-acetyl-6-sulfato)-β-cyclodextrin as chiral resolving agent. J Chromatogr A 816:233–241, 1998.

190. BA Williams, G Vigh. Dry look at the CHARM (charged resolving agent migration) model of enantiomer separations by capillary electrophoresis. J Chromatogr A 777: 295–309, 1997.

191. KA Connors. Population characteristics of cyclodextrin complex stabilities in aqueous solution. J Pharm Sci 84:843–848, 1995.

192. KA Connors. Prediction of binding constants of alpha-cyclodextrin complexes. J Pharm Sci 85:796–802, 1996.

193. E Schneiderman, AM Stalcup. Cyclodextrins: A versatile tool in separation science. J Chromatogr A 745:83–102, 2000.

194. B Koppenhoefer, U Epperlein, R Schlun, X Zhu, B Lin. Separation of enantiomers of drugs by capillary electrophoresis. V. Hydroxypropyl-α-cyclodextrin as chiral solvating agent. J Chromatogr A 793:153–164, 1998.

195. M Fillet, L Fotsing, J Crommen. Enantiomeric separation of uncharged compounds by capillary electrophoresis using mixtures of anionic and neutral β-cyclodextrin derivatives. J Chromatogr A 817:113–119, 1998.

196. P Dzygiel, P Wieczorek, JA Jonsson. Enantiomeric separation of amino acids by capillary electrophoresis with α-cyclodextrin. J Chromatogr A 793:414–418, 1998.

197. M Blanco, J Coello, H Iturriaga, S Maspoch, C Perez-Maseda. Separation of profen enantiomers by capillary electrophoresis using cyclodextrins as chiral selectors. J Chromatogr A 793:165–175, 1998.

198. LJ Jin, SFY Li. Comparison of chiral recognition capabilities of cyclodextrins for the separation of basic drugs in capillary zone electrophoresis. J Chromatogr B 708:257–266, 1998.

199. M Fillet, I Bechet, Ph Hubert, J Crommen. Resolution improvement by use of carboxymethyl-β-cyclodextrin as chiral additive for the enantiomeric separation of basic drugs by capillary electrophoresis. J Pharm Biomed Anal 14:1107–1114, 1996.

200. A Guttman, S Brunet, N Cooke. Capillary electrophoresis separation of enantiomers by cyclodextrin array chiral analysis. LC-GC 14:32–42, 1996.

201. F Bressolle, M Audran, T-N Pham, J-J Vallon. Cyclodextrins and enantiomeric separations of drugs by chromatography and capillary electrophoresis: Basic principles and new developments. J Chromatogr B 687:303–336, 1996.

202. AR Kranack, MT Bowser, P Britz-McKibbin, DDY Chen. The effects of a mixture of charged and neutral additives on analyte migration behavior in capillary electrophoresis. Electrophoresis 19:388–396, 1998.

203. DW Armstrong, UB Nair. Capillary electrophoretic enantioseparations using macrocyclic antibiotics as chiral selectors. Electrophoresis 18:2331–2342, 1997.

204. C Desiderio, S Fanali. Chiral analysis by capillary electrophoresis using antibiotics as chiral selector. J Chromatogr A 807:37–56, 1998.

205. KH Ekborg-Ott, GA Zientara, JM Schneiderheinze, K Gahm, DW Armstrong. Avoparcin, a new macrocyclic chiral run buffer additive for capillary electrophoresis. Electrophoresis 20:2438–2457, 1999.

206. Q Sun, SV Olesik. Chiral separation by simultaneous use of vancomycin as stationary phase chiral selector mobile phase additive. J Chromatogr B 745:159–166, 2000.

207. KL Sutton, RM Sutton, AM Stalcup, JA Caruso. A comparison of vancomycin and sulfated beta-cyclodextrin as chiral selectors for enantiomeric separations of selenoamino acids using capillary electrophoresis with UV absorbance detection. Analyst 125:231–234, 2000.

208. TJ Ward. Chiral separations. Anal Chem 72:4521–4528, 2000.

209. R Vespalec, HA Billiet, J Frank, P Bocek. Vancomycin as a chiral selector in capillary electrophoresis: An appraisal of advantages and limitations. Electrophoresis 17:1214–1221, 1996.

210. R Vespalec, P Bocek. Chiral separations in capillary electrophoresis. Electrophoresis 20:2579–2591, 1999.

211. MP Gasper, A Berthod, UB Nair, DW Armstrong. Comparison and modeling study of vancomycin, ristocetin A, and teicoplanin for CE enantioseparations. Anal Chem 68:2501–2514, 1996.

212. DW Armstrong, K Rundlett, GL Reid 3rd. Use of a macrocyclic antibiotic, rifamycin B, and indirect detection for the resolution of racemic amino alcohol by CE. Anal Chem 66:1690–1695, 1994.

213. K Rundlett, MP Gasper, EY Zhou, DW Armstrong. Capillary electrophoretic enantiomeric separations using the glycopeptide antibiotic, teicoplanin. Chirality 8:88–107, 1996.

214. R Kuhn, J Wagner, Y Walbroehl, T Bereuter. Potential and limitations of an optically active crown ether for chiral separation in capillary zone electrophoresis. Electrophoresis 15:828–834, 1994.

215. Y Walbroehl, J Wagner. Enantiomeric resolution of primary amines by capillary electrophoresis and high-performance liquid chromatography. J Chromatogr A 680: 253–261, 1994.

216. MG Schmid, G Gübitz. Capillary zone electrophoretic separation of enantiomers of dipeptides based on host-guest complexation with a chiral crown ether. J Chromatogr A 709:81–88, 1995.

217. H Nishi. Enantioselectivity in chiral electrophoresis with polysaccharides. J Chromatogr A 792:327–347, 1997.

218. A D'Hulst, N Verbecke. Chiral separation by capillary electrophoresis with oligosaccharides. J Chromatogr 698:257–287, 1992.

219. AM Abushoffa, BJ Clark. Resolution of enantiomers of oxamniquine by capillary electrophoresis and high-performance liquid chromatography with cyclodextrins and heparin as chiral selectors. J Chromatogr A 700:51–58, 1995.

220. Y Tanaka, S Terabe. Separation of the enantiomers of basic drugs by affinity capillary electrophoresis using a partial filling technique and α-1-acid glycoprotein as chiral selector. Chromatographia 44:119–128, 1997.

221. H Shiono, A Shibukawa, Y Kuroda, T Nakagawa. Effect of sialic acid residues of human alpha 1-acid glycoprotein on stereoselectivity in basic drug-protein binding. Chirality 9:291–296, 1997.

222. D Eberle, RP Hummel, R Kuhn. Chiral resolution of pantoprazole sodium and related sulfoxides by complex formation with bovine serum albumin in capillary electrophoresis. J Chromatogr A 759:185–192, 1997.

223. L Valtcheva, J Mohammad, G Pettersson, S Hjerten. Chiral separation of β-blockers by high-performance capillary electrophoresis based on non-immobilized cellulase as enantioselective protein. J Chromatogr 638:263–267, 1993.

224. GE Barker, P Russo, RA Hartwick. Chiral separation of leucovorin with bovine serum albumin using affinity capillary electrophoresis. Anal Chem 64:3024–3028, 1992.

225. J Wang, IM Warner. Combined polymerized chiral micelle and gamma-cyclodextrin for chiral separation in capillary electrophoresis. J Chromatogr A 711:297–304, 1995.

226. KA Agnew-Heard, MS Pena, SA Shamsi, IM Wagner. Studies of polymerized sodium N-undecylenyl-L-valinate in chiral micellar electrokinetic capillary chromatography of neutral, acidic, and basic compounds. Anal Chem 69:958–964, 1997.

227. EW Tsai, MM Singh, HH Lu, MA Brook. Application of capillary electrophoresis to pharmaceutical analysis. Determination of alendronate in dosage forms. J Chromatogr 626:245–250, 1992.

228. E Gonzalez, A Becerra, JJ Laserna. Direct determination of diuretic drugs in urine by capillary zone electrophoresis using fluorescence detection. J Chromatogr B 687:145–150, 1996.

229. CS Liu, XF Li, D Pinto, EB Hansen, CE Cerniglia, NJ Dovichi. On-line nonaqueous capillary electrophoresis and electrospray mass spectrometry of tricyclic antidepressants and metabolic profiling of aminotriptyline by *Cunninghamella elegans*. Electrophoresis 19:3183–3189, 1998.

230. M Hernandez, F Borull, M Calull. Determination of amoxicillin in plasma samples by capillary electrophoresis. J Chromatogr B 731:309–315, 1999.

231. YJ Heo, YS Whang, MK In, KJ Lee. Determination of enantiomeric amphetamines and selegiline in urine by capillary electrophoresis using modified beta-cyclodextrin. J Chromatogr B 741:221–230, 2000.

232. A Ramseier, J Caslavaska, W Thormann. Stereoselective screening for and confirmation of urinary enantiomers of amphetamine, methamphetamine, designer drugs, metadone and selected metabolites by capillary electrophoresis. Electrophoresis 20:2726–2738, 1999.

233. P Esseiva, E Lock, O Gueniat, MD Cole. Identification and quantification of amphetamine and analogues by capillary zone electrophoresis. Sci Justice 37:113–119, 1997.

234. Y Nakahara. Hair analysis for abused and therapeutic drugs. J Chromatogr B 733:161–180, 1999.

235. L Geiser, S Cherkaoui, JL Veuthey. Simultaneous analysis of some amphetamine derivatives in urine by nonaqueous capillary electrophoresis coupled to mass spectrometry. J Chromatogr A 895:111–121, 2000.

236. MLL de Moraes, B Polakiewicz, MF Mattua, MFM Tavares. Comparative evaluation of capillary electrophoresis and high-performance liquid chromatography for

the separation of cis-cis, cis-trans, and trans-trans isomers of atracurium besylate. J Capillary Electrophor 5:33–38, 1998.

237. S Pálmarsdótir, B Lindegård, P Deininger, L-E Edholm, L Mathiasson, J-A Jönsson. Supported liquid membrane technique for selective sample workup of basic drugs in plasma prior to capillary zone electrophoresis. J Capillary Electrophor 2: 185–189, 1995.

238. K Jinno, Y Han, M Nakamura. Analysis of anxiolytic drugs by capillary electrophoresis with bare and coated capillaries. J Capillary Electrophor 3:139–145, 1996.

239. MG Schmid, O Lecnik, U Sitte, G Gubitz. Application of ligand-exchange capillary electrophoresis to the chiral separation of alpha-hydroxy acids and beta-blockers. J Chromatogr A 875:307–314, 2000.

240. SE Lucangioli, VG Rodríguez, GC Fernández Otero, NM Vizioli, CN Carducci. Development and validation of capillary electrophoresis methods for pharmaceutical dissolution assays. J Capillary Electrophor 4:27–31, 1997.

241. SE Lucangioli, VG Rodríguez, GC Fernández Otero, CN Carducci. Determination of related impurities of bile acids in bulk drugs by cyclodextrin-modified micellar electrokinetic chromatography. J Capillary Electrophor 5:139–142, 1998.

242. EP Lai, E Dabek-Zlotorynska. Separation of theophylline, caffeine and related drugs by normal-phase capillary electrochromatography. Electrophoresis 20:2366–2372, 1999.

243. G Luo, Y Wang, AG Ewing, TG Strein. Chiral separation of a novel calcium antagonist using capillary electrophoresis with cyclodextrins and diode array detection. J Capillary Electrophor 1:175–180, 1994.

244. S Hillaert, W Van den Bossche. Determination of captopril and its degradation products by capillary electrophoresis. J Pharm Biomed Anal 21:65–73, 1999.

245. Y Mrestani, R Neubert, J Schiewe, A Hartl. Application of capillary zone electrophoresis in cephalosporin analysis. J Chromatogr A 690:321–326, 1997.

246. BR Thomas, XG Fang, X Cheng, RJ Tyrrell, S Ghodbane. Validated micellar electrokinetic capillary chromatography method for quality control of the drug substances hydrochlorothiazide and chlorothiazide. J Chromatogr B 657:383–394, 1994.

247. J Luksa, D Josic. Determination of cimetidine in human plasma by free capillary zone electrophoresis. J Chromatogr B 667:321–327, 1995.

248. B Toussaint, PH Hubert, UR Tjaden, J van der Greef, J Crommen. Enantiomeric separation of clenbuterol by transient isotachophoresis-capillary zone electrophoresis-UV detection new optimization technique for transient isotachophoresis. J Chromatogr A 871:173–180, 2000.

249. SPD Lalljie, J Vindevogel, P Sandra. Capillary electrophoresis as a fast screening method for the determination of clenbuterol in urine. J Capillary Electrophor 1: 241–243, 1994.

250. Z Juvancz, I Ürmös, I Klebovich. Capillary electrophoretic separation of clomiphene isomers using various cyclodextrins as additives. J Capillary Electrophor 3: 181–189, 1996.

251. K Persson-Stubberud, O Astrom. Separation of ibuprofen, codeine phosphate, their degradation products and impurities by capillary electrophoresis. I. Method devel-

opment and optimization with fractional factorial design. J Chromatography A 798: 307–314, 1998.

252. KP Stubberud, O Astrom. Separation of ibuprofen, codeine phosphate, their degradation products and impurities by capillary electrophoresis. II. Validation. J Chromatogr A 826:95–102, 1998.

253. M Korman, J Vindevogel, P Sandra. Separation of codeine and its byproducts by capillary zone electrophoresis as a quality control tool in the pharmaceutical industry. J Chromatogr 645:366–370, 1993.

254. X Gu, M Meleka-Boules, C-L Chen. Micellar electrokinetic capillary chromatography combined with immunoaffinity chromatography for identification and determination of dexamethasone and flumethasone in equine urine. J Capillary Electrophor 3:43–49, 1996.

255. BR Thomas, S Ghodbane. Evaluation of a mixed micellar electrokinetic capillary electrophoresis method for validated pharmaceutical quality control. J Liq Chromatogr 16:1983–2006, 1993.

256. J Sevcik, Z Stransky, BA Ingelse, K Lemr. Capillary electrophoretic enantioseparation of selegiline, methamphetamine and ephedrine using a neutral beta-cyclodextrin epichlorhydrin polymer. J Pharm Biomed Anal 14:1098–1094, 1996.

257. MT Ackermans, JL Beckers, FM Everaerts, IGJA Seleen. Comparison of isotachophoresis, capillary electrophoresis and high performance liquid chromatography for the determination of salbutamol, terbutaline sulphate and fenoterol hydrobromide in pharmaceutical dosage forms. J Chromatogr 590:341–353, 1992.

258. E De Lorenzi, C Galbusera, V Belloti, P Mangione, G Massolini, E Tabolotti, A Andreola, G Caccialanza. Affinity capillary electrophoresis is a powerful tool to identify transthyretin binding drugs for potential therapeutic use in amyloidosis. Electrophoresis 21:3280–3289, 2000.

259. F Tagliaro, R Valentini, G Manetto, F Crivellente, G Carli, M Marigo. Hair analysis by using radioimmunoassay, high-performance liquid chromatography and capillary electrophoresis to investigate chronic exposure to heroin, cocaine and/or ecstasy in applicants for driving licences. Forensic Sci Int 107:121–128, 2000.

260. A Karbaum, T Jira. Nonaqueous capillary electrophoresis: application possibilities and suitability of various solvents for the separation of basic analytes. Electrophoresis 20:3396–3401, 1999.

261. I Bjørnsdottir, J Tjørnelung, SH Hansen. Nonaqueous capillary electrophoresis in pharmaceutical analysis. J Capullary Electrophor 3:83–87, 1996.

262. T Buzinkaiova, J Sadecka, J Polonsky, E Vlaisicova, V Korinkova. Isotachophoresis analysis of some antidepressants. J Chromatogr 638:231–234, 1993.

263. CJ Sciacchitano, B Mopper, JJ Specchio. Identification and separation of five cephalosporins by micellar electrokinetic capillary chromatography. J Chromatogr 657: 395–399, 1994.

264. D Tsikas, A Hofrichter, G Brunner. Capillary isotachophoretic analysis of β-lactam antibiotics and their precursors. Chromatographia 30:657–662, 1990.

265. GN Okafo, P Cutler, DJ Knowles. Capillary electrophoresis study of the hydrolysis of a β-lactamase inhibitor. Anal Chem 67:3697–3701, 1995.

266. MG Quaglia, S Fanali, A Nardi, C Rossi, M Ricci. Separation of leucinostatins by capillary zone electrophoresis. J Chromatogr 593:259–263, 1992.

267. L Hernandez, NA Guzman, BG Hoebel. Bidirectional microdialysis in vivo shows differential dopaminergic potency of cocaine, procaine and lidocaine in the nucleus accumbens using capillary electrophoresis for calibration of drug outward diffusion. Psychopharmacology 105:264–268, 1991.

268. L Hernandez, NA Guzman, BG Hoebel. Differential dopaminergic potency of cocaine, procaine and lidocaine infused locally in the nucleus accumbens in vivo with calibration by capillary electrophoresis in vitro. NIDA Res. Monogr 105:355–356, 1991.

269. NA Guzman, MA Trebilcock, JP Advis. The use of a concentration step to collect urinary components separated by capillary electrophoresis and further characterization of collected analytes by mass spectrometry. J Liq Chromatogr 14:997–1015, 1991.

270. TJ O'Shea, SM Lunte. Selective detection of free thiols by capillary electrophoresis-electrochemistry using a gold/mercury amalgam microelectrode. Anal Chem 65:247–250, 1993.

271. SR Rabel, R Trueworthy, JF Stobaugh. Recent developments utilizing capillary electrophoresis with laser-induced fluorescence for the determination of 6-mercaptopurine metabolites. J High Resol Chromatogr 16:326–327, 1993.

272. G Vargas, J Havel, K Frgalova. Capillary zone electrophoresis determination of thyreostatic drugs in urine. J Capillary Electrophor 5:9–12, 1998.

273. JL Tsai, WS Wu, HH Lee. Qualitative determination of urinary morphine by capillary zone electrophoresis and ion trap mass spectrometry. Electrophoresîs 21:1580–1586, 2000.

274. KD Altria, NG Clayton, RC Harden, M Hart, J Hevizi, J Makwana, MJ Portsmouth. An inter-company cross-validation exercise on capillary electrophoresis testing of dose uniformity of paracetamol content in formulations. Chromatographia 39:180–184, 1994.

275. KD Altria, J Bestford. Main component assay of pharmaceuticals by capillary electrophoresis: Considerations regarding precision, accuracy, and linearity data. J Capillary Electroph 3:13–23, 1996.

276. LC Hsu, DJ Constable, DR Orvos, RE Hannah. Comparison of high-performance liquid chromatography and capillary zone electrophoresis in penciclovir biodegradation kinetic studies. J Chromatogr B 669:85–92, 1995.

277. T de Boer, R Bijma, K Ensing. Tuning of the selectivity in capillary electrophoresis by cyclodextrins illustrated by the separation of some structurally related phenothiazines. J Capillary Electrophor 5:65–71, 1998.

278. W Baeyens, G Weiss, G Van Der Weken, W Van Den Bossche, Analysis of pilocarpine and its trans epimer, isopilocarpine, by capillary electrophoresis. J Chromatogr 638:319–326, 1993.

279. K Persson, O Astrom. Fractional factorial design optimization of the separation of pilocarpine and its degradation products by capillary electrophoresis. J Chromatogr B 697:207–215, 1997.

280. S Kimakhe, S Bohic, C Larrose, A Reynaud, P Pilet, B Giumelli, D Heymann, G Daculsi. Biological activities of sustained polymyxin B release from calcium phosphate biomaterial prepared by dynamic compaction: an in vitro study. J Biomed Mater Res 47:18–27, 1999.

281. KD Altria. High and low injection volumes in CE for improved quantitative deter-
      mination of drug related impurities. Chromatographia 35:493–496, 1993.
282. R Gotti, S Furlanetto, V Andrisano, V Cavrini, S Pinzauti. Design experiments for
      capillary electrophoretic enantioresolution of salbutamol using dermatan sulfate. J
      Chromatogr A 875:411–422, 2000.
283. T You, X Yang, E Wang. Determination of sulfadiazine and sulfamethoxazole by
      capillary electrophoresis with end-column electrochemical detection. Analyst 123:
      2357–2360, 1998.
284. MT Ackermans, JL Beckers, FM Everaerts, H Hoogland, MJH Tomassen. Determi-
      nation of sulphonamides in pork meat extracts by capillary zone electrophoresis.
      J Chromatogr 596:101–109, 1992.
285. MC Ricci, RF Cross. Capillary electrophoresis separation of sulphonamides and
      dehydrofolate reductase inhibitors. J Microcol Sep 5:207–215, 1993.
286. Q Dang, L Yan, Z Sun, D Ling. Separation of sulphonamides and determination
      of the active ingredients in tablets by micellar electrokinetic capillary chromatogra-
      phy. J Chromatogr 603:259–266, 1992.
287. CL Ng, HK Lee, SFY Li. Determination of sulphonamides in pharmaceuticals by
      capillary electrophoresis. J Chromatogr 632:165–170, 1993.
288. PC Dabas, MC Vescina, CN Carducci. Determination of suramin by micellar elec-
      trokinetic chromatography with direct serum injection. J Capillary Electrophor 4:
      253–256, 1997.
289. LL Garcia, ZK Shihabi. Suramin determination by direct serum injection. J Liq
      Chromatogr 16:1279–1288, 1993.
290. LL Garcia, ZK Shihabi. Suramin determination by capillary electrophoresis. J Liq
      Chromatogr 16:2049–2057, 1993.
291. KC Chan, AB Alvarado, MT McGuire, GM Muschik, HJ Issaq, KM Snader. High-
      performance liquid chromatography and micellar electrokinetic chromatography of
      taxol and related taxanes from bark and needle extracts of Taxus species. J Chro-
      matogr 657:301–306, 1994.
292. M Petersson, K-G Wahlund, S Nilsson. Miniaturised on-line solid-phase extraction
      for enhancement of concentration sensitivity in capillary electrophoresis. J Chro-
      matogr A 841:249–261, 1999.
293. MFM Tavares, VL McGuffin. Separation and characterization of tetracycline anti-
      biotics by capillary electrophoresis. J Chromatogr 686:129–142, 1994.
294. YM Li, A Van Schepdael, E Roets, J Hoogmartens. Optimized methods for capil-
      lary electrophoresis of tetracyclines. J Pharm Biomed Anal 15:1063–1069, 1997.
295. Q Dang, L Yan, Z Sun, D Ling. Separation and simultaneous determination of
      the active ingredients in theophylline tablets by micellar electrokinetic capillary
      chromatography. J Chromatogr 630:363–369, 1993.
296. Z Aturki, E Camera, F La Torre, S Fanali. Direct chiral resolution of tiaprofenic
      acid in pharmaceutical formulations by capillary zone electrophoresis using cyclo-
      dextrins as chiral selectors. J Capillary Electrophor 2:213–217, 1995.
297. L Mateus, S Cherkaoui, P Christen, JL Veuthey. Capillary electrophoresis-diode
      array detection-electrospray mass spectrometry for the analysis of selected tropane
      alkaloids in plant extracts. Electrophoresis 20:3402–3409, 1999.
298. L Mateus, S Cherkaoui, P Christen, JL Veuthey. Capillary electrophoresis for the

analysis of tropane alkaloids: Pharmaceutical and phytochemical applications. J Pharm Biomed Anal 18:815–825, 1998.

299. M Eeva, JP Salo, KM Oksman-Caldentey. Determination of the main tropane alkaloids in transformed *Hyoscyamus muticus* plants by capillary zone electrophoresis. J Pharm Biomed Anal 16:717–722, 1998.

300. S Fanali, V Cotichini, R Porrà. Analysis of venlafaxine by capillary zone electrophoresis. J Capillary Electrophor 4:21–26, 1997.

301. KD Altria. Quantitative aspects of the application of capillary electrophoresis to the analysis of pharmaceuticals and drug related impurities. J Chromatogr 646: 245–257, 1993.

302. B Nickerson, B Cunningham, S Scypinski. The use of capillary electrophoresis to monitor the stability of a dual-action cephalosphorin in solution. J Pharm Biomed Anal 14:73–83, 1995.

303. MT Ackermans, FM Everaerts, JL Beckers. Determination of aminoglycoside antibiotics in pharmaceuticals by capillary zone electrophoresis with indirect UV detection coupled with micellar electrokinetic capillary chromatography. J Chromatogr 606:229–235, 1992.

304. C-X Zhang, Z-P Sun, D-K Ling, Y-J Zhang. Separation of tetracycline and its degradation products by capillary zone electrophoresis. J Chromatogr 627:281–286, 1992.

305. CM Boone, JP Franke, RA de Zeeuw, K Ensing. Evaluation of capillary electrophoretic techniques towards systematic toxicological analysis. J Chromatogr A 838: 259–272, 1999.

306. S Li, SG Weber. Determination of barbiturates by solid-phase microextraction and capillary electrophoresis. Anal Chem 69:1217–1222, 1997.

307. K Jinno, Y Han, H Sawada. Analysis of toxic drugs by capillary electrophoresis using polyacrylamide-coated columns. Electrophoresis 18:284–286, 1997.

308. Ph Morin, JC Archambault, P André, M Dreux, E Gaydou. Separation of hydroxylated and methoxylated flavonoids by micellar electrokinetic capillary chromatography. Determination of analyte partition coefficients between aqueous and sodium dodecyl sulfate micellar phases. J Chromatogr A 791:289–297, 1997.

309. MA Aramendia, I Garcia, F Lafont, JM Marinas. Determination of isoflavones using capillary electrophoresis in combination with electrospray mass spectrometry. J Chromatogr A 707:327–333, 1995.

310. GH Rolando, SM Laritza. Methods for the simultaneous analysis of water-soluble vitamins. Rev Mex Cienc Farm 30:9–14, 1999.

311. L Fotsing, M Fillet, I Bechet, P Hubert, J Crommen. Determination of six water-soluble vitamins in a pharmaceutical formulation by capillary electrophoresis. J Pharm Biomed Anal 15:1113–1123, 1997.

312. L Fotsing, M Fillet, I Becket, Ph Hubert, J Crommen. Determination of six water-soluble vitamins in a pharmaceutical formulation by capillary electrophoresis. J Pharm Biomed Anal 15:1113–1123, 1997.

313. L Fotsing, B Boulanger, P Chiap, M Fillet, P Hubert, J Crommen. Multivariate optimization approach for the separation of water-soluble vitamins and related compounds by capillary electrophoresis. Biomed Chromatogr 14:10–11, 2000.

314. S Pedersen-Bjergaard, KE Rasmussen, T Tilander. Separation of fat-soluble vita-

mins by hydrophobic interaction electrokinetic chromatography with tetradecylammonium ions as pseudostationary phase. J Chromatogr A 807:289–295, 1998.

315. O Naess, T Tilander, S Pedersen-Bjergaard, KE Rasmussen. Analysis of vitamin formulations by electrokinetic chromatography utilizing tetradecylammonium ions as the pseudostationary phase. Electrophoresis 19:2912–2917, 1998.

316. GNW Leung, HPO Tang, TSC Tso, TSM Wan. Separation of basic drugs with non-aqueous capillary electrophoresis. J Chromatogr A 738:141–154, 1996.

317. W Ding, JS Fritz. Separation of neutral compounds and basic drugs by capillary electrophoresis in acidic solutions using laurylpoly(oxyethylene) sulfate as an additive. Anal Chem 70:1859–1865, 1998.

318. MG Vargas, J Havel, J Patocka. Capillary zone electrophoretic determination of some drugs against Alzheimer's disease. J Chromatogr A 802:121–128, 1998.

319. B Yang, J Mo, X Yang, L Wang. Separation and determination of chlorphenamine maleate, acetaminophen and vitamin C of cold medicines by capillary electrophoresis with amperometric detection. Sepu 17:477–479, 1999.

320. X Liu, M Hou. Determination of compounds in common cold preparations by high performance capillary electrophoresis (HPCE). Yaowu Fenxi Zazhi 17:315–318, 1997.

321. M Katoka, M Imamura, K Nishijima, K Nishi. Simultaneous determination method for the active ingredients in cold medicines by capillary electrophoresis. Iyakuhin Kenkyu 27:45–56, 1996.

322. H Nishi, T Fukuyama, M Matsuo, S Terabe. Effect of surfactant structures on the separation of cold medicine ingredients by micellar electrokinetic chromatography. J Pharm Sci 79:519–523 1990.

323. J Palmer, S Atkinson, WY Yoshida, AM Stalcup, JP Landers. Charged chelate-capillary electrophoresis of endogenous corticosteroids. Electrophoresis 19:3045–3051, 1998.

324. H Nishi, T Fukuyama, M Matsuo, S Terabe. Separation and determination of lipophilic corticosteroids and benzothiazepin analogues by micellar electrokinetic chromatography using bile salts. J Chromatogr 513:279–295, 1990.

325. V Baeyens, E Varesio, J-L Veuthey, R Gurny. Determination of dexamethasone in tears by capillary electrophoresis. J Chromatogr B 692:222–226, 1997.

326. K Raith, E Althoff, J Banse, H Neidhardt, RH Neuber. Two examples of rapid and simple drug analysis in pharmaceutical formulations using capillary electrophoresis: Naphazoline, dexamethasone and benzalkonium in nose drops and nystatin in an oily suspension. Electrophoresis 19:2907–2911, 1998.

327. SM Berge, LD Bighley, DC Monkhouse. Pharmaceutical salts. J Pharm Sci 66:1–19, 1977.

328. PL Gould. Salt selection for basic drugs. Int J Pharmaceut 33:201–217, 1986.

329. WR Jones. Electrophoretic capillary ion analysis. In: JP Landers, ed. Handbook of Capillary Electrophoresis. Orlando, FL: CRC Press, 1994, pp 209–232.

330. WR Jones, PJ Jandik. Various approaches to the analysis of difficult sample matrices of anions using CE. J Chromatogr 608:385–393, 1992.

331. A Weston, PR Brown, PJ Jandik, AL Heclanberg, WR Jones. Optimization of detection sensitivity in the analysis of inorganic cations by CE using indirect photometric detection. J Chromatogr 608:395–402, 1992.

332. G Bondoux, P Jandik, WR Jones. New approach to the analysis of low levels of anions in water. J Chromatogr 602:79–88, 1992.

333. D Kaniansky, M Masar, J Marak, R Bodor. Capillary electrophoresis of inorganic anions. J Chromatogr A 834:133–178, 1999.

334. P Kuban, P Kuban, V Kuban. Simultaneous capillary electrophoretic separation of small anions and cations after complexation with ethylenediaminetetraacetic acid. J Chromatogr A 836:75–80, 1999.

335. T Hiissa, H Siren, T Kotiaho, M Snellman, A Hautojarvi. Quantification of anions and cations in environmental water samples. Measurements with capillary electrophoresis and indirect-UV detection. J Chromatogr A 853:403–411, 1999.

336. O Henin, B Barbier, A Brack. Determination of phosphate and pyrophosphate ions by capillary electrophoresis. Anal Biochem 270:181–184, 1999.

337. PN Bories, E Scherman, L Dziedzic. Analysis of nitrite and nitrate in biological fluids by capillary electrophoresis. Clin Biochem 32:9–14, 1999.

338. RC Williams, RJ Boucher. Analysis of potassium counter ion and inorganic cation impurities in pharmaceutical drug substance by capillary electrophoresis with conductivity detection. J Pharm Biomed Anal 22:115–122, 2000.

339. WG Kuhr, ES Yeung. Indirect fluorescence detection of native amino acids in capillary zone electrophoresis. Anal Chem 60:1832–1834, 1988.

340. ES Yeung, WG Kuhr. Indirect detection methods for capillary separations. Anal Chem 63:275A–282A, 1991.

341. B Baraj, M Martínez, A Sastre, M Aguilar. Simultaneous determination of Cr(III), Fe(III), Cu(II) and Pb(II) as UV-absorbing EDTA complexes by capillary zone electrophoresis. J Chromatogr A 695:103–111, 1995.

342. LA Colon, KJ Reynolds, R Alicea-Maldonado, AM Fermier. Advances in capillary electrochromatography. Electrophoresis 18:2162–2174, 1997.

343. AS Rathore, C Horvath. Capillary electrochromatography: Theories on electroosmotic flow in porous media. J Chromatogr A 781:185–195, 1997.

344. JJ Pesek, MT Matyska. Column technology in capillary electrophoresis and capillary electrochromatography. Electrophoresis 18:2228–2238, 1997.

345. AM Fermier, LA Colon. Capillary electrochromatography in columns packed by centripetal forces. J Microcol Sep 10:439–447, 1998.

346. TD Hatajik, PR Brown. Chiral separations of pharmaceuticals using capillary electrochromatography (CEC): An overview. J Capillary Electrophor 5:143–151, 1998.

347. KD Altria. Overview of capillary electrophoresis and capillary electrochromatography. J Chromatogr A 856:443–463, 1999.

348. D Wistuba, V Schurig. Enantiomer separation of chiral pharmaceuticals by capillary electrochromatography. J Chromatogr A 14:255–276, 2000.

349. G Gubitz, MG Schmid. Chiral separation by capillary electrochromatography. Enantiomer 5:5–11, 2000.

350. MM Robson, MG Cikalo, P Myers, MR Euerby, KD Bartle. Capillary electrochromatography: A review. J Microcol Sep 9:357–372, 1997.

351. MR Euerby, D Gilligan, CM Johnson, SCP Roulin, P Myers, KD Bartle. Applications of capillary electrochromatography in pharmaceutical analysis. J Microcol Sep 9:373–387, 1997.

352. MR Euerby, CM Johnson, KD Bartle. Practical experiences and applications of

capillary electrochromatography in the pharmaceutical industry. LC-GC 16:386–394, 1998.

353. SCP Roulin, KD Bartle, MR Euerby. Capillary electrochromatography and its potential in the pharmaceutical industry. J Pharm Pharmacol 50:127, 1998.

354. V Lopez-Avila, J Benedicto, C Yan. Determination of selected heterocyclic compounds containing nitrogen, oxygen, and sulfur by capillary electrochromatography. J High Resol Chromatogr 20:615–618, 1997.

355. S Terabe, K Otsuka, H Nishi. Separation of enantiomers by capillary electrophoretic techniques. J Chromatogr A 666:295–319, 1994.

356. G Gübitz, MG Schmid. Chiral separation principles in capillary electrophoresis. J Chromatogr A 792:179–225, 1997.

357. S Li, DK Lloyd. Direct chiral separations by capillary electrophoresis packed with an α1-acid glycoprotein chiral stationary phase. Anal Chem 65:3684–3690, 1993.

358. NW Smith, AS Carter-Finch. Electrochromatography. J Chromatogr A 892:219–255, 2000.

359. IS Krull, RL Stevenson, K Misty, M Swartz. Capillary Electrochromatography and Pressurized Flow Capillary Electrochromatography. An Introduction. New York: HNB Publishing, 2000, pp 94–184.

360. TD Hatajik, PR Brown. Chiral separations using capillary electrochromatography (CEC): An overview. J Capillary Electrophor 5:143–151, 1998.

361. JJ Pesek, MT Matyska. A new open-tubular approach to capillary electrochromatography. J Capillary Electrophor 4:213–217, 1997.

362. W Bertold, J Walter. Protein purification: Aspects of processes for pharmaceutical products. Biologicals 22:135–150, 1994.

363. RL Lundblad, RA Bradshaw. Applications of site-specific chemical modification in the manufacture of biopharmaceuticals. I. An overview. Biotechnol Appl Biochem 26:143–151, 1997.

364. GC Davis, RM Riggin. Characterization and establishment of specifications for biopharmaceuticals. Dev Biol Stand 91:49–54, 1997.

365. RG Werner, W Noe, K Kopp, M Schluter. Appropriate mammalian expression systems for biopharmaceuticals. Arzneimittelforschung 48:870–880, 1998.

366. JH Lupker. Residual host cell protein from continuous cell lines. Effects on the safety of protein pharmaceuticals. Dev Biol Stand 93:61–64, 1998.

367. JM Reichert. New Pharmaceuticals in the USA: Trends in development and marketing approvals 1995–1999. Trends Biotechnol 18:364–369, 2000.

368. G Walsh. Biopharmaceutical benchmarks. Nature Biotechnol 18:831–833, 2000.

369. PM Doran. Foreign protein production in plant tissue cultures. Curr Opin Biotechnol 11:199–204, 2000.

370. KR Reddy. Controlled-release, pegylation, liposomal formulations: New mechanisms in the delivery of injectable drugs. Ann Pharmacother 34:915–923, 2000.

371. JC Lee. Biopharmaceutical formulation. Curr Opin Biotechnol 11:81–84, 2000.

372. W Wang. Instability, stabilization, and formulation of liquid protein pharmaceuticals. Int J Pharm 185:129–188, 1999.

373. MC Lai, EM Topp. Solid-state chemical stability of proteins and peptides. J Pharm Sci 88:489–5000, 1999.

374. NB Finter. The safety of biopharmaceutical products—A personal point of view. Dev Biol Stand 88:133–137, 1996.

375. G Crotts, TG Park. Protein delivery from poly(lactic-coglycolic acid) biodegradable microspheres: Release kinetics and stability issues. J Microencapsul 15:699–713, 1998.

376. DK Pettit, WR Gombotz. The development of site-specific drug-delivery systems for protein and peptide biopharmaceuticals. Trends Biotechnol 16:343–349, 1998.

377. SD Putney, PA Burke. Improving protein therapeutics with sustained-release formulations. Nature Biotechnol 16:153–157, 1998.

378. WP Blackstock, MP Weir. Proteomics: Quantitative and physical mapping of cellular proteins. Trends Biotechnol 17:121–127, 1999.

379. KL Williams. Genomes and proteomes: Towards a multidimensional view of biology. Electrophoresis 20:678–688, 1999.

380. RD Unwin, MA Knowles, PJ Selby, RE Banks. Urological malignancies and the proteomic-genomic interface. Electrophoresis 20:3629–3637, 1999.

381. P Cash. Proteomics in medical microbiology. Electrophoresis 21:1187–1201, 2000.

382. C Rohlff. Proteomics in molecular medicine: Applications in central nervous systems disorders. Electrophoresis 21:1227–1234, 2000.

383. J Rosamond, A Allsop. Harnessing the power of the genome in the search for new antibiotics. Science 287:1973–1976, 2000.

384. M Qabar, J Urban, C Sia, M Klein, M Kahn. Peptide secondary structure mimetics: Applications to vaccines and pharmaceuticals. Farmaco 51:87–96, 1996.

385. A Friedler, N Zakai, O Karni, YC Broder, L Baraz, M Kotler, A Loyter, C Gilon. Backbone cyclic peptide, which mimics the nuclear localization signal of human immunodeficiency virus type 1 matrix, inhibits nuclear import and virus production in nondividing cells. Biochemistry 37:5616–5622, 1998.

386. R Kasher, DA Oren, Y Barda, C Gilon. Miniaturized proteins: the backbone cyclic proteinomimetic approach. J Mol Biol 292:421–429, 1999.

387. A Apffel, J Chakel, S Udiavar, S Swedberg, WS Hancock, C Souders, E Pungor Jr. Application of new analytical technology to the production of a well-characterized biological. Dev Biol Stand 96:11–25, 1998.

388. G Siuzdak. The emergence of mass spectrometry in biomedical research. Proc Natl Acad Sci USA 91:11290–11297, 1994.

389. CE Costello. Bioanalytic applications of mass spectrometry. Curr Opin Biotechnol 10:22–28, 1999.

390. F Regnier, A Amini, A Chakraborty, M Geng, J Ji, L Riggs, C Sioma, S Wang, X Zhang. Multidimensional chromatography and the signature peptide approach to proteomics. LC-GC 19:200–213, 2001.

391. J Leonil, V Gagnaire, D Molle, S Pezennec, S Bouhallab. Application of chromatography and mass spectrometry to the characterization of food proteins and derived peptides. J Chromatogr A 881:1–21, 2000.

392. JA Loo, DE DeJohn, P Du, TI Stevenson, RR Ogorzalek Loo. Application of mass spectrometry for target identification and characterization. Med Res Rev 19:307–319, 1999.

393. D Figeys, R Aebersold. High sensitivity analysis of proteins and peptides by capil-

lary electrophoresis-tandem mass spectrometry: Recent developments in technology and applications. Electrophoresis 19:885–892, 1998.

394. Y Shen, SJ Berger, GA Anderson, RD Smith. High-efficiency capillary isoelectric focusing of peptides. Anal Chem 72:2154–2159, 2000.

395. HG Lee. High-performance sodium dodecyl sulfate-capillary gel electrophoresis of antibodies and antibody fragments. J Immunol Methods 234:71–81, 2000.

396. T Liu, XX Shao, R Zeng, QC Xia. Analysis of recombinant and modified proteins by capillary zone electrophoresis coupled with electrospray ionization tandem mass spectrometry. J Chromatogr A 855:695–707, 1999.

397. MS Wong, TC Aw. A preliminary evaluation of serum proteins by capillary electrophoresis. J Capillary Electrophor 5:217–221, 1998.

398. MA Jenkins, S Ratnaike. Five unusual serum protein presentations found by capillary electrophoresis in the clinical laboratory. J Biochem Biophys Methods 41:31–47, 1999.

399. MA Jenkins. Three methods of capillary electrophoresis compared with high-resolution agarose gel electrophoresis for serum protein electrophoresis. J Chromatogr B 720:49–58, 1998.

400. F-TA Chen, RA Evangelista. Protein analysis by capillary electrophoresis. In: H Shintani, J Polonsky, eds. Handbook of Capillary Electrophoresis Applications. London: Blackie, 1997, pp 173–197.

401. WL Tseng, HT Chang. On-line concentration and separation of proteins by capillary electrophoresis using polymer solutions. Anal Chem 72:4805–4811, 2000.

402. T Miura, T Funato, S Yabuki, T Sasaki, M Kaku. Detection of monoclonal proteins by capillary electrophoresis using a zwitterion in the running buffer. Clin Chim Acta 299:87–99, 2000.

403. K Kubo, E Honda, M Imoto, Y Murishima. Capillary zone electrophoresis of albumin-depleted human serum using a linear polyacrylamide-coated capillary: Separation of serum alpha- and beta-globulins into individual components. Electrophoresis 21:396–402, 2000.

404. NA Guzman, J Moschera, CA Bailey, K Iqbal, AW Malick. Assay of protein drug substances present in solution mixtures by fluorescamine derivatization and capillary electrophoresis. J Chromatogr 598:123–131, 1992.

405. NA Guzman, J Moschera, K Iqbal, AW Malick. Effect of buffer constituents on the determination of therapeutic proteins by capillary electrophoresis. J Chromatogr 608:197–204, 1992.

406. NA Guzman, H Ali, J Moschera, K Iqbal, AW Malick, Assessment of capillary electrophoresis in pharmaceutical applications. J Chromatogr 559:307–315, 1991.

407. TM Phillips, BF Dickens. Analysis of recombinant cytokines in human body fluids by immunoaffinity capillary electrophoresis. Electrophoresis 19:2991–2996, 1998.

408. G-H Zhou, G-A Luo, Y Zhou, K-Y Zhou, X-D Zhang, L-Q Huang. Applications of capillary electrophoresis, liquid chromatography, electrospray-mass spectrometry and matrix-assisted laser desorption/ionization-time of flight-mass spectrometry to the characterization of recombinant human erythropoietin. Electrophoresis 19:2348–2355, 1998.

409. AD Tran, S Park, PJ Lisi, CT Huynh, RR Ryall, PA Lane. Separation of carbohy-

drate-mediated microheterogeneity of recombinant human erythropoietin by free solution capillary electrophoresis. Effects of pH, buffer type and organic additives. J Chromatogr 542:459–471, 1991.

410. F-TA Chen. Analysis of erythropoietin by capillary electrophoresis. In: H Shintani, J Polonsky, eds. Handbook of Capillary Electrophoresis Applications. London. Blackie, 1997, pp 198–206.

411. A Cifuentes, MV Moreno-Arribas, M de Frutos, JC Diez-Masa. Capillary isoelectric focusing of erythropoietin glycoforms and its comparison with flat-bed isoelectric focusing and capillary zone electrophoresis. J Chromatogr A 830:453–463, 1999.

412. A Bristow, E Charton. Assessment of suitability of a capillary zone electrophoresis method for determining isoform distribution of erythropoietin. Pharmaeuropa 11: 290–300, 1999.

413. P Dupin, F Galinou, A Bayol. Analysis of recombinant human growth hormone and its related impurities by capillary electrophoresis. J Chromatogr 707:396–400, 1995.

414. S Kundu, C Fenters. Isoelectric focusing of monoclonal antibodies by capillary electrophoresis. J Capillary Electrophor 2:273–277, 1995.

415. N Bihoreau, C Ramon, R Vincentelli, J-P Levillain, F Troalen. Peptide mapping characterization by capillary electrophoresis of human monoclonal anti-Rh(D) antibody produced for clinical studies. 2:197–202, 1995.

416. DJ Kroon, S Goltra, B Sharma. Analysis of monoclonal antibodies by sodium dodecyl sulfate-capillary gel electrophoresis with special reference to quantitation of half-antibody. J Capillary Electrophor. 2:34–39, 1995.

417. G Hunt, W Nashabeh. Capillary electrophoresis sodium dodecyl sulfate nongel sieving analysis of a therapeutic recombinant monoclonal antibody: A biotechnology perspective. Anal Chem 71:2390–2397, 1999.

418. MA Costello, C Woititz, J De Feo, D Stremlo, L-FL Wen, DJ Palling, K Iqbal, NA Guzman. Characterization of humanized anti-TAC monoclonal antibody by traditional separation techniques and capillary electrophoresis. J Liq Chromatogr 15:1081–1097, 1992.

419. NA Guzman, L Hernandez. A rapid procedure for the quantitative analysis of monoclonal antibodies by high performance capillary electrophoresis. In: TE Hugli, ed. Techniques in Protein Chemistry. San Diego, CA: Academic Press, 1989, pp. 456–467.

420. A Tran, H Parker, V Levi, M Kunitani. Analysis of recombinant human platelet-derived growth factor by reversed-charge capillary zone electrophoresis. Anal Chem 70:3809–3817, 1998.

421. K Tsuji. Evaluation of sodium dodecyl sulfate non-acrylamide polymer gel-filled capillary electrophoresis for molecular size separation of recombinant bovine somatotropin. J Chromatogr 652:139–147, 1993.

422. KC Chan, GM Muschik, HJ Issaq. Solid-state UV laser-induced fluorescence detection in capillary electrophoresis. Electrophoresis 21:2062–2066, 2000.

423. R Zhu, WT Kok. Postcolumn derivatization of peptides with fluorescamine in capillary electrophoresis. J Chromatogr A 814:213–221, 1998.

424. JP Advis, K Iqbal, AW Malick, NA Guzman. Capillary electrophoresis coupled to

fluorescence detection for the determination of multiple peptides. In: F de Pablo, CG Scanes, BD Weintraub, eds. Handbook of Hormonal Assay Techniques. New York: Academic Press, 1993, pp 127–144.

425. M Larsson, ES Lutz. Transient isotachophoresis for sensitivity enhancement in capillary electrophoresis-mass spectrometry for peptide analysis. Electrophoresis 21: 2959–2865 2000.

426. VL Ward, MG Khaledi. Nonaqueous capillary electrophoresis with laser induced fluorescence detection. J Chromatogr B 719:15–22, 1998.

427. G Okafo, D Tolson, S Monte, J Marchbank. Analysis of process impurities in the basic drug SB-253149 using capillary electrophoresis and on-line mass spectrometry detection. Rapid Commun Mass Spectrom 14:2320–2327, 2000.

428. X Cahours, H Dessans, P Morin, M Dreux, L Agrofoglio. Determination at ppb level of an anti-human immunodeficiency virus nucleotide drug by capillary electrophoresis-electrospray ionization tandem mass spectrometry. J Chromatogr A 895:101–109, 2000.

429. S Sentellas, L Puignou, E Moyano, MT Galceran. Determination of ebrotidine and its metabolites by capillary electrophoresis with UV and mass spectrometry detection. J Chromatogr A 888:281–292, 2000.

430. A Wang, F Fang. Applications of capillary electrophoresis with electrochemical detection in pharmaceutical and biomedical analyses. Electrophoresis 21:1281–1290, 2000.

431. S McWhorter, SA Soper. Near-infrared laser-induced fluorescence detection in capillary electrophoresis. Electrophoresis 21:1267–1280, 2000.

432. RB Taylor, S Toasaksiri, RG Reid. A literature assessment of sample pretreatments and limits of detection for capillary electrophoresis of drugs in biological fluids and practical investigation with some antimalarials in plasma. Electrophoresis 19: 2791–2797, 1998.

433. G Hempel. Strategies to improve sensitivity in capillary electrophoresis for the analysis of drugs in biological fluids. Electrophoresis 21:691–698, 2000.

434. ME Swartz, M Merion. On-line sample preconcentration on a packet-inlet capillary for improving the sensitivity of capillary electrophoresis of pharmaceuticals. J Chromatogr 632:209–213, 1993.

435. TM Phillips, JJ Chmielinska. Immunoaffinity capillary electrophoresis analysis of cyclosporin in tears. Biomed Chromatogr 8:242–246, 1994.

436. JH Beattie, R Self, MP Richards. The use of solid phase concentrator by capillary zone electrophoresis. Electrophoresis 16:322–328, 1995.

437. MA Strausbauch, BJ Madden, PJ Wettstein, JP Landers. Sensitivity enhancement and second-dimensional information from solid phase extraction-capillary electrophoresis of entire high-performance liquid chromatography fractions. Electrophoresis 16:541–548, 1995.

438. NA Guzman. Biomedical applications of on-line preconcentration-capillary electrophoresis using an analyte concentrator. Investigation of design options. J Liq Chromatogr 18:3751–3768, 1995.

439. K Ensing, A Paulus. Immobilization of antibodies as a versatile tool in hybridized capillary electrophoresis. J Pharm Biomed Anal 14:305–316, 1996.

440. LG Rashkovetsky, YV Lyubarskaya, F Foret, DE Hughes, BL Karger. Automated

microanalysis using magnetic beads with commercial capillary electrophoresis instrumentation. J Chromatogr A 781:197–204, 1997.

441. D Figeys, Y Zhang, R Aebersold. Optimization of solid phase microextraction-capillary zone electrophoresis-mass spectrometry for high sensitivity protein identification. Electrophoresis 19:2338–2347, 1998.

442. Q Yang, AJ Tomlinson, S Naylor. Membrane preconcentration CE. Anal Chem 71:183A–189A, 1999.

443. JC Waterval, G Hommels, J Teeuwsen, A Bult, H Lingeman, WJ Underberg. Quantitative analysis of pharmaceutically active peptides using on-capillary analyte preconcentration transient isotachophoresis. Electrophoresis 21:2851–2858, 2000.

444. P Cao, JT Stults. Mapping the phosphorylation sites of proteins using on-line immobilized metal affinity chromatography/capillary electrophoresis/electrospray ionization multiple stage tandem mass spectrometry. Rapid Commun Mass Spectrom 14:1600–1606, 2000.

445. MC Breadmore, M Macka, N Avdalovic, PR Haddad. On-capillary ion-exchange preconcentration of inorganic anions in open-tubular capillary electrochromatography with elution using transient-isotachophoretic gradients. 2. Characterization of the isotachophoretic gradient. Anal Chem 73:820–828, 2001.

446. DS Hage. Survey of recent advances in analytical applications of immunoaffinity chromatography. J Chromatogr B 715:3–28, 1998.

447. M Rozycka, N Collins, MR Stratton, R Wooster. Rapid detection of DNA sequence variants by conformation-sensitive capillary electrophoresis. Genomics 70:34–40, 2000.

448. JM Song, ES Yeung. Alternative base-calling algorithm for DNA sequenced based on four-label multicolor detection. Electrophoresis 21:807–815, 2000.

449. H Pang, V Pavski, ES Yeung. DNA sequencing using 96-capillary array electrophoresis. J Biochem Biophys Methods 41:121–132, 1999.

450. NJ Dovichi, J Zhang. DNA sequencing by capillary electrophoresis. Methods Mol Biol 162–163: 85–94, 2001.

451. K Brown. The Human Genome business today. Sci Am 283:50–55.

452. ES Razvi, LJ Leytes. High-throughput genomics. Modern Drug Discovery 3:40–42, 2000.

453. AM Ewards, CH Arrowsmith, B des Pallieres. Proteomics: New tools for a new era. Modern Drug Discovery 3:34–45, 2000.

454. G Ocvirk, T Tang, DJ Harrison. Optimization of confocal epifluorescence microscopy for microchip-based miniaturized total analysis systems. Analysts 123:1429–1434, 1998.

455. CL Colyer, T Tang, N Chiem, DJ Harrison. Clinical potential of microchip capillary electrophoresis systems. Electrophoresis 18:1733–1741, 1997.

456. D Schmalzing, S Buonocore, C Piggee. Capillary electrophoresis-based immunoassays. Electrophoresis 21:3919–3930, 2000.

457. GJ Bruin. Recent developments in electrokinetically driven analysis on microfabricated devices. Electrophoresis 21:3931–3951, 2000.

458. D Figeys. Lab-on-a-chip: A revolution in biological and medical sciences. Anal Chem 72:330A–335A, 2000.

459. J Khandurina, TE McKnight, SC Jacobson, LC Waters, RS Foote, JM Ramsey.

Integrated system for rapid PCR-based DNA analysis in microfluidic devices. Anal Chem 72:2995–3000, 2000.

460. CT Culbertson, SC Jacobson, JM Ramsey. Microchip devices for high-efficiency separations. Anal Chem 72:5814–5819, 2000.

461. B He, N Tait, F Regnier. Fabrication of nanocolumns for liquid chromatography. Anal Chem 70:3790–3797, 1998.

462. D Figeys, SP Gygi, G McKinnon, R Aebersold. An integrated microfluidics-tandem mass spectrometry system for automated protein analysis. Anal Chem 70:3728–3734, 1998.

463. Y Deng, J Henion, J Li, P Thibault, C Wang, DJ Harrison. Chip-based capillary electrophoresis/mass spectrometry determination of carnitines in human urine. Anal Chem 73:639–646, 2001.

464. ZH Fan, S Mangru, R Granzow, P Heaney, W Ho, Q Dong, R Kumar. Dynamic DNA hybridization on a chip using paramagnetic beads. Anal Chem 71:4851–4859, 1999.

465. J Xu, L Locascio, M Gaitan, CS Lee. Room-temperature imprinting method for plastic microchannel fabrication. Anal Chem 72:1930–1933, 2000.

466. LC Waters, SC Jacobson, N Kroutchinina, J Khandurina, RS Foote, JM Ramsey. Multiple sample PCR amplification and electrophoretic analysis on a microchip. Anal Chem 70:5172–5176, 1998.

467. Y Shi, PC Simpson, JR Scherer, D Wexler, C Skibola, MT Smith, RA Mathies. Radial capillary array electrophoresis microplate and scanner for high-performance nucleic acid analysis. Anal Chem 71:5354–5361, 1999.

468. P Gfrörer, J Schewitz, K Pusecker, E Bayer. On-line coupling of capillary separation techniques with 1H NMR. Anal Chem 73:315A–321A, 1999.

469. ME Lacey, AG Webb, JV Sweedler. Monitoring temperature changes in capillary electrophoresis with nanoliter-volume NMR thermometry. Anal Chem 72:4991–4998, 2000.

470. R Subramanian, WP Kelley, PD Floyd, ZJ Tan, AG Webb, JV Sweedler. A micro-coil NMR probe for coupling microscale HPLC with on-line NMR spectroscopy. Anal Chem 71:5335–5339, 1999.

471. DL Olson, ME Lacey, JV Sweedler. The nanoliter niche. Anal Chem 70:257A–264A, 1998.

472. DH Thomas, DJ Rakestraw, JS Schoeniger, V Lopez-Avila, J Van Emon. Selective trace enrichment by immunoaffinity capillary electrochromatography on-line with capillary zone electrophoresis-laser induced fluorescence. Electrophoresis 20:57–96, 1999.

473. JB Shear, EB Brown, WW Webb. Multiphoton-excited fluorescence of fluorogen-labeled neurotransmitters. Anal Chem 68:1779–1783, 1996.

474. JB Shear. Multiphoton-excited fluorescence in bioanalytical chemistry. Anal Chem 71:598A–605A, 1999.

8

Atomic Spectroscopy

Helen E. Taylor and Stephen G. Schulman
University of Florida, Gainesville, Florida

I. INTRODUCTION

A substantial number of pharmaceutically and clinically related problems require the detection and determination of small amounts of metal ions and other inorganic constituents of biological and xenobiotic substances (1–3). Some obvious examples are the detections of heavy metals and lithium in biological fluids and tissue samples in cases of suspected intoxication and the determination of potassium for purposes of quality control in intravenous solutions to be given to cardiac patients. Trace amounts of nonmetals such as selenium and iodine, which are associated with the functions of coenzymes or hormones, also must be analyzed in order to determine their roles in metabolic pathways.

Conventional chemical and instrumental analytical techniques as well as conventional separational methods frequently lack the sensitivity, selectivity, and efficiency desired in the analyses of pharmaceutically or biochemically significant elements. This is due to the fact that these elements often have low concentrations, complex matrices, or difficult chemistries. Also, in most circumstances, it is necessary to know only the identity and amount of a particular element in a pharmaceutical sample, rather than the diversity of chemical circumstances under which it exists in the sample. For these reasons, the various methods common to atomic spectroscopy are of importance in analytical pharmaceutical chemistry.

In all of the areas of atomic spectroscopy, the sample is subjected to sufficient thermal energy to cause complete vaporization and, ideally, complete decomposition of molecules into atoms. The absorption or emission of light by these vapor-state atoms may be measured and quantitatively related to the concentrations of the species which give rise to them (4,5). Moreover, the absorption

387

and emission spectra of atoms consist of irregularly spaced lines in the near-ultraviolet and visible regions of the electromagnetic spectrum. The pattern of absorption and emission lines is unique for each element and can be used for identification even when several absorbing or emitting elements are present in the sample. The fact that the sample is atomized eliminates the necessity for separation or lengthy chemical workup of the analyte of interest, although simple extraction from biological matrices is often practiced in order to reduce the number of possible interferants. Moreover, direct analysis of the element of interest circumvents the necessity of preparing chemical derivatives of the analyte for analysis, a convenience of no small consequence in the trace analyses of the alkali metal ions, which do not tend to form stable complexes for photometric analysis and have a very limited range of electrochemistry which can be exploited for analytical purposes.

Atomic spectroscopy can be divided into several broad classes based on the nature of the means of exciting the sample. One of these classes is generally known as atomic emission spectroscopy, in which excitation is thermally induced by exposing the sample to very high electric fields. Another class is known as flame emission spectroscopy or flame photometry, in which excitation is thermally induced by exposing the sample to a high-temperature flame. These methods differ from atomic absorption spectroscopy, in which the absorption of light from a radiation source by the atom is observed rather than the emission from the electronically excited atom.

In atomic emission spectroscopy, atomization of the sample is a multistep process. Historically, an electrical arc or spark discharge caused intense heating and ionization of the gas between two electrodes. The hot, nearly completely ionized gas, known as a ''plasma,'' would then vaporize, atomize, and electronically excite the sample, which was usually placed in a hollow in one of the electrodes. The electronic excitation of the sample atoms ultimately resulted in emission (akin to atomic fluorescence, but with many more lines produced) of visible or near-ultraviolet radiation which would be quantitated by the amount of darkening it caused on a photographic plate. Although the traditional arc and spark emission methods have greatly declined in popularity in the past several decades as a result of the ascendancy of flame methods, atomic emission spectroscopy is enjoying a resurgence of popularity owing to the recently developed technique of emission spectroscopy in inductively coupled plasmas. This method eliminates many of the limitations (especially reproducibility) of the arc and spark methods and is applicable to certain metalloids which are not easily amenable to analysis by atomic absorption or atomic fluorescence spectroscopy.

In flame emission spectroscopy, atomization is achieved by aspirating the sample, usually in a volatile solvent, into a controlled flame. The vaporized sample is atomized by collisions of the molecules of the sample with the energetic molecules comprising the flame-gas. The ''hot'' molecules of the flame-gas also collisionally effect electronic excitation of certain elements. Subsequent to excitation, these electronically excited atoms return to their ground state, emitting visible light in the process. The emitted light is observed visually in the flame and may be monitored photometrically and thereby used for analysis. Because the energy available for electronic excitation from the flame is rather low, only those elements with low-energy electronic transitions are amenable to analysis by this method. Elements such as the alkali metals, alkaline earths, copper, zinc, and mercury can be analyzed by flame emission spectroscopy.

The limitations of flame excitation imposed by the low energies (electronically speaking) available can be avoided by using the flame merely as the sample cell and exciting the sample in the flame with light of an appropriate frequency. If the frequency (or wavelength) of the exciting light satisfies the Planck frequency condition,

$$E_f - E_i = h\nu_a$$

where E_f and E_i are, respectively, the energies of the terminal and initial states involved in absorption, h is Planck's constant, and ν_a is the frequency of the exciting light, the light of frequency ν_a will be absorbed. The amount of light of frequency ν transmitted through (or absorbed by) the sample in the flame can be related to the concentration of the absorber. This forms the basis of atomic absorption spectroscopy. Alternatively, shortly after absorbing the light of frequency ν_a, the sample in the flame may emit visible or ultraviolet radiation of frequency $\nu_f < \nu_a$. The re-emission of light by the photo-excited atoms and its quantitative measurement forms the basis of atomic fluorescence spectroscopy. Both atomic absorption and atomic fluorescence are more general in their range of application than is flame emission spectroscopy.

The various areas of atomic spectroscopy will be discussed in more detail in the experimental and applications sections of this chapter. However, in order to better appreciate the ranges of applicability and limitation of the various atomic spectroscopic methods, it is in order to proceed next to a consideration of the features of atomic electronic structure which form the basis for atomic line spectra and to the processes which result in the production of atomic absorption or emission spectra.

II. THEORY

Analytical atomic spectroscopy is based on the absorption and emission of light by atoms (4,6–8). These processes originate with the promotion of atoms in their ground electronic states to electronically excited states and the return of electronically excited atoms to their ground electronic states. Because the frequencies as well as the intensities of the light absorbed and emitted are determined by the electronic configurations of the atoms in their ground and excited states, it will be useful to consider some of the features of atomic electronic structure and its influence on the observed optical spectra of atoms.

A. Electronic Structure of Atoms

1. The Hydrogen Atom

An atom consists of a positively charged nucleus and a number of negatively charged electrons. In a neutral atom, the total negative charge of all the electrons is equal to the total positive charge of the nucleus. The forces holding the atom together are electrostatic, consisting of attractions between each electron and the nucleus and repulsion between all of the atomic electrons. The classical theory of electrostatics, however, fails to account for the stability of atoms and for the characteristics of their spectra. According to the classical planetary theory, electrons should describe curved paths about the nucleus, thereby accelerating, continuously emitting radiation, losing energy, and eventually falling into the nucleus. The atom should, therefore, be an unstable entity.

Early in the twentieth century, the quantum mechanical approach was applied to the hydrogen atom by Bohr, Sommerfeld, DeBroglie, Heisenberg, Schrodinger, and Dirac in a successful attempt to rationalize its stability and the discrete nature of its absorption and emission spectra (9).

According to the quantum mechanical theory of the hydrogen atom, the single planetary electron of hydrogen can exist only in certain regions of space or orbitals, each orbital being defined by three quantum numbers, n, ℓ, and m. The physical significance of these quantum numbers is that most of the energy and radius of the hydrogenic atom in a state defined by occupation by the planetary electron of an orbital associated with a particular set of quantum numbers are determined by n, while the shape of the orbital, and therefore the average angular momentum, is determined by ℓ. The z component of the angular momentum of the atom in an external magnetic field, directed along the z axis, is governed by m. The latter property is related to the relative spatial attitudes of states having the same n and ℓ but different m.

The quantum mechanical treatment of the hydrogen atom dictates the possible values of n, ℓ, and m and their relationships to one another. All positive integral values are possible for n. For any orbital having a given value of n, the value of ℓ for the orbital can be

$$l = 0, 1, 2, \ldots, n - 1$$

Thus, associated with any value of n there are n different values of l. A set of orbitals all having the same value of n and l are said to belong to the same subshell, characterized by the azimuthal quantum number l. For any subshell the possible values of m are determined by the value of l for the subshell, and can be

$$m = -l, -l + 1, \ldots, 0, \ldots, l - 1, l$$

Hence for any value of l there are $2l + 1$ possible values of m, the magnetic quantum number. An orbital characterized by a set of the three quantum numbers n, l, and m is said to be a spatial orbital. Hydrogenic orbitals belonging to the . same shell are degenerate; i.e., they have the same energy. This is not directly observable. However, in a strong magnetic or electric field the degeneracy is removed and this phenomenon is observable through the splitting of certain spectral lines.

Hydrogenic subshells corresponding to $l = 0, 1, 2, 3, \ldots$ have been labeled $s, p, d, f, \ldots$, respectively. These letters are a legacy from old spectroscopic terminology. Spectral line progressions arising from transitions originating from orbitals of $l = 0, 1, 2, 3$ were called the *s*harp, *p*rincipal, *d*iffuse, and *f*undamental series. A subshell having $n = 1$ and $l = 0$ would be labeled $1s$, one with $n = 2$, $l = 0$ would be designated $2s$, one with $n = 2$, $l = 1$, $2p$, and so on. This system will be used throughout this chapter. The ground state of the hydrogen atom is then the $1s$ state, or the state produced by the electron occupying the $1s$ orbital.

By applying the principles of special relativity to the quantum mechanical treatment of the hydrogen atom, Dirac was able to derive not only the quantum numbers n, l, and m, but also a fourth quantum number, s, called the spin quantum number. This quantum number has been interpreted to be related to the angular momentum of the electron spinning on its own axis and can have only two values, $+\frac{1}{2}$ and $-\frac{1}{2}$. Thus, an electron occupying a hydrogenic spatial orbital of given n, l, and m can have either $s = +\frac{1}{2}$ or $s = -\frac{1}{2}$. Dirac's discovery of the spin quantum number was useful in the rationalization of the hyperfine structure (additional line splitting) which occurred in the spectral lines of the hydrogen atom in the presence of very strong magnetic fields and under optical magnification. The hyperfine splitting could not be rationalized on the basis of the existence of only three quantum numbers n, l, and m. In many-electron atoms the spin quantum number also plays a major role in the gross structure of spectral lines.

2. Many-Electron Atoms

Because of their greater complexity, the electronic structures and spectra of many-electron atoms cannot be rigorously determined by quantum mechanical methods. Rather, they are determined approximately by a "building up" process using the properties of the hydrogenic orbitals just discussed.

In the "building up" (*Aufbau*) process the ground state of a neutral atom of nuclear charge Z will be constructed by assigning Z electrons to the hydrogenic

orbitals in such a way that the electronic configuration of lowest potential energy,
will be obtained. In order to do this, it is necessary to know how many electrons
each orbital can accommodate and in what order the orbitals are filled.

The orbital capacities and order of filling of atomic orbitals are governed,
respectively, by the Pauli exclusion principle and Hund's rules of maximum angular momentum. In its simplest form, the exclusion principle states that no two
electrons in the same atom can have four identical quantum numbers. Hence, if
an orbital is specified by n, l, and m, it can accommodate a maximum of two
electrons, one with $s = +\frac{1}{2}$ and one with $s = -\frac{1}{2}$. A third electron would have
n, l, m, and s equal to one of the electrons already in that orbital and hence could
not occupy the orbital. For the shell with $n = 1$ there is only one orbital, with
$l = 0$ and $m = 0$. Thus, the $n = 1$ shell is closed to further occupation when it
contains two electrons. A closed $n = 1$ shell is designated $1s^2$. For the $n = 2$
shell there are four orbitals: one with $l = 0$ and $m = 0$ and three with $l = 1$ ($m
= -1, 0, 1$). The $n = 2$ shell is therefore closed when it contains eight electrons
and is designated $2s^2 2p^6$. The maximum number of electrons any shell can accommodate is $2n^2$. The maximum number any subshell can hold is $2(2l + 1)$.
The Pauli exclusion principle thus determines the electronic capacities and configurations of the closed shells as well as those of the closed subshells and orbitals.

Hund's rule of maximum multiplicity (maximum spin angular momentum)
enables prediction of the electronic configurations of partially filled shells. This
rule has its basis in classical electrostatics and states that in the case of degenerate
orbitals (orbitals of the same n and l), the configuration of minimum potential
energy will be achieved by allowing the electrons occupying these orbitals to
stay as far apart as possible. This means that in filling degenerate orbitals, each
orbital will accept one electron before spin pairing (double occupancy) will occur,
because the separate orbitals occupy different regions of space while two electrons in one orbital are close together, resulting in greater repulsive energy.

The combined application of Hund's rule and the Pauli exclusion principle
has been quite successful in predicting the ground-state electronic configurations
of the lighter elements. Thus, the ground electronic states of oxygen ($Z = 8$) and
lithium ($Z = 3$) are, respectively, $1s^2 2s^2 2p^4$ (with two paired electrons in one
and one unpaired electron in each of the other two $2p$ orbitals) and $1s^2 2s^1$, respectively. These configurations correlate well with the observed chemical properties
of these elements. The electronically excited states of the elements can be constructed by removing an electron from the highest occupied shell and prompting
it to an orbital in a higher unoccupied shell. For example, one excited state of
lithium could be represented by the configuration $1s^2 3p^1$, which is illustrated,
along with excited states for other alkali metals, in Fig. 1.

The above approach has not been, by itself, quite as successful in the assignments of the ground-state electronic configurations of the heavier elements, espe-

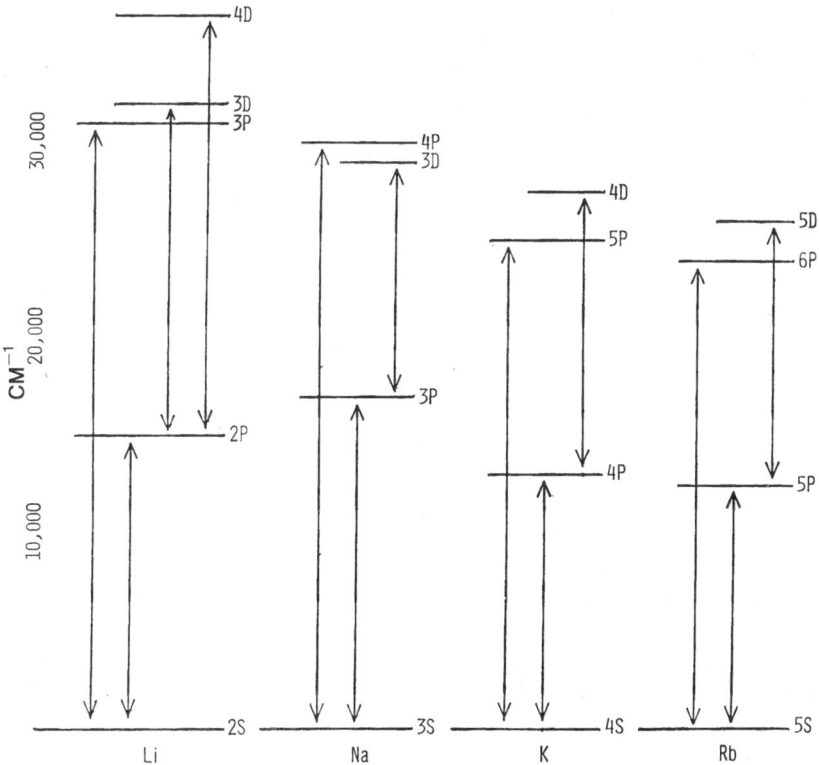

Fig. 1 Predominant atomic transitions in alkali metals.

cially those of the transition elements. For example, cobalt ($Z = 27$) would be assigned a configuration of $1s^2 2s^2 2p^6 3s^2 3p^6 3d^9$ (with one unpaired electron and four sets of paired electrons in the five $3d$ orbitals. This configuration is inconsistent with the chemical, spectroscopic, and magnetic properties of cobalt. From measurements of paramagnetic susceptibilities, it has been possible to assign to the ground state of cobalt a configuration of $1s^2 2s^2 2p^6 3s^2 3p^6 3d^7 4s^2$ (with three unpaired and four paired $3d$ electrons and two paired $4s$ electrons). This discrepancy is due to repulsion of the $3d$ electrons by the inner-shell electrons. This causes the $4s$ orbital to drop below the $3d$ orbitals and therefore become filled first. The principal failing of the hydrogenic orbitals is that they do not in any way take into account interelectronic repulsions. This means that in rationalizing the electronic structures and spectra of many-electron atoms, it is often necessary to blend empiricism with the oversimplified theory to obtain the correct answers.

3. Orbitals, States, and the Complexities of Atomic Electronic Structure

The relatively simple approach based on the orbital model of atomic electronic structure, discussed in the previous sections, suggests that the frequencies of the spectral lines observed in atomic spectra should be equal to the differences in energies between the orbital configurations involved in the transitions, divided by Planck's constant, and that all transitions between the same configurations should have the same energy. In practice, this approach serves to locate the approximate positions of spectral lines but does not account for the breadth or for the fine structure of these lines seen under high magnification or in the presence of an external magnetic field. Obviously, refinements to the orbital model are needed. These refinements are based on the fact that atomic electrons interact with each other, through their electric and magnetic properties, to perturb the hydrogenic orbitals.

Atomic electrons are moving charges and therefore act as small magnets. They produce an orbital magnetic moment, by virtue of their orbital motion, which is located at the center of the orbit (at the nucleus) and is directed at right angles to the orbital plane collinear with the orbital angular momentum vector associated with the occupied orbital. The electron may also be thought of as spinning on its own axis, producing a magnetic moment at the electron whose direction is perpendicular to the plane of rotation of the electron and is either "up" or "down" according to whether $s = +\frac{1}{2}$ or $s = -\frac{1}{2}$.

In many-electron atoms the individual electronic spin and orbital magnetic moments add vectorially (couple) to give a resultant atomic magnetic moment. There are two physically distinct ways in which individual electronic l and s values may add vectorially to give the resultant atomic angular momentum quantum number J. First, the l and s of each electron may couple strongly to give a resultant, one-electron, angular momentum quantum number j. The j values then are added vectorially for the atoms to give J. This is known as the $j-j$ coupling scheme. Alternatively, the individual l values of each electron may couple vectorially, to give a total orbital angular momentum quantum number L. Then the s values of each electron may be added vectorially to give a total spin angular momentum quantum number S. L and S may then be added vectorially to give J for the atom. This is known as the Russell-Saunders coupling scheme (10). While the values of J derived by either method of addition may be identical, the strong coupling between individual electronic l's and between individual s values, characteristic of the Russell-Saunders scheme, implies that L and S are good quantum numbers for the atom and consequently that orbital angular momentum and spin angular momentum are well-defined "constants of the motion" of atoms. The $j-j$ coupling scheme, however, does not define a total orbital angular momentum and a total spin angular momentum, but only a resultant atomic angular momentum. The two addition methods correspond to different physical situations, one in which L and S as well as J can be used to describe atomic electronic states and one in which L and S have no meaning and J is the only good angular momentum quantum number.

The importance of the coupling of electronic magnetic moments and angular momenta to the spectroscopy of atoms is twofold. First, just as the shells of hydrogenic atoms are split into different subshells and orbitals, each of which can determine the initial or terminal state in absorptive or emissive transitions and thereby contribute to the complexity of the spectrum, the angular momentum states of many electron atoms are split into several components in the presence of an external magnetic field. The existence of these component substates results in a greater variety of transitions (fine spectral structure) than would occur if the angular momentum states were not split. Second, the coupling of the angular momenta and magnetic moments in the Russell-Saunders coupling scheme indicates that the spin quantum number S for the atom is defined and that transitions between states of different S cannot occur (transition with a change in S violates the law of conservation of angular momentum). However, in the j–j coupling scheme, S is not defined rigorously and it is possible for transitions with change of spin to occur. Hence, for atoms whose angular momenta couple according to the j–j scheme, the splitting of spectral lines in the presence of a magnetic field should be more complex than for atoms whose angular momenta couple according to the Russell-Saunders scheme. Whether a given atom has its angular momentum and magnetic states defined by the Russell-Saunders or j–j coupling scheme depends very much on the size (nuclear charge) of the atom. If an atom has a nucleus of low charge, the individual electronic orbital angular momentum vectors, all located at the nucleus, will couple strongly, as will the individual electronic spin angular momenta located at the periphery of the electronic orbitals. The coupling of spin and orbital angular momenta will be secondary to the coupling of the l and s values, and Russell-Saunders coupling will describe the situation adequately. On the other hand, if a many-electron atom has a high nuclear charge, the orbital and spin angular momentum vectors will have a high probability of being found at the nucleus, and spin–orbital coupling will be appreciable for individual electrons contributing to the existence of a j–j coupling scheme. Usually, even when appreciable spin–orbital coupling occurs, it is more convenient to retain the formalism of the Russell-Saunders scheme and to incorporate into it an appropriate perturbation treatment to account for the spin–orbital coupling. This allows us to use the j–j coupling scheme to justify the appearance of spectral lines due to transitions involving spin changes while at the same time using the relative orderliness of the Russell-Saunders L and S terminology to classify all atomic spectral states.

The energy of an atom having a particular orbital configuration is determined by the principal quantum numbers of the occupied orbitals, the total orbital angular momentum L, the total spin angular momentum S, and the resultant atomic angular momentum J. The electrons which effectively contribute to L, S, and J are those which occupy the partially filled shells, which are the electrons responsible for the transitions seen in optical spectra. In evaluating the energy levels of atoms, all electrons in completely filled shells and subshells may be neglected, because they contribute nothing to the angular momentum of the atom (for filled subshells, $L = 0$, $S = 0$, and $J = 0$). The energy levels of the atom are then determined by the electronic configurations of the partially filled subshells. The definition of the atomic term which corresponds to a collection of energy levels of the atom all having the same atomic orbital angular momentum quantum number L and the same atomic spin angular momentum quantum number S is $^{2s+1}L_J$. In the same way that the one-electron orbital angular momentum quantum states $l = 0, 1, 2, 3, \ldots$ were denoted, conventionally, by $s, p, d, f,$ $\ldots$, the atomic orbital angular momentum quantum states corresponding to $L = 0, 1, 2, 3, \ldots$ are designated by $S, P, D, F, \ldots$, respectively. For example, sodium, having an outer-shell electronic configuration of $3S^1$, will have a term corresponding to $L = 0$ and $S = \frac{1}{2}$, which would be designated a ^2S term. Because J can have positive values of $L + S, L + S - 1, \ldots, L - S$, this term contains one energy level, corresponding to $J = \frac{1}{2}$, which is designated $3^2S_{1/2}$. The quantity $2S + 1$ is called the multiplicity of the term and is equal to the number of energy levels contained in the term when $L > S$. As it occurs in energy levels, the multiplicity gives the number of degenerate states contained in each energy level when $L > S$ for the level. These states become apparent only when the degeneracy is removed, as when spectra are taken in the presence of an external magnetic field. If $L < S$, the multiplicities of terms and energy levels have no significance other than the number of unpaired electrons plus one.

Hund's is rules can sometimes be used to predict the relative energies of terms derived from the same configuration. For example, an atomic p^2 configuration is split into ^1D, ^3P, and ^1S terms (Fig. 2). Because the ^3P term has the highest multiplicity (spin), it has the lowest energy. This term contains the ground state of the atom. Of the singlet terms, the ^1D term has the highest orbital angular momentum and, hence, lies at lower energy than the ^1S term. The ^3P term is composed of three energy levels; both singlet terms have only one energy level. The energy levels of the ^3P term are $^3P_2$, $^3P_1$, and $^3P_0$. To predict the relative energies of these levels, it is necessary to invoke the third of Hund's rules. This rule states that for less-than-half-filled subshells, the lower the value of J, the lower the energy of the level. For more-than-half-filled subshells, the higher the value of J, the lower the energy of the level. Hence, the $^3P_0$ level is lowest in energy, and contains the ground state, and the $^3P_2$ level is highest. For further

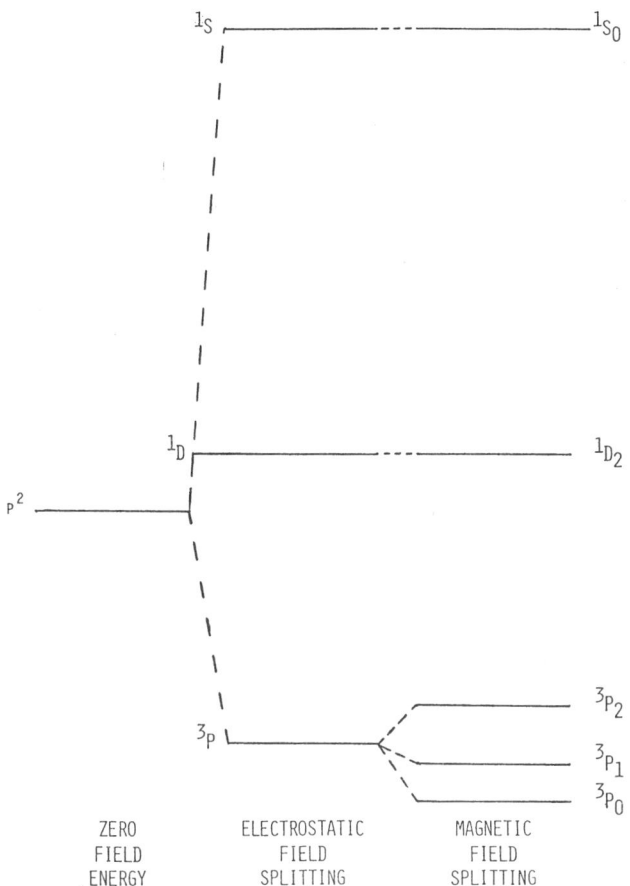

Fig. 2 Energy levels arising from the p^2 configuration.

information on atomic terms, levels, and states, the reader is advised to refer to the references at the end of this chapter (4–8).

B. Electronic Transitions In Atoms

1. Types of Atomic Transitions

Atomic electronic spectra can be broadly classified as either absorption or emission spectra. Absorption spectra arise from the absorption of light by atoms, the absorbed energy resulting in the elevation of the atom from the ground state to an electronically excited state. This entails raising an electron from an occupied orbital (in the valence shell) to an orbital which was unoccupied in the ground state of the atom. Under conditions of extremely high exciting light intensity, such as can be obtained with lasers, It is occasionally possible to produce light absorption resulting from excitation from a low to a higher excited state, but this process is of not much consequence in analytical atomic absorption spectroscopy. Emission spectra arise from the demotion of electronically excited atoms to lower excited states or to the ground state, a process often accompanied by the ejection of the lost energy as visible or ultraviolet light. Because the terminal state is often an intermediate excited state in the emission process, the emission spectrum of an atom contains many more lines than the absorption spectrum of the same atom.

There are several distinct mechanisms of atomic emission. Resonance emission occurs when atoms are excited from the ground electronic state to an excited state (resonance absorption), and then undergo rotational deactivation directly to the ground state. The emitted radiation is of the same wavelength as that absorbed. The most intense emission line for many atoms corresponds to the resonance emission process entailing transition between the ground and lowest excited states.

Normal direct-line emission occurs when an atom is in a higher excited state and undergoes a radiational transition to a lower excited state above the ground state. The radiation in this process will be of longer wavelength than the radiation absorbed.

Thermally assisted direct-line emission occurs when a long-lived state above the ground state is excited thermally (in a flame or plasma) and radiational excitation of the long-lived state results in further excitation to an upper excited state, which then undergoes a radiational transition to the ground state or a state lower in energy than the original long-lived excited state involved in the absorption process.

Normal stepwise line emission occurs when an atom is excited from the ground state to a higher excited state and then undergoes a transition to a lower excited state, from which a radiational transition to a lower state produces emission. The radiation emitted has a longer wavelength than that absorbed.

Thermally assisted stepwise line emission may occur in atoms radiationally excited to a metastable state above the ground state and then thermally excited to an upper excited state, which subsequently undergoes a radiational transition to the ground state or to an excited state of energy lower than that of the radiationally excited state. The emitted radiation has a shorter wavelength than the absorbed radiation.

Spectral lines are characterized by their positions in the electromagnetic spectrum and their intensities. These properties are dependent on the electronic structure and the environment of the absorbing and emitting atoms and are of prime importance in spectral analysis.

2. Positions of Spectral Lines

The frequencies at which atomic spectral lines are observed are determined by the differences in energy between the atomic states involved in the transitions. Energy levels for typical transitions in alkali metals are shown in Fig. 1. If, in the transition, the atom is initially in state i and finally in state f, the frequency of the line produced by that transition is

$$\nu = \frac{\varepsilon_f - \varepsilon_i}{h}$$

where ε_f and ε_i are the energies of the states designated f and i, respectively, and h is Planck's constant. For absorption spectra, ε_i corresponds to the energy of the lower state and ε_f to that of the upper state. For emission spectra, the opposite is true. Customarily, the positions of spectral lines are given by the reciprocals of the wavelengths of the lines. The wavenumber, $\bar{\nu}$, is related to the true frequency by

$$\bar{\nu} = \frac{\nu}{c}$$

where c is the speed of light and $\bar{\nu}$ is expressed in s^{-1} (or Hz). In the presence of an external magnetic field, the Zeeman effect will often be observed. Therein, a spectral line which was assumed to have originated from a single transition between two states will appear under magnification to be split into several closely spaced lines, clustered about the frequency of the transition in the absence of the external field (Fig. 2). Each line in this fine structure actually corresponds to a separate transition between the state from which the transition originated and one of the several substates created by the different possible interactions of the magnetic moment of the atom with the magnetic field.

3. Intensities of Spectral Lines

The observed intensity of a spectral line is determined by the rate of transition between the initial and final states corresponding to the frequency of that line. The rate of transition is, in turn, governed by the intensity of exciting radiation, the path length of exciting radiation through the sample, the concentration of potential absorbers or emitters in the sample, and the probability that a radiative transition will occur between the initial and final states of the absorber or emitter. In quantitative analysis, it is the concentration of absorbers or emitters in the sample that is of interest. These factors will be discussed further in the experimental section. However, it is in order to consider here the probability of radiative transition, which is a function of atomic electronic structure and the nature of the transition.

The probability of radiative transition between two electronic states is described quantum mechanically by a term called the transition moment integral, R_{if}. Although the details of the quantum mechanical derivation of R_{if} are beyond the scope of this treatment, it would be approximately correct to visualize R_{if} as the change in electronic dipole moment of the atom undergoing electronic transition (i.e., if e is the electronic charge and $\bar{r}_{if}$ is the vector along which the displacement of electronic charge in the atom occurs as a result of the transition, then $R_{if} = e\bar{r}_{if}$). The spectroscopic significance of R_{if} is that the molecular probability (or intensity) or radiative transition is proportional to R_{if}^2. The greater the value of the electronic dipole moment change in the electronic transition, the greater will be the inherent intensity of the transition.

For absorptive transition, as in atomic absorption spectrophotometry, it can be shown that the measurable integrated inherent intensity of an absorption line, expressed as absorbance A, per unit concentration of absorber c and optical path length d in the flame, is given by

$$\frac{A}{cd} = \int_0^\infty \varepsilon_{\bar{v}_a} \, d\bar{v}_a = 1.085 \times 10^{38} \bar{v}_a |R_{if}|^2$$

where $\bar{v}_a$, the reciprocal of the wavelength of the line, is measured in cm^{-1}; $\varepsilon_{\bar{v}_a}$, the absorptivity at any given value of $\bar{v}_a$ in the spectral line of interest, is usually expressed in terms of cm^2/mg (when c is expressed as mg/cm^3 and d in cm) and R_{if} in esu-cm. For Gaussian-shaped absorption lines,

$$\int_0^\infty \varepsilon_{\bar{v}} \, d\bar{v} \approx \varepsilon_m \, \Delta\bar{v}_{1/2}$$

where ε_m is the absorptivity at the line maximum and $\Delta\bar{v}_{1/2}$ is the line width at half-maximum absorption. It may seem surprising that we speak of the width or half-width of a spectral line, since the discussion up to the present has almost implied that the spectral lines are infinitesimal in width. However, real spectral lines have finite widths owing to several physical phenomena. The widths of atomic spectral lines (ca. 2 cm^{-1}, at ~3,000 K) are due to the following.

a. *The natural width.* The uncertainty principle of quantum mechanics requires that, for a spectroscopic transition,

$$\Delta E \, \Delta t \approx \frac{h}{2\pi}$$

where ΔE and Δt are, respectively, the uncertainties in the energy and the time of transition. For transitions between the ground state and an electronically excited state (as in atomic absorption),

$$\Delta E\, \tau_R \approx \frac{h}{2\pi}$$

where τ_R is the radiative lifetime of the upper state. The smaller τ_R is, the greater will be ΔE. For allowed electronic transitions, $\tau_R \approx 1 \times 10^{-9}$ s, so that it may be expected that the mean value of ΔE is

$$\Delta E = \frac{h}{2\pi \times 10^{-9}}$$

But, since $E = hc\bar{v}$, $\Delta E = hc\,\Delta\bar{v}$, where c is the speed of light, or $\Delta\bar{v} = 1/(2\pi c \times 10^{-9}) = 5.3 \times 10^{-2}$ cm^{-1}. This value of $\Delta\bar{v}$ can be considered a mean value for the minimum possible half-bandwidth of an atomic spectral line (this value will vary slightly, depending on the actual value of τ_R).

b. *Doppler broadening.* Atoms in rapid motion toward or away from the spectrometer detector will absorb or emit radiation which is slightly shifted to higher or lower frequencies by comparison with the spectral lines of atoms moving laterally with respect to the detector. This is comparable to the change in pitch of a siren on a vehicle moving toward or away from an observer (the Doppler effect). Because the high- and low-frequency components lie on either side of the unaffected spectral line, the latter appears to be broadened.

c. *Collisional broadening* (Holtzmark and Lorentz broadening). Broadening of spectral lines can also occur from collisions of absorbing or emitting atoms with foreign species (Lorentz broadening) or other atoms of the same substance (Holtzmark broadening). Collisional broadening is characterized by broadening of the spectral line, shift of the line maximum to the red (lower frequencies), and distortion of the red side of the spectral line (the line becomes asymmetric). The broadening itself is related to the changes in velocity, which occur in absorbing or emitting atoms as a result of inelastic collisions. The atomic energy levels of the emitting or absorbing atoms are perturbed as a result of interaction with ground-state species. The nature of the perturbation is such as to compress the energy levels of the excited atom, resulting in lower-energy transitions and, therefore, a red shift in the spectral lines. Collisional broadening can be recognized by the skewing of the spectral line toward the red side of the spectrum.

d. *Stark and Zeeman broadening.* In gaseous systems, where ions are produced, broadening of atomic spectral lines can occur due to the splitting of atomic energy levels by the electric fields resulting from high concentrations of ions (the Stark effect). This type of broadening, however, is significant only in high-energy arcs, sparks, and plasma discharges, where ionization occurs on a large scale. In atomic fluorescence spectroscopy, the sources, e.g., hollow cathode discharge lamps, electrodeless discharge lamps, lasers, or flames, as well as the cells, flames, or furnaces used, do not exhibit appreciable Stark broadening. Similarly, Zeeman broadening due to the splitting of spectral lines by magnetic fields is usually insignificant under conditions common to analytical atomic spectroscopy because of the absence of strong magnetic fields.

For the most part, it is Doppler broadening and, to a lesser extent, Holtz-mark broadening that have the greatest influence on the half-widths of the spectral lines of interest in analytical atomic spectroscopy.

For emissive transitions, as in atomic fluorescence, flame emission, and plasma emission, the transition moment R_{if} can be related to the rate constant or probability of emission k_e by

$$k_e = 3.14 \times 10^{29} \bar{\nu}_e^3 |R_{if}|^2 \frac{g_i}{g_f}$$

where $\bar{\nu}_e$, the reciprocal of the wavelength of the emission line, is expressed in cm^{-1} and R_{if} in esu-cm; g_i/g_f is the ratio of the spin multiplicities of the initial and final states of the transition, e.g., $g_i/g_f = 1$ for transitions between singlet and singlet states and $g_i/g_f = 3$ or $1/3$ for transitions between singlet and triplet states. The significance of k_e is that the observed intensity of emission is proportional to the quantum yield of emission ϕ_e, which, in turn, depends on k_e:

$$\phi_e = k_e \tau^0 = \frac{k_e}{k_e + k_d}$$

where τ^0 is the mean lifetime of the excited state from which emission arises and k_d is the sum of the first order of pseudo-first-order rate constants which compete with emission for the deactivation of the excited state from which emission arises.

Under certain circumstances the change in the electronic dipole moment accompanying electronic transition between certain electronic states of an atom will be zero. The transition should then have zero intensity and is then said to be forbidden. When R_{if} is nonzero, the transition is said to be allowed. Whether or not a given transition is allowed (the allowedness or forbiddenness of a transition) depends on the space and spin properties (angular momenta) of the atomic orbitals involved in the transition and can be derived rigorously only by using quantum mechanics. The permissible values of the angular momentum quantum numbers of the electronic states involved in the allowed transitions (more precisely, the permissible values of charges of these quantum numbers) give rise to a set of optical selection rules. In terms of the principal and azimuthal quantum numbers

n and l of the optical electron and the orbital, total and spin angular momentum quantum numbers L, J, and S, the optical selection rules for atoms are

$$\Delta n = 0, 1, 2, \ldots$$
$$\Delta l = \pm 1$$
$$\Delta L = 0, \pm 1$$
$$\Delta J = 0, \pm 1$$
$$\Delta S = 0$$

The selection rules dictate that of all the possible transitions that could be conceived, only a fraction will actually be observed. However, it is in order to note that for atoms of high atomic number, where j–j coupling predominates, the selection rule $\Delta S = 0$ is relaxed so that transitions which are spin-forbidden in the Russell-Saunders scheme may actually appear weakly in the spectrum.

In order to apply the optical selection rules to a simple example, consider the spectrum of the hydrogen atom. The ground state of hydrogen has an electron in the $1s$ orbital. In this state, $L = 0$ and $S = -\frac{1}{2}$. The only possible value of J, therefore, is $\frac{1}{2}$, and the ground state of hydrogen is $1\,^2S_{1/2}$. If the electron is excited to the $n = 2$ shell, the possible values of L are 1 and 0 and the only possible value of S is still $\frac{1}{2}$. For $L = 1$, $J = \frac{3}{2}$ and $J = \frac{1}{2}$. For $L = 0$, $J = -\frac{1}{2}$. Hence, the possible energy levels are $2\,^2S_{1/2}$, $2\,^2P_{1/2}$, and $2\,^2P_{3/2}$. There are no interelectronic repulsions in hydrogen. The $2S$ and $2P$ terms should, therefore, have the same energy. Actually, the $^2P_{3/2}$ and the $^2P_{1/2}$ levels are slightly split, due to their different resultant angular momenta. If the electron occupies the third shell, then $S = \frac{1}{2}$ and $L = 2, 1, 0$. For $L = 2$, $J = \frac{5}{2}$ and $\frac{3}{2}$; for $L = 1$, $J = \frac{3}{2}$ and $\frac{1}{2}$; and for $L = 0$, $J = \frac{1}{2}$. The possible energy levels are then $3\,^2S_{1/2}$, $3\,^2P_{3/2}$, $3\,^2P_{1/2}$, $3\,^2D_{5/2}$, and $3\,^2D_{3/2}$. It can be seen that for all terms with $L = 0$, there are two slightly separated energy levels. Now, remembering the selection rules $\Delta S = 0$, $L = \pm 1$, and $J = 0, \pm 1$, only the absorptions $1\,^2S_{1/2} \rightarrow 2P_{3/2}$, $1\,^2S_{1/2} \rightarrow 2\,^2P_{1/2}$, $1\,^2S_{1/2} \rightarrow 3\,^2P_{3/2}$, $1\,^2S_{1/2} \rightarrow 3\,^2P_{1/2}$ are allowed. Because of the close spacings of the $^2P_{3/2}$ and $^2P_{1/2}$ levels, these transitions will appear as two doublets—that is, two pairs of slightly separated lines, with the pairs being appreciably separated from one another. In the emission spectra, these same pairs of doublets would appear, but also the $3\,^2D_{5/2} \rightarrow 2\,^2P_{3/2}$, $3\,^2D_{3/2} \rightarrow 2\,^2P_{1/2}$ doublets would be seen. In hydrogen, resonance transitions involving terms with the same L but different n form series with the spacing between lines being proportional to $1/(n_i^2) - 1/(n_j^2)$. This is the principal difference between hydrogen spectra and alkali metal spectra. Although hydrogen and the alkali metals all have one optical electron and a $^2S_{1/2}$ ground-state energy level, the excited states formed by occupation of different subshells of the same shell in the alkali metals do not have the same energy. This is due to the stronger attraction of the nucleus for electrons of the lowest values of l in the same shell, due to more effective screening of electrons of higher *l* from the nucleus by inner-shell electrons. This has the effect of destroying the regular spacing between resonance transitions involving energy levels of different *n* but the same *L*, because the effect of inner-shell electron screening is not constant but falls off with increasing *n*. Other than in this respect, the spectra of the alkali metals are similar to those of hydrogen. For example, they both show doublet fine structure in the allowed transitions.

The spectra of the other elements are somewhat more complex, due to the large number of terms that arise as a result of the occurrence of several unpaired electrons in their atoms, and will not be discussed in this chapter.

III.   EXPERIMENTAL

Atomic spectroscopy, in general, is effected by vaporizing and atomizing the analytical sample, which is then excited by a suitable means. The manner in which the sample is excited is the major point of difference among the various types of atomic spectroscopic techniques. Excitation may be produced by an electrical arc or spark, a chemical flame, or an inductively coupled argon plasma, all of which also facilitate vaporization and atomization as well. After excitation, the excited atoms undergo relaxation, producing atomic emission. Alternatively, the absorption of electromagnetic radiation by the atomic vapor after excitation with a high-intensity light source forms the basis of atomic absorption spectroscopy. The absorbed light may be subsequently emitted as atomic fluorescence.

Among the various types of atomic spectroscopy, only two, flame emission spectroscopy and atomic absorption spectroscopy, are widely used and accepted for quantitative pharmaceutical analysis. By far the majority of literature regarding pharmaceutical atomic spectroscopy is concerned with these two methods. However, the older method of arc emission spectroscopy is still a valuable tool for the qualitative detection of trace-metal impurities. The two most recently developed methods, furnace atomic absorption spectroscopy and inductively coupled plasma (ICP) emission spectroscopy, promise to become prominent in pharmaceutical analysis. The former is the most sensitive technique available to the analyst, while the latter offers simultaneous, multielemental analysis with the high sensitivity and precision of flame atomic absorption.

These methods have yet to be extensively utilized in pharmaceutical analysis, primarily due to their novelty and their high cost. Lastly, atomic fluorescence spectroscopy has been largely ignored but may ultimately find use in pharmaceutical analysis due the simplicity of its design and limits of detection which are occasionally somewhat lower than with atomic absorption spectroscopy.

The relative sensitivities, precisions, and characteristics of the various techniques are summarized in Table 1. Although less than full advantage has been made of several, some have been applied directly to pharmaceutical analysis. The

Table 1 Comparison of Atomic Spectroscopic Methods

Method	Type of analysis	Sensitivity	Precision	Sample size	Advantages	Disadvantages
Emission						
Arc	Qualitative	Moderate	10%	1–10 mg	Solid or liquid samples	Matrix interference
Spark	Qualitative & quantitive	Low	5%	0.1 mL	Solid or liquid samples	Matrix interference
Inductively coupled plasma	Qualitative & quantitive	High	1–2%	0.1 mL	High linear dynamic range	High cost
					Determination of re-fractory elements	
Flame	Quantitive	Moderate	1–4%	5 mL	Determination of alakli and alkaline earth metals	Spectral inter-ference
Absorption						
Flame	Quantitive	High	1–2%	5 mL	High specificity	Some elements are refractory
Furnace	Quantitive	Very high	5%	1–10 μL	Solid or liquid samples	Matrix interference
					High specificity	
Fluorescence	Quantitive	High	1–2%	5 mL	Low background inter-ference	Quenching effects
					High specificity	

problems and benefits of these techniques must be examined in greater detail before a consideration of their pharmaceutical utility can be addressed.

A. Arc/Spark Emission Spectroscopy

Excitation of atoms can be induced by an electrical arc or spark discharge, with subsequent emission of electromagnetic radiation (11–13). In a typical experiment (Fig. 3), a sample is placed on one of a pair of graphite electrodes, usually the anode. The electrode containing the sample of interest is referred to as the sample electrode; the other is the counter electrode. When an electrical discharge is allowed to pass across these electrodes, a high-temperature arc (4000–6000 K) or spark (10,000 K) will produce volatilization and dissociation of the sample, accompanied by ionization of atoms with low ionization potentials. The atoms are then collisionally excited by the electrons of the discharge to higher electronic states. Upon relaxation, each element in the sample emits a series of characteristic spectral lines which are proportional in intensity to the elemental concentration of the sample. The pattern of the lines provides a "fingerprint" for simultaneous, quantitative identification of up to 69 elements.

The DC arc is produced by a direct current, flowing through two electrodes in a shorted circuit. The arc is adjusted with a variable resistor to obtain the desired potential. The sample, which is ionized at the anode, then drifts toward the cathode, forming a cathodic layer of excited atoms and ions. This layer of intense luminosity produces a multitude of emission lines and is the optimum location for the measurement of atomic emission. Because the arc can localize on one part of the electrode, the latter is often rotated rapidly to produce a more even arc. The nonuniformity of the DC arc produces a substantial variation in line intensities, limiting the method to qualitative identification. The apparatus is simple, inexpensive, and sensitive.

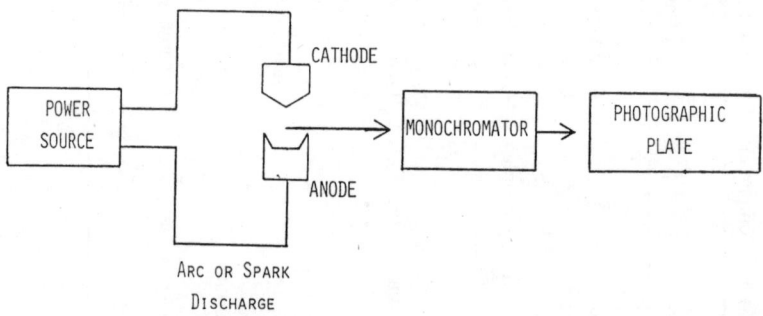

Fig. 3 Instrumental design of an arc/spark emission spectrometer.

Replacement of the direct current by an alternating current produces a more homogenous arc and thus somewhat more reproducible line intensities. The arc is, however, produced less than half of the time, due to the alternating current, resulting in fewer emission lines and a loss in sensitivity. The normal commercial spectrograph allows the user to adjust the operating conditions in order to attain those necessary for optimum analyses, thus the emission source used may be anything from a completely DC arc to an AC arc, as well as a completely AC spark (13).

Spark emission spectroscopy is the most reproducible but the least sensitive of these techniques. A charge buildup in a capacitor causes an intermittent sparking between the two electrodes. This sparking produces short peaks of approximately 1000–2000 pulses during a 10- to 100-μs time period (13). Although the electrodes remain cold, the temperature of the spark discharge reaches 10,000 K, causing not only very efficient atomization, but a high degree of ionization as well. An element often appears in several states of ionization, producing spectra much more complex than arc emission spectra. The quantitative precision of the spark source is several times that achievable with an arc, even though it is less sensitive qualitatively.

The electrodes for arc/spark spectroscopy are usually made from high-purity graphite with a pointed cathode and the anode cupped for sample containment. Samples may be solid, liquid, or in solution, thus eliminating many sample preparation procedures necessary for other methods, as well the errors which may be introduced by this handling. Solid samples may be placed into the cup, or solutions may be dried in the cup. An alternative design allows the anode to turn slowly while a portion of it dips into the analyte solution, continually feeding fresh sample onto the electrode surface. The sample is often intimately mixed with a high-purity graphite powder to minimize matrix effects; however, matrix effects still produce significant variations in signal levels. Standards used for calibration should, therefore, be as similar to the analytical sample as possible. Barring even this problem, quantitative measurements are still somewhat imprecise due to the instability of the electrical discharges.

Variable volatility of the analyte mixture can also be a problem. Various elements in a sample are removed at different rates, causing the temperature of the flame to vary with time. For this reason, the sample is often mixed with a spectroscopic buffer such as lithium carbonate, in order to prevent the composition and temperature of the arc or spark from changing as the analysis proceeds. This procedure is also effective in lowering the temperature of the discharge to prevent ionization of alkali and alkaline earth metals.

Detection of emission for qualitative analysis usually employs a photographic plate (13,14). The emission passes through a dispersive prism or grating, and impinges on the plate. The entire spectrum of the sample is thus displayed with one measurement, providing a permanent record of the results. Since the spectrum may be recorded for a substantial time, the photographic plate actually integrates the signal. The fluctuations of the source are thereby averaged, giving a truer analysis.

The degree of blackening, or density, of the photographic plate facilitates quantitative measurement. By measuring the intensity of light transmitted through a spectral line, I, and the intensity through an unexposed portion of the plate, I_0, the density, D, can be calculated:

$$D = \log_{10} \frac{I_0}{I}$$

Unfortunately, the density is a complex function of the exposure time. If the exposure is too short, the contrast of the spectral line to the unexposed portion will be poor; if the exposure is too long, the transmission of the line will be too small to measure. The exposure time should therefore be chosen to be between these extremes. The density is also a function of development time, the spectral transition measured, and differences between individual plates. The use of an internal standard substantially improves the precision and accuracy of the method.

The increased reproducibility of spark emission is more conducive to quantitative analysis than is arc emission. Spark emission spectrometers often employ a more sophisticated detection system. Rather than impinging on a photographic plate, the dispersed radiation passes onto an array of photomultiplier tubes positioned at preset wavelengths. The photomultiplier is more accurate and faster to use in quantitative measurements than film (12). Such an instrument is called a direct reader and will be discussed further in relation to inductively coupled plasma emission spectroscopy.

B. Inductively Coupled Plasma Emission Spectroscopy

The major disadvantage of arc/spark emission spectroscopy is the instability of the excitation source. This problem can be virtually eliminated by the use of a plasma torch. The most common commercially available method uses an inductively coupled plasma (ICP), which is also called RF plasma, to excite the sample (13–19). The resulting spectrometers (Fig. 4) can simultaneously measure up to 60 elements with high sensitivity and an extraordinarily wide linear dynamic range.

The RF plasma is formed by passing a stream of argon gas between the middle and innermost of three concentric quartz tubes. The argon gas is ionized as it passes through the magnetic field of a radio-frequency induction coil. The resulting argon plasma reaches temperatures estimated to range between 5000

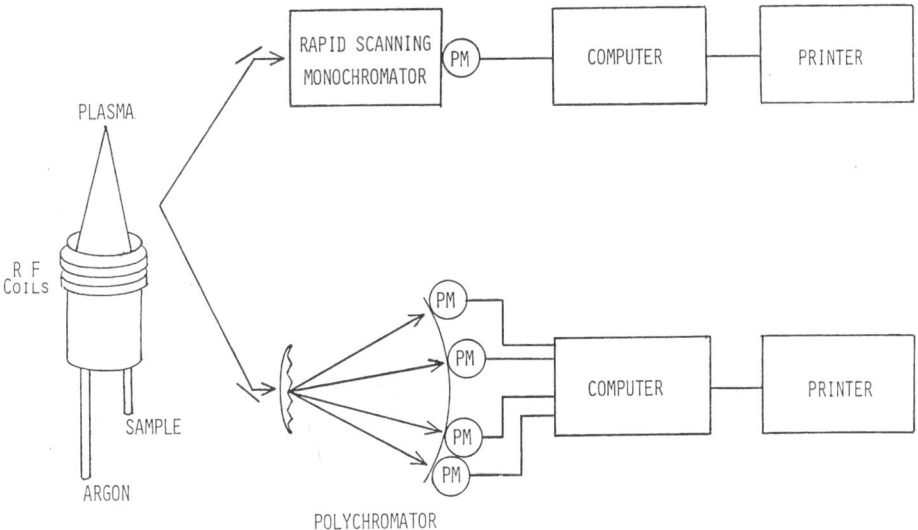

Fig. 4 Instrumental design of an inductively coupled plasma spectrometer utilizing either a monochromator (top) or polychromator (bottom) system of light dispersion. Photomultiplier tubes are denoted by PM.

and 15,000 K. An additional stream of argon is passed between the middle and outside tubes in a helical pattern, to cool the middle tube and protect it from the hot plasma inside. The sample usually does not travel through the hottest part of the plasma, but through the "doughnut" hole, thus realistically attaining temperatures of only about 9000–10,000 K (13).

The analyte is placed into the plasma, usually by utilizing a crossflow nebulizer (18), whereby the sample solution is introduced as a fine mist into a stream of argon at a right angle to its flow. The aerosol of argon and analyte is then carried through the innermost quartz tube and into the argon plasma. The high-temperature plasma heats the sample, producing essentially complete vaporization and atomization of the analyte. Several refractory elements which are difficult to atomize by flame atomic absorption spectroscopy can be quantitated at much lower concentrations by inductively coupled plasma spectroscopy, due to this extensive atomization.

Excepting refractory elements, the sensitivity of inductively coupled plasma spectroscopy is comparable with that of flame atomic absorption spectroscopy (Table 2). Additionally, interactions that occur between the atmosphere and an arc or flame are absent due to the inertness of the argon plasma. Since ioniza-

Table 2 Limits of Detection (ppm)

Element	Emission				Absorption		Fluorescence		Blood Levels[h]
	DC arc[a]	Spark[b]	ICP[c]	Flame[d]	Flame[d]	Furnace[e]	Lamp[f]	Laser[g]	
Ag	0.0006	0.2	0.001	0.008	0.001	0.00001	0.0001	0.004	0.01
Al	0.05	0.05	0.01	0.05	0.01	0.0001	0.1	0.0006	0.4
As	0.1	5.	0.015	10.	0.03	0.0002	0.1		0.01
Au	0.05	0.1	0.04	2.	0.02	0.0001	0.003		(0.001)
B	0.07	0.4	0.002	0.05	2.5	0.02			(0.1)
Ba	0.005	0.02	0.0002	0.002	0.02	0.00006		0.008	0.08
Be	0.0006	0.0002	0.0005	1.	0.002	0.000001	0.01		<0.004
Bi	0.03	0.1	0.05	20.	0.05	0.0003	0.005	0.003	(0.01)
Ca	0.01	0.05	0.00007	0.0002	0.002	0.0004	0.02	0.00008	50.
Cd	0.02	0.1	0.001	0.8	0.001	0.000003	0.000001	0.008	(0.006)
Co	0.01	0.05	0.002	0.03	0.002	0.00006	0.005	1	0.001
Cr	0.01	0.05	0.002	0.004	0.002	0.00004	0.005	0.001	0.03
Cs				0.6	0.05				0.001
Cu	0.0003	0.01	0.002	0.01	0.004	0.00004	0.0005	0.001	1.
Fe	0.03	0.5	0.001	0.03	0.004	0.000008	0.03		1.
Ga	0.02	0.02	0.014	0.06	0.05	0.0001	0.01	0.0009	
Ge	0.02	0.1	0.15	0.4	0.1	0.0003	0.1		
Hf	1.	0.25	0.005	20.	20.			100.	
Hg	0.07	1.	0.2	10.	0.5	0.0005			0.003
K	0.3		0.08	0.00005	0.003	0.0001			40.
La	0.005	0.02	0.006	6.	2.				
Li	2.		0.0002	0.00002	0.001	0.0006			0.03
Lu	0.009	10.	0.008	1.	3.		0.1	3.	
Mg	0.007	0.05	0.0007	0.07	0.003	0.00007	0.0001	0.0002	20.
Mn	0.003	0.01	0.0005	0.008	0.0008	0.00001	0.001	0.00004	0.008
Mo	0.006	0.03	0.005	0.2	0.03	0.0003	0.5	0.01	0.004
Na	0.005	0.1	0.0002	0.0005	0.0008	0.000004		0.0001	1400.

Nb	5.	0.1	0.01	1.	3.	0.0002	0.003	1.5	0.08
Ni	0.02	0.05	0.005	0.02	0.005	0.0001		0.002	0.03
P	0.15	4.	0.03	1.	100.	0.0003		150.	100.
Pb	0.04	0.1	0.015	0.1	0.01	0.00005	0.01	0.5	0.05
Pd	0.005	0.02	0.007	0.05	0.01	0.0004	0.04		
Pt	0.04	0.4	0.02	4.	0.05	0.001			
Rb	~30.		0.008	0.008	0.005	0.0001			0.2
Rh	0.02	0.05	0.003	0.03	0.02	0.0008	3.0	0.15	
Sb	0.07	2.	0.2	0.06	0.03	0.0002	0.05		0.003
Sc	0.2	0.01	0.0008	0.8	0.1	0.006		0.01	
Se			0.015	100.	0.1	0.0001	0.04		0.01
Si	0.1	0.02	0.01	3.	0.1	0.007	0.6		8.
Sn	0.05	0.3	0.3 0.1	0.05	0.0004	0.05		0.03	
Sr	0.00003	0.002	0.00002	0.0005	0.005		0.03	0.003	
Ta	30.	0.3	0.07	4.	3.				
Te	60.	4.	0.08	2.	0.05	0.0001	0.005		0.03
Ti	0.0001	0.01	0.001	0.2	0.1	0.004	4.	0.002	0.04
Tl	0.07	0.8	0.2	0.02	0.02	0.00007	0.008	0.004	
U	~10	2.	0.075	5.	20.				(0.0008)
V	0.02	0.02	0.002	0.1	0.02	0.0005	0.07	0.03	0.01
W	0.03	0.4	0.002	0.6	3.				
Y	0.0009	0.8	0.0002	0.03	0.3				
Zn	0.01	0.5	0.001	10.	0.001	0.000002	0.00002		10.
Zr	0.004	0.01	0.002	5.	4.				0.4

a Refs. 11 and 12.
b Refs. 11 and 92.
c Refs. 13 and 15.
d Ref. 21.
e Ref. 26.
f Refs. 13 and 26.
g Refs. 26 and 33.
h Refs. 64 and 65. Values are normal levels for human blood except for values in parentheses, which are for serum.

tion does not occur appreciably in the argon plasma, it is unnecessary to use an easily ionizable metal such as lithium as a spectroscopic buffer in order to inhibit the analyte ionization.

The emission spectrum of the plasma is often very rich, providing a multitude of lines potentially available for analysis. However, this can actually complicate the analysis, due to the presence of overlapping lines emanating from other elements in the sample. The choice of the analytical wavelength to be used for each element must thus be made with a foreknowledge of all components which may be expected to be present in the sample.

Emission from the plasma is measured either by a polychromator (17) or monochromator system (18,19). The polychromator system is also used by direct-reading arc/spark emission spectrometers. In the polychromator system, a Rowland circle is used (13). This is an optical setup in which the emitted radiation impinges upon a concave diffraction grating which disperses it in a radial manner. The light intensity at any particular wavelength is measured by placing a photomultiplier tube at the proper position on an arc (16). A different phototube is thus required for each element to be analyzed. Each phototube charges a separate capacitor, with the resultant charge being equal to the integrated line intensity. Such an arrangement can quantitate as many as 60 different elements simultaneously. The analysis is not only multielemental but also rapid and conservative of the sample. The greatest objection to the polychromator system is that the elements to be analyzed and the analytical wavelengths to be used are preset for a particular instrument by the manufacturer. Adjustment for the addition of other elements is a complex and expensive exercise. This type of system is certainly excellent for routine analysis.

More flexibility can be achieved by the utilization of a rapid-scanning monochromator. An instrument of this type employs a stepper motor to scan all emission wavelengths between 180 and 900 nm in a matter of seconds. By coupling a microprocessor to the system, experimental parameters may be individually optimized for each element as the wavelength scan progresses. The analytical wavelength for each element can also be optimized with respect to interfering lines within the emission spectrum of the sample. The disadvantage of the monochromator is that both measurement time and sample volume must be increased in order to achieve sensitivity and accuracy comparable to a polychromator. For the determination of N elements in the sample, the monochromator system will require a minimum of N times longer analysis time and a sample volume of N times more. The choice is therefore between the speed and efficiency of the polychromator and the extreme versatility of the monochromator.

The intensity of the emission is directly proportional to the analyte concentration, provided that all instrumental parameters remain constant throughout the analysis. Normally, however, several factors may change slightly, reducing the accuracy of the measurement. Variations in the sample matrix or sample viscosity are common problems. In order to circumvent the difficulties, an element that is not native to the sample but that has an emission spectrum near that of a sample constituent is quantitatively added to the sample as an internal standard. Deviations in the intensity of the resulting emission will then reflect abnormalities present in the analysis. A computer is then used to normalize the instrumental response of all elements being analyzed in relation to the internal standard. Measurements made by inductively coupled plasma spectroscopy, therefore, exhibit high reproducibility.

Perhaps the most outstanding feature of the inductively coupled plasma spectrometer is its remarkable linear dynamic range. Components differing in concentration by as much as six orders of magnitude can be quantitated without dilution of the sample. This result is due to the plasma being "optically thin." The concentration of analyte in the plasma is isolated in the center, with little being present at the edges of the plasma. Self-absorption of the emitted radiation is therefore almost completely eliminated, accounting for the surprising linearity of response. This can diminish sample preparation time and the possibility of making dilution errors. More important, trace elements or impurity concentrations can be measured at the same time that the analysis for the major constituents is performed.

C. Flame Emission Spectroscopy

The electronic excitation of atoms by a flame, followed by the emission of ultraviolet and visible light, forms the basis for flame or atomic emission spectroscopy (20–23). It may also be referred to as flame photometry (13). Typically, a sample solution is drawn into a nebulizer, which sprays a fine mist of the solution into a 2000–3200 K flame (Fig. 5). The flame evaporates the solvent and volatizes the sample, whereupon the analyte can be atomized. The flame can then electronically excite or ionize the atomized analyte. After excitation, the atoms can undergo emission of light, which is then passed through a narrow bandpass filter, a prism, or a grating monochromator. A photocell or photomultiplier is used for detection, with the emission intensity being proportional to the concentration of the analyte in the sample. Flame emission is more precise than arc/spark emission and produces simpler emission spectra, due to the lower temperatures of the flame. A flame is more stable than an arc, but emission spectra are sensitive to changes in its temperature. Using flame emission spectroscopy, the concentrations of alkali and alkaline earth metals in clinical solutions can be easily and conveniently measured.

The flame used for excitation is produced by a mixture of a fuel gas and an oxidant gas which is ignited at the surface of a burner. The burning of the gas mixture takes place at a considerable rate, called the burning velocity. The gas flow rate must exceed this velocity in order to avoid blowback into the mixing

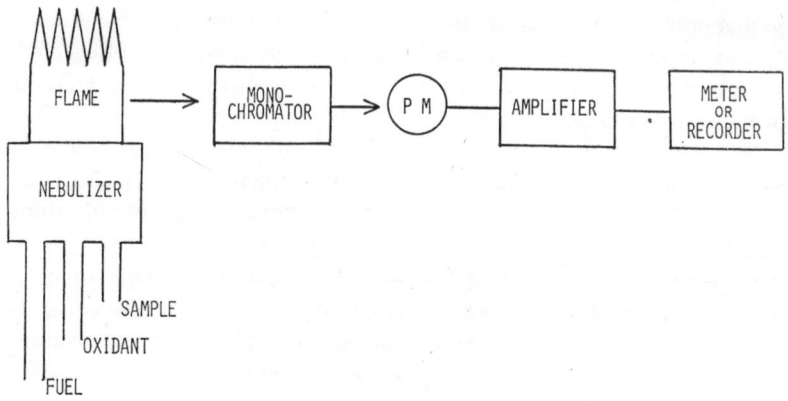

Fig. 5 Instrumental design of a flame emission spectrometer.

chamber and subsequent explosion. Thus, the gas mixture should have a low burning velocity but be capable of relatively high flame temperatures. A commonly used fuel gas, acetylene, coupled with either air or nitrous oxide as an oxidant, provides a moderate burning velocity of 270–500 cm/s with a flame temperature of 2450–2950 K. Such mixtures produce relatively safe and efficiently atomizing flames which are also relatively quiet.

The structure of the flame itself consists of two major zones, the primary reaction zone and a secondary reaction or postheating zone. The primary reaction zone, the area of the flame just above the burner surface, is the region where combustion, atomization, and excitation occurs. Some typical flame combustion products formed in this region are CO, CO_2, H_2, N_2, and H_2O molecules, as well as $O\cdot$, $H\cdot$, $OH\cdot$, and $C\cdot$ radical species. The secondary reaction zone is a much cooler region where the flame gases mix with atmospheric components that may include impurities and emission interferants. Between these two zones lies a smaller, but important, intermediate region, where little reaction occurs. In this region of the flame, the fraction of the atoms in the ith excited state, α_i, is controlled only by the prevalent temperature and can be represented by the Boltzmann distribution:

$$\alpha_i = \frac{N_i}{\sum_{j=0}^{N} N_j} = \frac{g_i e^{-(E_i - E_0)k_B T}}{\sum_{j=0}^{N} g_j e^{-(E_i - E_0)k_B T}}$$

where N_i is the number of electrons in the ith excited state, $(E - E_0)$ is the difference in energy between the ith excited state and the ground state, k_B is the

Boltzmann constant, and g_i is a statistical weight for the ith excited state. The summation over all possible states is the electronic partition function. If the flame temperature is constant throughout the analysis, the signal level will be subject only to the amount of sample in this region. Thus the intermediate zone is usually aligned with the optical path and is of most importance for analytical measurements. However, this alignment of the optical path should also be optimized for the particular element to be quantitated.

The analyte is placed into the flame by drawing a sample solution into a nebulizer, where the sample is mixed with the carrier gas as a fine mist and borne with the gas into the flame. The rate of flow of the gas controls the rate of nebulization and hence the rate of sample atomization. The solvent can be aqueous or organic and should be chosen judiciously, since several solvent effects may be evident during analysis. Solvents or organic sample substituents will often increase the temperature of the flame. A change in the solvation of the analyte or the presence of other solutes in the sample, which interact with the analyte, may alter the degree of atomization. The solvent should also be selected to minimize viscosity, since highly viscous solvents are poorly nebulized.

The addition of a spectroscopic buffer such as lithium carbonate has a stabilizing effect on the flame temperature and decreases the ionization of elements with low ionization potentials as well. At a constant flame temperature, the intensity of a given emission line is directly proportional to the concentration of analyte according to the equations given for emission spectrometry in the theoretical section. In practice, a standard curve is usually prepared in order to assess linearity between intensity and concentration. Alternatively, the method of standard addition (24) can be utilized.

The magnitude of analyte concentration is an important consideration. At high concentration, a considerable amount of material will be present in the cooler, exterior portions of the flame. Emission originating at the flame interior can be subsequently absorbed by atoms that reside at the edges, before it can exit the flame. This process, called ''self-absorption'' and in the extreme case ''self-reversal'' (5), gives rise to a nonlinear relationship between concentration and emission. If the range of analyte concentration is not known, it is recommended that an analysis be made using several dilutions of the same sample in order to ensure the absence of this effect.

D. Atomic Absorption Spectroscopy

Atomic absorption (21,22,25–29) differs from atomic emission spectroscopy in that the quantitative measure of an element is made by observing the absorption of light passing through an atomized sample instead of the emission from the thermally excited atom (Fig. 6). The precision of measurement is greater than arc, spark, or flame emission and comparable to inductively coupled plasma spec-

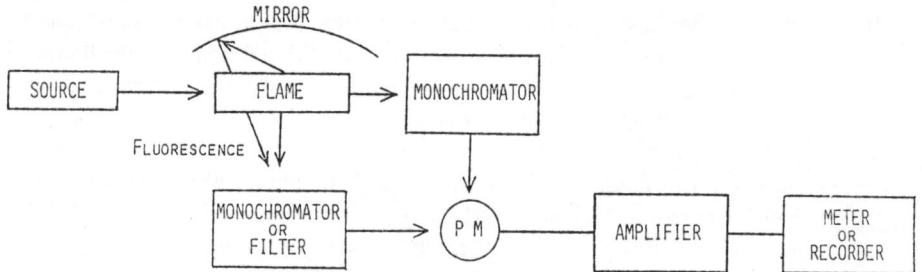

Fig. 6 Instrumental design of a spectrometer for the measurement of atomic absorption or atomic fluorescence. The flame may be replaced with a graphite furnace.

trometry. The sensitivity is comparable to the latter, but the cost of an atomic absorption spectrometer is lower by approximately an order of magnitude. It is the most widely used technique for quantitative analysis by atomic spectroscopy.

Atomization of the sample is usually facilitated by the same flame aspiration technique that is used in flame emission spectrometry, and thus most flame atomic absorption spectrometers also have the capability to perform emission analysis. The previous discussion of flame chemistry with regard to emission spectroscopy applies to absorption spectroscopy as well. Flames present problems for the analysis of several elements due to the formation of refractory oxides within the flame, which lead to nonlinearity and low limits of detection. Such problems occur in the determination of calcium, aluminum, vanadium, molybdenum, and others. A high-temperature acetylene/nitrous oxide flame is useful in atomizing these elements. A few elements, such as phosphorous, boron, uranium, and zirconium, are quite refractory even at high temperatures and are best determined by nonflame techniques (Table 2).

The graphite furnace is an alternative means of atomizing the sample. A small sample of perhaps only a few microliters is dried inside a graphite tube which is flushed with argon. The sample is heated by passing a moderate current through the graphite so that the organic portion of the sample will be ashed and matrix effects diminished. When a heavy current is passed through the tube, the temperature increases rapidly to over 2500 K, causing the sample to be efficiently vaporized. The tube is aligned so that the vapor within will lie along the optical path of the source. The time that the vapor remains within the optical path, called the residence time, is substantially longer than in flames and, coupled with the efficiency of vaporization, gives rise to markedly increased sensitivity. This sensitivity, plus the capability to handle extremely small sample volumes, makes flameless atomic absorption spectroscopy one of the most valuable techniques for quantitation in biological samples (1).

After the sample has been atomized by a flame or a graphite furnace, an intense light source is passed through the vapor. Nearly all of the atoms in the vapor are in their electronic ground state according to the conditions of a Boltzmann distribution and can thus absorb the impinging radiation. The light which is unabsorbed by the vapor passes through a monochromator and then to a photomultiplier tube, as shown below. The intensity of this transmitted light, I, is electronically compared to the transmission of light in the absence of the sample, I_0, to obtain the absorbance due to the analyte.

Excitation sources used in atomic absorption spectroscopy are usually hollow cathode lamps or electrodeless discharge tubes, both of which produce high-intensity line excitation. Continuum sources, which emit a continuous level of energy over a large spectral region, are also used, though less frequently, The choice of the spectral source will affect the sensitivity and linearity of the analysis (5,30).

Line sources are capable of producing the best linear relationship between instrument response and concentration. For optimum linear response, the half-width of the source line used for excitation should be less than the half-width of the absorption line of the sample. This requirement is met by most line sources, since at temperatures of the flame, absorption lines of the sample undergo substantial Doppler and collisional broadening whereas the corresponding source lines remain narrow.

The hollow cathode is the most frequently used atomic absorption line source. A cupped cathode made of the element to be quantitated and a tungsten anode are positioned in a glass tube which is filled with an inert gas at reduced pressure. The end of the tube is sealed with an optically transparent quartz window. When an electrical potential is struck between the electrodes, the inert gas at the anode is ionized and moves toward the cathode. The element in the cup is "sputtered" into the gas and excited by the discharge to higher electronic states. The lamp emits intense lines due to resonance radiation. The emission will also show lines characteristic of the electrode itself as an impurity. When feasible, the electrode may be made of the element to be analyzed, thereby avoiding this possible interference. Lamps are available for over 60 different elements and are readily obtainable,

The electrodeless discharge tube is somewhat different in that the discharge is produced by microwave radiation. A few grams of the element in question are placed in a resonance cavity in a low-pressure argon atmosphere. The discharge excites the atoms within the cavity, producing high-purity emission lines uncontaminated by the emission lines of an electrode. These lamps emit more intense light than the hollow cathode lamps, but are successful for only a few elements. The electrodeless discharge is used whenever a high-purity spectral line source is desired and when a more intense emission source is required. With either type of source, a disadvantage of atomic absorption spectroscopy is that a different source lamp must be used for each element. A few hollow cathode lamps have been developed with multielement capability, but these are of limited utility.

A major difficulty encountered with atomic absorption techniques is the presence of incompletely absorbed background emission from the source and scattered light from the optical system. As the background becomes more intense relative to the absorption of the analyte, the precision of the measurement decreases dramatically. For this reason, several background correction techniques have been implemented. A commonly used method is the method of proximity, which was discussed in relation to inductively coupled plasma spectroscopy.

Another strategy involves alternating the line source with a continuum source such as a deuterium or tungsten lamp. The absorption of the continuum by the analyte is much less than that of the background absorption, due to the relative narrowness of the absorption line. The absorption of the continuum can then be electronically subtracted from the line-source absorption, thereby providing a background correction at the same wavelength as used for the measurement of the analyte. For proper operation it is necessary to ensure that the continuum and line sources have precisely the same optical path alignment and precisely balanced intensities at the wavelength of interest.

A powerful background correction technique involves the Zeeman effect (31,32). The atomized vapor is placed within a magnetic field and the light from the source is passed through a polarizer. The polarized light cannot be absorbed by the atomic vapor when the plane of polarization is perpendicular to the applied magnetic field. However, the absorption of background light will be independent of the orientation. By modulation of the magnetic field or by rotation of the polarizer, the intensity of the background light may be measured and electronically subtracted from the total measured intensity. The error of background correction s reduced from ~3% with continuum source correction to ~0.05% by the use of the Zeeman effect.

E. Atomic Fluorescence Spectroscopy

Atomic fluorescence spectroscopy (27,33,34) is based on the remission of light by atoms, in a vapor, which have been excited by a high-intensity light source. This fluorescence is omnidirectional and distinctive in energy of the emitting atoms. The instrumental design (Fig. 6) is similar to an atomic absorption spectrometer with the exception of the detector, which is placed at a right angle to the optical excitation path in order to avoid background interference from the source.

Line sources for atomic fluorescence spectroscopy can be the hollow cathode lamps or electrodeless discharge lamps discussed previously. The source should have the highest possible output intensity since, as in molecular fluorescence spectroscopy, the intensity of fluorescence is directly proportional to the intensity of the exciting light. High-intensity dye lasers have therefore been utilized to achieve greater sensitivities for many elements (34). Other elements which must be excited at wavelengths where laser efficiency is reduced are better determined using conventional sources.

Continuum sources can also be utilized advantageously in atomic fluorescence spectroscopy, since they provide the capability of exciting a large number of elements with a single source. The output intensities of these sources are significantly lower than those of line sources, and they should be used only when sensitivity is not critical in the analysis.

As in atomic absorption spectrometry, a flame is usually used as the analytical cell, although a graphite furnace may also be utilized. The interferences found in flames which were discussed in relation to flame emission, such as the formation of refractory oxides and other molecular species, will also affect atomic fluorescence and should be considered accordingly. Furthermore, scattering of the fluorescence within the flame is a serious source of background interference. The most important problem associated with flame atomic fluorescence is the quenching of fluorescence arising from collisional deactivation of the excited atoms in the vapor by other species in the flame. This effect is especially evident with nitrogen and hydrocarbons, and care should be given to minimize their presence in the flame. Replacement of the nitrogen in an air/hydrogen mixture with argon produces a low quenching flame.

Because the atomic fluorescence is measured at a right angle to the source, spectral interferences are minimal and a simple cutoff filter may often be used to isolate the emission line. The intensity of the fluorescence is directly proportional to the analyte concentration. As the analyte concentration within the flame becomes large, self-absorption of resonance fluorescence becomes significant, as it does in flame emission spectroscopy. Under these conditions, the linearity of the instrumental response breaks down and a calibration curve must be used or the analyte solutions diluted accordingly.

IV. APPLICATIONS

The analytical measurement of elemental concentrations is important for the analysis of the major and minor constituents of pharmaceutical products. The use of atomic spectroscopy in this regard has been the subject of several reviews (2,3,35,36). Metals are major constituents of several pharmaceuticals such as dialysis solutions, lithium carbonate tablets, antacids, and multivitamin–mineral tablets. For these substances, spectroscopic analysis is an important tool. It is indispensable for the determination of trace-metal impurities in pharmaceutical products and the qualitative and quantitative analysis of metals, essential and toxic, in biological fluids and tissues (37). Beyond this, several drugs which do not contain metallic components themselves may be analyzed indirectly by spectroscopic methods utilizing metal complexation or precipitation reactions.

The elements Na, K, Li, Mg, Ca, Al, and Zn are among the most common elements subjected to pharmaceutical analysis and coincidentally are also among the elements most readily determinable by flame emission. Although it is not as versatile as other methods, flame emission spectroscopy exhibits sensitivity greater than or approximately equal to that of flame absorption spectroscopy for the above elements (Table 2). Where precision of the analysis is critical, however, the analyst should consider the alternative of atomic absorption spectroscopy.

Sodium and potassium levels are difficult to analyze by titrimetric or colorimetric techniques but are among the elements most easily determined by atomic spectroscopy (2,38) (Table 2). Their analysis is important for the control of infusion and dialysis solutions, which must be carefully monitored to maintain proper electrolyte balance. Flame emission spectroscopy is the simplest and least expensive technique for this purpose, although the precision of the measurement may be improved by employing atomic absorption spectroscopy. Both methods are approved by the U.S. (39), British (40), and European (41) Pharmacopeias and are commonly utilized. Sensitivity is of no concern, due to the high concentrations in these solutions; furthermore, dilution of the sample is often necessary in order to reduce the metal concentrations to the range where linear instrumental response can be achieved. Fortunately, the analysis may be carried but without additional sample preparation because other components, such as dextrose, do not interfere.

Lithium carbonate tablets are also readily analyzed by flame emission spectroscopy after first dissolving them in mineral acids (2,42). Both lithium and sodium levels must be carefully controlled in these tablets. Clinically, lithium levels in serum should be maintained between 3.5 and 7 ppm and may be monitored easily and rapidly by either emission or absorption spectroscopy (43, 44).

Calcium and magnesium compounds are often added to dialysis and infusion solutions and are also used extensively, along with aluminum compounds, in antacid preparations. After dissolution of insoluble compounds in mineral acid, homogeneous solutions of these metals are easily determined by either emission or absorption spectroscopy (2,45–47). An investigation into aluminum contamination in infusion solutions for parenteral nutrition used both ICP and graphite furnace atomic absorption spectroscopy to determine the aluminum concentrations (48). Unfortunately, phosphates or other oxyanions which are often present in these products interfere by the formation of stable oxy-metal compounds in the aerosol, which are difficult to atomize in the flame. This complication can be eliminated simply by the excessive addition of chemical releasing agents which compete for the component ions of the oxy-metal compound. Thus, lanthanum or strontium salts which form stable oxy-metal compounds themselves are commonly added to free the analyte metal. Additionally, chelating agents such as ethylenediaminetetraacetic acid (EDTA) may be added to chelate the analyte metal and prevent the formation of these compounds. The metal may also be extracted into a nonaqueous solvent before the analysis. Where interferences are a major factor, the method of standard addition (24) is recommended for the improvement of accuracy.

Zinc is present in a number of pharmaceuticals, the most important of which is life-sustaining insulin. Many topical preparations contain zinc as the oxide, sulfate, or stearate as an astringent or antipruritic. Some foot powders contain the antifungicidal zinc undecenoate, and zinc pyrithione is used in antidandruff shampoos. After dissolving them in acid, the topical products can be easily analyzed by either atomic emission or atomic absorption spectroscopy (49), since they contain a relatively high concentration of zinc. However, atomic absorption is approximately four orders of magnitude more sensitive than atomic emission for the determination of zinc (Table 2) and offers superior precision for the analysis of injectable insulin (50), where zinc concentrations can be as low as 4 ppm (39).

Most other metals present in pharmaceuticals are present in sufficient concentrations that high sensitivity is not imperative and they may therefore be determined by flame atomic absorption spectroscopy. These products are extremely variable in composition but nonetheless yield easily to this type of analysis, which is generally unaffected by compounding agents such as binders or expanders. Thus, the elements Na, K, Mg, Ca, Mn, Fe, Co, Cu, Zn, and Mo are among those determinable by flame (51–53) and, recently, furnace (54) atomic absorption in multivitamin–mineral tablets. Chemical interactions between some metals dictate the use of an internal standard when several elements are present simultaneously. It should be noted here that a spark emission or ICP spectrometer equipped with an appropriate polychromator would have the advantage of simultaneous and therefore more rapid analysis in these multielemental products. These techniques have probably not been fully utilized in this regard.

A number of organomercurials (3,55,56) can be quantitated after digestion of the sample to release the mercury. Gold has been analyzed in the form of the antiarthritic sodium aurothiomalate (57). Antacid products containing Si, Ti, and Bi (47) and As in arsenamide (58) have all been determined by atomic absorption. Atomic absorption is also invaluable for the analysis of Pt in the antineoplastic drug, cis-diaminodichloroplatinum(II) (3): Although in most cases these metals are determinable by gravimetry, titrimetry, polarography, colorimetry, etc., atomic absorption spectroscopy is often more efficient of time and sample, more precise, and/or more accurate.

Investigational pharmaceutical techniques also have been observed using ICP spectroscopic techniques. Boron neutron capture therapy is such an example, where malignant cells can be preferentially loaded with ^{10}B and irradiated with thermal neutrons, thus killing the cancerous cells (59). This can be especially useful in inoperable cancers such as cranial tumors. A number of boron compounds have been observed in various biological fluids and tissue samples using ICP atomic emission spectroscopy to measure and determine the selectivity of ^{10}B-containing molecules as pharmaceuticals (60–65).

The determination of trace metal impurities in pharmaceuticals requires a more sensitive methodology. Flame atomic absorption and emission spectroscopy have been the major tools used for this purpose. Metal contaminants such as Pb, Sb, Bi, Ag, Ba, Ni, and Sr have been identified and quantitated by these methods (59,66–68). Specific analysis is necessary for the detection of the presence of palladium in semisynthetic penicillins, where it is used as a catalyst (57), and for silicon in streptomycin (69). Furnace atomic absorption may find a significant role in the determination of known impurities, due to higher sensitivity (Table 2). Atomic absorption is used to detect quantities of known toxic substances in the blood, such as lead (70–72). If the exact impurities are not known, qualitative as well as quantitative analysis is required, and a general multielemental method such as ICP spectrometry with a rapid-scanning monochromator may be utilized. Inductively coupled plasma atomic emission spectroscopy may also be used in the analysis of biological fluids in order to detect contamination by environmental metals such as mercury (73), and to test serum and tissues for the presence of aluminum, lead, cadmium, nickel, and other trace metals (74–77).

Perhaps the most difficult problem concerning atomic spectroscopy for the pharmaceutical analyst is the determination of metals at the parts-per-billion level in biological samples (25,37,78). Two basic prerequisites must be met for this type of analysis. The sensitivity of the method must be maximal and, by virtue of clinical limitations, the required sample size must be minimal for the analysis of minute metal concentrations.

The first requirement can be easily fulfilled by the preconcentration of the analyte before the analysis. Preconcentration has been applied to sample preparation for flame atomic absorption (25) and, more recently, for ICP (79,80) spectroscopy. However, preconcentration is not completely satisfactory, because of the increased analysis time (which may be critical in clinical analysis) and the increased chance of contamination or sample loss. Most important, however, a larger initial sample size is necessary. The apparent solution is a more sensitive technique. Table 2 lists concentrations of various metals in whole blood or serum (81,82) in comparison to limits of detection for the various atomic spectroscopy techniques. In many cases, especially for the toxic heavy metals, only flameless atomic absorption using a graphite furnace can provide the necessary sensitivity and accommodate a sample of only a few microliters (Table 1). The determination of therapeutic gold in urine and serum (83,84), chromium in serum (85), skin (86) and liver (87), copper in semen (88), arsenic in urine (89), manganese in animal tissues (90), and lead in blood (91) are but a few examples in analyses which have utilized the flameless atomic absorption technique.

An interesting application of atomic spectroscopy to pharmaceutical analysis is the indirect determination of organic drugs (2). Most often this involves reacting an analytical reagent with a drug which contains a specific functional group in such a way that the resultant compound or the excess analytical reagent can be measured by atomic spectroscopy. The determination of methamphetamine, for example, can be facilitated by the precipitation of the bismuth complex of the drug. A known quantity of $BiCl_3$ is added to a solution of methamphetamine, followed by a determination of the unreacted bismuth in solution by atomic absorption spectroscopy (92). Isoniazid has been determined by the extraction of its cupric chelate into methylisobutyl ketone and subsequent measurement of the copper content by aspiration of the copper complex directly into the flame of an atomic absorption spectrometer (93). Another application involves the determination of antihistamines containing a quaternary ammonium ion functional group at a 1 μM concentration (94,95). The antihistamine is first extracted by the formation of an ion pair with an excess of an ionic detergent such as sodium dioctylsulfosuccinate. The excess of the latter is then determined by the addition of copper, extraction of the copper complex into an organic phase, and determination of the copper content of the complex by atomic absorption spectrometry. Other drugs analyzed by indirect methods include alkaloids (96–99), barbiturates (100), benzylpenicillin (101), diazepines (102), noscapine (103), phenothiazines (104), and various compounds which contain ionic or covalent halogen (105), aldehyde (106), or diol (107,108) functional groups. Many of these determinations have been performed in the presence of possible interferents such as binders, salts, or other drugs.

The indirect analysis suffers from a basic problem in that while many sample impurities do not affect the analysis, the chemical reaction is not specific for a particular drug. Thus, drugs of the same structural classification, drug metabolites or decomposition products of a drug, will often react in the same manner as the drug of analytical interest and therefore lead to misleading results. If the purity of the drug can be established by an independent procedure, however, the indirect analysis of organic pharmaceutical products may be performed accurately and rapidly by indirect atomic absorption spectrometry.

REFERENCES

1. FW Sunderman Jr. In: DT Forman, RW Mattoon, eds. Clinical Chemistry. Washington, DC: American Chemical Society, 1975.
2. JHM Miller. Am Lab 10:41, 1978.
3. F Rousselet, F Thuillier. Prog Anal Atomic Spectrosc 1:353, 1979.
4. JD Winefordner, SG Schulman, TC O'Haver. Luminescence Spectroscopy in Analytical Chemistry. New York: Wiley-Interscience, 1972.

5. AP Thorne. Spectrophysics. London: Chapman & Hall, 1974.
6. EU Condon, GH Shortley. The Theory of Atomic Spectra. New York: Cambridge University Press, 1951.
7. C Sandorfy. Electronic Spectra and Quantum Chemistry. Englewood Cliffs, NJ: Prentice-Hall, 1964.
8. HG Kuhn. Atomic Spectra. New York: Longmans, 1969.
9. WR Hindmarsh. Atomic Spectra, with Selections from Early Papers in Spectroscopy. New York: Pergamon Press, 1967.
10. HN Russell, FA Saunders. The coupling of the momenta of electrons. Astrophys J 61:38, 1925.
11. KI Zil'bershtein. Spectrochemical Analysis of Pure Substances, transl. JH Dixon. New York: Crane Russak, 1977.
12. V Svoboda, I Kleinmann. Anal Chem 40:1534, 1968.
13. JW Robinson. Atomic Spectroscopy. New York: Marcel Dekker, 1990.
14. VA Fassel, RN Kniseley. Anal Chem 46: IIIOA, 1974.
15. CC Butler, RN Kniseley, VA Fassel. Anal Chem 47:825, 1975.
16. AF Ward. Am Lab 10:79, 1978.
17. VA Fassel. Science 202:183, 1978.
18. HL Kahn, SB Smith, RG Schleicher. Am Lab 11:65, 1979.
19. MA Floyd, VA Fassel, RK Winge, JM Katzenberger, AP D'Silva. Anal Chem 52: 431, 1980.
20. JA Dean. Flame Photometry. New York: McGraw-Hill, 1960.
21. R Maurodineaneau, H Boiteux. Flame Spectroscopy. New York: Wiley, 1965.
22. GD Christian, FJ Feldman. Appl Spectrosc 25:660, 1971.
23. N Omenetto, NN Hatch, LM Fraser, JD Winefordner. Spectrochim Acta 288:65, 1973.
24. DA Skoog, DM West. Fundamentals of Analytical Chemistry. 3rd ed. New York: Holt, Rinehart and Winston, 1976, p 577.
25. GD Christian, FJ Feldman. Atomic Absorption Spectroscopy: Applications in Agriculture, Biology and Medicine. New York: Wiley-Interscience, 1970.
26. J Ramirez-Munoz. Atomic-Absorption Spectroscopy. New York: Elsevier, 1968.
27. JD Winefordner, JJ Fitzgerald, N Omenetto. Appl Spectrosc 29:369, 1975.
28. M Slavin. Atomic Absorption Spectroscopy. 2nd ed. New York: Wiley, 1978.
29. BV L'vov. Atomic Absorption Spectrochemical Analysis, transl. JH Dixon. London: Adam Hilger, 1970.
30. JD Winefordner, V Svoboda, LJ Cline. CRC Crit Rev Anal Chem 1: 233, 1970.
31. T Hadeishi, RD McLaughlin. Science 174:404, 1971.
32. PR Liddell, KG Brodie. Anal Chem 52:1256, 1980.
33. V Sychra, V Svoboda, I Rubeska. Atomic Fluorescence Spectroscopy. London: Van Nostrand Reinhold, 1975.
34. SJ Weeks, H Haraguchi, JD Winefordner. Anal Chem 50:360, 1978.
35. RV Smith. Am Lab 5:27, 1973.
36. E Vauder Eeckhout, P DeMoerloose. Pharm Weekbl 106:749, 1971.
37. GH Morrison. CRC Crit Rev Anal Chem 8:287, 1979.
38. GF Hazebroncq. In: 3rd International Congress of Atomic Absorption and Atomic Fluorescence Spectrometry. London: Adam Hilger, 1971.

39. US Pharmacopeia. XXIV. Rockville, MD. The United States Pharmacopeial Convention, 2000.
40. British Pharmacopeia. London: Her Majesty's Stationery Office, 1998.
41. European Pharmacopeia. Maisonneuve, 1997.
42. J Sagel. Pharm Weekbl 111:897, 1976.
43. Remington's. The science and practice of pharmacy. 20th ed. Easton, PA: Mack Publishing, 2000.
44. R Nafissy. Acta Med Iran 19:82–88, 1977.
45. RV Smith, MA Nessen. J Pharm Sci 60:907, 1971.
46. BA Dalrymple, CT Kenner. J Pharm Sci 58:604, 1969.
47. A Stahlavska, K Propokova, M Tuzar. Pharmazie 29:140, 1974.
48. S Recknagel, P Brattler, A Chrissafidou, HJ Gramm, J Kotwas, U Rosick. Infusionsther-Transfusionsmed 21:266–273, 1994.
49. RR Moody, RB Taylor. J Pharm Pharmacol 24:848, 1972.
50. K Szivos, L Polos, L Bezur, E Pungor. Acta Pharm Hung 43:90, 1973.
51. YS Chae, JP Vacik, WH Shelver. J Pharm Sci 62:1838, 1973.
52. SA El-Kinawy, MI Walash, MS Abou-Bakr, and IZ Diaie. J Drug Res Egypt 7: 151, 1975.
53. Y Kidani, K Takeda, H Koike. Jpn Anal 22:719, 1973.
54. PO Kosonen, A Salonen, A Nieminen. Finn Chem Lett 33:136, 1978.
55. RD Thompson, TJ Hoffman. J Pharm Sci 64:1863, 1975.
56. IT Calder, JHM Miller. J Pharm Pharmacol 28:25P, 1976.
57. F Rousselet, V Courtois, ML Girard. Analysis 3:132, 1975.
58. JR Leaton. J Assoc Offic Anal Chem 53:237, 1970.
59. WH Thomas, YK Lee. Acta Pharm Suec 11:495, 1974.
60. W Porschen, J Marx, F Dallacker, H Muckter, T Bohmel, R Fairchild. Radiat Environ Biophys 26:209–218, 1987.
61. K Ishiwata, M Shiono, K Kubota, K Yoshino, J Hatazawa, T Ido, C Honda, M Ichihashi, Y Mishima. Melanoma Res 2:171–179, 1992.
62. SL Kraft, PR Gavin, CE DeHaan, CW Leathers, WF Bauer, DL Miller, RV Dorn III. Proc Natl Acad Sci USA 89:11973–11977, 1992.
63. V Gregoire, AC Begg, R Huiskamp, R Verrijk, H Bartilink. Radiotherm Oncon 27:46–54, 1993.
64. SL Kraft, PR Gavin, CW Leathers, CE DeHaan, WF Bauer, DL Miller, RV Dorn III, ML Griebenow. Cancer Res 54:1259–1263, 1994.
65. K Haselberger, H Radner, G Pendl. Cancer Res 54:6318–6320, 1994.
66. A Kovar, W Lantenshlaeget, R Seidel. Dt Apothz 115:1855, 1975.
67. J Pawlaczyk, M Makowska. Acta Polon Pharm 36:59, 1979.
68. J Mohay, M Veress, G Szasz. Magy Kem Foly 85:465, 1979.
68. RJ Hurtubise. J Pharm Sci 63:1128, 1974.
70. DI Bannon, C Murashchik, CR Zapf, MR Farfel, JJ Chisolm Jr. Clin Chem 40: 1730–1734, 1994.
71. B Berney. Milbank Q 71:3–39, 1993.
72. PL Ooi, KT Goh, BH Heng, CT Sam, KH Kong, U Rajan. Rev Environ Health 9: 207–213, 1991.
73. F Buneaux, A Buisine, S Bourdon, R Bourdon. J Anal Toxicol 16:99–101, 1992.

74. A Franzblau, L Rosenstock, DL Eaton. Environ Res 46:15–24, 1988.
75. J Moon, TJ Smith, S Tamaro, D Enarson, S Fadl, AJ Davidson, L Weldon. Sci Total Environ 54:107–125, 1986.
76. E Berman, In: EL Crove, ed. Applied Atomic Spectroscopy, Vol II. New York: Plenum Press, 1978.
77. TP Moyer, GV Mussman, DE Nixon. Clin Chem, 37:709–714, 1991.
78. DE Nixon, TP Moyer, P Johnoson, JT McCall, AB Ness, WH Fjerstad, MB Wehde. Clin Chem 32:1660–1665, 1986.
79. WJ Haas, VA Fassel, F Grabau IV, RN Kniseley, WL Sutherland. Adv Chem Ser 1972:91, 1979.
80. RM Barnes, JS Genna. Anal Chem 51:1065, 1979.
81. PL Altman, DS Dittmer, eds. Biology Data Book. Vol III. 2nd ed. Bethesda, MD: Federation of American Societies for Experimental Biology, 1974, pp 1751–1753.
82. J Versieck, R Cornelis. Anal Chim Acta 116:217, 1980.
83. JV Dunckley, FA Staynes. Ann Clin Biochem 14:53, 1977.
84. JV Dunckley, DM Grennan, DG Palmer. J Anal Toxicol 3:242, 1979.
85. E Gvaf-Harsanyi, FJ Langmyhr. Anal Chim Acta 116:105, 1980.
86. S Liden, E Lundberg. J Invest Dermatol 72:42, 1979.
87. SS Chao, EE Pickett. Anal Chem 52:335, 1980.
88. M Arroyo. Rev Clin Esp 151:211, 1978.
89. M Ishizaki, M Fujiki, S Yamaguchi. Sangyo Igaku 21:234, 1979.
90. DI Paynter. Anal Chem 51:2086, 1979.
91. G Carelli, V Rimatori, B Sperduto. Med Lav 70:313, 1979.
92. T Mitsui, Y Fuiimura, T Suzuki. Bunseki Kagaku 24:244, 1975.
93. Y Kidani, K Inagaki, T Saotome, H Koike. Bunseki Kagaku 22:896, 1973.
94. J Alary, J Rochat, A Villet, A Coeur. Ann Pharm Fr 34:345, 1976.
95. A Villet, J Alary, A Coeur. Bull Trav Soc Pharm Lyon 21:31, 1977.
96. SI Simon, DF Boltz. Microchem J 20:468, 1975.
97. T Mitsui, Y Fujimura. J Hyg Chem 21:183, 1975.
98. T Minamikawa, K Matsumura, A Kamei, M Yamakawa. Bunseki Kagaku 20:1011, 1971.
99. T Minamikawa, N Yamagishi. Bunseki Kagaku 22:1058, 1973.
100. T Mitsui, Y Fujimara. Bunseki Kagaku 24:575, 1975.
101. Y Kidani, K Nakamura, K Inagaki, H Koike. Bunseki Kagaku 24:575, 1975.
102. J Alary, A Villet, A Coeur. Ann Pharm Fr 34:419, 1976.
103. T Minamikawa, K Matsumura. Yakugaki Zasshi 96:440, 1976.
104. J Alary, A Villet, A Coeur. Ann Pharm Fr 35:439, 1977.
105. Y Kidani, H Takemura, H Koike. Bunseki Kagaku 22:187, 1973.
106. PJ Oles, S Siggia. Anal Chem 46:911, 1974.
107. PJ Oles, S Siggia. Anal Chem 46:2197, 1974.
108. B Tan, P Melius, MV Kilgore. Anal Chem 52:602, 1980.

9
Luminescence Spectroscopy

John H. Miyawa and Stephen G. Schulman
University of Florida, Gainesville, Florida

I. INTRODUCTION

Luminescence spectroscopy is an analytical method derived from the emission of light by molecules which have become electronically excited subsequent to the absorption of visible or ultraviolet radiation. Due to its high analytical sensitivity (concentrations of luminescing analytes $\sim 1 \times 10^{-9}$ moles/L are routinely determined), this technique is widely employed in the analysis of drugs and metabolites. These applications are derived from the relationships between analyte concentrations and luminescence intensities and are therefore similar in concept to most other physicochemical methods of analysis. Other features of luminescence spectral bands, such as position in the electromagnetic spectrum (wavelength or frequency), band form, emission lifetime, and excitation spectrum, are related to molecular structure and environment and therefore also have analytical value.

Luminescence spectroscopy may be divided into two major areas: fluorescence spectroscopy and phosphorescence spectroscopy. The differences between the two are based mostly on the time frames on which the phenomena of fluorescence and phosphorescence occur, phosphorescence decaying much more slowly (often taking several seconds) than fluorescence subsequent to excitation. Slight differences between the instrumentation used to observe fluorescence and that to observe phosphorescence take advantage of the temporal distinction between the two luminescence phenomena. Chemiluminescence is a form of fluorescence differing only in the fact that a chemical reaction as opposed to incident light generates the excited state.

The origin, nature, and measurement of molecular fluorescence, phosphorescence, and chemiluminescence, as well as the dependence of these luminescences on molecular structure, reactivity, and interactions with the environment, will be considered.

II. THE EMISSION OF LIGHT BY MOLECULES

A. Postabsorption Disposition of Excitation Energy

The electronic excitation of molecules occurs as the result of the absorption of near-ultraviolet or visible light. Absorption consists of the interaction of the electric field of the exciting light with the π or nonbonded electrons of the absorbing molecule. This interaction causes energy to be absorbed from the light wave, a process that occurs in $\sim 10^{-15}$ s. The intensity of light absorbed (I_a) is related to the Lambert-Beer Law by

$$I_a = I_0 - I = I_0(1 - 10^{-\varepsilon cl}) \tag{1}$$

where I_0, I, c, and l are, respectively, the intensities of light incident upon and transmitted through the sample, the molar concentration of absorber in the sample and the optical depth of the sample. ε is the molar absorptivity of the absorber at the nominal analytical wavelength. The energy of the light absorbed (per mole) is equal to the difference in energy, E, between the ground and excited states of the absorbing molecules and is given by the Planck frequency relation,

$$E = Nh\nu \tag{2}$$

where h is Planck's constant, ν is the frequency of absorbed light and N is Avogadro's number.

It should be noted that most molecules of interest to the pharmaceutical scientist, which will absorb near UV or visible light, are derived from benzene or naphthalene or related heterocycles and exhibit two or three absorption bands corresponding to excitation from the ground electronic state to two or three excited states.

Subsequent to excitation to an electronically excited singlet state, in which the molecule of interest may be vibrationally excited as well, the process of return to the ground electronic state begins. The loss of excess vibrational energy, known as vibrational or thermal relaxation, takes place in a stepwise fashion, each step taking 10^{-14} to 10^{-13} s and assisted by inelastic collisions with solvent molecules to which excess thermal energy is lost. Because of the overlap between higher and lower electronically excited states, there is also an efficient radiationless pathway for the demotion of the excited molecule from higher to lower electronically excited singlet states. This process is called internal conversion. In aliphatic molecules with a high degree of vibrational freedom, vibrational relaxation and internal conversion may return the excited molecule to the ground electronic state radiationlessly within 10^{-12} s after excitation.

Under this circumstance, fluorescence does not occur. However, in aromatic and a few other highly conjugated molecules, the degree of vibrational freedom is restricted, resulting in a very inefficient thermal mechanism of return to the ground state. In this case, the excited molecule may, as the result of internal conversion and vibrational relaxation, rapidly arrive in the lowest vibrational level of the lowest electronically excited singlet state. After the relatively long period of 10^{-11}–10^{-7} s, it may change spin and go from the singlet to the triplet state, or by emitting the difference in energy between the ground and lowest excited singlet states in the form of near-ultraviolet or visible fluorescence. The frequency of the fluorescent light is related to its energy by Eq. (2). However, because of vibrational relaxation in the excited state (Fig. 1) subsequent to absorption, and in the ground state subsequent to fluorescence, the energy of fluorescence is lower (and the wavelength of fluorescence therefore longer) than that of absorption to the lowest excited singlet state even though the same electronic states are involved in both transitions. Moreover, in very rigid molecules, such as aromatic hydrocarbons, the longest-wavelength absorption band and the fluorescence band will appear as mirror images of one another when plotted on an abscissa that is linear in frequency or energy (Fig. 2).

It is important, from the analytical point of view, to keep in mind that fluorescence almost always occurs from the lowest excited singlet state. This means that only one fluorescence band may be observed for any given molecule, even though it will usually have several absorption bands. Therefore, the observation of several fluorescence bands in a solution of a supposedly pure sample suggests either the occurrence of a chemical reaction or the presence of impurities.

If all molecules absorbing light emitted fluorescence, the observed intensity of fluorescence would be given by I_a in Eq. (1). However, several processes may compete with fluorescence for deactivation of the lowest excited singlet state. Consequently, the intensity of fluorescence I_f is obtained by multiplying I_a by the fraction of molecules φ_f which are deactivated by fluorescence. Hence, φ_f

$$I_f = \varphi_f I_a \tag{3}$$

is called the quantum yield of fluorescence or the fluorescence efficiency, is always unity or a fraction, and is related to the rates of fluorescence (k_f) and competitive deactivating processes (k_d) such as internal conversion and crossover to the triplet state,

$$\varphi_f = \frac{k_f}{k_f + \Sigma k_d} \tag{4}$$

The greater the numbers or rates of processes competing with fluorescence, the lower the value of φ_f and, hence, the lower the intensity of fluorescence. On most commercial instrumentation, φ_f must be greater than 1×10^{-4} for fluorescence to be observable. From Eqs. (1) and (3) it follows that

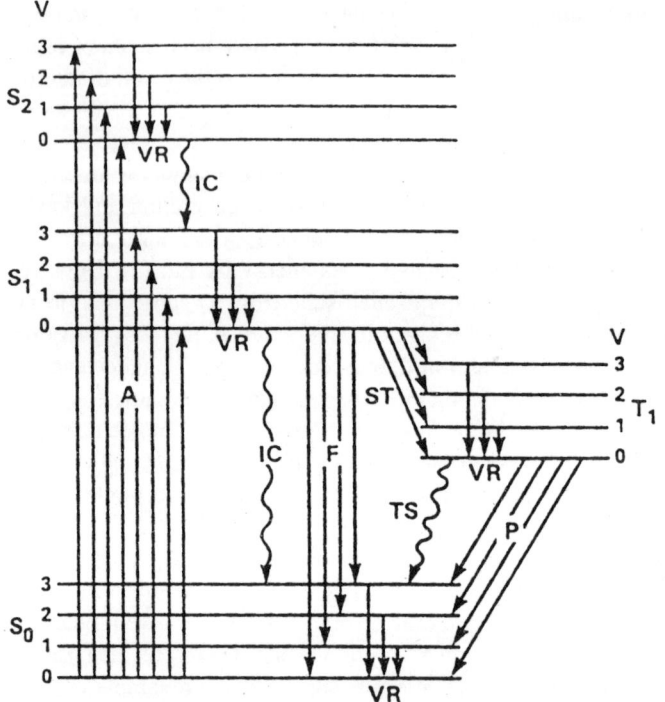

Fig. 1 Photophysical processes of conjugated molecules. Electronic absorption (A) from the lowest vibrational level ($v = 0$) of the ground state (S_0) to the various vibrational levels ($v = 0, 1, 2, 3$) of the excited singlet states (S_1 and S_2) is followed by rapid, radiationless internal conversion (IC) and vibrational relaxation (VR) to the lowest vibrational level ($v = 0$) of S_1. Competing for deactivation of the lowest excited singlet state S_1 are the radiationless internal conversion and singlet–triplet intersystem crossing (ST) as well as fluorescence (F). Fluorescence is followed by vibrational relaxation (VR) in the ground state. Intersystem crossing (ST) is followed by vibrational relaxation (VR) in the triplet state (T_1). Phosphorescence (P) and nonradiative triplet–singlet intersystem crossing (ST) return the molecule from the triplet state (T_1) to the ground state (S_0). Vibrational relaxation in S_0 then thermalizes the "hot" ground-state molecule.

$$I_f = \varphi_f I_0 \left(1 - 10^{-\varepsilon cl}\right) \tag{5}$$

indicating that fluorescence intensity is not linear but rather exponential in its variation with absorber concentration. However, Eq. (5) may be expanded, in series, to

$$\varphi_f I_0 \left(1 - 10^{-\varepsilon cl}\right) = \left(2.3\varepsilon cl - \frac{(2.3\varepsilon cl)^2}{2!} - \frac{(2.3\varepsilon cl)^3}{3!} - \ldots\right)\varphi_f I_c \tag{6}$$

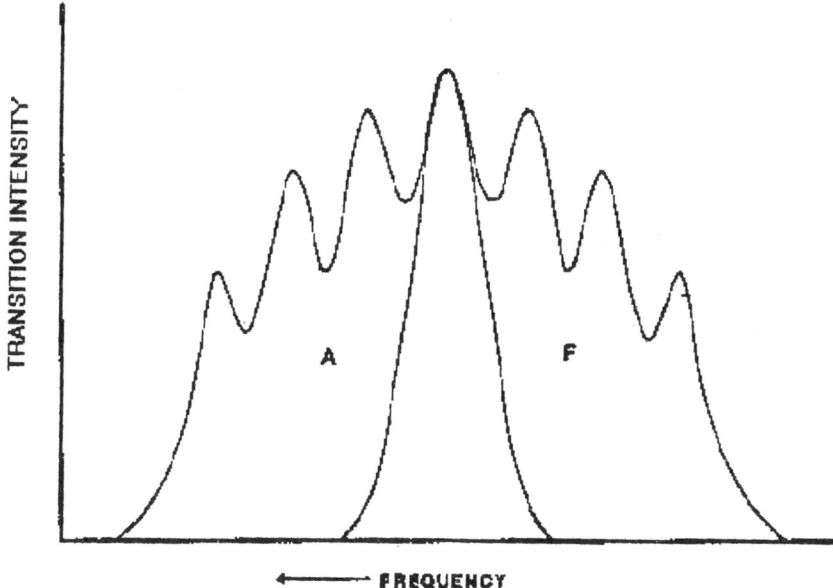

Fig. 2 Illustrating the mirror image relationship between absorption (A) and fluorescence (F), extant when the vibrational spacing are identical in ground and electronically excited states.

For values of $\varepsilon cl < 0.2$, the higher terms in Eq. (6) may be neglected (to within 2.3% error) by comparison with the first, so that

$$I_f = 2.3 \; \varphi_f I_0 \varepsilon cl \tag{7}$$

Thus, at very low absorbance I_f is linear with the analyte concentration and under constant experimental conditions Eq. (8) may be employed analytically in the relative form

$$\frac{I_{fu}}{I_{fs}} = \frac{C_u}{C_s} \tag{8}$$

where the subscripts u and s refer to unknown and standard solutions, respectively. However, it must be borne in mind that at higher absorbances Eq. (7) breaks down and correction for higher terms in Eq. (6) or the preparation of a calibration curve must be employed for analytical purposes.

Another important property of fluorescing molecules is the lifetime of the lowest excited singlet state (τ). If the mean rate of fluorescence is the number of fluorescence events per unit time, the mean lifetime of the excited state is the reciprocal rate, or the mean time per fluorescent event. The greater the number and the faster the processes that compete with fluorescence for deactivation of the lowest excited singlet state, the shorter will be τ. Fluorescence lifetime measurement is a valuable technique in the analysis of multicomponent samples containing analytes with overlapping fluorescence bands.

Among the processes which compete with fluorescence is intersystem crossing from the lowest excited singlet to the lowest excited triplet state. Intersystem crossing is a radiationless process which results in a change in the spin angular momentum of the excited molecule. Although it diminishes the quantum yield of fluorescence, it results in population of the lowest triplet state. The return of molecules from the lowest triplet state to the ground state may be accompanied by a long-duration light emission known as phosphorescence.

Phosphorescence, like fluorescence, is most often observed in molecules having rigid molecular skeletons and having large energy separations between the lowest excited triplet state and the ground singlet state (i.e., aromatic molecules). Because triplet states are long-lived (10^{-4}–10 s), chemical and physical processes in solution compete effectively with phosphorescence for deactivation of the lowest excited triplet state. Except for the shortest-lived phosphorescences, collisional deactivation by solvent molecules, quenching by paramagnetic species (e.g., oxygen), photochemical reactions, and certain other processes preclude the observation of phosphorescence in fluid media. Rather, phosphorescence is normally studied in glasses, at liquid-nitrogen temperature, or, more recently, with analytes adsorbed on filter paper or other solid substrates, at ambient temperatures, where collisional processes cannot completely deactivate the triplet state.

Because fluorescence normally originates only from the lowest excited singlet state and phosphorescence only from the lowest triplet state, only one fluorescence band and one phosphorescence band may be observed from any given molecular species. Since phosphorescence is precluded in fluid solutions, only a single emission band, that due to fluorescence, may be observed for a single molecular species in solution. However, when fluorescence and phosphorescence are observed for the same molecule, the band occurring at longest wavelengths will be the phosphorescence band because the lowest triplet state always lies lower in energy than the lowest excited singlet state. Phosphorescence is usually distinguished from fluorescence by taking advantage of the differences in the mean lifetimes of the two processes. For very-long-lived phosphorescences (>0.1 s), electronic chopping devices are employed to distinguish between fluorescence and phosphorescence.

Phosphorescence is analytically useful in much the same way as fluorescence. This is especially so in the study of metabolites that phosphoresce but do not fluoresce. For example, p-nitrophenol, one of the metabolites of parathion, does not fluoresce, yet it can be detected and determined by its intense phosphorescence at concentrations down to about 10^{-10}. However, where both luminescences can be employed analytically, fluorescence is usually the method of choice, because it is not restricted to rigid media. The intensity of phosphorescence I_p, for low sample absorbance ($\varepsilon cl < 0.02$) is

$$I_p = 2.3\ \varphi_{st}\varphi_p\ I_0\ \varepsilon cl \tag{9}$$

where φ_p and φ_{st} are, respectively, the quantum yields of phosphorescence and intersystem crossing.

B. Chemical Structural Effects on Luminescence Spectra

Most luminescence spectra of pharmaceutical interest arise from functionally substituted aromatic molecules. Consequently, the compounds of interest in this chapter are those derived from drugs possessing aromatic rings, such as benzene, naphthalene, or anthracene, or their heteroaromatic analogs pyridine, quinoline, acridine, etc. As is the case for absorption spectra, the luminescence spectra of these substances may often be understood in terms of the electronic interactions between the simple aromatic structures and their substituents.

Understanding of the spectra–structure relationships can be valuable in assessing the feasibility of a luminescence method for a given compound. In this section will be considered the influence of exocyclic and heterocyclic substituents, in aromatic rings, on the intensities and positions of the luminescence spectra of aromatic molecules.

1. Chemical Structure and Luminescence Intensity

The intensity of fluorescence observable from a given molecular species depends on the molar absorptivity of the transition excited and the quantum yield of fluorescence. The molar absorptivity determines the number of molecules ultimately populating the lowest excited singlet state. For most aromatic molecules, the π, π^* absorption bands lying in the near-ultraviolet and visible regions of the spectrum have molar absorptivities of 1000–10,000, so that the proper choice of the transition to excite can affect the intensity of fluorescence by about one order of magnitude.

Far more important is the quantum yield of fluorescence, which may affect the intensity of fluorescence over about four orders of magnitude and may determine whether fluorescence is observable at all. The quantum yield of fluorescence is dependent on the rates of processes competing with fluorescence for the deactivation of the lowest excited singlet state.

The current state of spectroscopic theory does not permit the quantitative prediction of how efficiently nonradiative deactivation processes compete with fluorescence from molecule to molecule. However, some qualitative generalizations are possible.

Aromatic molecules containing lengthy aliphatic side chains generally tend to fluoresce less intensely than those without side chains. This is brought about by the introduction of a large number of vibrational degrees of freedom by the aliphatic moieties.

The greater vibrational freedom introduces vibrational coupling between the ground and lowest excited singlet states and thus reduces the quantum yield of fluorescence by providing an efficient pathway for internal conversion. In unsubstituted aromatic molecules, the rigidity of the aromatic ring results in wide separation of the ground and lowest excited singlet states. In general, molecular rigidity and high quantum yield of fluorescence are closely related to one another. These arguments are also valid for phosphorescence. Rigid molecules tend to phosphoresce more intensely.

The fluorescence of aromatic molecules is quenched (diminished in intensity) partially or completely by heavy-atom substituents such as $-As(OH)_2$, Br, and I, and by certain other groups such as $-NH_2$, $-CHO$, $-COR$, and nitrogen in six-membered heterocyclic rings (e.g., quinoline). Each of these substituents has the ability to cause mixing of the spin and orbital electronic motions of the aromatic system. Spin–orbital coupling destroys the concept of molecular spin as a well-defined property of the molecule and thereby enhances the probability or rate of singlet $\rightarrow$ triplet intersystem crossing.

This process favors population of the lowest triplet state at the expense of the lowest excited singlet state and thus decreases the fluorescence quantum yield. However, the efficient population of the lowest triplet state favors a high yield of phosphorescence. Consequently, aromatic arsenates, nitro-compounds, bromo- and iodo-derivatives, aldehydes, ketones, and N-heterocyclics tend to fluoresce very weakly or not at all. However, most of them phosphoresce quite intensely. In the heavy atom-substituted aromatics, spin–orbital coupling results from the high nuclear charge of the heavy atom substituent, which causes the π electrons of the aromatic system to spend more time, on the average, near the heavy nuclei. This situation causes juxtaposition of the spin and orbital angular momentum vectors at the heavy nucleus and thereby favors their coupling. In the aldehyde, ketone, and N-heterocyclic molecules, however, the phenomenon which causes high electron density at an atomic nucleus, and therefore spin–orbital coupling, is somewhat different. In these molecules, the oxygen atoms of the carbonyl groups and the heterocyclic nitrogen atoms have nonbonded electron pairs in orbitals which are sp^2 hybridized. The promotion of a nonbonded electron to a vacant π^* orbital usually occurs at lower energy and with lower molar absorptivity than the lowest-energy $\pi \rightarrow \pi^*$ transition. Hence, in molecules having nonbonded electrons, the n, π^* state is the lowest excited singlet state.

The greater degree of s character in the sp^2 hybridized nonbonding orbital compared to the $p\pi$ orbitals of the aromatic system results in a higher degree of spin–orbital coupling in molecules having nonbonded electrons than in those that do not. These factors combine to result in a high efficiency of intersystem crossing in molecules having functional groups with nonbonded electrons. Hence, these molecules tend to fluoresce weakly and phosphoresce strongly. A rather interesting distinction between molecules having heavy-atom substituents and those having nonbonded electrons is that, upon going to a highly polar or hydrogen-bonding solvent from a weakly interacting solvent, the quantum yield of fluorescence of the latter group of compounds increases and the quantum yield of phosphorescence decreases. This is due to the binding of the nonbonded electrons, which stabilizes them and causes the lowest π, π^* state to drop below the n, π^* state. This diminishes the role of the n, π^* state in populating the triplet. Strongly interacting solvents do not have a dramatic effect on the fluorescence yields of heavy-atom-substituted aromatic molecules.

2. Chemical Structure and Position of the Luminescence Maximum

The energies of the ground and excited states of fluorescing and phosphorescing molecules are affected by molecular structure. This is reflected in the positions (energies) of the fluorescence and phosphorescence maxima.

According to Eq. (2), the greater the separation between the ground and excited states, the greater will be the frequency and the shorter will be the wavelength of luminescence. This separation depends on the energy difference between the highest occupied and lowest unoccupied molecular orbitals and the repulsion energy between the electronic configurations corresponding to the ground and excited states. It should be recalled that these factors and Eq. (2) also defined the effects of molecular structure on electronic absorption spectra. It will suffice to say that the discussions of structural effects on absorption maxima are also applicable to fluorescence and phosphorescence maxima and need not be repeated here.

C. Environmental Effects on Luminescence Spectra

1. Solvent Effects

The solvents in which fluorescence spectra are observed play a major role in determining the spectral positions and intensities with which fluorescence bands occur. In some cases, the solvent may determine whether or not fluorescence is observed at all. The effects of the solvent on the fluorescence spectra are determined by the nature and degree of the interactions of the solvent molecules with the ground and lowest electronically excited singlet state of the fluorescing solute molecules. The influence of the solvent on the appearance of phosphorescence is much smaller than that on fluorescence.

Solvent interactions with solute molecules are electrostatic in nature and may be classified as dipolar or hydrogen bonding. The position of the fluorescence band maximum in one solvent, relative to that in another, depends on the relative separations between the ground and excited states in either solvent and therefore the relative strengths of ground- and excited-state solvent stabilization.

This subject has already been discussed at length for absorption spectra, and the relationships between solvent polarity and luminescence maxima are qualitatively similar. The influences of solvent hydrogen-bonding and polar properties on luminescence intensities, however, are rather different from those on absorption spectra.

Hydrogen bonding in the lowest excited singlet state occasionally results in the loss of fluorescence intensity in molecules whose lowest excited singlet states are of the intramolecular charge transfer type. This is observed as a decrease in fluorescence quantum yield upon going from hydrocarbon to hydrogen-bonding solvents. Many aryl amines and phenolic compounds demonstrate this behavior, which is due to internal conversion enhanced by coupling of the vibrations of the molecule of interest to those of the solvent. It can be avoided as an analytical interference by making fluorescence measurements in non-hydrogen-bonding media.

Molecules having lowest excited singlet states of the n, π^* type phosphoresce, but they rarely fluoresce in hydrocarbon solvents (aprotic, nonpolar) because the n, π^* singlet state is efficiently deactivated by intersystem crossing. However, in polar, hydrogen-bonding solvents, such as ethanol or water, these molecules become fluorescent and their phosphorescence efficiencies decrease. This results from the stabilization, relative to the ground state, of the lowest singlet π, π^* state, by dipole–dipole interaction and the destabilization of the lowest singlet n, π^* state by hydrogen-bonded interaction. If both of these interactions are sufficiently strong, the π, π^* state drops below the n, π^* state in the strongly solvated molecule, thereby permitting intense fluorescence. Quinoline and 1-naphthaldehyde, for example, do not fluoresce in cyclohexane but do so in water.

Solvents containing atoms of high atomic number (e.g., alkyl iodides) also have a pronounced effect on the intensity of fluorescence of solute molecules. However, this effect is not directly related to the polarity or hydrogen-bonding properties of the "heavy-atom solvent." Atoms of high atomic number in the solvent cage of the solute molecule enhance spin–orbital coupling between the lowest excited singlet state and the lowest triplet state of the solute. This favors the radiationless population of the lowest triplet state at the expense of the lowest excited singlet state. Thus, in "heavy-atom solvents," fluorescence is usually less intense than in solvents of low molecular weight. In frozen solutions the "heavy atom effect" favors the enhancement of phosphorescence intensity. Thus, in analytical practice, heavy-atom solvents are to be avoided in fluorometry and to be desired in phosphorimetry.

2. The Influence of pH

The influences of pH on luminescence spectra are derived from the dissociation of acidic functional groups or the protonation of basic functional groups, associated with the aromatic portions of fluorescing molecules.

The spectral shifts observed when basic groups are protonated or when acidic groups are dissociated are greater in magnitude, but qualitatively similar in direction, to the shifts resulting from interaction of lone or nonbonded electron pairs with hydrogen-bond donor solvents or from interaction of acidic hydrogen atoms with hydrogen bond acceptor solvents, respectively. Thus, the protonation of electron-withdrawing groups such as carboxyl, carbonyl, and pyridinic nitrogen results in shifts of the luminescence spectra to longer wavelengths, while the protonation of electron-donating groups such as the amino group produces spectral shifts to shorter wavelengths. The protolytic dissociation of electron-donating groups such as hydroxyl, sulfhydryl, or pyrrolic nitrogen produces spectral shifts to longer wavelengths, while the dissociation of electron-withdrawing groups such as carboxyl produces shifting of the fluorescence and phosphorescence spectra to shorter wavelengths.

In some molecules, notably the carbonyl derivatives of benzene, the presence of nonbonded electrons obviates the occurrence of fluorescence even in strongly hydrogen-bonding solvents such as water. However, protonation of the functional group possessing the nonbonded electron pair, or dissociation of electron-donor groups in the molecule, raises the n, π^* lowest excited singlet state above the lowest excited singlet π, π^* state and thereby allows fluorescence to occur. Benzaldehyde, for example, does not fluoresce as the neutral molecule but does so, moderately intensely, as the cation, in concentrated sulfuric acid. Tranquilizers derived from butyrophenone (e.g., haloperidol) fluoresce in their protonated forms in concentrated sulfuric acid. However, these compounds are nonfluorescent in aqueous media.

One of the more interesting aspects of acid–base reactions of fluorescent or potentially fluorescent molecules is derived from the occurrence of protonation and dissociation during the lifetime of the lowest excited singlet state, a classical and often-observed example of photochemistry originating from the lowest excited singlet state.

The lifetimes of molecules in the lowest excited singlet state are typically of the order of 10^{-11}–10^{-7} s. Typical rates of proton transfer reactions are 10^{11} s^{-1} or less. Consequently, excited-state proton transfer may be much slower, much faster, or competitive with radiative deactivation of the excited molecules.

If excited-state proton transfer is much slower from fluorescence, the fluorescence intensity-versus-pH curve will be exactly like the absorbance-versus-pH curve (i.e., no excited-state proton transfer occurs). If excited-state proton transfer is much faster than fluorescence, the fluorescence intensity-versus-pH curve will reflect the acid–base equilibrium in the lowest excited singlet state and the dissociation constant taken from the fluorescence intensity-versus-pH curve will be that of the excited-state reaction. Equilibrium in the excited state is a rare phenomenon and will not be dealt with further here.

If the rate of proton transfer in the excited state is comparable to the rates of deactivation of acid and conjugate base by fluorescence, the variations of the fluorescence intensities of acid and conjugate base with pH will be governed by the kinetics of the excited state proton-transfer reactions. In general, the pH region over which the emissions of both conjugate acid and conjugate base are observed will be much wider (Fig. 3) than, say, $pK_a -2 < pH < pK_a + 2$ because in some cases it will be possible to excite the conjugate acid and in others the conjugate base exclusively, and yet in both cases see emission from the conjugate acid and base. This obviously represents a potential interference in analytical fluorimetry. It is worth noting that buffer ions act as proton donors and acceptors with excited, potentially fluorescent molecules. In solutions containing high concentrations of buffer ions, the latter may induce excited-state proton transfer in molecules which would not ordinarily enter into this process in water. Consequently, the intelligent application of fluorimetry in buffered aqueous solutions requires the use of very dilute buffers and, therefore, a compromise between optimal buffer capacity and fluorimetric sensitivity.

Because phosphorescence is normally observed only in rigid media, proton transfer in the lowest triplet state is rarely observed, in real time, in the pH dependence of the phosphorescence spectrum.

3. The Influence of High Solute Concentrations

The discussion of luminescence has, up to the present, been based on the properties of dilute solutions in which the analyte molecules were presumed not to interact with one another. It has already been established that at high absorbance at the wavelength of excitation, deviations from linearity of the fluorescence intensity-versus-concentration relationship may occur because of the exponential variation of luminescence intensity with concentration. However, over a wide range of solute concentrations, solute–solute interactions may also account for loss of luminescence intensity with increasing solute concentration.

Several types of excited-state solute–solute interaction are common. The aggregation of excited solute molecules with ground-state molecules of the same type produces an excited polymer or "excimer," which, by virtue of the coupling of the aromatic systems of the excited and unexcited molecules, may either not

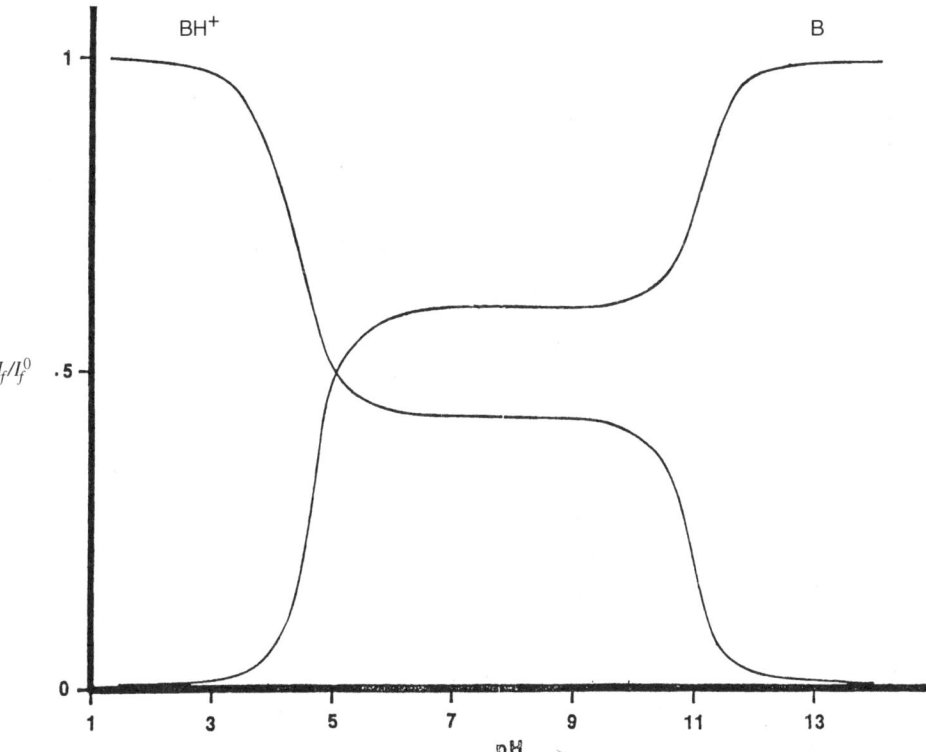

Fig. 3 pH dependences of the relative fluorescence intensities (I_f/I_f^0) of a base (B) and its conjugate acid (BH+) resulting from B becoming a stronger base in the lowest excited singlet state ($pK_{BH}+ = 11.5$) but having insufficient time, prior to the fluorescence of B, for complete protonation. Acid–base equilibrium is therefore not truly attained during the lifetime of the lowest excited singlet state.

luminesce or may luminesce at lower frequency than the monomeric excited molecule. The excimer fluorescence occurs at the expense of monomer fluorescence and, if not anticipated, can cause serious errors in fluorimetric analysis. Because excimer fluorescence takes place in the excited state, it is demonstrable in the fluorescence spectrum. However, after fluorescence, the deactivated polymer rapidly decomposes. Hence, the absorption spectrum does not reflect the presence of the excited-state complexes. Occasionally, excited-state complex formation may occur between two different solute molecules. The phenomenon is comparable to excimer formation; however, the term "exciplex" has been coined to described a heteropolymeric excited-state complex. Excimer and exciplex formation are usually observed only in fluid solution because diffusion of the excited species is necessary to form the excited complexes. One concentration effect that is observed in molecules in fluid or rigid media is resonance energy transfer. Energy transfer entails the excitation of a molecule which, during the lifetime of the excited state, passes its excitation energy off to a nearby molecule. This process falls off in probability as the inverse sixth power of the distance between donor and acceptor and can occur between molecules which are separated by up to 10 nm. If the energy donor is a chemical species distinct from the energy acceptor, and is of analytical interest, the analytical interference due to loss of luminescence intensity from the former species should be obvious. Because the mean distance between molecules decreases with increasing concentration, energy transfer is favored by increasing concentration of the acceptor. For energy transfer to occur between two dissimilar molecules, the fluorescence spectrum of the energy donor must overlap the absorption spectrum of the energy acceptor.

At very high absorber concentrations, ground-state aggregation effects may come into play and may ultimately result in the decrease of luminescence intensity with increasing absorber concentration. Moreover, when very highly absorbing solutions are studied, the loss in fluorescence intensity with increasing absorber concentration may also be due to absorption of all of the exciting light before it completely traverses the sample. Fluorescence is then observed from only part of the sample cell. If the absorption spectrum and the fluorescence spectrum of the solute overlap, diminution of fluorescence intensity may result from the reabsorption of part of the emitted radiation. This is called trivial reabsorption.

Fluorescence may be quenched (diminished in intensity or eliminated) due to the deactivation of the lowest excited singlet state of the analyte by interaction with other species in solution. The mechanisms of quenching appear to entail internal conversion, intersystem crossing, electron transfer, and photodissociation as modes of deactivation of the excited analyte–quencher complexes.

Quenching processes may be divided into two broad categories. In dynamic or diffusional quenching, interaction between the quencher and the potentially fluorescent molecule takes place during the lifetime of the excited state. As a result, the efficiency of dynamic quenching is governed by the ratio constant of the quenching reaction, the lifetime of the excited state of the potential fluorescer, and the concentration of the quenching species. Interaction between quencher and excited analyte results in the formation of a transient excited complex which is nonfluorescent and may be deactivated by any of the usual radiationless modes of deactivation of excited singlet states. Because interaction occurs only after excitation of the potentially fluorescing molecule, the presence of the quenching species has no effect on the absorption spectrum of the analyte. Many aromatic molecules—for example, the aminobenzoic and hydroxybenzoic acids and their metabolites—are dynamically quenched by halide ions such as Cl^- and Br^-.

Static quenching is characterized by complexation in the ground state between the quenching species and the molecule, which, when alone excited, should eventually become a potential fluorescing molecule. The complex is generally not fluorescent, and although it may dissociate in the excited state to release the fluorescing species, this is, at most, only partially effective in producing fluorescers because radiationless conversion to the ground state may be much faster than photodissociation. As a result, the ground-state reaction diminishes the intensity of fluorescence of the potentially fluorescent species. The quenching of the fluorescence of doxorubicin by complexation with double-helical DNA is an example of static quenching.

The dependence of the relative quantum yield of fluorescence (ϕ/ϕ_0) on quencher concentration [Q] may be derived from steady-state kinetics and is

$$\frac{\phi}{\phi_0} = \frac{1 - \alpha}{1 + k_D \tau [Q]} \tag{10}$$

where τ is the mean lifetime of the excited state in the absence of quenching (i.e., when [Q] = 0), k_D is the rate constant of complexation of the potential fluorescing species by the quencher Q in the excited state, and α is the fraction of complex, formed in the ground state, that is excited by direct absorption.

Although quenching of fluorescence is most often regarded as an analytical interference, it can also form the basis for analytical methodology. This approach makes use of the fact that if the intense fluorescence of a reference compound is statically quenched by quantitative reaction with a nonfluorescent analyte of interest, the difference between the fluorescence intensities of the reference compound before and after quenching by the analyte is proportional to the concentration of the analyte. Although this technique has not been employed to the extent that direct fluorimetry has, it is nearly as sensitive as the direct fluorimetry of the reference fluorophore. Moreover, it is as sensitive as most fluorimetric methods in which a fluorescing derivative of the analyte must be generated chemically. Dynamic fluorescence quenching is not particularly useful from the point of view of trace analysis, because the occurrence of quenching during the lifetime of the lowest excited singlet state requires that quencher concentrations be $> 10^{-4}$ M.

III. INSTRUMENTATION

Fluorescence spectrophotometers consist of a light source, to provide the excitation energy, a device for selecting the excitation wavelength, a sample compartment, a second device for selecting the emission wavelength, a photodetector, and a data acquisition and recording device to determine the intensity of fluorescence at any given wavelength (Fig. 4).

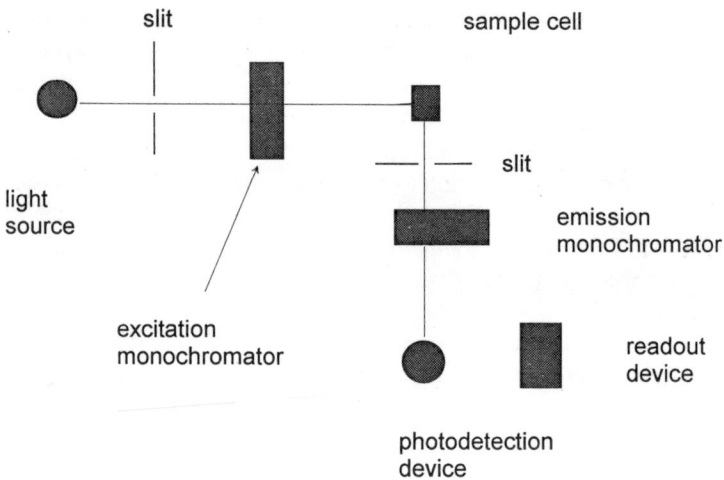

Fig. 4   Schematic diagram of a spectrofluorimeter.

A.   Light Sources

Gas discharge lamps are the most commonly used light sources. These lamps consist of two electrodes in a carrier gas at high pressure, across which a high potential difference is applied. The carrier gas employed is frequently the inert gas xenon. The high potential difference ionizes the carrier-gas atoms, which then accelerate toward the cathode, generating, through collisions, other excited states which emit radiation upon returning to their ground states. The high pressure enhances the probability of intermolecular collisions, broadening the energy distribution of the excited gas atoms and, in the process, transforming an otherwise sharp atomic line spectrum into a continuum. This continuum, however, is not uniform in intensity and falls off sharply below 300 nm. The lamps, however, do not provide a stable light output profile with time. This profile often changes as a result of changes in the discharge conditions, variations in the power supply, and sputtering of electrode material onto the quartz envelope.

The high pressure mercury vapor lamp has also found use as a light source. It is characterized by the presence of high-intensity emission lines at 253.7, 296.5, 302.2, 312.2, 313.3, 365.0, 404.1, 435.8, 516.1, 577.0, and 579.0 nm. The mercury vapor lamp has a more stable light output profile than the xenon arc lamp, generates intense line emissions, and is less expensive than the xenon arc lamp. Mercury vapor lamps, however, emit little light at wavelengths other than those of the mercury line emission spectrum, limiting the choice of useful excitation wavelengths. In the less expensive fluorimeters a low-pressure mercury vapor lamp is commonly employed because some of the lines are sufficiently intense to excite samples.

Incandescent lamps have also been used as light sources in fluorimeters. Tungsten wire filaments are typically employed, in which the metal wire is electrically heated in an inert gas atmosphere. As a result of the heating, the hot tungsten filament emits light, which is usually in the visible region of the electromagnetic spectrum. This limited range of emission, coupled with the relatively lower intensity of emission, limits the usefulness of incandescent lamps. The lamps are stable and inexpensive, but are useful only for the excitation of molecules which absorb light in the visible range.

Tunable dye lasers have also found use as excitation sources. Their ability to generate high-intensity, narrow-line-width electromagnetic radiation has proved desirable. However, this advantage is offset by their high cost and complexity, as well as by the fact that high excitation energy enhances light scatter, photodecomposition, and heat generation within the sample cell. The latter disadvantage has, as a result, limited lasers to principal use as light sources for the analysis of inorganics. They are, however, also useful as line sources in fluorescence detectors for flowing streams.

B. Wavelength Selection Devices

The wavelength selection devices employed in fluorimetry are either filters or grating monochromators. The filters are either absorption or interference based. Absorption filters are constructed of either tinted glass or gelatin containing dyes sandwiched between glass. These filters absorb electromagnetic radiation above or below a cutoff wavelength. Interference filters, on the other hand, consist of two partially transmitting films of silver separated by a transparent spacer film. The filters transmit light of wavelength corresponding to the optical separation of the spacer film as well as integral multiples of this principal wavelength. All other wavelengths are eliminated by destructive interference.

The two types of filters are relatively inexpensive, and collectively the filters provide a narrow to broad (approx 1–100 nm) bandpass with peak transmissions of up to 90%. However, the filters are not very useful in scanning instruments.

Monochromators consist of an entrance slit, a dispersion device, and an exit slit. The dispersion device is usually a diffraction grating, though a prism may still be used in the older devices. The gratings are preferred, as they are less expensive, have uniform resolution, a linear dispersion throughout the ultraviolet–visible range, and a better light flux throughput than the prisms. The gratings are either transmission or reflectance based. Transmission gratings have a large number (600–1200 lines/mm) of parallel transparent and opaque lines arranged alternately on the grating. Monochromatic light impingent on the incident plane is transmitted through the closely spaced transparent lines, generating an interference pattern. The reflectance grating is similar to the transmission grating except that the grooves are on a reflective rather than a transparent surface. Generally, the greater the number of lines per millimeter on the grating, the higher is the resolution and the better the separation of polychromatic light into its component wavelengths. A drawback of gratings is that several orders of spectra, which contribute to spectral interferences, are obtained. The higher orders of spectra can be reduced, however, by the use of filters in the optical path.

The types of slits and their widths also influence the resolution of fluorescence spectrophotometers. The slits serve to focus incident light onto the dispersion devices and the photodetectors. The slits may be fixed or adjustable, the latter being more commonly used despite the fact that they are more expensive. Adjustable slits may be unilateral or bilateral, depending on whether only one or both of two beveled blades are moved with respect to each other in adjusting slit width. Bilateral slits are preferred over unilateral slits, as they maintain a constant line upon changing slit width. As a rule, the smaller the slit width, the better is the resolution and, hence, the analytical selectivity. However, this is associated with a decrease in intensity of transmitted light and, as a result, sensitivity. Conversely, the larger the slit width, the better is the sensitivity and the poorer the resolution, which may be significant in multicomponent analysis. There are usually two monochromators in fluorescence spectrophotometers, an excitation monochromator and an emission monochromator.

C. Sample Compartment

The sample compartment has its internal surfaces painted a flat black and is covered during measurement to minimize stray light. The sample compartment is normally positioned so that the excitation and emission monochromators are at right angles to each other, so as to minimize interference from stray excitation light. However, for strongly absorbing solutions or solid analytes, a frontal surface configuration is adopted in which the fluorescence emitted at 30° to the incident light path is measurable. The fluorescence cells are constructed of either quartz or silica, as these materials are able to transmit light of wavelength as low as 200 nm up to well into the near-infrared. Occasionally, cheaper glass or plastic cells are employed, where the excitation wavelength range is above 330 nm. The sample cells are normally rectangular, with a horizontal cross-sectional area of 1 cm^2 for room-temperature fluorescence spectroscopy. However, for low-temperature work including phosphorimetry, cylindrical cells are employed so as to preclude the possibility of cracking at edges. Sample compartments of the more recent spectrophotometers usually contain accessories to permit low-temperature fluorescence or phosphorescence work. When the fluorescence spectrophotometer is employed as a detector for liquid chromatography, the fluorescence cell employed is a flow-through cell with entrance and exit ports which can be connected to the chromatographic system. Similarly, the compartment can be customized to house polarized prisms or to provide for the maintenance of constant temperatures.

D. Photodetectors

The most widely used photodetectors are the photomultiplier tubes. These tubes generate an electrical signal upon exposure to light. The photomultiplier tube consists of an evacuated glass or quartz tube containing a photocathode, a series of dynodes across which are maintained successively increasing potential differences, and an anode. Light photons strike the photocathode, causing it to emit photoelectrons. These photoelectrons are then accelerated and amplified across the series of dynodes, eventually striking the anode to yield a measurable current. The sensitivity of the detector depends primarily on the composition of the cathode surface. This is usually cesium, selenium, tellurium, or a mixed alkali metal alloy. These metals have relatively low ionization potentials, easily losing valence electrons on exposure to light. Most photomultiplier tubes have an optimal spectral response in the region 300–500 nm, which falls off dramatically beyond 500 nm. At longer wavelengths, special red-sensitive photomultiplier tubes have been developed for use.

For the measurement of low light intensities, photon counting has been employed. In photon counters the photoelectron pulses at the anode of the photomultiplier tube are counted using a high-speed electronic counter. One pulse accounts for each electron ejected from the photocathode, and the mean pulse count rate is proportional to the light intensity. These counters, however, have a limiting count rate, beyond which they do not discriminate between signals. This restricts their utility to the measurement of low light intensities, where they exhibit less drift.

Imaging detectors have more recently been developed for use as detectors in spectrofluorimeters. The image detectors, more commonly known as vidicons, are essentially video camera tubes. These may be photoemissive or photoconductive. In the image orthicon the target surface containing a photosensitive material is biased by a scanning electron beam, resulting in a charge separation on the photosensitive surface. Incident light discharges this bias to an extent proportional to the amount of light striking each pixel. Upon return of the scanned beam, the bias is replaced by the generation of a signal. On the other hand, in the return vidicon the photoelectrons arising from the photocathode are focused onto a secondary electron emitter which in turn releases electrons which are collected onto a nearby mesh. A low-velocity electron beam is then scanned across the secondary electron emitter, discharging the positive charge present in some regions of the emitter. Over the other regions of the emitter surface the electrons are simply scattered and collected by an anode located at the base of the tube, giving rise to a signal. To improve sensitivity, intensified vidicons have been developed. In these intensified vidicons the electrons emerging from the photocathode are accelerated, in an orderly fashion, to the photoconductive target. Depending on target construction, these intensified vidicons are called either secondary electron conductor (SEC) intensified vidicons (MgO, KCl, MgF, or Ag) or silicon-intensified (SIT) vidicons (silicon diode arrays).

E. Data Acquisition Devices

The electronic signal obtained from the photodetector is usually electronically amplified, measured using some sort of galvanometer, and presented in either analog or digital form. The signal may alternatively be recorded on a strip-chart recorder, to supply a permanent record of the spectrum. The limitation of these recorders, however, is the relatively long response time of the pen. This limits the speed at which accurate spectra can be obtained. Oscilloscopes have been developed, however, in which elaborate electrical circuits have been employed to obtain spectra with a much shorter response time.

F. Time-Resolved Fluorimetry

With the advances in electrooptical technology, it has become possible to excite a potentially fluorescent sample with a pulsed lamp which emits its radiation at nanosecond intervals. The thyraton pulsed lamps, for example, emit radiation in bursts of 2–10 ns duration with about 0.2 ms between pulses, and the pulsed lasers generate even shorter-duration pulses. A fluorescent sample excited with a pulsed source does not fluoresce continuously. Rather, each pulse generates a batch of excited molecules which then emit the energy as fluorescence in an exponential fashion. The next pulse then excites a batch of the molecules which then also decay in an exponential fashion. In pulse fluorimetry the fluorescence from the sample, excited by the pulsed source, is represented, after detection, as a function of time on a fast sampling oscilloscope. However, in time-correlated single-photon counting, it is represented on an $X–Y$ plotter used in conjunction with a multichannel pulse analyzer.

Evidently, fluorescers with decay times much longer than the lamp pulse characteristics can be analyzed in much the same way as radioactive decay curves. A semilogarithmic plot of fluorescence intensity against time is linear, with a slope proportional to the decay time and the ordinate intercept providing a quantitative measure of the amount of fluorophore. If the lamp pulse time and the decay time of the fluorophore are comparable, the fluorophore's decay characteristics can be obtained by computerized deconvolution of the spectrum obtained.

The pulsed-source (time-resolved) method thus effects spectroscopic separation of the emission of several fluorescing species by making use of differences in their decay times rather than their fluorescence intensities. This is useful where strongly overlapping fluorescence spectra complicate simultaneous quantitative analysis without chemical or mechanical separation.

The pulse fluorimetric method, however is applicable only when one species has a much longer decay time than other species in the solution, it is much less useful when several species in solution have similar decay times. The pulse fluorimetric approach has only limited application to the resolution of more than two fluorescers under any circumstances.

Time-resolved fluorimetry is also useful for the elimination of interferences from stray light due to Rayleigh and Raman scatter. The latter phenomena occur on a time scale of $10^{-14}–10^{-13}$ s and, as they have a much shorter duration than lamp or laser pulses, the light associated with them can be eliminated from the signal that ultimately reaches the detector. Time-correlated single-photon counting is superior in its ability to resolve multiple fluorescence from the same solution.

Phase fluorometry is another useful fluorometric technique for the determination of substances with overlapping fluorescence spectra. In phase fluorometry the phase angle between the lamp pulse and the emission of fluorescent light allows discrimination between fluorescences of different origins. Phase-sensitive optics and electronics are rather complicated and, as in the case of time-resolved fluorometry, will be mentioned here only in passing. For further information on these areas, the reader is referred to the bibliography.

G. Fiber-Optic Fluorescence Spectroscopy

In the fiber-optic fluorimeter a fiber-optic cable replaces the sample compartment and cell and a covalently bound fluorophore and perhaps a chemically selective membrane at the distal end of the cable comprise a fluorescence sensor. This device has found extensive use in environmental and biological analysis. In the fiber optic instrument, light from a xenon lamp, laser, or light-emitting diode travels along an optically conducting fiber to its end, where activation of a covalently bound fluorophore causes fluorescence. Fluorescence sensors based on direct fluorescence or competitive binding are available, but most are based on fluorescence quenching of the sensor fluorophore by the analyte.

Instrumentation consists of a light source, optical filters, the fiber-optic cable, a sensing zone (the fluorophore and ancillary membranes or reagents), and a detector. Lasers, xenon lamps, hydrogen, deuterium, mercury, and halogen lamps, and light-emitting diodes (LEDs) have been used as excitation sources. Fiber-optic fluorescence spectrometers may be operated in a continuous or pulsed excitation mode. The latter mode allows for the application of time-resolved fluorometry.

The material of which the optical fiber is made determines the excitation wavelength range used. Fused silica, glass, and plastic fibers are the most common fiber materials. Silica can be used from the ultraviolet range down to 220 nm, but the fibers are expensive. Glass is suitable for use in the visible region and is lower in cost than silica. Plastic fibers are even less expensive but are limited to use above 450 nm.

Fiber-optic sensors are less expensive, more rugged, and smaller than electrodes; in the future we may see the former replacing the latter in various areas of analytical and clinical chemistry. Fields of application include environmental monitoring, process control, remote spectroscopy in high-risk areas with radioactive, explosive, biological, or other hazards, titrimetry, and in-vivo bioanalysis.

IV. ANALYTICAL ASPECTS

Fluorescence spectroscopy is used mainly as a quantitative analytical tools. The intensity of light absorbed, I_{abs}, is related to the concentration of analyte and optical parameters of the measurement system by the Beer-Lambert law. If all of the molecules that absorbed light fluoresced, I_{abs} would also be the intensity of fluorescence. However, the processes of internal conversion, intersystem crossing, and quenching compete with fluorescence so that the intensity of fluorescence is given by Eq. (11),

$$I_f = \phi_f I_0 (I - 10^{-\varepsilon c l}) \tag{11}$$

where I_f is the intensity of fluorescence, I_0 the intensity of light absorbed in creating the excited state, ϕ_f the quantum yield of fluorescence (the fraction of excited molecules that actually fluoresce), ε the molar absorptivity, l the optical pathlength of the cell, and C the molar concentration. A correction factor should be included in the equation to account for the fact that only a fraction of the fluorescent light emitted by the sample is collected by the detector. However, since I_0 is never measured absolutely, fluorimetric intensity measurements are only relative and the correction factor can be assumed to be included in I_0.

A. Quantitation and Sensitivity

As the intensity of light absorbed is exponentially related to concentration, the intensity of fluorescence can be related directly to concentration by a Maclaurin series,

$$I_f = \phi_f I_0 (2.303 \ A \ - \ \frac{(2.303A)^2}{2!} + \frac{(2.303A)^3}{3!} - \ldots)$$ (12)

where A is the absorbance, given by $A = \varepsilon C l$.

The relationship between fluorescence and concentration is virtually linear at very low concentrations where the absorbance is less than 0.02 and is approximated by

$$I_f = 2.303\phi_f I_0 A = 2.303\phi_f I_0 \ \varepsilon c l.$$ (13)

Therefore, at low concentrations, for a given instrument, at set wavelengths of measurement, the intensity of fluorescence is proportional to concentration. In this linear range, analytical sensitivity is highest. At higher concentrations, the fluorescence intensity increases exponentially with increasing concentration (Fig. 5). Despite the linear relationship, a calibration curve should always be obtained for any quantitative analysis. This is simply because a number of parameters other than concentration can influence the linearity of this relationship. At the higher analyte concentrations, inner filter effects are observed when the excitation and emission optics are perpendicular (the most common experimental arrangement), as a result of analyte absorption of the incident radiation in regions of the cell where the emanating fluorescence is not received by the detector.

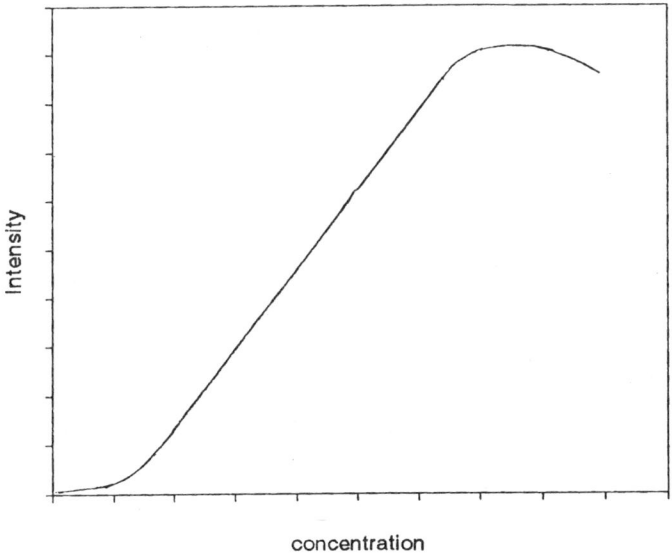

concentration

Fig. 5 Fluorescence–concentration plot illustrating nonlinear relation at the extremes of concentration.

There are two inner filter effects, the prefilter and postfilter effects. Of these, the prefilter effect is more common and more important. The prefilter effect arises from the fact that part of the sample cell is shielded from the photodetector by parts of the cell holder or by optical baffles (Fig. 6). The incident exciting light of intensity I_0 falls on the sample and traverses the pathlength l_1 before reaching the part of the cell visible to the detector. Hence the intensity of exciting light entering the part of the cell visible to the detector is

$$I'_0 = I_0 10^{-\varepsilon cl_1} \tag{14}$$

If the length of the path of light in the cell which is visible to the detector is l_2, the fluorescence intensity I_f seen by the detector is

$$I_f = \phi_f I_0 (1 - 10)^{-\varepsilon cl_2} = \phi_f 10^{-\varepsilon cl_1}(1 - 10^{-\varepsilon cl_2}) \tag{15}$$

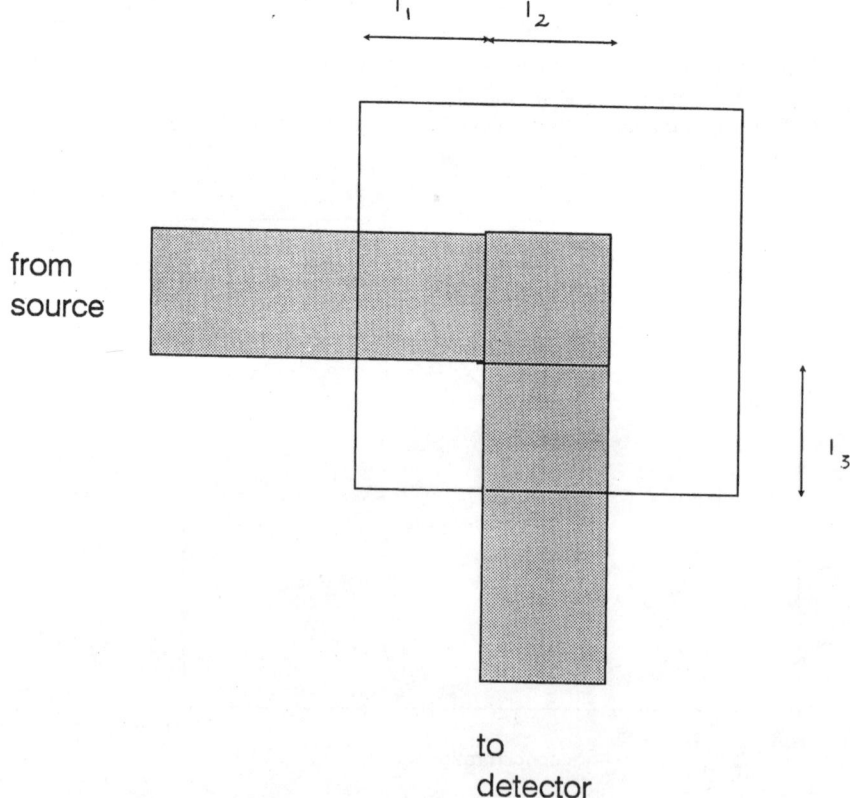

Fig. 6   Schematic diagram of light passage through cuvette.

Now, $1 - 10^{-\varepsilon c l_2}$ has the dependence on C seen in Fig. 5, having an approximately linear region at low concentration and a slowly increasing region at higher concentration. According to Eq. (11), which holds when front surface or reflective illumination is employed, the term $1 - 10^{-\varepsilon c l_2}$ will eventually become unity when C is very high, so that

$$I_f = \phi_f I_0 \tag{16}$$

Under this circumstance I_f is independent of concentration and functions as what is called a "quantum counter." However, when Eq. (15) is applicable (perpendicular illumination and detection), as the term $(1 - 10^{-\varepsilon c l_2})$ slowly increases with increasing absorber concentration, the $10^{-\varepsilon c l_1}$ term decreases. As a result, the increase in fluorescence is slower, reaching a maximum and then decreasing with increasing absorber concentration. The slower rate of change of I_f with C is associated with decreased analytical sensitivity. This prefilter effect is insignificant when the absorbance $\varepsilon c l_2$ is small enough (approx ≤ 0.02); $\varepsilon C l_1$ will then be even smaller and the fluorescence intensity-concentration relationship is linear as in Eq. (13). It is therefore important to adjust the concentration of the sample or excitation wavelength of the instrument so that analysis is performed in the linear region of the fluorescence-versus-concentration curve.

A postfilter effect arises from absorption by the sample of light emitted by the sample as the fluorescent light crosses the path of length l_3 (Fig. 6) through the sample cell, on the way to the detector. In this region, analytical sensitivity is adversely affected only when there is overlap of the absorption and fluorescence spectra. The postfilter factor for decrease in fluorescence intensity is given by $10^{-\varepsilon' c l_3}$, where ε' is the molar absorptivity in the absorption spectrum over the range of wavelength where the absorption and fluorescence spectra overlap. The postfilter effect is less common, as it requires overlap of the absorption and fluorescence spectra which most molecules of analytical interest do not demonstrate. A solution to the problem is to ensure that excitation is effected at short enough wavelengths in the absorption spectrum and fluorescence is monitored at long enough wavelengths in the emission spectrum, so that the postfilter effect is precluded.

The analytical sensitivity, defined as the rate of change of fluorescence intensity with analyte concentration, is decreased at high analyte concentration as the result of exponential dependence of the fluorescence intensity. However, at very low concentrations the linearity of the fluorescence signal with concentration and sensitivity are also lost. In this case, it is usually due to the fact that background noise, e.g., stray light, instrumental noise, etc., became comparable in intensity to the analytically significant signal, causing the latter to appear to change slowly with concentration. Eventually, the fluorescence is lost in the noise and the "limit of detection is reached," although the detection limit is formally defined as that signal which is three times the background noise signal. The quantum yield of fluorescence can be influenced by a variety of factors. These factors include the temperature and viscosity of the solution, which in turn influence the rate of competing deactivation processes. Similarly, solutes present in the analytical solutions that are able to influence the efficiency of other deactivation processes will alter the relationship between fluorescence intensity and concentration. For example, hydrogen ion concentration is able to influence the quantum yields of fluorescence for various substances as a result of proton transfer in the lowest excited singlet state. Not only does hydrogen ion concentration influence fluorescence, other dissolved solutes such as oxygen that are able to quench fluorescence will also influence the relationship between fluorescence intensity and concentration. Quenching of the fluorescence signal by dissolved oxygen can limit sensitivity in quantitative analyses. Indeed, improved analytical sensitivity has often been achieved by removal of oxygen from the sample, usually by flushing nitrogen or helium through the sample prior to analysis. In differing solvents, fluorescent analytes are known to exhibit differing fluorescence efficiencies. The intensity of incident light directly influences the measured fluorescence. Generally, the intensity of incident light should be high enough to permit high analytical sensitivity but not high enough to initiate photochemical decomposition within the cell, a feature that limits the use of powerful light sources such as lasers. Alterations in the intensity of incident light usually involve alterations in the slit width. These changes, however, also compromise the selectivity of the analytical system. Hence a compromise among the three factors often has to be made.

Despite these apparent limitations, fluorescence methods are employed for the determination of a wide variety of compounds. The selectivity of these analyses arises from the choice of both excitation and emission wavelengths, whereas the sensitivity of the analyses arises from the fact that absolute as opposed to relative measurements of light emission are made. This can be compared to ultraviolet–visible spectroscopy, where the ratio of incident to transmitted light is determined. Fluorescence measurements also have the advantage of a wide linear range of analysis.

Deliberate modification of the microenvironment of the fluorescing species has also been employed as a means of improving analytical sensitivity. A simple approach has involved selection of an appropriate solvent system that will enhance fluorescence. This has often involved the use of acids, bases, salts, or buffers to modify hydrogen ion concentration or ionic strength in aqueous media. A more recent approach involves modification of the immediate environment of the fluorescent species by complexation with an appropriate molecule. The cyclodextrins, for example, have been reported to enhance the intensity of fluorescence from a number of fluorescers. This enhancement arises from the ability to form inclusion complexes with appropriately sized molecules, shielding the excited singlet-state species from the nonradiative deactivation processes. Micellar systems have also been employed to enhance the intensity of fluorescence from fluorescent substances. The micellar environment has been reported to increase fluorescent lifetimes and quantum yields significantly. The analyte must preferentially associate with the micelle rather than the bulk solvent for this effect to manifest, affording an element of selectivity. With the availability of double-beam instruments, analytical sensitivity has been improved, particularly where there is significant background fluorescence, as with biological samples. The double-beam instruments have also opened up the possible application of pH and reagent-induced difference spectrofluorimetry as a means of analyzing fluorescent substances in the presence of known interferents.

Just as in ultraviolet–visible spectrophotometry, derivative techniques have also been applied to the analysis of fluorescent substances in multicomponent preparations. The advantages here are reported to be the enhanced spectral resolution and amplification of signal that manifests with substances exhibiting narrower spectral bands. Derivative spectrofluorimetric techniques have been applied, for example, to the determination of chlorpromazine sulfoxide in chlorpromazine hydrochloride preparations.

Synchronous scanning techniques have also been applied to the quantitative analysis of fluorescent substances. Synchronous scanning involves scanning both the excitation and emission monochromators simultaneously, while maintaining a constant wavelength interval between them. The technique has been employed in the analysis of multicomponent preparations. The technique is reported to simplify the spectra of multicomponent samples and reduce the bandwidths of fluorescence spectra. The equation relating the measured fluorescence to concentration is given by

$$I_s = Kcl \, \text{Ex}(\lambda - \delta\lambda)\text{Em}(\lambda) \qquad (17)$$

where I_s is the synchronous luminescence intensity, c is the concentration of measured species, l is the pathlength, $\delta\lambda = \lambda_{em} - \lambda_{ex}$, K is a geometric constant, and Ex and Em are the intensity distribution patterns of the excitation and emission, respectively. The synchronous scanning technique, either alone or in conjunction with derivative spectroscopy, has been applied to the quantitative analysis of organic substances. Appropriate selection of the wavelength interval for the synchronous scan has been reported to be the basis for selectivity in multicomponent analysis. However, loss of spectral information associated with the technique, as well as the possibility of wavelength coincidences, are two possible sources of error. Both can be circumvented by the use of multidimensional spectrofluorimetry to select an optimum wavelength interval, as well as the use of derivative spectrophotometry to improve analytical selectivity. The combination of techniques has been employed to the simultaneous determination of a variety of mixtures of drug substances and their metabolites as well as to mixtures of metal ion complexes.

Improved spectral resolution can also be achieved by use of time-resolved spectrofluorimetry. The selectivity here arises from the differing decay characteristics of overlapping fluorescence curves. The use of a pulsed light source whose pulse width is shorter than the lifetime of the lowest excited singlet state of the fluorophore, coupled with a detection system that generates a plot of the fluorescence intensity as a function of time, would facilitate selection of the appropriate decay curve for use in the quantitation of a specific analyte in the mixture. Despite the overlapping fluorescence spectra, a semilogarithmic plot of fluorescence intensity versus decay time can yield several straightline segments, each of which corresponds to a different fluorescent species in the sample. Extrapolation of each line segment to the intensity axis would facilitate calculation of concentration of the corresponding fluorescent species.

B. Interferences

1. Light Scattering

Solvents containing carbon–hydrogen and oxygen–hydrogen bonds, when excited with strong radiation, produce Raman bands shifted slightly to the long-wavelength side of the exciting light. These bands can be mistaken for part of the analyte fluorescence in qualitative fluorimetric analysis and cause serious interference in quantitative fluorimetric measurements when the Raman band of the solvent overlaps a weak fluorescence band of the analyte. Unfortunately, most analytically useful solvents contain groups which give rise to Raman scatter. It should be noted that Raman bands are usually weak and do not interfere in fluorimetric measurements at low instrumental sensitivity.

Raman bands can be distinguished from analyte emission by the facts that their wavelengths vary with a change in excitation wavelength and their bandwidth, at constant emission slit width, varies with the width of excitation slit. Interference from Raman bands, when present, can be minimized by using a cutoff filter before the emission monochromator to remove all wavelengths below and including the Raman band. Elastic scattering of the exciting light, called Rayleigh scatter, can also cause stray-light interference. This can be minimized by exciting the sample at wavelengths much shorter than the wavelengths at which fluorescence is monitored or by use of cutoff filters.

2. Luminescent Impurities

The fluorescent impurities present in solvents and those present in reagents used for the preparation of fluorophores and buffers are common sources of interference in fluorimetric analysis, especially at low analyte concentration. These interferences can be kept to a minimum by using pure solvents and reagents. Many of these chemicals are now available from commercial suppliers in a degree of purity which is designed for spectral studies. Some solvents can be purified by single or repeated distillation in ultra-clean glassware. Water can be purified by first distilling it from alkaline permanganate solution, followed by a second distillation. Deionization is also an effective way of purifying water. Purified solvents should be stored in clean glass bottles with aluminum-lined caps.

One serious form of interference is contamination of glassware and cells with luminescent impurities. This contamination can be minimized by soaking all the glassware in a concentrated nitric acid bath for about 24 h, rinsing with tap water, and distilled water, and then drying in an oven. Cells should be cleaned in a similar fashion when suspected of contamination. The repeated use of the nitric acid bath for cleaning cells should be avoided, as this may damage them. For routine work, the contamination of cells can be minimized by rinsing at least three times with the sample to be measured. If an ultrasonic cleaner is available, the cells, during rinsing, should be placed in the ultrasonic cleaner for about 30 s. Prolonged exposure of square cells to ultrasonic vibrations should be avoided, as this could damage the joints. Cells should be wiped with lint-free soft tissue paper each time they are used. The use of soap or detergents should be avoided, as these contain highly fluorescent components. Plastic laboratory ware usually contains fluorescent impurities which can be leached into solvents. The use of such plastic ware is not recommended for fluorimetric applications.

Nonluminescent impurities may also interfere by quenching analyte fluorescence. They may be removed much as luminescent impurities are removed.

C. Fluorescence Immunoassay

Quantitative fluorimetric analysis on real samples usually requires a separation of the fluorescent analyte from its matrix because of the wide occurrence of fluorescent and quenching impurities. Occasionally, simple solvent extraction will do the trick. More often, chromatographic separation is necessary, and the fluorimeter becomes merely a detector for the chromatographic eluates. Recently, however, the application of immunochemistry to analytical chemistry has permitted in-situ analysis on a scale never before possible. Although the earliest immunochemical analyses utilized radioisotopes, the most popular ones used today employ fluorescent probes or labels.

In fluorescent immunoassay (FIA), a fluorescent molecule is covalently bound to a molecule of analytical interest. The labeled analyte interacts reversibly with an antibody (usually derived from a rabbit or goat) which is specific for the analyte and will bind with about equal affinity to the labeled or unlabeled analyte. The labeled analyte usually shows different fluorescent properties according to whether it is bound by the antibody or free to diffuse in solution. The serial additional of different concentrations of the analyte to the labeled analyte–protein complex causes displacement of the labeled analyte and consequent alteration of the measured fluorescence intensity to a degree depending on the total concentration of nonlabeled analyte added. This allows the construction of a calibration curve from which quantitative analysis can be performed when an unknown amount of analyte is added to the labeled analyte–antibody complex. If there is sufficient change in the fluorescence of the labeled analyte upon dissociation from its antibody complex, the analysis may be performed in situ and is called a homogenous immunoassay. If separation of the bound labeled analyte from the unbound material is necessary, either because of insufficient differences in fluorescence or because of interfering substances present, the analysis is called a heterogeneous immunoassay.

A fluorescent label to be used in homogeneous immunoanalysis should fulfill several requirements. Since the fluorescence signal must often be measured in a "dirty" matrix, the probe should have a high fluorescence quantum yield and the excitation and emission maxima of the probe should occur at wavelengths longer than those of fluorescent impurities in the matrix.

Currently, the most popular fluorescent labels for FIA are those derived from the long-wavelength, strongly absorbing and emitting xanthene dyes fluorescein and rhodamine B. The isothiocyanates or isocyanates of these fluorophores are commercially available and can be used to label primary and secondary aliphatic amines in aqueous solutions by simple procedures. Consequently, they can be used to label a wide variety of organic molecules. Even those compounds which do not have indigenous alkyl amino groups can often be labeled by introducing bridging groups using reagents such as carbodiimides which are amenable to coupling with the isothiocyanato or isocyanato functions.

Heterogenous fluorescence immunoassays can be carried out with the aid of the same separation procedures used in radioimmunoassay. The more expedient homogenous fluorescence immunoassays require quenching, enhancement, polarization, or shifting of the fluorescence of the label upon binding of the labeled analyte to its antibody. Occasionally, a second antibody, directed at the anti analyte antibody, will be used in a "double-antibody" method to precipitate the bound labeled and unlabeled analyte or to alter the optical properties of the label in such a way as to make the analysis more sensitive.

Homogeneous fluorescence immunoassays can often be effected even when there is no obvious change in the intensity or spectral position of fluorescence of the labeled analyte upon binding to the antibody. If light-polarizing polymer films are used to polarize the exciting light and analyze the fluorescence of the sample excited by polarized light, it will generally be observed that the intensity of the fluorescence reaching the detector will be considerably less in those samples where a greater amount of antibody binding of the labeled analyte is extant. This is a result of the higher degree of polarized fluorescence emitted from the labeled analyte which is bound to the slowly rotating antibody. The polarized emission is more efficiently attenuated by the analyzer polarizing film than the unpolarized light emitted by the rapidly rotating, labeled analyte molecules which are not bound to macromolecules. This is the basis of fluorescence polarization immunoassay. The decrease in fluorescence intensity measured with decreasing labeled analyte binding occurs as a result of increasing unlabeled analyte concentration and can be used to construct a calibration curve from which the concentrations of unknown analyte samples can be determined when their polarized fluorescence is measured. Fluorescence polarization immunoassay probably accounts for most of the fluorescent immunoassays currently performed.

V. CHEMILUMINESCENCE

The term chemiluminescence has generally referred to the luminescent phenomena associated with a variety of chemical reactions. Chemiluminescence can be more specifically described as the electromagnetic emission that arises from the exothermic oxidation of an organic compound. Generally, the exothermic oxidation of the organic compound yields an energy-rich product that is luminescent because the molecule is either rigid or so small that it is unable to quickly dissipate the energy of the exothermic reaction internally. Chemiluminescence is distinguished from the more efficient bioluminescence by the fact that in bioluminescence, visible light is produced from an enzymically controlled reaction involving the chemical components of a living system. For a molecule to exhibit chemiluminescence, it must be able to form an electronically excited species through a chemical reaction at ordinary temperatures. The energy must be generated in a single reaction step, and the molecule receiving this energy must have a limited number of accessible vibrational energy states which would otherwise act as an energy sink. If the excited state is emissive, it can chemiluminesce directly. However, it can also transfer the energy to another molecule which, following excitation, then emits the energy as light.

The luminescence exhibited by luminol was one of the earliest cases of chemiluminescence to be studied. Luminol (5-amino-2,3-dihydrophthalazine-1,4-dione) is an aminophthalic hydrazide that is able to exist in several tautomeric forms. Other significant derivatives of luminol include isoluminol, aminobutylethylnaphthalhydrazide, and diazoluminol (Fig. 7). Luminol and its derivatives are generally oxidized in alkaline media to form an excited aminophthalate derivative, which then releases the energy as light. In protic media, a catalyst and a source of oxygen are required.

The 1,2-dioxetanes are another important group of chemiluminescent compounds. These compounds are oxidatively cleaved thermally in a concerted fashion to yield two carbonyl moieties, one of which is excited. The mechanism of this cleavage has been described as a chemically initiated electron-exchange

luminol   isoluminol derivatives   benzoperylene derivative

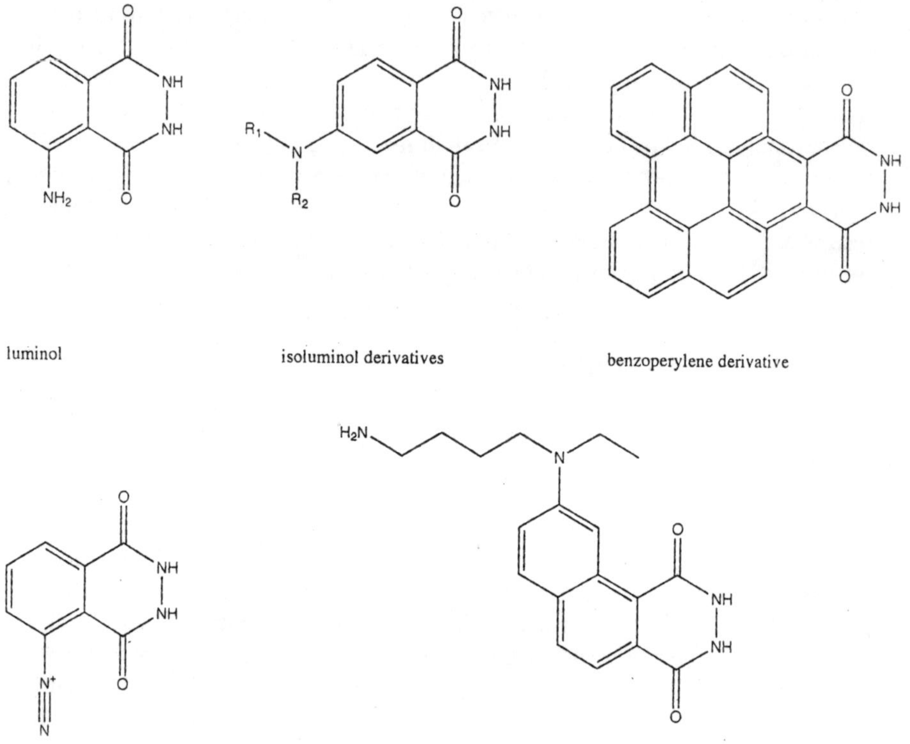

Fig. 7 Chemical structures of luminol derivatives exhibiting chemiluminescence.

chemiluminescence (CIEEL) (Fig. 8). Substituents substantially affect the conditions necessary to effect decomposition. A phenolic substituent, for example, reportedly permits triggering of the reaction in organic solvents by the addition of base. Similarly, siloxy substituents facilitate triggering of the reaction by fluoride (F) ions. Derivatives triggered by enzymes have also been developed, the such as (3-(2′-spiroadamantane)-4-methoxy-4-(3″-phosphoryloxy) phenyl-1,2-dioxetane sodium salt (AMPPD), a substrate for alkaline phosphatase, and its galactopyranosyl derivative (3-(2′-spiroadamantane)-4-methoxy-4-(3″-β-D-galactopyranosyloxyphenyl)-1,2-dioxetane sodium salt (AMPGD), a substrate for β-D-galactosidase.

Acridinium compounds form a third group of chemiluminescent compounds. These are commonly acridinium esters, in which intramolecular displacement of the alcohol derivative generates a four-membered dioxetanone ring which upon opening releases energy (Fig. 9). These have been modified by altering the leaving group, to yield more efficient luminescers.

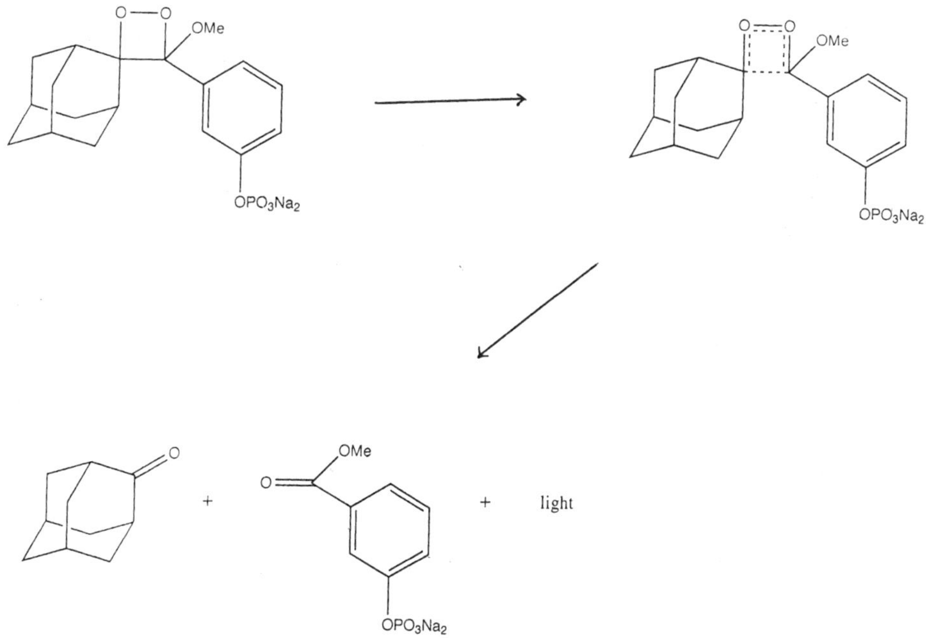

Fig. 8 Chemiluminescent reaction scheme for AMPPD.

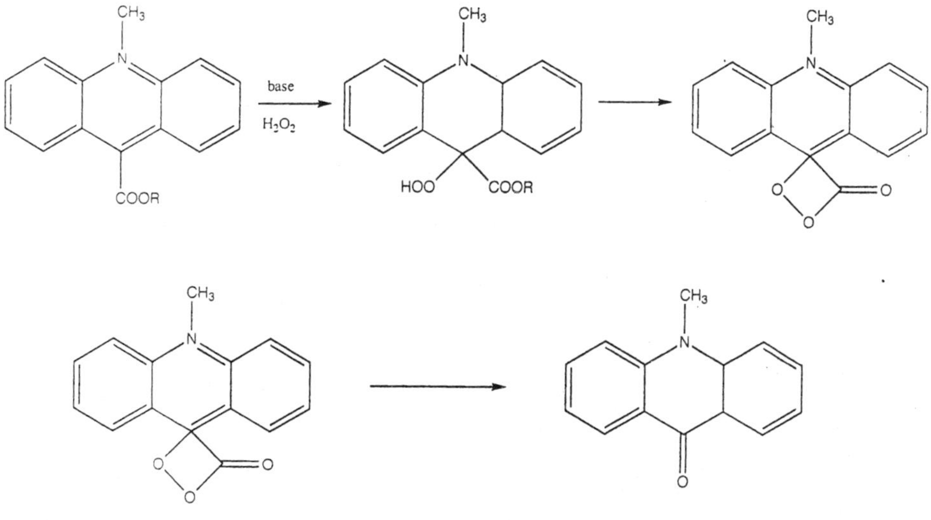

Fig. 9 Chemiluminescent reaction scheme for acridinium esters.

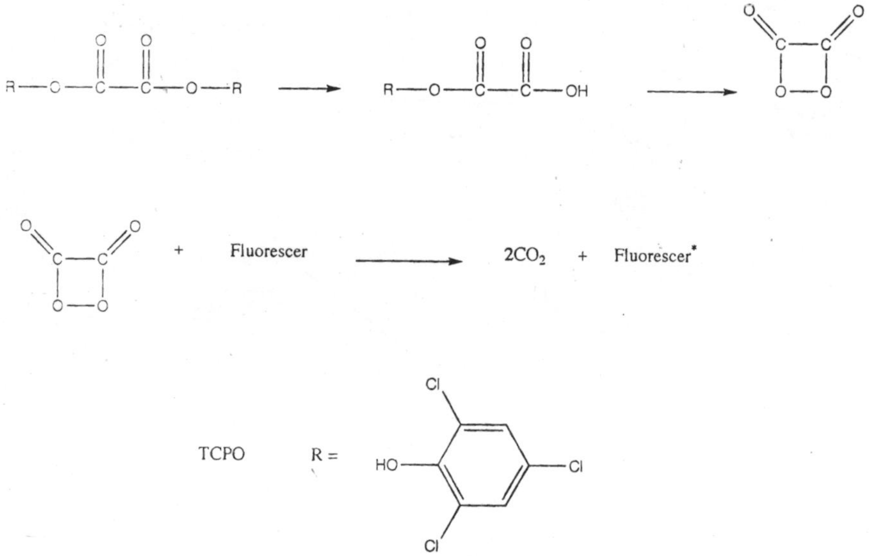

Fig. 10 Chemiluminescent reaction scheme for oxalates.

A variety of leaving groups, including various phenols, thiols, etc., have been developed, and it has been shown that the leaving group should have a pK_a < 11 for reasonable light yields. Oxalate esters are the most efficient chemiluminescent reagents. The esters in themselves are not chemiluminescent, but following reaction with an oxidant, transfer the energy to an added fluorescent molecule (Fig. 10). Of these compounds, bis(2,4,6-trichlorophenyl) oxalate (TCPO) has been well studied. In the presence of a suitable fluorescer it has been reported to achieve an efficiency ϕ_{CL} as high as 27%. Electronegative substituents attached to the oxalate group have been shown to yield efficient chemiluminescing molecules. In aqueous systems the oxalates are less efficient luminescers, partly as a result of the hydrolysis they undergo.

A. Reaction Mechanism

The chemiluminescent reactions involve a simple oxidation that results in the formation of a high-energy species. Cleavage of this high-energy species then results in the transfer of the energy to an electron-rich moiety in some other part of the molecule or to some other electron-rich species in solution. In the case of luminol the aromatic ring residue accepts this excitation energy, whereas for the oxalates the energy arising from cleavage of the four-membered ring is transferred to a fluorophore in solution.

The oxidative reagent employed is usually alkaline hydrogen peroxide. In basic solutions of hydrogen peroxide the hydroperoxide anion, which stems from the initial reduction of oxygen, is believed to be a critical reactant. By nucleophilic attack of an electron-deficient site on the luminol molecule, it generates a peroxy species whose cleavage yields the excited aminophthalate ion and nitrogen. The former then loses its energy as visible light.

Peroxide oxidation in the other compounds, on the other hand, generates a strained four-membered dioxetane or dioxetanone ring which, upon cleavage, releases its energy to the rest of the molecule or a fluorescent species present in the reaction medium. The light emitted in this case is from the fluorescent species as it returns to the ground energy state.

Other peroxide oxidants have been employed; even molecular oxygen under strongly basic solutions will generate the hydroperoxide. Other nucleophilic species reported in the literature include a peroxynitrite species which has also been reported to facilitate luminescence.

B. Application of Luminol Chemiluminescence to Analysis in Solution

The intensity of chemiluminescence arising from the chemiluminescent reaction can be described by the following equation:

$$I_{CL} = \phi_{CL} \cdot \frac{dC}{dt}$$

where ϕ_{CL} is the efficiency of the chemiluminescence and dC/dt is the rate of the chemical reaction. The efficiency of chemiluminescence depends on how efficiently excited states are generated from the molecular reaction and on how efficiently the excited states luminesce, i.e.,

$$\phi_{CL} = \phi_{exc} \cdot \phi_{lum}$$

where ϕ_{exc} is the excitation efficiency and ϕ_{lum} is the luminescence efficiency. Both the excitation and luminescence efficiency can be influenced by a variety of reaction conditions such as solvents employed, concentration, pH, and purity of reagents.

The intensity of luminescence can be used as the basis for determination of any species whose concentration influences the rate or efficiency of the chemiluminescent reaction. In applying chemiluminescence to analysis of the species, the reaction conditions should be adjusted so that the analyte of interest is the limiting reagent in the system and all other reactants are in excess. To obtain precise measurements, the chemiluminescence reaction should be initiated in a controlled and reproducible manner, largely because the emission intensity varies with time as the reactants are consumed. Chemiluminescent analyses are reported to have several advantages, which include good sensitivity, a wide linear dynamic range, low detection limits down to the femto- or atto-mole range, and the requirement for simple instrumentation.

The chemiluminescent signal is transient, hence the measurement of emitted light intensity is time dependent. The signal is therefore either recorded at a specific time after mixing, or by integrating the light emission plot during the entire time period or during a specific fraction of time when light is emitted. Commonly the reactants are rapidly mixed and the emission intensity measured as a function of time after mixing. The Intensity–time profile obtained is typically a frontally skewed sigmoid. The initial part of the intensity–time profile is influenced by the method of mixing employed, while the general shape of the curve depends on the kinetics of the reaction as well as on any changes in quantum yield with time.

Most chemiluminescence reactions are reported to have low efficiencies, less than 10%, which has restricted their usefulness for analyses. The duration of the reactions is influenced by the reaction conditions and may occur rapidly, within 1 s, or last longer than 24 h. In the development of chemiluminescent assay methods the two basic factors that influence the intensity of chemiluminescence (i.e., efficiency and rate) should be considered. The efficiency of the reaction influences both analytical sensitivity and detection limits, while the reaction kinetics determine both the precision and sample throughput.

Any substance able to quantitatively influence the light output can be determined by chemiluminescence. In fact, the luminol reaction has been used to determine a number of compounds that are able to interact with the oxidant initiating the chemiluminescent reaction. The substances so quantitated have not necessarily been the analyte of interest but rather compounds related to the analyte. These substances have been divided into three categories.

The first category is those measured species that consist of one of the reagents that is consumed in the course of the reaction. This case is exemplified by the determination of hydrogen peroxide using the luminol system and was applied to the determination of hydrogen peroxide in irradiated water as early as 1955.

The second category of reactions entails the analysis of compounds that are able to generate one of the chemiluminescent reactants. An example is the indirect determination of glucose by treatment with the enzyme glucose oxidase. The enzyme oxidizes the glucose to gluconic acid and hydrogen peroxide, the hydrogen peroxide so formed then being assayed by the luminol reaction:

Glucose $\rightarrow$ gluconic acid $+$ H_2O_2

luminol $+$ H_2O_2 $\rightarrow$ aminophthalate $+$ $h\nu$ $+$ N_2

A method for the determination of vitamin B_{12} by means of the luminol–hydrogen peroxide system has been developed. The method depends on the release of bound cobalt in vitamin B_{12} by acidification of the vitamin. The cobalt so released is then permitted to quantitatively catalyze the oxidation of luminol by hydrogen peroxide.

The coupling of reactions is not always possible, as there is often the problem of incompatibility of conditions for all the reactions. This has, however, not impeded use of the technique for the analysis of a range of biochemicals.

The third category consists of those substances able to modify the primary chemiluminescent reaction. The analysis of a wide range of substances falls into this category. This includes the analyses of metal ions such as arsenic (Ar^{3+}), cobalt (Co^{2+}), and Nickel (Ni^{2+}), nonmetallic inorganics such as the gases oxygen and nitrogen dioxide, the halide ions, and a number of nitro-, amino-, or hydroxy-group-containing organics. These substances are reported to have either an excitatory or inhibitory influence on the chemiluminescence exhibited by luminol. More recently a number of drug substances have been reported to enhance a variety of otherwise chemiluminescent reactions. For example, quinine has been determined using its ability to enhance the chemiluminescence exhibited by the oxidation of sulfite by cerium(IV). Similarly, the determination of acetaldehyde by monitoring the chemiluminescence emission from the luminol hexacyanoferrate(III) reaction in the presence of xanthine oxidase has been described. Xanthine oxidase catalyses the oxidation of xanthine and hypoxanthine in the presence of molecular oxygen to yield uric acid, H_2O_2, and the superoxide anion radical. The latter anion radical is reported to induce light emission from luminol more efficiently than H_2O_2. The steroidal hormones hydrocortisone and betamethasone, as well as the antihistamine promethazine, have also been determined from their ability to enhance the chemiluminescence of the cerium(IV)–sulfite system. The chemiluminescent oxidation of the antimycobacterial isoniazid by N-bromosuccinimide has also been applied to the determination of isoniazid. The penicillins, penicillin G, penicillin V, and the cephalosporin cephalothin, have also been determined using their ability to enhance luminol chemiluminescence.

C. Instrumentation

The instrumentation employed for chemiluminescence measurements basically consists of a mixing device and a detection system. Three approaches have been used to measure the intensity of emitted light.

The first approach involves the use of a static measurement system in which the mixing of reagents is performed in a vessel held in front of the detector. The chemiluminescent reagent is added to the analyte in a cuvette held in a dark enclosure, and the intensity of the light emitted is measured through an adjacent photomultiplier tube. In this static system the mixing is induced by the force of the injection. This simple reagent addition system without a mixing device is of limited usefulness in measuring fast chemiluminescent reactions with precision. The procedure is also rather cumbersome, requiring separate cuvettes for each measurement as well as repeated opening of the light-tight apparatus, which necessitates special precautions to protect the photomultiplier tube.

The second approach is the two-phase measurement system. In this system the chemiluminescent reagents are immobilized on a solid support such as filter paper, and the analyte is permitted to interact with the immobilized reagent by diffusion or convection. The light emitted is measured using a microliter plate reader or by contact printing with photographic detection. The chemiluminescent intensity of these systems is influenced by both the kinetics of the reaction and the efficiency of the mass transfer processes bringing the reactants together. The principal advantages of this system lie in the conservation of reagents and the convenience of measurement.

The third approach involves the use of flow measurement systems. The flow injection approach has been described as the most successful of the methods. It involves injection of the analyte into a stream of appropriate pH, remote from the detector, and the chemiluminescent reagent flows in another stream. The two streams meet at a T-junction inside a light-tight enclosure, then flow through a flat coil placed immediately in front of a photomultiplier tube. This compact assembly provides more rapid and reproducible mixing, resulting in reproducible emission intensities and permitting rapid sample throughput. The design of the mixing device and the means of retaining the emitting solution in view of the detector are important considerations. Mixing is reported as being most effective at a T-piece or Y-junction, although some workers have used a conventional FIA system in which sample is injected into the surrounding flowing reagent to achieve mixing. This approach is reproducible but reported as not producing rapid mixing. In flow measurement systems the chemiluminescent signal has to be measured during the mixing; as a result, only a section as opposed to the whole intensity–time curve is measured. An additional feature of these systems is that the shape of the curve is now dependent on both the kinetics of the chemiluminescent reaction as well as the parameters of the flow system. Chemical variables such as reagent concentrations, pH, etc., and physical parameters such as flow rate, reaction coil length, sample size, and the limitations of experimental apparatus all affect the performance of flow injection procedures. The effects of these variables on the observed analytical signal are not necessarily independent, as interactions can and probably do occur.

Most of the chemiluminescence determinations reported have involved the determination of compounds able to quench the luminescence of luminol or other chemiluminescent systems. A smaller number have been reported for substances that enhance the chemiluminescent signal. A variety of determinations fall into the latter group of compounds that are able to sensitize the chemiluminescent reactions. The chemiluminescent systems have not necessarily involved the luminol peroxide system alone, but have included among others the cerium(IV)–sulfite system in which the luminescence arises from the oxidation of sulfite by cerium(IV), hexacyanoferrate(III), the peroxodisulfate system, and the bromine-based oxidative systems.

D. Chemiluminescent Immunoassays

Chemiluminescence offers a useful alternative to both fluorescence and radioactivity, as it does not require an excitation source, there is less light scattering, and the problem of source instability is absent. The radiation hazard is also absent from chemiluminescent molecules. Most chemiluminescent reagents and their conjugates are stable and can be applied to both homogeneous and heterogenous assays. The chemiluminescent label may be attached to an enzyme or may serve either as a substrate or as a cofactor for the enzyme.

The acylhydrazides are the most frequently used chemiluminescent labels in chemiluminescent immunoassays, as they are easily conjugated and can be used for a range of immunoassay types. The acylhydrazides that have found use in luminescence analysis are those in which it has been possible to covalently link to an antigen or antibody without significantly compromising luminescence. Those compounds that have been used include isoluminol, diazoluminol, aminobutylethylnaphthalhydrazide, and a benzoperylene derivative of luminol. The reagents have also been used in determinations of enzymes that generate H_2O_2, such as the peroxidase enzyme, glucose dehydrogenase, and xanthine oxidase enzymes.

The 1,2-dioxetanes have mainly been used to develop substrates for the determination of enzymes, the more significant examples being (3-(2′-spiroadamantane)-4-methoxy-4-(3″-phosphoryloxy) phenyl-1,2-dioxetane sodium salt (AMPPD), a substrate for alkaline phosphatase, and its galactopyranosyl derivative, (3-(2′-spiroadamantane)-4-methoxy-4-(3″-β-D-galactopyranosyloxyphenyl)-1,2-dioxetane sodium salt (AMPGD), a substrate for β-D-galactosidase.

Only one acridinium phenyl carboxylate has found significant use as a chemiluminescent label. 4-(2-succinimidyloxycarbonylethyl)-phenyl-10-methyl-acridinium-9-carboxylate fluorosulphonate has been employed for the determination of albumin and thyroxine and α-fetoprotein.

In contrast, the oxalate esters have hardly found use as chemiluminescent labels in immunoassays, despite their greater chemiluminescence efficiency. This is probably a consequence of the need to have a fluorescer in the medium with the oxalate ester to generate a measurable light signal.

The use of enhancers will probably extend the usefulness of chemiluminescent reagents as labels in immunoassays, making chemiluminescence immunoassays a part of routine clinical analysis.

BIBLIOGRAPHY

1. CA Parker. Photoluminescence of Solutions. Amsterdam: Elsevier, 1968.
2. JD Winefordner, SG Schulman, TC O'Haver. Luminescence Spectrometry in Analytical Chemistry. New York: Wiley Interscience, 1972.
3. G Guilbault. Practical Fluorescence. 2nd ed. New York: Marcel Dekker, 1990.
4. JR Lakowicz. Principles of Fluorescence Spectroscopy. New York: Plenum Press,
   1983.
5. JN Demas. Excited State Lifetime Measurements. New York: Academic Press, 1983.
6. SG Schulman, ed. Molecular Luminescence Spectroscopy: Methods and Applications.
   New York: Wiley-Interscience, Part I, 1985, Part II, 1988; Part III, 1993.
7. A Sharma, SG Schulman. Introduction to Fluorescence Spectroscopy. New York: Wiley-Interscience, 1999.
8. AM Garcia Campaña, WRG Baeyens, eds. Chemiluminescence in Analytical Chemistry. New York: Marcel Dekker, 2001.

10
Solid-State Nuclear Magnetic Resonance Spectroscopy

David E. Bugay
SSCI, Inc., West Lafayette, Indiana

I. INTRODUCTION

Nuclear magnetic resonance (NMR) spectroscopy has become a fundamental technique for pharmaceutical analysis (1–8). Typically, when one thinks of NMR spectroscopy, solution-phase studies, such as structure elucidation, immediately come to mind (1). Within a drug discovery setting, structure elucidation is the one of the primary uses of NMR. It is ironic to think that although approximately 80% of all pharmaceutical products are solid-state formulations, solid-state NMR is relatively new to pharmaceutical analysis. Based on the myriad of uses of NMR spectroscopy as applied to pharmaceutical analysis, this chapter focuses on the use of solid-state NMR for the analysis of pharmaceuticals.

Since the advent of solid-state NMR in the 1970s, an increasing number of applications of solid-state NMR to the study of pharmaceuticals have been published. Probably the most popular use of solid-state NMR has been in the study of polymorphism (9,10). Traditionally, polymorphism has been investigated by X-ray diffraction and thermal analysis techniques. Various solid-state NMR methodologies and pulse sequences have shed new light on the understanding of polymorphism. An extensive review of scientific advances within this aspect of drug development will be reviewed in this work. Additionally, solid-state NMR has been utilized for studying drug–excipient interactions, drug–membrane interactions, chirality, structural characterization, and molecular mobility—a key aspect of solid-state reactions.

This chapter provides a basis of theory for solid-state NMR and a brief review of experimental aspects. In the discussion section, the various applications of solid-state NMR as applied to pharmaceuticals are outlined. Qualitative characterization techniques are summarized as well as quantitative methodology used for the study of the active pharmaceutical ingredient (API) or the final drug product.

II. SOLID-STATE NMR THEORY

Conventional utilization of solution-phase NMR data acquisition techniques on solid samples yields broad, featureless spectra (Fig. 1A). The broad nature of the signal is due primarily to dipolar interactions, which do not average out to zero in the solid state, and chemical shift anisotropy (CSA), which again is a function of the fact that our compound of interest is in the solid state. Before one describes the two principal reasons for the broad, featureless spectra of solid-state materials, it is important to understand the main interactions that a nucleus with a magnetic moment experiences when situated within a magnetic field. In addition, manifestations of these interactions in the solid-state NMR spectrum need to be discussed.

A. Zeeman Interaction

The principal interaction that a nucleus experiences when placed in an externally applied magnetic field (B_0, tesla), is the Zeeman interaction (H_Z), which describes the interaction between the magnetic moment of the nucleus and B_0. The nuclear magnetic moment (μ, ampere meter^2) is proportional to the nuclear spin quantum number (I) and the magnetogyric ratio (γ, radian tesla^{-1} second^{-1}), Eq. (1) (h = Planck's constant). Thus, the Zeeman interaction occurs only with nuclei which possess a spin greater than zero, and yields $2I + 1$ energy levels of

$$\mu = \frac{\gamma I h}{2\pi} \tag{1}$$

$$H_Z = -\gamma \hbar B_0 I_Z \tag{2}$$

separation $v_0 = \gamma B_0/2\pi$. Equation (2) describes the Hamiltonian, where v_0 is the corresponding Larmor frequency ($\hbar = h/2\pi$). The interaction is linear with the applied magnetic field, thus giving the impetus to manufacture higher-magnetic-

---

Fig. 1 Solid-state ^{13}C NMR spectra of didanosine (VIDEX-ddI) acquired by: (A) conventional solution-phase pulse techniques; (B) high-power proton decoupling only; (C) high-power proton decoupling and magic angle spinning (5 kHz); (D) high-power proton decoupling combined with magic angle spinning (MAS) and cross-polarization (CP). In each case, 512 scans were accumulated. (From Ref. 11.)

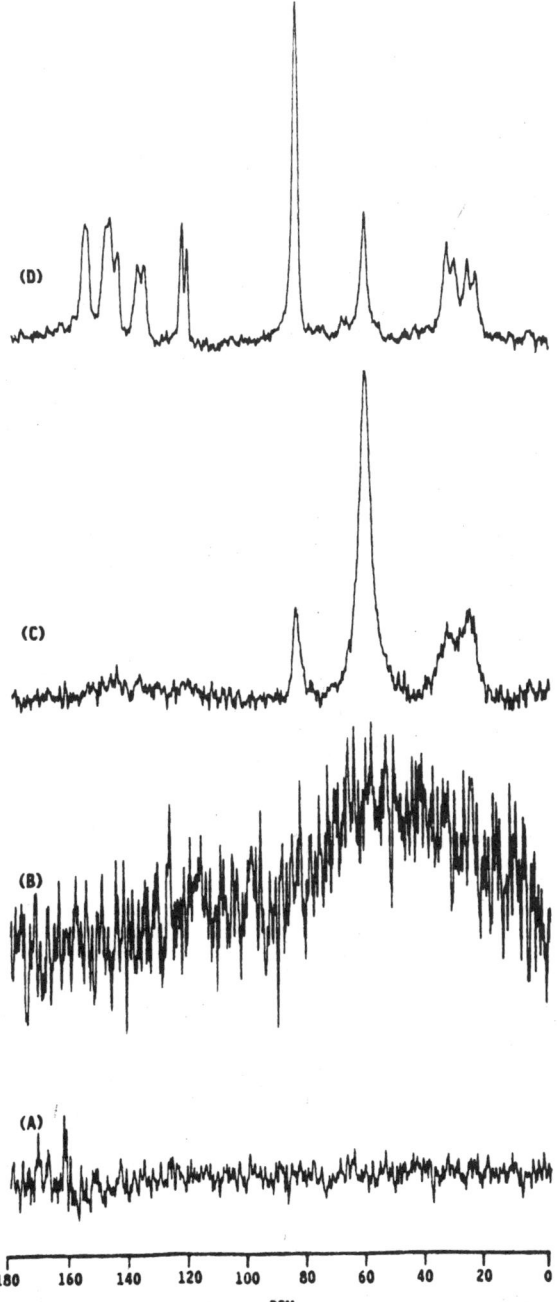

field spectrometers since a larger separation of energy levels leads to a greater population difference, a subsequent increase in the signal-to-noise ratio (S/N), and increased spectral resolution. Since the Zeeman interaction incorporates the magnetogyric ratio, a constant for each particular nucleus, the resonant frequency for each nucleus is different at a specific applied magnetic field (Table 1). The magnitude of the Zeeman interaction for a ^{13}C nucleus in a 5.87-T magnetic field is 62.86 MHz. Small perturbations to the Zeeman effect are produced by other interactions, such as dipole–dipole, quadrupolar, shielding, and spin–spin coupling. Typically, these small perturbations are less than 1% (<600 kHz) of the Zeeman interactions, which range in frequency from 10^6 to 10^9 Hz.

B. Dipole-Dipole Interaction

Dipole–dipole interaction is the direct magnetic coupling of two nuclei through space. This interaction may involve two nuclei of equivalent spin or nonequivalent spin, and is dependent on the internuclear distance and dipolar coupling tensor. Additionally, the total interaction, labeled H_D, is the summation of all possible pairwise interactions (homo- and heteronuclear). It is important to note that the interaction is dependent on the magnitude of the magnetic moments, which is reflected in the magnetogyric ratio, and on the angle (θ) that the internuclear vector makes with B_0. Therefore, this interaction is significant for spin-$\frac{1}{2}$ nuclei with large magnetic moments, such as ^{1}H and ^{19}F. Also, the interaction decreases rapidly with increasing internuclear distance (r), which generally corresponds to contributions only from directly bonded and nearest-neighbor nuclei. Equation (3) describes the dipolar interaction for a pair of nonequivalent, isolated spins I and S. Since the dipolar coupling tensor, D, contains a $(1 - 3 \cos^2 \theta)$ term, this interaction is dependent on the orientation of the molecule. In

$$H_D^{IS} = \frac{\gamma_I \gamma_S \hbar^2}{r^3} I \cdot \hat{D} \cdot S \tag{3}$$

solution-phase studies where the molecules are tumbling rapidly, the $(1-3 \cos^2 \theta)$ term is integrated over all angles of θ and subsequently disappears. In solid-state NMR, the molecules are fixed with respect to B_0, thus the $(1-3 \cos^2 \theta)$ term does not approach zero. This leads to broad resonances in the solid-state NMR spectrum, since dipole–dipole interactions typically range from 0 to 10^5 Hz in magnitude.

In the case of pharmaceutical solids which are dominated by carbon and proton nuclei, the dipole–dipole interactions may be simplified. The carbon and proton nuclei may be perceived as "dilute" and "abundant" based on their isotopic natural abundance, respectively (Table 1). Homonuclear ^{13}C–^{13}C dipolar inter-

Table 1 NMR Frequency Listing

Isotope	Spin	Natural abundance	Sensitivity rel.[a]	Sensitivity abs.[b]	NMR frequency (MHz) at a field (T) of 2.35	5.87	11.74
^{1}H	½	99.98	1.00	1.00	100.00	250.00	500.00
^{2}H	1	1.5×10^{-2}	9.65×10^{-3}	1.45×10^{-6}	15.35	38.38	76.75
^{10}B	3	19.58	1.99×10^{-2}	3.90×10^{-3}	10.75	26.87	53.73
^{11}B	3/2	80.42	0.17	0.13	32.08	80.21	160.42
^{13}C	½	1.11	1.59×10^{-2}	1.76×10^{-4}	25.14	62.86	125.72
^{14}N	1	99.63	1.01×10^{-3}	1.01×10^{-3}	7.22	18.06	36.12
^{15}N	½	0.37	1.04×10^{-3}	3.85×10^{-6}	10.13	25.33	50.66
^{17}O	5/2	3.7×10^{-2}	2.91×10^{-2}	1.08×10^{-5}	13.56	33.89	67.78
^{19}F	½	100.00	0.83	0.83	94.08	235.19	470.39
^{23}Na	3/2	100.00	9.25×10^{-2}	9.25×10^{-2}	26.45	66.13	132.26
^{25}Mg	5/2	10.13	2.67×10^{-3}	2.71×10^{-4}	6.12	15.30	30.60
^{27}Al	5/2	100.00	0.21	0.21	26.06	65.14	130.29
^{29}Si	½	4.7	7.84×10^{-3}	3.69×10^{-4}	19.87	49.66	99.33
^{31}P	½	100.00	6.63×10^{-2}	6.63×10^{-2}	40.48	101.20	202.40
^{33}S	3/2	0.76	2.26×10^{-3}	1.72×10^{-5}	7.67	19.17	38.35
^{35}Cl	3/2	75.53	4.70×10^{-3}	3.55×10^{-3}	9.80	24.50	48.99
^{37}Cl	3/2	24.47	2.71×10^{-3}	6.63×10^{-4}	8.16	20.39	40.78
^{39}K	3/2	93.1	5.08×10^{-4}	4.73×10^{-4}	4.67	11.67	23.33
^{41}K	3/2	6.88	8.40×10^{-5}	5.78×10^{-6}	2.56	6.40	12.81
^{43}Ca	7/2	0.145	6.40×10^{-3}	9.28×10^{-6}	6.73	16.82	33.64
^{67}Zn	5/2	4.11	2.85×10^{-3}	1.17×10^{-4}	6.25	15.64	31.27
^{79}Br	3/2	50.54	7.86×10^{-2}	3.97×10^{-2}	25.05	62.63	125.27
^{81}Br	3/2	49.46	9.85×10^{-2}	4.87×10^{-2}	27.01	67.52	135.03
^{127}I	5/2	100.00	9.34×10^{-2}	9.34×10^{-2}	20.00	50.02	100.04
^{195}Pt	½	33.8	9.94×10^{-3}	3.36×10^{-3}	21.50	53.75	107.50

[a] Determined for an equal number of nuclei at a constant field.
[b] Product of the relative sensitivity and natural abundance.
Source: From Ref. 11.

actions essentially do not exist because of the low concentration of ^{13}C nuclei (natural abundance of 1.1%). On the other hand, $^{1}H–^{13}C$ dipolar interactions contribute significantly to the broad resonances, but this heteronuclear interaction may be removed through simple high-power proton decoupling fields, similar to solution-phase techniques.

C. Chemical Shift Interaction

The three-dimensional magnetic shielding by the surrounding electrons is an additional interaction that the nucleus experiences in either the solution or the solid state. This chemical shift interaction (H_{CS}) is the most sensitive interaction to changes in the immediate environment of the nucleus and provides the most diagnostic information in a measured NMR spectrum. The effect originates from the small magnetic fields that are generated about the nucleus by currents induced in orbital electrons by the applied field. These small perturbations upon the nucleus are reflected in a change in the magnetic field experienced by the nucleus. Therefore, the field at the nucleus is not equal to the externally applied field and hence the difference is the nuclear shielding, or chemical shift interaction [Eq. (4)]. It is important to note the orientation dependence of the shielding constant, σ, and the fact that shielding is proportional to the applied field, hence the need for chemical shift reference materials such as tetramethylsilane.

$$H_{CS} = \gamma_I \, \hbar I \cdot \hat{\sigma} \cdot B_0 \qquad\qquad (4)$$

Solution-state NMR spectra yield "average" chemical shift values which are characteristic of the magnetic environment for a particular nucleus. The average signal is due to the isotropic motion of the molecules in solution. In other words, B_0 "sees" an average orientation of a specific nucleus. For solid-state NMR, the chemical shift value is also characteristic of the magnetic environment of a nucleus, but normally, the molecules are not free to move. It must be kept in mind that the shielding will be characteristic of the nucleus in a particular orientation of the molecule with respect to B_0. Therefore, a specific functional group oriented perpendicular to the magnetic field will give a sharp signal characteristic of this particular orientation (Fig. 2A). Analogously, if the functionality is orientated parallel to B_0, then a sharp signal characteristic of that orientation will be observed (Fig. 2B). For most polycrystalline pharmaceutical samples, a random distribution of all orientations of the molecule will exist. This distribution produces all possible orientations and is thus observed as a very broad NMR signal (Fig. 2C). The magnitude of the chemical shift anisotropy is typically between 0 and 10^5 Hz.

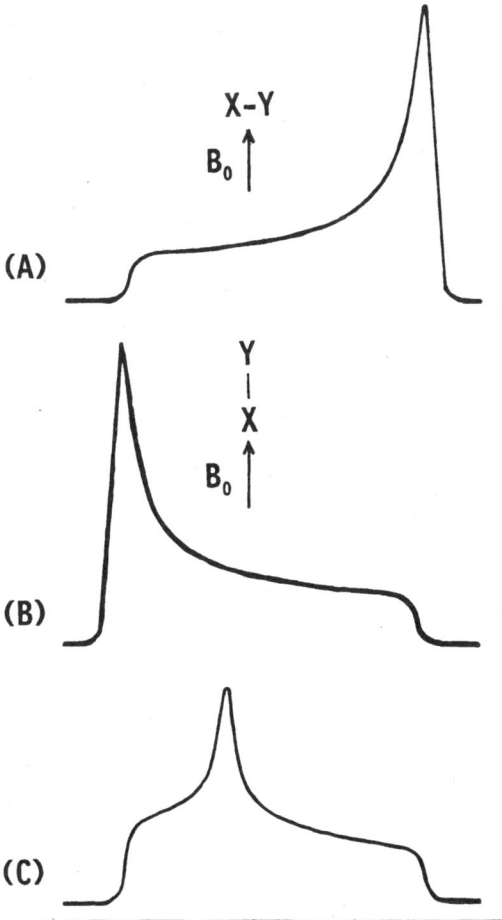

Fig. 2 Schematic representation of the ^{13}C NMR signal of a single crystal containing the functional group A-B, orientated (A) perpendicular to the applied field, and (B) parallel to the applied field. The lineshape in (C) represents the NMR signal of a polycrystalline sample with a random distribution of orientations yielding the chemical shift anisotropy pattern displayed. (From Ref. 11.)

D. Spin–Spin Couplings and Quadrupolar Interactions

Two additional interactions experienced by the nucleus in the solid state are spin–spin couplings to other nuclei and quadrupolar interactions, which involve nuclei of spin greater than $1/2$. Spin–spin (H_{SC}), or J coupling [Eq. (5)], originates from indirect coupling between two spins by means of their electronic surroundings and are several orders of magnitude smaller (possibly $0–10^4$ Hz, typically only several kilohertz) than dipole interactions. Although J coupling interactions occur in both the solid and the solution state, these couplings are typically used in solution-phase work for conformational analysis (12). Quadrupolar interactions (H_Q) arise from dipolar coupling of quadrupolar nuclei, which display a non-

$$H_{SC} = I \cdot \hat{J} \cdot S \tag{5}$$

spherically symmetrical field gradient, with nearby spin-$1/2$ nuclei [Eq. (6)]. The magnitude of the interaction is dependent on the relative magnitudes of the quadrupolar and Zeeman interactions and may completely dominate the spec-·trum ($0–10^9$ Hz), but since pharmaceutical compounds are primarily hydrocarbons, quadrupolar interactions typically do not interfere.

$$H_Q = I \cdot \hat{Q} \cdot I \tag{6}$$

Summarizing the interactions, the isotropic motions of molecules in the solution state yield a discrete average value for the scalar spin–spin coupling and chemical shift interactions. For the dipolar and quadrupolar terms, the average obtained is zero, and the interaction is not observed in solution-phase studies. In sharp contrast, interactions in the solid state are orientation-dependent, subsequently producing a more complicated spectrum, but one that contains much more information. In order to yield highly resolved, "solution-like" spectra of solids, a combination of three techniques is used: dipolar decoupling (13), magic-angle spinning (14,15), and cross-polarization (16). The remainder of this section discusses the three techniques and their subsequent effect on measured spectra of pharmaceutical compounds which are dominated by carbon nuclei.

E. Dipolar Decoupling

Discussed earlier was the fact that high-powered proton decoupling fields can eliminate the heteronuclear $^1H–^{13}C$ dipolar interactions that may dominate a solid-state NMR spectrum. The concept of decoupling is familiar to the solution-phase NMR spectroscopist, but needs to be expanded for solid-state NMR studies. In solution-phase studies, the decoupling eliminates the scalar spin–spin coupling, not the dipole–dipole interactions (this averages out to zero due to isotropic motions of the molecules). Irradiation of the sample at the resonant frequency of the nucleus to be decoupled (B_2 field) causes the z component of the spins to flip rapidly compared to the interaction one wishes to eliminate. Scalar interactions usually require 10 W of decoupling power or less. In pharmaceutical solids work, decoupling is used primarily to remove the heteronuclear dipolar interactions between protons and carbons. The magnitude of the dipolar interaction ($\sim$50 kHz) usually requires decoupling fields of 100 W and subsequently removes both scalar and dipole interactions. Even with the use of high-power decoupling, broad resonances still remain, principally due to chemical shift anisotropy (Fig. 1B).

F. Magic-Angle Spinning

Molecules in the solid state are in fixed orientations with respect to the magnetic field. This produces chemical-shift anisotropic powder patterns for each carbon atom, since all orientations are possible (Fig. 2). It was shown as early as 1959 that rapid sample rotation of solids narrowed dipolar-broadened signals (14). A number of years later, it was recognized that spinning could remove broadening caused by CSA yet retain the isotropic chemical shift (15).

The concept of magic-angle spinning arises from the understanding of the shielding constant, σ [Eq. 4]. This constant is a tensor quantity and, thus, can be related to three principal axes:

$$\sigma_{zz} = \sum_{i=1}^{3} \sigma_i \cos^2 \theta_i \qquad (7)$$

where σ_i is the shielding at the nucleus when B_0 aligns along the ith principal axis, and θ_i is the angle this axis makes with B_0. Under conditions of mechanical spinning, this relationship becomes time dependent and a ($3 \cos^2 \theta - 1$) term arises. By spinning the sample at the so-called magic-angle of 54.7°, or 54°44′, this term becomes zero and thus removes the spectral broadening due to CSA (Fig. 1C), provided the sample rotation (kHz) is greater than the magnitude of the CSA (kHz).

CSA may range from 0 to 20 kHz, so our spin rates must exceed this value or spinning sidebands are observed. Figure 3A displays the solid-state ^{31}P-NMR spectrum of fosinopril sodium (Monopril), a novel ACE inhibitor (17), acquired under proton decoupling and static spinning conditions. At a relatively low spin rate of 2.5 kHz, broadening due to CSA is removed and the center band and sidebands appear (Fig. 3B). As the sample is spun at higher rates (Figs. 3C and 3D), the sidebands become less intense and move out from the center band. Even at a spin rate of 6 kHz (maximum spin rate for the probe/instrument used), the CSA is not completely removed (Fig. 3E). Although fast enough spin rates may not be achieved to remove CSA totally for specific compounds, slower than optimal rates will still narrow the resonances. Spinning sidebands, at multiples of

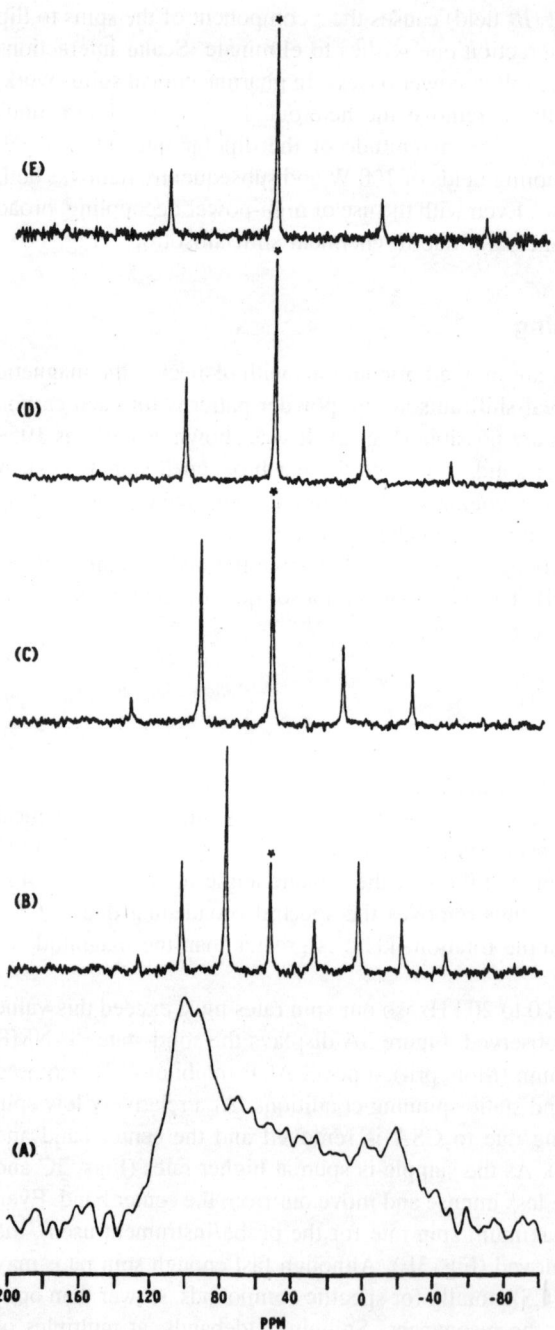

the spin rate, will complicate the spectrum, but they can be easily identified by varying the spin rate and observing which signals change in frequency. In addition, specific pulse sequences may be used to minimize the spinning sidebands (18). While increasing the magnetic field strength increases the signal-to-noise ratio (Zeeman interaction), it also increases the CSA, since this interaction is field-dependent [Eq. (4)]. Utilizing today's high-field spectrometers (>4.7 T), spinning sidebands may exist but can be identified and used to gain additional information or be potentially eliminated if necessary.

The techniques of magic-angle spinning and heteronuclear dipolar decoupling produce solid-state NMR spectra which approach the linewidths and appearance of solution-phase NMR spectra. Unfortunately, there is an inherent lack of sensitivity in the general NMR experiment due to the nearly equivalent population of the two spin states for spin-$\frac{1}{2}$ nuclei. In addition, the sensitivity of the experiment is decreased with pharmaceutical compounds, since they are composed primarily of carbon atoms where the ^{13}C observable nuclei have a natural abundance of only 1%. The long relaxation times of specific carbon nuclei also pose a problem since quick, repetitive pulsing cannot occur. The technique of cross-polarization provides a means of both signal enhancement and reduction of long relaxation times.

G. Cross-Polarization

The concept of cross-polarization as applied to solid-state NMR was implemented by Pines et al. (16). A basic description of the technique is the enhancement of the magnetization of the rare spin system by transfer of magnetization from the abundant spin system. Typically, the rare spin system is classified as ^{13}C nuclei and the abundant system as ^{1}H spins. This is especially the case for pharmaceutical solids, and the remaining discussion of cross-polarization focuses on these two spin systems only.

Figure 4 describes the cross-polarization (CP) pulse sequence and the behavior of the ^{1}H and ^{13}C spin magnetizations during the pulse sequence in terms of the rotating frame of reference (19). Step 1 of the sequence involves rotation of the proton magnetization onto the y' axis by application of a 90° pulse (rotating-frame magnetic field B_{1H}). Subsequent spin-locking occurs along y' by an on-resonance pulse along y' for a specific period of time, t. At this point, a high

---

Fig. 3 Solid-state ^{31}P NMR spectra of fosinopril sodium acquired under single pulse, high-power proton decoupling and various conditions of magic-angle spinning: (A) static; (B) 2.5 kHz; (C) 4.0 kHz; (D) 5.0 kHz; (E) 6.0 kHz. The isotropic chemical shift is designated by an asterisk. (From Ref. 11.)

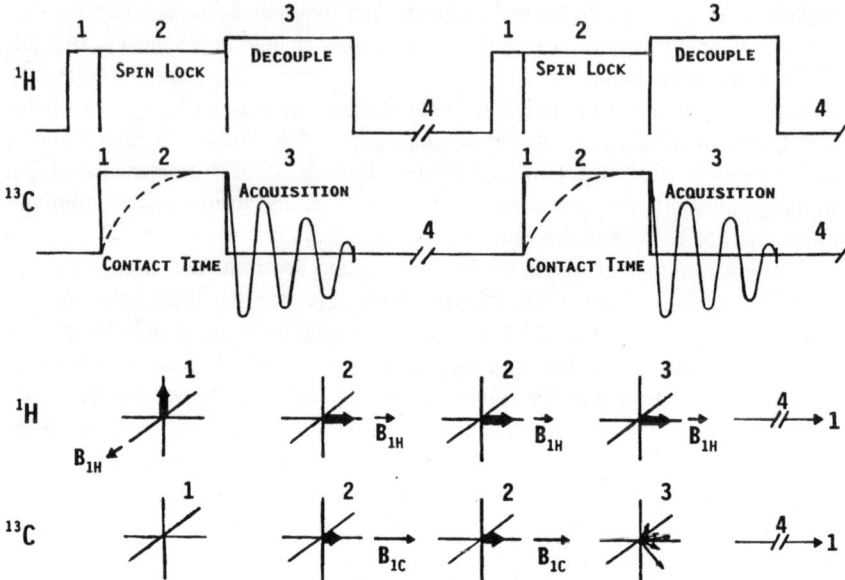

Fig. 4 The top diagram represents the pulse sequence for the cross-polarization experiment, whereas the bottom diagram describes the behavior of the ¹H and ¹³C spin magnetizations during the sequence. (From Ref. 11.)

degree of proton polarization occurs along B_{1H}, which will decay with a specific time referred to as $T_{1\rho H}$. As the ¹H spins are locked along y', an on-resonance pulse, B_{1C}, is applied to the ¹³C spins. The ¹³C spins are also "locked" along y' and decay with the time $T_{1\rho C}$, (step 2). By correctly choosing the magnitude of the spinlocking fields B_{1H} and B_{1C}, the Hartmann-Hahn (20) condition [Eq. (8)] will be satisfied and transfer of magnetization will occur from the ¹H spin

$$\gamma_{1H}B_{1H} = \gamma_{13C}B_{1C} \tag{8}$$

reservoir to the ¹³C spin reservoir. Once this equality is obtained by the correct spin-locking fields, the dilute ¹³C spins will take on the characteristics of the more favorable ¹H system. In the end, a maximum magnetization enhancement equal to the ratio of the magnetogyric ratios may be achieved (γ_H/γ_C). Once the carbon magnetization has built up during this "contact period"·(t), B_{1C} is switched off and the carbon-free induction decay recorded (step 3). During this data acquisition period, the proton field is maintained on for heteronuclear decoupling of the ¹H–¹³C dipolar interactions. Step 4 involves a standard delay period in which no pulses occur and the two spin systems are allowed to relax back to their equilibrium states.

The entire CP pulse sequence provides two advantages for the solid-state NMR spectroscopist: significant enhancement of the rare spin magnetization, and reduction of the delay time between successive pulse sequences since the rare spin system takes on the relaxation character of the abundant spin system. Signal enhancement (magnetization enhancement) for less sensitive nuclei is immediately apparent from the CP process. Less obvious is the fact that the rare spin system signal is not dependent on the recycle time in step 4 (regrowth of carbon magnetization), but on the transfer process in step 2 and the relaxation behavior of the proton spin system in step 4. Since the single spin lattice relaxation time of the proton system is typically much shorter than the carbon system (1's to 10's versus 100's of seconds), the delay time is much shorter. This corresponds to a greater number of scans per unit time, yielding better S/N.

Throughout the cross-polarization pulse sequence, a number of competing relaxation processes are occurring simultaneously. Recognition and understanding of these relaxation processes are critical in order to apply CP pulse sequences for quantitative solid-state NMR data acquisition or ascertaining molecular motions occurring in the solid state.

H. Relaxation

Familiar to most chemists is the notion of spin-lattice relaxation (21). Labeled as T_1, the spin-lattice relaxation time is defined as the amount of time for the net magnetization (M_z) to return to its equilibrium state (M_0) after a spin transition is induced by a radiofrequency pulse [Eq. (9)]. The "lattice" term originates from the idea that the spin system gives up energy to its surroundings as

$$\frac{\partial M_z}{\partial_t} = \frac{-1}{T_1}(M_z - M_0) \tag{9}$$

it tries to reestablish spin equilibrium. The process of transferring spin energy to other modes of energy may be classified as relaxation mechanism. For T_1 in solids, all of the following mechanisms may contribute: dipole–dipole (T_{1DD}), spin-rotation (T_{1SR}), quadrupolar (T_{1Q}), scalar (T_{1SC}), and chemical shift anisotropy (T_{1CSA}). The simple inversion–recovery pulse sequence may be used to measure T_1 times for solid state samples (22). In the use of simple high-powered decoupling, single-pulse sequences for quantitative NMR studies, the inversion–recovery pulse sequence may be used to determine T_1's of interest. A multiple of five times T_1 may then be incorporated as the recycle time between successive pulses, assuring sufficient time for the magnetization to return to equilibrium. In this way, the NMR signal observed is truly representative of the number of nuclei producing it.

To understand the cross-polarization process, two other rate processes must be defined: (a) spin-lattice relaxation in the rotating frame ($T_{1\rho}$) and (b) the cross-polarization relaxation time (T_{CH}). The spin-lattice relaxation in the rotating frame characterizes the decay of magnetization in a field B_1, which is normally much smaller than the externally applied field B_0. In steps 1 and 2 of the CP sequence (Fig. 4), the carbon and proton spin systems are locked by the application of fields B_{1H} and B_{1C} and each system decays with its characteristic time. Figure 5 represents a thermodynamic model for the relaxation processes during CP. During contact between the two spin systems, magnetization is transferred at the rate T_{CH}. Since competing relaxation processes are occurring, the following conditions must be met to obtain a spectrum by CP: $T_{1C} > T_{1H} \geq T_{1\rho H} >$ CP time $> T_{CH}$. It is apparent that for quantitative NMR studies of solids by CP, the individual relaxation processes must be measured to assure that the signal is truly proportional to the amount of species present.

With an understanding of magnetic interactions in the solid state, inherent relaxation processes, and experimental techniques to overcome these difficulties,

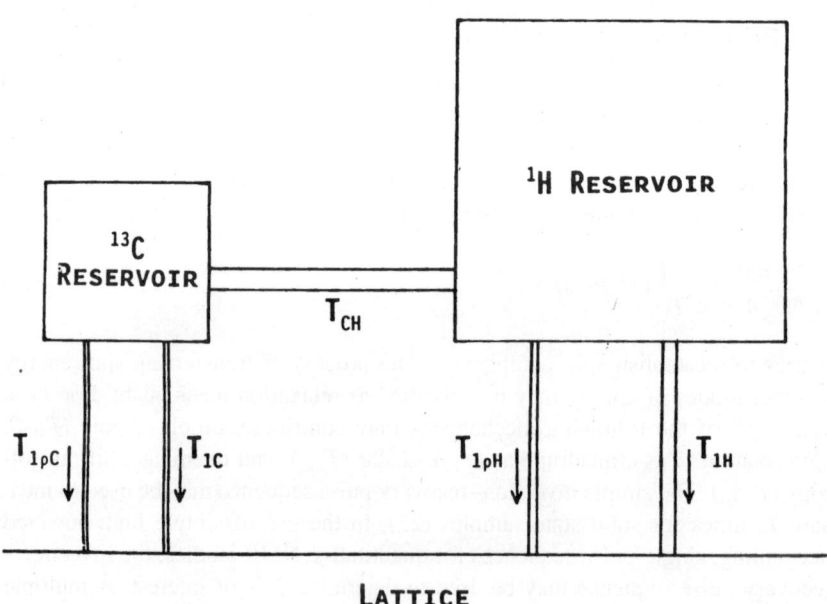

Fig. 5 Thermodynamic model of the cross-polarization sequence and representation of the competing relaxation processes. (From Ref. 11.)

Schaefer et al. acquired the first "liquid-like" NMR spectrum of a solid by CP/ MAS techniques (23).

Before a review of recent applications of solid-state NMR to the pharmaceutical sciences, a brief Experimental section is presented making reference to a number of important pulse sequences.

III. EXPERIMENTAL ASPECTS

Manufacturers of today's modern NMR spectrometers normally offer a number of different models of instruments that are capable of measuring solid-state NMR spectra. The basic components of the solid-state spectrometer are the same as the solution-phase instrument: data system, pulse programmer, observe and decoupler transmitters, magnetic system, and probes. In addition, high-power amplifiers are required for the two transmitters and a pneumatic spinning unit to achieve the necessary spin rates for MAS. Normally, the observe transmitter for ^{13}C work requires broadband amplification of approximately 400 W of power for a 5.87-T, 250-MHz instrument. The amplifier should have triggering capabilities so that only the radiofrequency (rf) pulse is amplified. This will minimize noise contributions to the measured spectrum. So that the Hartmann-Hahn condition may be achieved, the decoupler amplifier must produce an rf signal at one-fourth the power level of the observe channel for carbon work.

Although a series of standard referencing compounds has not been established for solid-state NMR as opposed to solution-phase studies (24), a number of unofficial standard samples are currently being used (25). Table 2 lists a series of compounds, their intended use, and appropriate references further detailing their intended uses and limitations. Additionally, like solution-phase NMR, a myriad of solid-state NMR pulse sequences exist that selectively investigate the sample of interest. A number of these pulse sequences are described within this Experimental section with further discussion outlined in the Applications section of this work.

The single pulse excitation/magic-angle spinning (SPE/MAS) pulse sequence is a means of obtaining solid-state NMR spectra of very sensitive nuclei such as phosphorus-31. The sequence is an adaptation of the solution-phase, gated decoupling NMR experiment (32). In the SPE/MAS sequence, high-power proton decoupling occurs during the signal acquisition period. This decoupling period primarily removes the heteronuclear dipolar coupling that typically broadens solid-state NMR spectra. This pulse sequence is normally used to acquire decoupled spectra on heteronuclear atoms (signified as X in the pulse sequence) that possess high sensitivity (such as the ^{31}P spectra presented in Fig. 3). MAS is used throughout the sequence to minimize broadening due to CSA.

Table 2 Reference Samples for Solid-State NMR

Compound	Intended use	Reference
Adamantane ($C_{10}H_{16}$)	External referencing for ^{13}C spectra, optimizing Hartmann-Hahn match, linewidth measurement, sensitivity measurement	(26)
Hexamethylbenzene ($C_{12}H_{18}$)	Optimizing Hartmann-Hahn match	(26)
Glycine ($C_2H_5O_2N$)	External referencing for ^{13}C spectra, sensitivity measurement, magic angle setting for ^{13}C experimentation	(26)
Ammonium phosphate ($NH_4H_2PO_4$) Monobasic	External referencing for ^{31}P spectra	(26)
Potassium bromide (KBr)	Magic angle setting for ^{13}C experimentation	(27)
ZNP[a], $Zn[S_2P(OC_2H_5)_2]_2$	Magic angle setting for ^{31}P experimentation	(28)
DMPPO[b], $C_8H_{11}OP$	Magic angle setting for ^{31}P experimentation	(29)
Samarium acetate tetrahydrate $[(CH_3CO_2)_3SmC4H_2O]$	^{13}C CP/MAS chemical shift thermometer	(30)
TTAA[c]	^{15}N CP/MAS chemical shift thermometer	(31)

[a] ZNP: zinc(II) bis (O,O'-diethyldithiophosphate).
[b] DMPPO: dimethylphenylphosphine oxide.
[c] TTAA: 1,8-dihydro-5,7,12,14-tetramethyldibenso (b,i)-1,4,8,11-tetraazacyclotetradeca-4,6,11,13-tetraene-$^{15}N_4$ (tetramethyldibenzotetraaza(14)annulene).
Source: Adapted from Ref. 11.

As demonstrated in Fig. 3, even with high-speed MAS, spinning sidebands do occur. These sidebands may be confused with actual resonances in the NMR spectrum and be misinterpreted. Dixon et al. have developed a pulse sequence which suppresses spinning sidebands typically observed in NMR spectra acquired with high-field magnets operating at 5 T and above (18). This pulse sequence has been labeled TOSS, *to*tal *s*uppression of *s*idebands. After initial cross-polarization, Hahn spin echoes are generated with a series of pulses on the X channel. The timing of these pulses are programmed such that the sidebands are inverted, but the isotropic peak is unaffected. Upon addition of spectra with inverted and normal sideband intensities, the sidebands cancel and the isotropic peak remains.

Another pulse sequence that is helpful for the assignment of resonances in a qualitative CP/MAS NMR analysis is the delayed decoupling, or dipolar dephasing experiment. This pulse sequence is simply the CP/MAS experiment with a variable delay time inserted between the spin-locking period and the start of decoupling/acquisition. Nonprotonated carbons are selectively observed by this pulse sequence (33), due to the inherent weaker $^{13}C-^{1}H$ dipolar interaction for carbons with nonbonded protons. During the variable delay time (usually programmed between 50 and 100 µs), directly protonated carbons dephase rapidly and their signal disappears. Subsequently, a simplified spectrum is measured in which only quaternary carbons are observed.

Quantitative analysis is a very important component of pharmaceutical analysis. In a number of the publications describing the use of solid-state NMR, the majority of the work has dealt with qualitative studies with brief references to the possibility of quantitative analysis. Under proper data acquisition conditions, solid-state NMR is a quantitative technique that typically provides sufficient selectivity and sensitivity. An excellent guide to the utilization of magic angle spinning and cross-polarization techniques for quantitative solid-state NMR data acquisition has been outlined by Harris (34).

As discussed in the Theory section, in order to acquire solid-state NMR spectra in which the signal intensities truly reflect the nuclei producing it, data acquisition parameters such as recycle time, pulse widths, cross-polarization time, Hartmann-Hahn match, and decoupling power must be explicitly determined for each chemical system. If these mechanisms, primarily relaxation, are not understood and improper acquisition parameters inserted into the pulse sequence, the quantitative nature of solid-state NMR can be compromised. In Harris's article, the problems of solid-state NMR acquisition techniques as applied to quantitative measurements are addressed. Additionally, errors in the setting of the magic angle, Hartmann-Hahn match, and cross-polarization mixing time are discussed in relation to obtaining quantitative NMR results. To this end, a series of pulse sequences is presented here in order to determine the time factor of various relaxation mechanisms and to program this information into the quantitative data acquisition pulse sequence.

From previous discussions it is known that competing rate processes occur in the cross-polarization NMR experiment. So that the measured NMR signal truly represents the number of nuclei producing it, explicit measurements must be made for these various mechanisms, namely, T_{CP} and $T_{1\rho H}$. For SPE experiments, T_1 must be measured for the nucleus of interest.

During the Hartmann-Hahn condition, proton magnetization is being transferred to the carbon spins at the rate of $1/T_{CH}$ *and* to the lattice at the rate of $1/T_{1\rho H}$. In order for CP to occur, $1/T_{CH} < 1/T_{1\rho H}$ ($1/T_{1\rho C}$ is being ignored, since it is greater than $1/T_{1\rho H}$). The ^{13}C magnetization will achieve a maximum value at a specific contact time, which also signifies optimum sensitivity for the CP experiment. For any quantitative analysis studies, the optimal contact time should be determined for the chemical system of interest. Normally, the contact time is measured by varying the contact time in the CP/MAS pulse sequence and measuring the subsequent signal intensity. By plotting the contact time versus the signal intensity on a logarithmic scale, the optimal contact time for each resonance may be determined (35). Once the contact time is determined for the distinct resonance observed in the NMR spectrum, this value, or a compromised/average value, may be inserted into the quantitative CP/MAS pulse sequence.

In addition to measuring T_{CH} for the chemical system in question, the proton $T_{1\rho}$ value must be determined since the repetition rate of a CP experiment is dependent on the recovery of the proton magnetization. Common convention states that a delay time between successive pulses of $1-5 \times T_1$ must be used. One advantage to solids NMR work is that a common proton $T_{1\rho}$ value will be measured, since protons communicate through a spin-diffusion process. The most common way to measure the proton $T_{1\rho}$ is to apply a proton 90° pulse, wait a variable delay time, and then apply the normal CP pulse sequence. By measuring the ^{13}C intensity versus the variable delay time, the proton $T_{1\rho}$ time may be approximated (35).

Once the proton $T_{1\rho}$ and T_{CH} values are determined for the chemical system, physical mixtures of the two entities can be generated (calibration samples). Subsequent acquisition of the solid-state NMR spectra under quantitative conditions yields signal intensities representative of the amount of each material present. Measurement of the relative signal intensities may then lead to determination of the amount of one species present in an unknown mixture.

In some cases, CP is not necessary to obtain a suitable solid-state NMR spectrum. In these cases, the SPE/MAS sequence may be used and for quantitative analysis only the X-nucleus T_1 time needs to be determined. The standard inversion–recovery experiment can be used to measure this value, keeping in mind that MAS and high-power decoupling is still necessary. As before, once the X-nucleus T_1 time is determined, $1-5 \times T_1$ may be inserted as the delay time between successive pulses for quantitative data acquisition.

IV. PHARMACEUTICAL APPLICATIONS

Solid-state NMR has and continues to be used in a myriad of pharmaceutical applications, but unfortunately, all of the different applications cannot be discussed in this chapter. Instead, applications in drug development, as opposed to drug discovery, are presented herein. Aboul-Enein has outlined some of the general uses of solid-state NMR for pharmaceutical research (36). Phosphorus-31 cross-polarization/magic-angle spinning (CP/MAS) NMR has been used to study the dehydration of disodium clodronate (37). A fast rise in temperature reveals that disodium clodronate loses lattice water and only one ^{31}P resonance is measured, whereas a slow increase in temperature converts the crystalline form to an anhydrous form which displays two nonequivalent phosphorus atoms (resonances).

Solid-state NMR has been used extensively for characterizing the structures of N-desmethylnefopam HCl (38), patellin (39), erythromycin A dihydrate (40), and the amorphous nature of ursodeoxycholic acid (41). The conformations of 3′-amino-3′-deoxythymidine (42), gramicidin A (43), and amiodarone HCl (44) have been confirmed by solid-state NMR.

Due to the specificity of solid-state NMR, it is an ideal technique to study inclusion complexes, drug–excipient interactions, or the effect of moisture on the drug substance or formulation. Studies on inclusion complexes include gliclazide-β-cyclodextrin (45) and hydrocortisone butyrate (46), whereas Makriyannis utilized NMR for investigating drug–membrane interactions (47). Other interactions studied by solid-state NMR include polyethylene glycol with griseofulvin (48) and the moisture-induced interaction with trospectomycin sulfate, which affects the 3′-gem-diol ↔ 3′-keto equilibrium in either the drug substance or freeze-dried formulation (49).

A. Polymorphism

The most widely used application of solid-state NMR in drug development is in the area of polymorphism (50,51). Pharmaceutical solids can exist in a number of solid forms, each having different properties of pharmaceutical importance, including stability and bioavailability; the number and properties of these forms are largely unpredictable and vary considerably from case to case. Pharmaceutical solids can be divided into crystalline and amorphous solids based on the presence of crystalline structure. This classification is usually based on X-ray powder diffraction and/or microscopic examination. Crystalline solids can then be further classified into: crystalline polymorphs, forms having the same chemical composition but different crystal structures and therefore different densities, melting points, solubilities, and many other important properties; and solvates, forms containing solvent molecules within the crystal structure, giving rise to unique differences in solubility, response to atmospheric moisture, loss of solvent, etc. Sometimes a drug substance may be a desolvated solvate, which is formed when solvent is removed from a specific solvate while the crystal structure is essentially retained—again, many important properties are unique to such a form.

Different physical forms of a drug substance can display radically different solubilities, which directly affects the dissolution and bioavailability characteristics of the compound. In addition, the chemical stability of one form, as compared to the other, may vary. The physical stability is also crucial. During various processing steps (grinding, mixing, tablet pressing, etc.), the physical form of the drug substance may be compromised, subsequently leading to dissolution problems. For these reasons, the full characterization of polymorphic systems is critical to numerous entities (preformulation/physical pharmacy group, chemical process development, regulatory/patent office, and analytical laboratories) within drug development companies.

The use of solid-state NMR for the investigation of polymorphism is easily understood based on the following model. If a compound exists in two true polymorphic forms, labeled as A and B, each crystalline form is crystallographically different. This means, for instance, that a carbon nucleus in form A may be situated in a slightly different molecular environment as compared to the same carbon nucleus in form B. Although the connectivity of the carbon nucleus is the same in each form, the local environment may be different. Since the local environment may be different, this potentially leads to a different chemical-shift interaction for each carbon, and ultimately, a different isotropic chemical shift for the same carbon atom in the two different polymorphic forms. If one is able to obtain pure material for the two forms, analysis and spectral assignment of the solid-state NMR spectra of the two forms can lead to the origin of the crystallographic differences in the two polymorphs. Solid-state NMR is thus an important tool in the multidisciplinary approach to polymorphism.

Solid-state NMR analysis of different solid forms can provide crucial information which often resolves conflicting results between other methods such as infrared, X-ray powder diffraction, and thermal methods (52). In addition, a dictate of formulation technology is that the physical form of the drug substance, after being defined and verified, should not change once the product has been manufactured. Solid-state NMR provides a powerful method for comparing the physical form of the drug substance after pharmaceutical processing or manufacturing. Furthermore, solid-state NMR provides a method for the analysis of mixtures of crystal forms in both the pure drug substance and in dosage forms (52).

There are a number of significant advantages to the use of solid-state NMR for the study of polymorphism. As compared to diffuse reflectance IR, Raman, and X-ray powder diffraction techniques, solid-state NMR is a bulk technique in which the intensity of the measured signal is not a function of the particle size of the material. In addition, NMR is an absolute technique under proper data acquisition procedures, meaning that the intensity of the signal is directly proportional to the number of nuclei producing it, thus permitting quantitative analysis. In some polymorphic investigations, a single crystal of sufficient quality may not be available for X-ray crystallography (structure determination). By proper assignment of the NMR spectrum, though, the origin of the polymorphism may be inferred by differences in the resonance position for identical nuclei in each polymorphic form.

B. Drug Substance Qualitative Analysis

The majority of applications of solid-state NMR used in the investigation of pharmaceutical polymorphs are performed in conjunction with other analytical techniques. Byrn et al. have reported differences in the solid-state NMR spectra for different polymorphic forms of benoxaprofen, nabilione, and pseudo-polymorphic forms of cefazolin (53). Although single-crystal X-ray crystallography was initially used to study the polymorphs, the solid-state ^{13}C CP/MAS NMR spectrum of each form was distinctly different. These studies focused primarily on the bulk drug material, although a granulation of benoxaprofen was studied. It was concluded that solid-state NMR could be used to differentiate the form present in the granulation even in the presence of excipients. In further studies, the crystalline forms of prednisolone *tert*-butylacetate (54), cefaclor dihydrate (55), and glyburide (56) were studied. The five crystal forms of prednisolone *tert*-butylacetate were again determined by single-crystal X-ray crystallography, and solid-state NMR was used to determine the effect of crystal packing on the ^{13}C chemical shifts of the different steroid forms. Although conformational changes were observed in the ester side chain by X-ray crystallography, no major differences were noted in the NMR spectra, indicating that the environment remains relatively unchanged. Significant chemical shift differences were noted for carbonyl atoms involved in hydrogen bonding. This theme is consistent in the NMR study of cefaclor dihydrate. Again, the effects of hydrogen bonding were discernible by solid-state NMR. The study of glyburide was principally concerned with the structural conformation of the molecule in the solution and solid state. The solution conformation was determined by ^{1}H and ^{13}C NMR, and in the solid by single-crystal X-ray crystallography, IR, and solid-state ^{13}C NMR. The solid-state NMR results suggested that this method would be useful for comparing solid- and solution-state conformations of molecules.

In 1989, Etter published an excellent paper on the complementary use of solid-state NMR and X-ray crystallography to study the structure of pharmaceutical solids (57). Crystallographic effects such as polymorphism, multiple molecules per asymmetric unit cell, disorder, intra- and intermolecular H-bonding, tautomerism, and solvation were all investigated by solid-state NMR.

In a series of publications by Harris and Fletton, solid-state NMR has been used to investigate the structure of polymorphs. The majority of studies involve X-ray and IR techniques and also address the possibility of quantitative measurements of polymorphic mixtures. The three pseudo-polymorphic forms of testosterone were examined by IR and ^{13}C CP/MAS NMR (58). The two forms of molecular spectroscopy were able to differentiate the forms, but NMR had the ability to investigate nonequivalent molecules in a given unit cell. A series of doublet resonances was noted for a series of different carbon atoms. This implied that the specific carbon atom within the molecule may resonate at two different frequencies, depending on the crystallographic site of the molecule. In addition to the hydrogen-bonding explanation of the crystallographic splittings, the use of NMR to determine quantitatively the amount of each pseudo-polymorph present in a mixture was addressed. In the study of androstanolone (59), high-quality ^{13}C NMR spectra were obtained by CP/MAS techniques and allowed for characterization of the anhydrous and monohydrate forms. Again, crystallographic splittings were noted for the two forms and were related to hydrogen bonding. An identical approach to study pharmaceutical polymorphic structure was used in the investigation of the two polymorphs of 4′-methyl-2′-nitroacetanilide (60). An additional study of cortisone acetate (61,62) by solid-state NMR revealed differences in the NMR spectra for the six crystalline forms. Further multidisciplinary approaches to the physical characterization of pharmaceutical compounds are detailed in separate studies on cyclopenthiazide (63), the excipient lactose (64), fursemide (65), losartan (66), fosinopril sodium (67), and leukotriene antagonists MK-679, MK-571 (68), L-660,711 (69), captopril (70), diphenhydramine HCl (71), mofebutazone (72), phenylbutazone (72), oxyphenbutazone (72), cimetidine (73), and the iron chelator 1,2-dimethyl-3-hydroxy-4-pyridone (74). In each case, solid-state NMR was used in conjunction with other techniques such as DSC, IR, X-ray diffraction, microscopy, and solubility/dissolution studies to characterize the polymorphic systems fully.

In further studies of polymorphism by solid-state NMR, conversion from one solid-state form to another by UV irradiation (75) and variable temperature techniques (76) are outlined. In the first study, NMR was employed to follow the chemical transformation within the organic crystals of p-formyl-$trans$-cinnamic acid (p-FCA). A photoreactive β-phase may be crystallized from ethanol, whereas a photostable γ-phase is produced from acetone. After irradiation of the β-phase with UV radiation, and subsequent acquisition of the solid-state ^{13}C NMR spectrum, the photoproduct was easily identified by NMR. The second conversion study (76) investigated the four forms of p-amino-benzenesulphonamide sulphanilamide (α, β, γ, and δ). The first three forms were fully investigated by solid-state ^{13}C NMR and X-ray crystallography techniques. Subsequent variable-temperature studies monitored the interconversion of the α- and β-forms to the γ-form. Coalescence of some NMR signals in the γ-form also suggested that phenyl ring motion occurred within the crystal. Conclusions from the study indicated that solid-state NMR had the ability to differentiate pharmaceutical polymorphs, determine asymmetry in the unit cell, and investigate molecular motion within the solid state.

C. Drug Substance Quantitative Analysis

In a number of the publications describing the use of solid-state NMR for polymorphic characterization, the majority of the work has dealt with qualitative studies, with brief references to the possibility of quantitative analysis. Quantitative solid-state ^{13}C CP/MAS NMR has been used to determine the relative amounts of carbamazepine anhydrate and carbamazepine dihydrate in mixtures (77). The ^{13}C-NMR spectra for the two forms did not appear different, although sufficient S/N for the spectrum of the anhydrous form required long accumulation times. This was determined to be due to the slow proton relaxation rate for this form. Utilizing the fact that different proton spin-lattice relaxation times exist for the two different pseudopolymorphic forms, a quantitative method was developed. The dihydrate form displayed a relatively short relaxation time, permitting interpulse delay times of only 10 s to obtain full-intensity spectra of the dihydrate form while displaying no signal due to the anhydrous form. By utilizing an internal standard (glycine) and the differences in the relaxation rate of the two forms, the peak area of the dihydrate could be measured and related through a calibration curve to the amount of anhydrous and dihydrate content in mixtures of carbamazepine.

D. Drug Product Analysis

A series of papers has been published on the solid-state NMR spectra of a number of analgesic drugs. Jagannathan recorded the solid-state ^{13}C NMR spectrum of acetaminophen in bulk and dosage forms (78). From the solution-phase NMR spectrum, assignments of the solid-state NMR resonances could be made in addition to explanations for the doublet structure of some resonances (dipolar coupling). Spectra of the dosage product from two sources indicated identical drug substance, but different levels of excipients. The topic of drug–excipient interactions was addressed by solid-state ^{13}C NMR in the investigation of different commercially available aspirin samples (79,80). In each commercial aspirin product, the only difference in the measured NMR spectrum was due to variations in the excipients, indicating that there were no interactions between the drug and the excipients under dry blending conditions. After lyophilization of two of the products, one aspirin sample did show a different NMR spectrum, indicating possible interaction during lyophilization or conversion to a different solid-state form during processing.

Saindon et al. used solid-state NMR to show that different brands of prednisolone tablets contained different polymorphs (81). For the listed drugs, the NMR analysis could discern the active component except for the low-dose enalapril maleate tablets. Excipients obscure some drug resonances which appear at the same chemical shift value, but overall, have little effect on resonances at other chemical shifts. Similar results were obtained for tablets of another closely related HMG-CoA reductase inhibitor, simvastatin.

E. Mobility

Molecular mobility can be studied by solid-state NMR and X-ray crystallography. Solid-state NMR offers several approaches to studying the molecular mobility of solids. These include: (a) the study of processes which result in peak coalescence of solid-state NMR resonances using variable temperature solid-state NMR, (b) determination of the T_1 relaxation of individual carbon atoms using variable temperature solid-state NMR, (c) use of interrupted decoupling to detect methylene and possible methine groups with unusual mobility, and (d) comparison of solid-state MAS spectra measured with and without cross-polarization.

The rate of relaxation of the carbon spins toward equilibrium is characterized by the spin-lattice relaxation time, T_1 (82). In this process, the excess energy from the spin system is given to the surroundings or lattice. Interactions of randomly fluctuating magnetic fields at the Lamor frequency of the nucleus stimulates these transitions. Such fields arise from motions of other nuclear magnetic moments such as protons. Spin-lattice relaxation is thus most efficient when these motions are near the Lamor frequency. Studies of T_1 can give valuable information about fast motions in the megahertz range, such as methyl group rotations. Variable-temperature solid-state NMR spectroscopy has been shown to be a powerful technique for the study of molecular motion in cortisone (83), cholesterol (84), proteins (85), polymers (86–90), and amino acids (91).

In addition, solid-state NMR has been used to study protein hydration and stability (92) and the activation energies of spinning methyl groups (93). The activation energy for the spinning of methyl groups in triethylphosphine oxide was reported to be 7.9 kJ/mol by T_1 studies and 8.69 kJ/mol by $T_{1\rho}$ studies (88). Relaxation studies of *peri*-substituted naphthalenes were used to determine the barrier to methyl rotation when substituents at the peri position were —H, —CH$_3$, —Cl, and —Br (94). The observed barriers were —H, 9.7 kJ/mol; —CH$_3$, 13.5 kJ/mol; —Cl, 15.2 kJ/mol; and —Br, 18.4 kJ/mol. The barriers follow the sequence of the van der Waals radii and indicate that if a large group abuts the methyl group, it can restrict methyl rotation. Rotation barriers for methyl groups in several amino acids have also been reported (93).

F. Chirality

The issue of chirality continues to be an important issue in drug development. With the advent of solid-state NMR, enantiomeric purity of a sample may be measured directly. A number of reports in the literature demonstrate the ability of ^{13}C or ^{31}P solid-state NMR to differentiate between optically pure material and racemic species (95). In one report, tartaric acid was studied by ^{13}C CP/MAS NMR (96). For each of the optically pure (2R, 3R), racemic, and *meso*-tartaric acid material studied, two molecules exist per asymmetric unit cell. Since the carbonyls and α-carbon atoms in a single molecule are not symmetry related, one would expect two resonances for each carbon (crystallographic splitting). The ^{13}C NMR spectra showed two resonances for each carbon, but also demonstrated a distinct isotropic chemical shift value for each resonance upon comparison to each other. This data demonstrated the potential for solid-state NMR to differentiate between racemic and enantiomeric crystallites.

A series of racemic and optically pure organophosphorus samples were studied by solid-state ^{31}P NMR (97). Analogous to the tartaric acid work, the two optical isomers ($+$ and $-$) showed the same ^{31}P NMR spectrum that were distinct from the spectrum of the racemic material. In addition, the issue of quantitative analysis of optical purity was addressed. The conformational analysis of DL-, L-, and D-methionine was studied by solid-state ^{13}C NMR (98). Based on the NMR studies, each crystalline form of DL-methionine consists of a single conformer, which agrees with X-ray diffraction data.

G. Future Direction

In recent years, two-dimensional solid-state NMR spectroscopy (99,100) has been applied to pharmaceutical systems. The overall goal of these studies is to obtain structural information. Such studies are particularly important and useful when the crystal structure cannot be determined.

Raftery's laboratory has used two-dimensional (2D) solid-state NMR to study the red, orange, and yellow conformational polymorphs of ROY (5-methyl-2-(2-nitrophenyl)amino]-3-thiophenecarbonitrile (101). The separation of side-bands by order experiment (2DTOSS) separates the isotropic and anisotropic chemical shifts over two dimensions, allowing analysis of the chemical-shift anisotropy patterns using magic-angle spinning in the 1500-Hz range (102). The chemical shift anisotropy of the C-3 carbon atom (the atom attached to the CN) could be easily distinguished using this pulse sequence and was used to probe differences in chemical environment between the different color polymorphs. Raftery and co-workers showed that two-dimensional solid-state NMR can easily distinguish differences between conformational polymorphs and is a powerful method for analyzing such systems. Raftery's group is now combining information from the 2DTOSS experiments with ab-initio or density functional methods to attempt to obtain quantitative structural information. This combined approach along with other 2-D solid-state NMR experiments, promises to provide structural information on polymorphs whose crystal structure cannot be determined.

Munson and co-workers have used 2D, high-speed ^{13}C CP/MAS NMR to study uniformly ^{13}C-labeled aspartame (103). Aspartame exists in three hydrates (two hemihydrates and a dihemihydrate). Two of these hydrates apparently contain three molecules per asymmetric unit, giving up to three resolvable resonances for each carbon atom. This complex crystal packing prevented assignments of the resonances using techniques based on the number of attached protons or J couplings. Typical MAS experiments at spinning rates of 7 kHz and proton decoupling powers of 63 kHz gave broadened spectra due to broadening by dipolar decoupling. However, increasing the spinning rate to 28 kHz and the decoupling to 263 kHz gave narrow line spectra that allowed the assignment of the crystallographically inequivalent sites. For one of the forms, peak assignments could be made for all three molecules in the asymmetric unit of the crystal.

As with the 2DTOSS studies, Munson and co-workers showed that solid-state NMR can provide important information on the structures of polymorphs and solvates whose crystal structure is not known. As we seek more detailed structural information of drugs in order to understand more about solid-state behavior, these 2D techniques will no doubt become more and more important. Additional 2D techniques have been utilized to study DNA complexes (104) and perform distance measurements (105), specifically using the rotational-echo double-resonance (REDOR) sequence (106,107).

Regulatory documentation is now making specific reference to solid-state NMR. A flow chart approach to the physical characterization of pharmaceutical solids was first published in 1995 (108). In the flow chart approach to determining the number of polymorphic forms, spectroscopy, and specifically solid-state NMR, is outlined as a recommended technique. The U.S. Food and Drug Administration (FDA) has also recognized the need for solid-state NMR characterization of drug substance or product. In the most recent International Committee of Harmonization (ICH) guideline for the setting of specifications for drug substance and product, spectroscopy (IR, Raman, and solid-state NMR) is referred to in decision tree four: investigating the need to set acceptance criteria for polymorphism in drug substances and drug products (109). Worldwide regulatory authorities now recognize the importance of solid-state NMR and how the technique is intimately tied to the drug development process.

V. SUMMARY

The multinuclear technique of solid-state NMR has been applied to various aspects of drug development with tremendous success. Although the technique ideally lends itself to the structure determination of drug compounds in the solid state, it is anticipated that in the future, solid-state NMR will be used routinely for method development and problem-solving activities in the analytical/materials science/physical pharmacy area of the pharmaceutical sciences. During the past few years, an increasing number of publications have emerged in which solid-state NMR has become an invaluable technique. With the continuing development of solid-state NMR pulse sequences and hardware improvements (increased sensitivity), solid-state NMR will provide a wealth of information for the characterization of pharmaceutical solids.

REFERENCES

1. PA Mirau. A strategy for NMR structure determination. J Magn Reson 96:480–490, 1992.
2. AH Beckelt. Chirality and its importance in drug development. Biochem Soc Trans 19:443–446, 1991.
3. O Kaplan, JS Cohen. Magnetic resonance as a non-invasive tool to study metabolism in pharmacological research. Trends Pharmacol Sci 11:398–399, 1990.
4. FMH Jeffrey, A Rajagopal, CR Malloy, AD Sherry. ^{13}C-NMR: A simple yet comprehensive method for analysis of intermediary metabolism. Trends Biochem Sci 16:5–10, 1991.
5. V Stoven, JY Lallemand, D Abergel, S Bovaziz, MA Delsuc, A Ekondzi, E Guittet, S Laplante, R LeGoas, T Malliavin, A Mikov, C Reisdorf, M Rogin, C van Heijenoort, Y Yang. Insight into protein nuclear magnetic resonance research. Biochimie 72:531–535, 1990.
6. JPM van Duynhoven, PJM Folkers, CWJM Prinse, BJM Harmsen, RNH Konings, CW Hilbers. Assignment of the ^{1}H NMR spectrum and secondary structure elucidation of the single-stranded DNA binding protein encoded by the filamentous bacteriophage IKe. Biochemistry 31:1254–1262, 1992.
7. GM Clore, AM Gronenborn. Determination of structures of larger proteins in solution by three- and four-dimensional heteronuclear magnetic resonance spectroscopy. In: GM Clore, AM Gronenborn, eds. NMR of Proteins. Boca Raton, FL: CRC Press, 1993, pp. 1–32.
8. JK Baker, CW Myers. One-dimensional and two-dimensional ^{1}H- and ^{13}C-nuclear magnetic resonance (NMR) analysis of vitamin E raw materials or analytical reference standards. Pharm Res 8:763–770, 1991.
9. A Burger, R Ramberger. On the polymorphism of pharmaceuticals and other molecular crystals. I. Theory of thermodynamic rules. Mikrochim Acta [Wien] II:259–271, 1979. A Burger, R Ramberger. On the polymorphism of pharmaceuticals and other molecular crystals. II. Applicability of thermodynamic rules. Mikrochim Acta [Wien] II:273–316, 1979.
10. J Haleblian, W McCrone. Pharmaceutical applications of polymorphism. J Pharm Sci 58:911–929, 1969.
11. DE Bugay. Solid state nuclear magnetic resonance spectroscopy—theory and pharmaceutical applications. Pharm Res 10:317–327, 1993.
12. Karplus MJ. Vicinal proton coupling in nuclear magnetic resonance. J Am Chem Soc 85:2870–2871, 1963.
13. TC Farrar, ED Becker. Pulse and Fourier Transform NMR. New York: Academic Press, 1971.

14. ER Andrew, A Bradbury, RG Eades. Removal of dipolar broadening of nuclear magnetic resonance spectra of solids by specimen rotation. Nature 183:1802–1803, 1959.
15. ER Andrew. The narrowing of NMR spectra of solids by high-speed specimen rotation and the resolution of chemical shift and spin multiplet structures for solids. Prog Nuclear Magn Reson Spectrosc 8:1–39, 1972.
16. A Pines, MG Gibby, JS Waugh. Proton-enhanced nuclear induction spectroscopy. A method for high resolution NMR of dilute spins in solids. J Chem Phys 56:1776–1777, 1972. A Pines, MG Gibby, JS Waugh. Proton-enhanced NMR of dilute spins in solids. J Chem Phys 59:569–590, 1973.
17. A Salvetti. Newer ACE inhibitors, a look at the future. Drugs 40:800–828, 1990.
18. WT Dixon, J Schaefer, MD Sefcik, EO Stejskal, RA McKay. Total suppression of sidebands in CPMAS carbon-13 NMR. J Magn Reson 49:341–345, 1982.
19. CA Fyfe. Solid State NMR for Chemists. Guelph, Ontario, Canada: CFC Press, 1983.
20. SR Hartmann, EL Hahn. Nuclear double resonance in the rotating frame. Phys Rev 128:2042–2053, 1962.
21. RJ Abraham, J Fisher, P Loftus. Introduction to NMR Spectroscopy. New York: Wiley, 1988:84–86.
22. TC Farrar, ED Becker. Pulse and Fourier Transform NMR. New York: Academic Press, 1971:20–22.
23. J Schaefer, EO Stejskal, R Buchdahl. High-resolution carbon-13 nuclear magnetic resonance study of some solid, glassy polymers. Macromolecules 8:291–296, 1975.
24. P Granger. Multinuclear NMR referencing. Appl Spectrosc 42:1–3, 1988.
25. WL Earl, DL VanderHart. Measurement of ^{13}C chemical shifts in solids. J Magn Reson 48:35–54, 1982.
26. Bruker Instruments. The CP/MAS Accessory Product Description Manual. Billerica, MA: Bruker Instruments, 1987.
27. JS Frye, GE Maciel. Setting the magic angle using a quadrupolar nuclide. J Magn Reson 48:125–131, 1982.
28. A Kubo, CA McDowell. Setting of the magic angle for ^{31}P MAS NMR using zinc(II) bis(O, O'-diethyldithiophosphate). J Magn Reson 92:409–410, 1991.
29. RC Crosby, JF Haw. Dimethylphenylphosphine oxide, a compound for setting the magic angle in ^{31}P CP/MAS NMR. J Magn Reson 82:367–368, 1989.
30. GC Campbell, RC Crosby, JF Haw. ^{13}C chemical shifts which obey the curie law in CP/MAS NMR spectra. The first CP/MAS NMR chemical-shift thermometer. J Magn Reson 69:191–195, 1986.
31. B Wehrle, F Aguilar-Parrilla, H-H Limbach. A novel ^{15}N chemical-shift NMR thermometer for magic angle spinning experiments. J Magn Reson 87:584–591, 1990.
32. RJ Abraham, J Fisher, P Loftus. Introduction to NMR Spectroscopy, New York: Wiley, pp. 113–116, 1988.
33. SJ Opella, MH Frey. Selection of nonprotonated carbon resonances in solid-state nuclear magnetic resonance. J Am Chem Soc 101:5854–5856, 1979.
34. RK Harris. Quantitative aspects of high-resolution solid-state nuclear magnetic resonance spectroscopy. Analyst 10:649–655, 1985.
35. D Michel, F Engelke. Cross-Polarization, Relaxation Times and Spin-Diffusion in Rotating Solids. In: P Diehl, E Fluck, H Günther, R Kosfeld, J Seelig, eds. NMR

Basic Principles and Progress, Vol. 32. New York: Springer-Verlag, 1994, pp. 69–125.

36. Y Aboul-Enein. Applications of solid-state nuclear magnetic resonance spectroscopy to pharmaceutical research. Spectroscopy 5:32–33, 1990.

37. JT Timonen, E Pohjala, H Nikander, TT Pakkanen. A ^{31}P CP-MAS NMR study on dehydration of disodium clodronate tetrahydrate. Pharm Res 15:110–115, 1998.

38. R Glaser, S Geresh, J Blumenfeld, D Donnell, N Sugisaka, M Drouin, A Michel. Solution and solid-state structures of N-desmethylnefopam hydrochloride, a metabolite of the analgesic drug. J Pharm Sci 82:276–281, 1993.

39. TM Zabriskie, MP Foster, TJ Stout, J Clardy, CM Ireland. Studies on the solution- and solid-state structure of patellin 2. J Am Chem Soc 112:8080–8084, 1990.

40. GA Stephenson, JG Stowell, PH Toma, RR Pfeiffer, SR Byrn. Solid-state investigations of erythromycin A dihydrate: Structure, NMR spectroscopy, and hygroscopicity. J Pharm Sci 86:1239–1244, 1997.

41. E Yonemochi, Y Ueno, T Ohmae, T Oguchi, S-I Nakajima, K Yamamoto. Evaluation of amorphous ursodeoxycholic acid by thermal methods. Pharm Res 14:798–803, 1997.

42. T Kovacs, L Parkanyi, I Pelczer, F Cervantes-Lee, KH Pannell, PF Torrence. Solid-state and solution conformation of 3′ amino-3′-deoxythymidine precursor to a noncompetitive inhibitor of HIV-1 reverse transcriptase. J Med Chem 34:2595–2600, 1991.

43. RR Ketchem, W Hu, A Cross. High-resolution conformation of gramicidin A in a lipid bilayer by solid-state NMR. Science 261:1457–1460, 1993.

44. RC Rao, F Lefebvre, G Commenges, C Castet, L Miguel, G Maire, M Gachon. Conformation of amiodarone HCl: Solution and solid state ^{13}C NMR study. Pharm Res 11:1088–1092, 1994.

45. JR Moyano, MJ Arias-Blanco, JM Gines, F Giordano. Solid-state characterization and dissolution characteristics of gliclazide-β-cyclodextrin inclusion complexes. Int J Pharm 148:211–217, 1997.

46. IK Chun, DS Yun. Inclusion complexation of hydrocortisone butyrate with cyclodextrins and dimethyl-β-cyclodextrin in aqueous solution and in solid state. Int J Pharm 96:91–103, 1993.

47. A Makriyannis, T Mavromoustakos. Studies on drug-membrane interactions using solid state NMR, small angle X-ray diffraction, and differential scanning calorimetry. Epitheor Klin Farmakol Farmakokinet, Int Ed 3:95–114, 1989.

48. M Alden, J Tegenfeldt, ES Saers. Structures formed by interactions in solid dispersions of the system polyethylene glycol griseofulvin with charged and non-charged surfactants added. Int J Pharm 94:31–38, 1993.

49. MD Likar, RJ Taylor, PE Fagerness, Y Hiyama, RH Robins. The 3′-keto-diol equilibrium of trospectomycin sulfate bulk drug and freeze-dried formulation: solid-state carbon-13 cross-polarization magic angle spinning (CP/MAS) and high-resolution carbon-13 nuclear magnetic resonance (NMR) spectroscopy studies. Pharm Res 10:75–79, 1993.

50. DE Bugay. Magnetic Resonance Spectrometry. In: HG Brittain, ed. Physical Characterization of Pharmaceutical Solids. New York: Marcel Dekker, pp. 93–125, 1995.

51. SR Byrn, B Tobias, D Kessler, J Frye, P Sutton, P Saindon, J Kozlowski. Relationship between solid state NMR spectra and crystal structures of polymorphs and solvates of drugs. Trans Am Crystallogr Assoc 24:41–54, 1989.

52. DE Bugay. Solid state nuclear magnetic resonance spectroscopy—theory and pharmaceutical applications. Pharm Res 10:317–327, 1993.

53. SR Byrn, G Gray, RR Pfeiffer, J Frye. Analysis of solid-state carbon-13 NMR spectra of polymorphs (benoxaprofen and nabilone) and pseudopolymorphs (cefazolin). J Pharm Sci 74:565–568, 1985.

54. SR Byrn, PA Sutton, B Tobias, J Frye, P Main. Crystal structure, solid-state NMR spectra, and oxygen reactivity of five crystal forms of prednisolone *tert*-butylacetate. J Am Chem Soc 110:1609–1614, 1988.

55. H Martinez, SR Byrn, RR Pfeiffer. Solid-state chemistry and crystal structure of cefaclor dihydrate. Pharm Res 7:147–153, 1990.

56. SR Byrn, AT McKenzie, MMA Hassan, AA Al-Badr. Conformation of glyburide in the solid state and in solution. J Pharm Sci 75:596–600, 1986.

57. MC Etter, GM Vojta. The use of solid-state NMR and X-ray crystallography as complementary tools for studying molecular recognition. J Mol Graphics 7:3–11, 1989.

58. RA Fletton, RK Harris, AM Kenwright, RW Lancaster, KJ Packer, N Sheppard. A comparative spectroscopic investigation of three pseudopolymorphs of testosterone using solid-state i.r. and high-resolution solid-state NMR. Spectrochim Acta 43A: 1111–1120, 1987.

59. RK Harris, BJ Say, RR Yeung, RA Fletton, RW Lancaster. Cross-polarization/ magic-angle spinning NMR studies of polymorphism: androstanolone. Spectrochim Acta 45A:465–469, 1989.

60. RA Fletton, RW Lancaster, RK Harris, AM Kenwright, KJ Packer, DN Waters, A Yeadon. A comparative spectroscopic investigation of two polymorphs of 4′-methyl-2′-nitroacetanilide using solid-state infrared and high-resolution solid-state nuclear magnetic resonance spectroscopy. J Chem Soc Perkin Trans 2, 11:1705–1709, 1986.

61. RK Harris, AM Kenwright, BJ Say, RR Yeung, RA Fletton, RW Lancaster, GL Hardgrove Jr. Cross-polarization/magic-angle spinning NMR studies of polymorphism: Cortisone acetate. Spectrochim Acta 46A:927–935, 1990.

62. EA Christopher, RK Harris, RA Fletton. Assignments of solid-state ^{13}C resonances for polymorphs of cortisone acetate using shielding tensor components. Solid State Nuclear Magn Reson 1:93–101, 1992.

63. JJ Gerber, JG van derWatt, AP Lötter. Physical characterisation of solid forms of cyclopenthiazide. Int J Pharm 73:137–145, 1991.

64. HG Brittain, SJ Bogdanowich, DE Bugay, J DeVincentis, G Lewen, AW Newman. Physical characterization of pharmaceutical solids. Pharm Res 8:963–973, 1991.

65. C Doherty, P York. Frusemide crystal forms; solid state and physicochemical analyses. Int J Pharm 47:141–155, 1988.

66. K Raghavan, A Dwivedi, GC Campbell Jr, E Johnston, D Levorse, J McCauley, M Hussain. A spectroscopic investigation of losartan polymorphs. Pharm Res 10: 900–904, 1993.

67. HG Brittain, KR Morris, DE Bugay, AB Thakur, ATM Serajuddin. Solid-state

NMR and IR for the analysis of pharmaceutical solids: polymorphs of fosinopril sodium. J Pharm Biomed Anal 11:1063–1069, 1994.

68. RG Ball, MW Baum. A spectroscopic and crystallographic investigation of the structure and hydrogen bonding properties of the chiral leukotriene antagonist MK-679 as compared to its racemate MK-571. J Org Chem 57:801–803, 1992.

69. EB Vadas, P Toma, G Zografi. Solid-state phase transitions initiated by water vapor sorption of crystalline L-660, 711, a leukotriene D_4 receptor antagonist. Pharm Res 8: 148–155, 1991.

70. HY Aboul-Enein, RA Dommisse, HO Desseyn. Solid state NMR of captopril and its zinc comples. Spectrosc Lett 23:491–495, 1990.

71. R Glaser, K Maartmann-Moe. X-ray crystallography studies and CP-MAS ^{13}C NMR spectroscopy on the solid-state stereochemistry of diphenhydramine hydrochloride, an antihistaminic drug. J Chem Soc Perkin Trans 2:1205–1210, 1990.

72. M Stoltz, DW Oliver, PL Wessels, AA Chalmers. High-resolution solid state carbon-13 nuclear magnetic resonance spectra of mofebutazone, phenylbutazone, and oxyphenbutazone in relation to X-ray crystallographic data. J Pharm Sci 80: 357–362, 1991.

73. DA Middleton, CS Duff, F Berst, DG Reid. Cross-polarization magic-angle spinning ^{13}C NMR characterization of the stable solid state forms of cimetidine. J Pharm Sci 86:1400–1402, 1997.

74. H-K Chan, S Venkataram, DJW Grant, Y-E Rahman. Solid state properties of an oral iron chelator, 1,2-dimethyl-3-hydroxy-4-pyridone and its acetic acid solvate. I: Physicochemical characterization intrinsic dissolution rate and solution thermodynamics. J Pharm Sci 80:677–685, 1991.

75. KDM Harris, JM Thomas. Probing polymorphism and reactivity in the organic solid state using ^{13}C NMR spectroscopy: Studies of *p*-formyl-*trans*-cinnamic acid. J Solid State Chem 93:197–205, 1991.

76. L Frydman, AC Olivieri, LE Diaz, B Frydman, A Schmidt, S Vega. A ^{13}C solid-state NMR study of the structure and the dynamics of the polymorphs of sulphanilamide. Mol Phys 70:563–579, 1990.

77. R Suryanarayanan, TS Wiedmann. Quantitation of the relative amounts of anhydrous carbamazepine ($C_{15}H_{12}N_2O$) and carbamazepine dihydrate ($C_{15}H_{12}N_2O\cdot2H_2O$) in a mixture by solid-state nuclear magnetic resonance (NMR). Pharm Res 7:184–187, 1990.

78. NR Jagannathan. High-resolution solid-state carbon-13 nuclear magnetic resonance study of acetaminophen: A common analgesic drug. Curr Sci 56:827–830, 1987.

79. C Chang, LE Díaz, F Morin, DM Grant. Solid state ^{13}C NMR study of drugs: Aspirin. Magn Res Chem 24:768–771, 1986.

80. LE Díaz, L Frydman, AC Olivieri, B Frydman. Solid state NMR of drugs: Soluble aspirin. Anal Lett 20:1657–1666, 1987.

81. PJ Saindon, NS Cauchon, PA Sutton, C-J Chang, GE Peck, SR Byrn. Solid-state nuclear magnetic resonance (NMR) spectra of pharmaceutical dosage forms. Pharm Res 10:197–203, 1993.

82. E Fukushima, SBW Roeder. Experimental Pulse NMR: A Nuts and Bolts Approach. Reading, MA: Addison-Wesley, 1981.

83. ER Andrew, M Kempka. Proton NMR study of molecular motion in solid cortisone. Solid State Nuclear Magn Reson 2:261–264, 1993.

84. ER Andrew, B Peplinska. NMR study of solid cholesterol. Mol Phys 70:505–512, 1990.

85. S Yoshioka, Y Aso, S Kojima. Determination of molecular mobility of lyophilized bovine serum albumin and γ-globulin by solid-state ^{1}H NMR and relation to aggregation-susceptibility. Pharm Res 13:926–930, 1996.

86. MD Poliks, J Schaefer. Microscopic dynamics in chloral polycarbonate by cross-polarization magic-angle spinning carbon-13 NMR. Macromolecules 23:2682–2686, 1990.

87. J Schaefer, MD Sefcik, EO Stejskal, RA McKay, WT Dixon, RE Cais. Molecular motion in glassy polystyrenes. Macromolecules 17:1107–1118, 1984.

88. J Schaefer, MD Sefcik, EO Stejskal, RA McKay. Carbon-13 $T_{1\rho}$ experiments on solid polymers having tightly spin-coupled protons. Macromolecules 17:1118–1124, 1984.

89. AN Garroway, WM Ritchey, WB Moniz. Some molecular motions in epoxy polymers: A carbon-13 solid-state NMR study. Macromolecules 15:1051–1063, 1982.

90. J Schaefer, EO Stejskal, R Buchdahl. Magic-angle carbon-13 NMR analysis of motion in solid glassy polymers. Macromolecules 1982:384–405, 1977.

91. A Naito, S Ganapathy, K Akasaka, CA McDowell. Spin-relaxation of carbon-13 solid amino acids using the CP-MAS technique. J Magn Reson 54:226–235, 1983.

92. F Separovic, YH Lam, X Ke, H-K Chan. A solid-state NMR study of protein hydration and stability. Pharm Res 15:1816–1821, 1998.

93. ER Andrew, WS Hinshaw, MG Hutchins, ROI Sjöblom. Proton magnetic relaxation and molecular motion in polycrystalline amino acids. I. Aspartic acid, cystine, glycine, histidine, serine, tryptophan, and tyrosine. Mol Phys 31:1479–1488, 1976.

94. F Imashiro, K Takegoshi, S Okazawa, J Furukawa, T Terao, A Saika, A Kawamori. NMR study of molecular motion in perisubstituted naphthalenes. J Chem Phys 78:1104–1111, 1983.

95. ZJ Li, MT Zell, EJ Munson, DJW Grant. Characterization of racemic species of chiral drugs using thermal analysis, thermodynamic calculation, and structural studies. J Pharm Sci 88:337–346, 1999.

96. HDW Hill, AP Zens, J Jacobus. Solid-state NMR spectroscopy. Distinction of diastereomers and determination of optical purity. J Am Chem Soc 101:7090–7091, 1979.

97. KV Anderson, H Bildsøe, HJ Jakobsen. Determination of enantiomeric purity from solid-state ^{31}P NMR of organophosphorus compounds. Magn Res Chem 28:S47–S51, 1990.

98. LE Díaz, F Morin, CL Mayne, DM Grant, C-J Chang. Conformational analysis of DL-, L-, and D-methionine by solid-state ^{13}C NMR spectroscopy. Magn Reason Chem 24:167–170, 1986.

99. DP Burum, A Bielecki. An improved experiment for heteronuclear-correlation 2D NMR in solids. J Magn Reson 94:645–652, 1991.

100. A Bielecki, DP Burum, DM Rice, FE Karasz. Solid-state two-dimensional ^{13}C-^{1}H correlation (HETCOR) NMR spectrum of amorphous poly (2,6-dimethyl-p-phenylene oxide) (PPO). Macromolecules 24:4820–4822, 1991.

101. J Smith, E MacNamara, D Raftery, T Borchardt, S Byrn. Application of two-dimensional ^{13}C solid-state NMR to the study of conformational polymorphism. J Am Chem Soc 120:11710–11713, 1998.
102. SF deLacroix, JJ Titman, A Hagemeyer, HW Spiess. Increased resolution in MAS NMR spectra by two-dimensional separation of sidebands by order. J Magn Res 97:435–443, 1992.
103. MT Zell, BE Padden, DJW Grant, MC Chapeau, I Prakash, EJ Munson. Two-dimensional high-speed CP/MAS NMR spectroscopy of polymorphs. 1. Uniformly ^{13}C-labeled aspartame. J Am Chem Soc 121:1372–1378, 1999.
104. C-L Juang, RA Santos, P Tang, W-J Chien, M-H Wann, R Bernstein, J Hong, GS Harbison. Structure and dynamics of DNA and drug DNA complexes by one and two-dimensional solid-state NMR. J Biomol Struct Dyn 8:18–22, 1991.
105. TP Jarvie, JS Bader, GT Went. Method and apparatus for distance measurements with solid-state NMR and usefulness for drug design. U.S. Patent Number 97 43627, 1997.
106. DD Beusen, LM McDowell, U Slomczynska, J Schaefer. Solid-state nuclear magnetic resonance analysis of the conformation of an inhibitor bound to thermolysin. J Med Chem 38:2742–2747, 1995.
107. TP Jarvie, GT Went, KT Mueller. Simultaneous multiple distance measurements in peptides via solid-state NMR. J Am Chem Soc 118:5330–5331, 1996.
108. S Byrn, R Pfeiffer, M Ganey, C Hoiberg, G Poochikian. Pharmaceuticals solids: A strategic approach to regulatory considerations. Pharm Res 12: 945–954, 1995.
109. Q6A Specifications: Test Procedures and Acceptance Criteria for New Drug Substances and New Drug Products: Chemical Substances. Fed Reg 62, No. 227 62890–62910, 1997.

11
Vibrational Spectroscopy

David E. Bugay
SSCI, Inc., West Lafayette, Indiana

W. Paul Findlay
Purdue University, West Lafayette, Indiana

I. INTRODUCTION

Typically, most chemists think of infrared (IR) spectroscopy as the only form of vibrational analysis for a molecular entity. In this framework, IR is routinely used as an identification assay for various intermediates and final bulk drug substance, and also as a quantitative technique for solution-phase studies. Full vibrational analysis of a molecule must also include Raman spectroscopy. Although IR and Raman spectroscopy are complementary techniques, widespread use of the Raman technique in pharmaceutical investigations has been limited. Before the advent of Fourier transform techniques and lasers, experimental difficulties tended to limit the use of Raman spectroscopy. However, over the last 20 years a renaissance of the Raman technique has been seen, due mainly to instrumentation development.

The use of vibrational spectroscopy in pharmaceutical applications is very diverse. Traditional techniques are used for bulk drug substance characterization including structure elucidation, routine compound identification, and solid-state characterization such as polymorphism (1–4). In addition, hyphenated techniques such as liquid chromatography-infrared (LC-IR) (5) and thermogravimetric-infrared (TG-IR) analysis (6,7) have complemented the analytical and pharmaceutical scientist's arsenal of instrumental techniques. Drug–excipient interactions, drug and/or excipient interaction with storage vessels, particulate identification, contaminant analysis, and mapping of pharmaceutical tablets are just a few of the microscopy-related applications of IR and/or Raman spectroscopy. IR and Raman spectroscopy have also branched out into biopharmaceutical applications such as peptide secondary structure determination (8), and more recently, reaction monitoring systems for chemical process engineers (9) and on-, or at-line systems for monitoring chemical processes, manufacturing operations, or clinical packaging procedures (10). Within any of these examples, it must be recognized that vibrational spectroscopy, whether it be IR or Raman, may not be the ultimate analytical technique. A multidisciplinary approach to pharmaceutical methods development and problem solving must be utilized, with vibrational spectroscopy as a major component.

Whichever form of vibrational analysis is utilized, it is imperative that the scientist thoroughly understand: (a) the theoretical concepts of IR or Raman spectroscopy, (b) limitations of the instrumentation, (c) appropriate sampling techniques, and (d) the ability to process and interpret the spectral data correctly. In some cases Raman spectroscopy may be the more appropriate technique but cost-prohibitive. IR may be applicable, but limitations of the technique must be appreciated by the scientist. The reverse scenario may be true for a different problem. In any case, a thorough understanding of IR and Raman spectroscopic techniques will allow one to study properly the scientific challenge at hand. Since the advent of Fourier transform techniques and advances in laser technology and computer systems, a myriad of sampling techniques are available for the vibrational spectroscopist. Analysis can now be performed on virtually any type of sample, such as single crystals, bulk material, slurries, creams, particulates, films, solutions (aqueous and organic), oils, gas-phase samples, and on-process streams through the use of fiber-optic probes. Additional advantages of vibrational analysis include: (a) typically nondestructive in nature, with the ability to recover the material for further characterization, (b) quantitative technique under proper sampling conditions, and (c) complementary to other characterization techniques.

This chapter attempts to present to the reader the theory, sampling techniques, and major applications of IR and Raman spectroscopy in the pharmaceutical fields in a practical format. Extensive references are included throughout the text, and the reader is encouraged to consult these for further information.

II. VIBRATIONAL SPECTROSCOPY THEORY

A brief description of IR and Raman theory will be presented so that a common understanding of the techniques is available to the reader. A complete description of the underlying theory of IR and Raman spectroscopy is outside the scope of this chapter, but can be obtained from the literature (11–15).

A. Infrared

1. Mid-Infrared

All molecules of pharmaceutical interest absorb some form of electromagnetic radiation. Within the electromagnetic spectrum (Fig. 1), infrared energy is a small portion which is typically divided into three regions, the near-, mid-, and far-IR regions with their respective energy/frequency limits. In IR spectroscopy, the energy unit wavenumber (cm^{-1}) is typically used. Wavenumber is the reciprocal of the IR wavelength expressed in centimeters.

When a broad-band source of IR energy irradiates a sample, the absorption of IR energy by the sample results from transitions between molecular vibrational and rotational energy levels. A vibrational transition may be approximated by treating two atoms bonded together within a molecule as a harmonic oscillator. Based on Hooke's law, the vibrational frequency between these two atoms may be approximated as:

$$v = \frac{1}{2\pi} \sqrt{\frac{k}{\mu}} \tag{1}$$

where μ is the reduced mass of the two atoms, $\mu = (m_1 m_2)/(m_1 + m_2)$, and k is the force constant of the bond (dynes/cm). Quantum mechanical analysis of the harmonic oscillator reveals a series of equally spaced vibrational energy levels

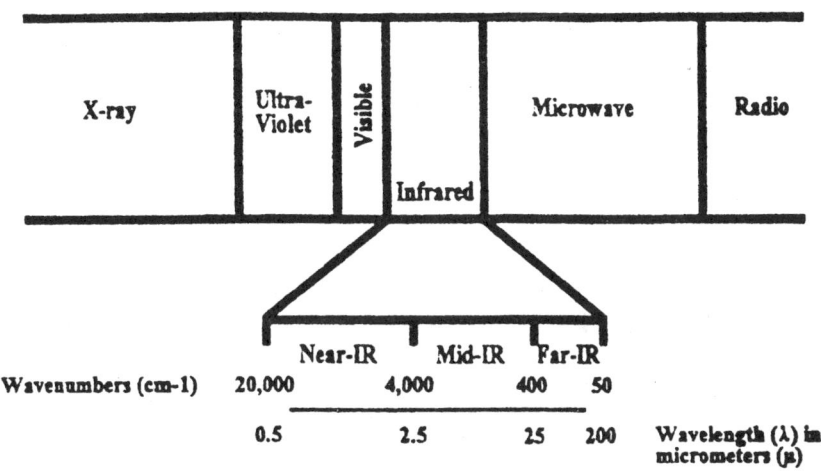

Fig. 1 A portion of the electromagnetic spectrum, comparing infrared energy to other forms of radiation.

(defined by the vibrational quantum number v, where $v = 0, 1, 2, 3, \ldots$) that are expressed as:

$$E_v = \left(v + \frac{1}{2}\right)hv_0 \tag{2}$$

where E_v is the energy of the vth level, h is Planck's constant, and v_0 is the fundamental vibrational frequency. These energy levels may be graphically described in a Jablonski energy-level diagram (Fig. 2). It must be noted that the fundamental vibrational frequencies in a polyatomic molecule do not necessarily correspond to the vibrations of single pairs of atoms, but rather to those of a group of atoms. The absorption of IR energy by a molecule corresponds to ap-

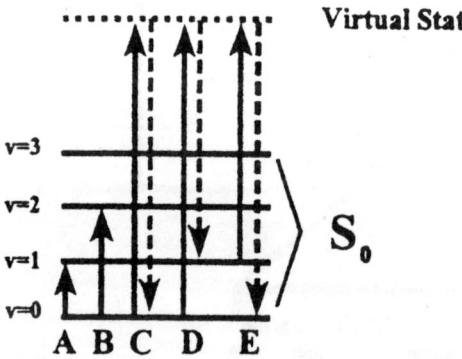

Fig. 2 Jablonski energy-level diagram illustrating possible transitions, where solid lines represent absorption processes and dashed lines represent scattering processes. Key: A, IR absorption; B, near-IR absorption of an overtone; C, Rayleigh scattering; D, Stokes Raman transition; E, anti-Stokes Raman transition. S_0 is the singlet ground state, S_1 the lowest singlet excited state, and v represents vibrational energy levels within each electronic state.

proximately 2–10 kcal/mole, which in turn equals the stretching and bending vibrational frequencies of most bonds in covalently bonded molecules. Thus, there is a correlation between IR spectroscopy and the ability to probe the vibrational motion of a molecule.

The number of fundamental vibrational modes of a molecule is equal to the number of degrees of vibrational freedom. For a nonlinear molecule of N atoms, $3N - 6$ degrees of vibrational freedom exist. Hence, there are $3N - 6$ fundamental vibrational modes. Six degrees of freedom are subtracted from a nonlinear molecule since: (a) three coordinates are required to locate the molecule in space, and (b) an additional three coordinates are required to describe the orientation of the molecule based on the three coordinates defining the position of the molecule in space. For a linear molecule, $3N - 5$ fundamental vibrational modes are possible, since only 2 degrees of rotational freedom exist. Thus, in a total vibrational analysis of a molecule by complementary IR and Raman techniques, $3N - 6$ or $3N - 5$ vibrational frequencies should be observed. It must be kept in mind that the fundamental modes of vibration of a molecule are described as transitions from one vibration state (energy level) to another [$v = 1$ in Eq. (2), Fig. 2]. Sometimes, additional vibrational frequencies are detected in a vibrational spectrum. These additional absorption bands are due to forbidden transitions that occur [$v \geq 2, 3, \ldots$ in Eq. (2), Fig. 2] and are described in the section on near-IR theory. Additionally, not all vibrational bands may be observed, since some fundamental vibrations may be too weak to observe or give rise to overtone and/or combination bands (vide infra).

For a fundamental vibrational mode to be IR active, a change in the molecular dipole must take place during the molecular vibration. This is described as the IR selection rule. Atoms which possess different electronegativity and are chemically bonded change the net dipole of a molecule during normal molecular vibrations. Typically, antisymmetric vibrational modes and vibrations due to polar groups are more likely to exhibit prominent IR absorption bands.

A percent transmission (%T) IR spectrum may be calculated by Eq. (3):

$$\text{Percent transmission} = \frac{I}{I_0} \times 100 \tag{3}$$

where I equals the intensity of transmitted IR radiation, and I_0 equals the intensity of irradiating IR energy. In this case, the percent transmission IR spectrum is represented by wavenumber (cm^{-1}) on the abscissa and %T on the ordinate (Fig. 3). IR spectra may also be represented in absorbance units [Eq. (4)],

$$\text{Absorbance} = \log\left(\frac{I_0}{I}\right) = abc \tag{4}$$

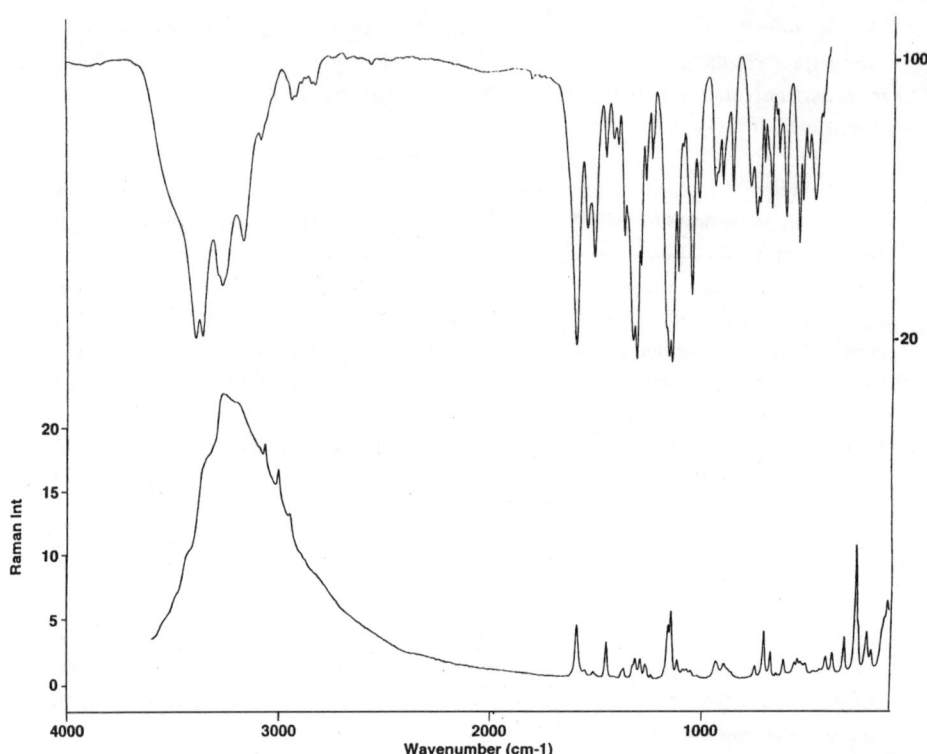

Fig. 3 FTIR (upper) and FT-Raman (lower) spectra of hydrochlorothiazide. The left ordinate scale is representative of the Raman intensity, whereas the right ordinate scale represents IR transmission units.

where a is absorptivity, b is the sample cell thickness, and c is concentration. The representation of IR spectra in absorbance units permits quantitative analysis.

As previously mentioned, IR spectroscopy is typically used for the identification of a molecular entity. This approach arises from the fact that the vibrational frequency of two atoms may be approximated from Eq. (1). If one assumes that the force constant (k) for a double bond is 10×10^5 dynes/cm, Eq. (1) allows one to approximate the vibrational frequency for C=C:

$$v = 4.12\sqrt{\frac{k}{\mu}} = 4.12\sqrt{\frac{10 \times 10^5}{[12 * 12/12 + 12]}} = 1682 \text{ cm}^{-1} \tag{5}$$

Fairly good agreement exists between the calculated value of 1682 cm^{-1} and the experimental value of 1650 cm^{-1}. Direct correlation does not exist because Hooke's law assumes that the vibrational system is an ideal harmonic oscillator and, as mentioned before, the vibrational frequency for a single chemical moiety in a polyatomic molecule corresponds to the vibrations from a group of atoms. Nonetheless, based on the Hooke's law approximation, numerous correlation tables have been generated that allow one to estimate the characteristic absorption frequency of a specific functionality (13). It becomes readily apparent how IR spectroscopy can be used to identify a molecular entity, and subsequently physically characterize a sample or perform quantitative analysis.

2. Near-Infrared

The near-IR region of the electromagnetic spectrum is generally from 750 to 2500 nm (13,300 to 4000 cm^{-1}). Typically, wavelength (in nanometers) is represented on the abscissa of a near-IR spectrum. Absorption bands in the near-IR region of the spectrum arise from overtones or combinations of fundamental vibrational motions, in addition to combinations of overtones. Overtone absorption bands are the result of forbidden transitions arising from the ground vibrational energy level and where $v > 1$ in Eq. (2). Since overtone bands arise from forbidden transitions, they are typically 10–1000 times less intense than their corresponding fundamental absorption bands. The majority of overtone bands in the near-IR arise from R—H stretching modes (O—H, N—H, C—H, S—H). Due to the large mass difference between the two atoms, large-amplitude vibrations arise with high anharmonicity and large dipole moments. The large anharmonicity causes the frequency of the overtone (or combination) bands to be slightly less than the sum of the frequency of the participating bands. This is mathematically described in Eq. (6), where v_0 is the fundamental vibrational frequency, v_n is

$$v_v = v\, v_0(1 - v\, x) \tag{6}$$

the frequency of the $(n - 1)$st overtone, v is the vibrational quantum number ($=$ 1 for fundamental, $= 2$ for the first overtone, $= 3$ for the second, etc.), and x is the anharmonicity factor, which measures the deviation of the potential function from the parabolic function. Combination bands arise from the simultaneous changes in the energy of two or more vibrational modes and are observed at a frequency given by Eq. (7) ($v =$ frequency of transition contributing to the combination band, $n =$ integer). Again, since combination bands are forbidden transitions, the intensity of these

$$v = n_1 v_1 + n_2 v_2 + n_3 v_3 \ldots \tag{7}$$

bands within a near-IR spectrum are from 10–1000 times less intense than the original fundamental vibrational bands. Typically, in pharmaceutical analysis, near-IR spectroscopy is used for the detection of water (strong O—H combination band).

3. Far-Infrared

The far-IR region may be thought of as an extension of the mid-IR, since this region of the spectrum is typically used to investigate the fundamental vibrational modes of heavy atoms. The region extends from 400 to 10 cm^{-1} and the low-frequency molecular vibrations arise from the vibrational motion of heavy atoms such as metal-ligands. In addition, this region of the spectrum is used to investigate intermolecular vibrations of crystalline materials. The motion of molecules relative to one another gives rise to lattice modes, sometimes referred to as external modes. These low-frequency modes are very sensitive to changes in the conformation or structure of a molecule and are found from 200 to 10 cm^{-1}. Although this spectral region is important for the study of polymorphism, very few literature reports have been published. From a practical standpoint, investigations in the far-IR require nontraditional optical windows such as polyethylene or polypropylene, since alkali halide windows do not transmit in this region. The spectral quality (signal-to-noise ratio: S/N) is also compromised due to relatively poor sources and insensitive detectors. From a pharmaceutical sciences perspective, this IR spectral region is primarily used for solid-state investigations of lattice modes.

B. Raman

When a compound is irradiated with monochromatic radiation, most of the radiation is transmitted unchanged but a small portion is scattered. If the scattered radiation is passed into a spectrometer, a strong Rayleigh line is detected at the unmodified frequency of radiation used to excite the sample. In addition, the scattered radiation also contains frequencies arrayed above and below the frequency of the Rayleigh line. The *differences* between the Rayleigh line and these weaker Raman line frequencies correspond to the vibrational frequencies present in the molecules of the sample. For example, we may obtain a Raman line at ± 2980 cm^{-1} on either side of the Rayleigh line and thus the sample possesses a vibrational mode of this frequency. The frequencies of molecular vibrations are typically 10^{12}–10^{14} Hz. A more convenient unit, which is proportional to frequency, is wavenumber (cm^{-1}), since fundamental vibrational modes lie between 3600 and 50 cm^{-1}.

The Raman lines are generally weak in intensity, approximately 0.001%

of the source, and hence their detection and measurement are difficult. Raman bands at wavenumbers less than the Rayleigh line are called Stokes lines, while anti-Stokes lines occur at greater wavenumbers than the source radiation. Generally, the anti-Stokes lines are less intense than the Stokes lines because these transitions arise from higher vibrational energy levels containing fewer molecules, as described by the Boltzmann-distribution (Fig. 2). Hence, the Stokes portion of the spectrum is generally used. The abscissa of the spectrum is usually labeled as wavenumber shift or Raman shift (cm^{-1}), and the negative sign (for Stokes shift) is dispensed with (Fig. 3).

C. Comparison of Raman and Infrared Spectroscopies

Both infrared and Raman are vibrational spectroscopic techniques, and the Raman scattering spectrum and infrared absorption spectrum for a given species often resemble one another quite closely. There are, however, sufficient differences between the types of chemical groups that are infrared and Raman active to make the techniques complementary rather than competitive. This is illustrated in Fig. 3, where the infrared and Raman spectra of hydrochlorothiazide are shown.

Briefly, a vibrational mode is infrared active when there is a change in the molecular dipole moment during the vibration, whereas a vibrational mode is Raman active when there is a change in polarizability during the vibration. Consequently, asymmetric modes and vibrations due to polar groups (e.g., C=O, N—H) are more likely to be strongly infrared active, whereas symmetric modes and homopolar bonds (e.g., C=C or S—S) tend to be Raman active. The complementarity of the two techniques in terms of molecular characterization for a range of antibacterial agents and related compounds has recently been demonstrated (16).

III. EXPERIMENTAL

A. Infrared Spectroscopy

Infrared spectroscopy instrumentation is almost as widely varying as the applications. Mid-infrared instrumentation today consists almost exclusively of Fourier transform instruments. Although dispersive mid-IR instruments are still used, only a few manufacturers still produce these instruments. Dispersive mid-IR instruments have found a niche in the process monitoring field. Near-infrared instrumentation, on the other hand, is still dominated by dispersive instruments. Recently, some manufacturers have begun to offer near-IR instruments with Fourier transform mechanics and optics.

1. Sampling Techniques

Sometimes, one of the greatest challenges for the IR spectroscopist is sample preparation. Since the development of the Fourier transform infrared spectrophotometer and its inherent signal-to-noise and throughput advantages, an abundance of sampling techniques for pharmaceutical analysis have been developed. In this Experimental section, the general configuration of the IR spectrophotometer has been overlooked so that a brief review of IR sampling techniques can be outlined. Although it is extensive, this section does not review all IR sampling techniques, just those widely used for pharmaceutical problem solving and methods development. A full description of the components of an IR spectrophotometer may be reviewed in the classic Griffiths and DeHaseth book (17).

a. Alkali Halide Pellet

The classic IR sampling technique is the alkali halide pellet preparation (18). This technique involves mixing the solid-state sample of interest with an alkali halide (typically KBr or KCl) at a 1–2% w/w sample/alkali halide ratio. The mixture is pulverized into a finely ground homogeneous mixture, placed into a die (typically stainless steel), and subjected to approximately 10,000 psi of pressure for a period of time to produce a glass pellet. The pellet (with the sample finely dispersed throughout the glass) may then be placed into the IR spectrophotometer for spectral data acquisition.

From the traditional view, this sampling technique is used for the preparation of samples for chemical identity testing and is commonly incorporated into a regulatory submission. The advantage of this technique is that only a small amount of sample is required (usually 1 mg), and a high-quality spectrum can be obtained in a matter of minutes. Disadvantages exist, such as solid-state transformation of the sample due to the pressure requirements to form the glass pellet and possible halide exchange between KBr (or KCl) and the sample of interest (19,20). These are critical disadvantages whenever IR spectroscopy is utilized for pharmaceutical polymorph investigations. Quantitation of mixtures has been attempted with fair success utilizing the alkali halide pellet sampling technique (21). Due to the aforementioned disadvantages, it is suggested that this IR sampling technique be used only for simple compound identification assays.

b. Mineral Oil Mull

Another classical sampling technique for solids is the mineral oil mull preparation (18). In this technique, a small amount of sample (~1 mg) is placed into an agate mortar. To this, a small amount of mineral oil is added and the sample and oil mixed to an even consistency. The mixture is then placed onto an IR optical window and sampled by the IR spectrophotometer. One advantage to this technique is that there is no likelihood of solid-state transformations due to mixing and/or grinding, and hence is a good technique for the qualitative identification of pharmaceutical polymorphs. Unfortunately, the mineral oil has a number of intense spectral contributions (2952, 2923, 2853, 1458, and 1376 cm^{-1}), which may overlap important absorption bands corresponding to the sample of interest. Typically, this technique is used for qualitative identification assays whenever the alkali halide pellet technique is inappropriate.

c. Diffuse Reflectance

The diffuse reflectance (DR) technique is probably one of the most important solid-state sampling techniques for pharmaceutical problem solving and methods development (22). Sometimes referred to as DRIFTS (diffuse reflectance infrared Fourier transform spectroscopy), this technique is used extensively in the mid- and near-IR spectral regions. The technique involves irradiation of the powdered sample by an infrared beam (Fig. 4). The incident radiation undergoes absorption, reflection, and diffraction by the particles of the sample. Only the incident radiation that undergoes diffuse reflectance contains absorptivity information about the sample. A number of significant advantages exist for diffuse reflectance analysis. Samples may be investigated neat, or diluted within a nonabsorbing matrix such as KBr or KCl (usually at a 1–5% w/w active-to-nonabsorbing matrix material ratio). Macro and micro sampling cups are usually provided with the diffuse reflectance accessory, and approximately 400 and 10 mg of sample are required for each cup, respectively. The sample is also 100% recoverable, so that other

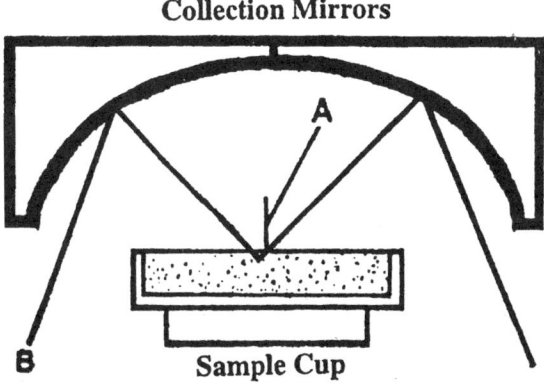

Fig. 4 Schematic representation of the diffuse reflectance sampling accessory. Key: A, blocker device to eliminate specular reflectance; B, path of IR beam.

solid-state investigations may take place on the same material. Through the use of an environmental chamber attachment, variable-temperature and -humidity DR experiments can be performed (23). By varying the temperature of the sample, information about temperature-mediated crystal form transformations and the nature of the interaction of a solvate with the parent molecule can be determined.

The DR technique lends itself to polymorph studies since the technique is noninvasive, the polymorph character remains intact due to limited sample handling, and the technique is quantitative (4). One disadvantage to diffuse reflectance IR is that it is a particle size-dependent technique (22). Development of quantitative polymorph assays require that the particle size of each component be limited to a specific range, including both components of a mixture, and the nonabsorbing matrix if the mixture is not sampled neat. It must also be kept in mind that for a quantitative assay, all calibration, validation, and subsequent samples to be assayed must fall within the particle size range; otherwise, significant prediction errors may arise.

d. Microspectroscopy

The first linkage between a microscope and an IR spectrophotometer was reported in 1949 (24). Today, every manufacturer of IR spectrophotometers offers an optical/IR microscope sampling accessory. The use of optical microscopy and IR microspectroscopy is a natural course of action for any solid-state investigation. Optical microscopy provides significant information about a sample, such as its crystalline or amorphous nature, particle morphology, and size. Interfacing the microscope to an IR spectrophotometer provides unequivocal molecular identification of one particular crystallite—hence the tremendous benefit of IR microspectroscopy for the identification of particulate contamination in bulk or formulated drug products.

The IR microspectroscopy sampling technique is the ultimate sampling technique, since only one particle is required for analysis. Due to the limitation of diffraction, typically, the particles of interest must be greater in size than 10 $\times$ 10 μm. The sample of interest is placed on an IR optical window and the slide is placed onto the microscope stage and visually inspected. Once the sample of interest is in focus, the field of view is apertured down to the sample. Depending on sample morphology, thickness, and transmittance properties, a reflectance and/or transmittance IR spectrum may be acquired by the IR microscope accessory. Obvious advantages for the technique exist, such as nondestructive sampling, reflectance and/or transmittance measurements, minimal sample requirements, and the ability to monitor solid-state transformations with the interface of a hot stage important for polymorphism studies (25).

Variable-temperature studies may also be performed with FT-IR microspectroscopy. Through the use of an environmentally controlled chamber, small samples can be observed both optically and spectroscopically while changing the environment about the sample. This is particularly useful when observing temperature-mediated crystal form changes that also display morphological changes.

e. Thermogravimetric/Infrared Analysis

In the thermogravimetric IR sampling technique, a thermogravimetric (TG) analyzer is interfaced to an IR spectrophotometer such that the evolved gas from the sample/TG furnace is directed to an IR gas cell. This IR sampling technique lends itself to the identification and quantitation of residual solvent content for a pharmaceutical solid (6), and also for the investigation of pharmaceutical pseudopolymorphs (7).

Analogous to standard thermogravimetric analysis procedures, the sample of interest for TG/IR analysis is placed into a TG sample cup (approximately 10 mg) and introduced into the TG furnace. The TG balance then monitors the weight loss of the sample as a function of temperature (typical heating rate of 10 °C/min). From the TG data, the amount of weight loss over a specific temperature range can be used to infer possible residual solvent content of a sample. Unfortunately, only the percent weight loss can be calculated from the TG experiment. An unequivocal identification of the evolved gas cannot be made, hence the significant advantage of interfacing the TG apparatus to an IR spectrophotometer.

Although residual solvent content of a sample can be identified and quantified by TG/IR analysis, gas chromatography techniques usually outperform TG/IR experiments because of lower detection limits and ease of quantitation. One significant advantage of the TG/IR technique is the investigation of pharmaceutical pseudo-polymorphs. Thermal analysis techniques such as TG analysis and differential scanning calorimetry (DSC) typically can determine the presence of pseudo-polymorphism based on correlating TG weight loss and detection of DSC endotherms at the same temperatures. Unfortunately, the solvent of crystallization cannot be identified by the thermal analysis techniques. IR analysis of the evolved gas at the same temperature as the TG weight loss/DSC endotherm can provide identification of the solvent of crystallization (7).

f. Attenuated Total Reflectance

Attenuated total reflectance (ATR) IR spectroscopy is a sampling technique for either solid, liquid, or gel-like samples (26). The basic premise of the technique involves placing the sample in contact with an infrared transmitting crystal with a high refractive index. The infrared beam is directed through the crystal, penetrating the surface of the sample, and displaying spectral information of that surface. An advantage of this technique is that it requires very little sample preparation; simply place the sample in contact with the crystal. Sufficient cleaning of the crystal between samples is necessary to avoid cross-contamination. It should also be noted that ATR does not sample the bulk of a material; penetration by the IR radiation is only a few micrometers or tenths of micrometers.

g. Photoacoustic

The photoacoustic effect was first discovered by Alexander Graham Bell in the early 1880s (27), but not applied to FTIR spectroscopy until a century later (28,29). Significant advantages of FTIR photoacoustic spectroscopy (PAS) include: (a) spectra may be acquired on opaque materials (commonly found in pharmaceutical formulations), (b) minimal sample preparation is necessary, and (c) depth profiling is possible.

The PAS phenomenon involves the selective absorption of modulated IR radiation by the sample. The selectively absorbed frequencies of IR radiation correspond to the fundamental vibrational frequencies of the sample of interest. Once absorbed, the IR radiation is converted to heat and subsequently escapes from the solid sample and heats a boundary layer of gas. Typically, this conversion from modulated IR radiation to heat involves a small temperature increase at the sample surface ($\sim 10^{-6}$ °C). Since the sample is placed into a closed cavity cell which is filled with a coupling gas (usually helium), the increase in temperature produces pressure changes in the surrounding gas (sound waves). Due to the fact that the IR radiation is modulated, the pressure changes in the coupling gas occur at the frequency of the modulated light, as well as the acoustic wave. This acoustical wave is detected by a very sensitive microphone and the subsequent electrical signal is Fourier processed and a spectrum produced.

Depth profiling of a solid sample may be performed by varying the interferometer moving mirror velocity (modulated IR radiation). By increasing the mirror velocity, the sampling depth varies and surface studies may be performed. Limitations do exist, but the technique has proven to be quite effective for solid samples (30). In addition, unlike diffuse reflectance sampling techniques, particle size has a minimal effect on the photoacoustic measurement.

h. Fiber Optics

The development of fiber optics in infrared spectroscopy grew out of the need to analyze hazardous materials from a remote location. Solids can be analyzed with a diffuse reflectance arrangement of optical fibers, while liquids can be analyzed with an ATR arrangement of optical fibers on the end of the probe. Much of the early development of IR spectroscopy using fiber-optic probes was done in the near-IR spectral region, due to the limitations of the materials available for use as optical fibers. With the development of chalcogenide fibers, the use of fiber optics has extended into the mid-IR spectral region. Transmission measurements with chalcogenide fibers are suitable over the range of 4000 to 1000 cm^{-1} (with absorptions due to hydrogen-containing impurities in the fiber at 2250–2100 cm^{-1}). Normal use of chalcogenide fibers is limited to approximately 2 m in length (31).

B. Raman Spectroscopy

The Raman effect was discovered in 1928, but the first commercial Raman instruments did not start to appear until the early 1950s. These instruments did not use laser sources, but used elemental sources and arc lamps. In 1962, laser sources started to become available for Raman instruments, and the first commercial laser Raman instruments appeared in 1964–1965. The first commercial FT-Raman instruments were available starting in 1988, and by the next year, FT-Raman microscopy was possible (32). Due to the various complexities when one compares dispersive Raman spectrometers with FT-based systems (33), only sampling techniques will be discussed here.

1. Sampling Techniques

Sampling techniques for Raman spectroscopy are relatively general, since the only requirement is that the monochromatic laser beam irradiate the sample of interest and the scattered radiation be focused on the detector. The sampling discussion outlined here is applicable to both types of spectrometers (dispersive/FT).

a. General Techniques

Raman spectroscopy may be performed on very small samples (e.g., a few nanograms), and sample preparation is simple; powders do not need to be pressed into disks or diluted with KBr, they just need to be irradiated by the laser beam. Solid samples are often examined in stainless steel or glass sample holders generally requiring ~25–50 mg of material. Typically, liquid samples are analyzed in quartz or glass cuvettes, which may have mirrored rear surfaces to improve the signal intensity. Glass is a very weak Raman scatter and so many samples (liquid and solid) can be simply analyzed in a bottle, or in for example, a nuclear magnetic resonance (NMR) tube, although fluorescence from some glasses can be problematic. Water is a good solvent for Raman studies, since the Raman spectrum of water is essentially one broad, weak band at 3500 cm^{-1}.

The complete Stokes Raman spectrum covering shifts in the range 10–3500 cm^{-1} can be obtained and the intensity of Raman scattering is directly proportional to the concentration of the scattering species, an important factor for quantitative analysis. However, the Raman effect is relatively weak and hence a material needs to be present to a level of about 1% for accurate assessments, whereas IR can be used to detect materials to a level of approximately 0.01%. Fluorescence can also be problematic in Raman studies, but is typically due to the NMR sample tubes being utilized, or impurities within the sample of interest. Data massaging techniques can sometimes blank out Raman spectral contributions due to fluorescent materials.

Variable temperature studies in Raman spectroscopy provide a wealth of information. Because a Raman spectrum typically covers a wavelength range that extends beyond the range normally associated with mid-IR spectroscopy (10–400 cm^{-1}), information about lattice vibrations of organic compounds is readily available. By varying the temperature of a sample, the lattice energies of the compound are changed, allowing for interpretation of the nature of the crystal lattice. In addition, information similar to that obtained in IR variable-temperature studies (crystal form changes and the nature of solvate association) can be achieved through variable-temperature Raman investigations.

b. Microspectroscopy

Analogous to IR microspectroscopy work, Raman spectra can be acquired on small amounts of material through the use of a Raman microprobe (34). In a similar fashion to the IR microscope, a sample for Raman microanalysis is first viewed optically through the microscope. Once optical measurements are made, a Raman spectrum may be obtained on the material. The amount of material irradiated for Raman microanalysis is defined by the size of the objective. On some systems, a 100 × objective achieves a sampling size as small as 1 µm. Since a high intensity of monochromatic radiation from the laser is focused on a small amount of sample, sample degradation by the laser must be monitored (35). Otherwise, the Raman microprobe is ideal for investigating polymorphism (single crystals), particulate contamination, and small amounts of samples. Using apparatus similar to those used for IR microspectroscopy, variable-temperature studies can be performed with a Raman microprobe. Variable-temperature Raman microprobe studies provide the same type of information as described in the previous section, with the added benefits of working with small samples and being able to observe optically any morphological changes.

c. Fiber Optics

Fiber optics have been used in Raman spectroscopy since the early 1980s (36,37). In 1986, Archibald et al. demonstrated the use of a fiber-optic probe with a FT-Raman system (38). Today, much of the research in the use of fiber optics in FT-Raman spectroscopy centers around fiber (and fiber bundle) design. The number, type, and arrangement of the fibers in a fiber bundle are all factors that are varied to produce fiber bundles for different applications. Fiber systems include single-fiber, where the laser excitation and collected scattered radiation travel along the same fiber; and multifiber, where laser excitation is transmitted along one (or multiple) fibers, and the scatter is transmitted to the detector along different fibers. The arrangements of the fibers in a multifiber system can also vary. Two examples include an arrangement in which one excitation fiber is surrounded by several collection fibers or in which several excitation and collection fibers are randomly mixed in a bundle.

IV. SELECTED APPLICATIONS

The remainder of this chapter focuses on the myriad of vibrational spectroscopy applications in the pharmaceutical field. Although these applications are not comprehensive (due to the limited manuscript space), the reader should appreciate the numerous applications of IR and Raman spectroscopy for pharmaceutical analysis.

A. Spectral Libraries

Spectral libraries play a key role in the work of the modern-day spectroscopist (39). Applications include contaminant analysis/identification, identity testing of bulk drug substances and intermediates, and as a tool for the structural elucidation of new chemical entities. A list of some of the types of commercially available spectral libraries is found in Table 1. Although few FT-Raman spectral libraries are presently available (40), with the growing interest in the technique, many libraries are in development. In addition, the creation of spectral libraries of proprietary compounds in the pharmaceutical analytical laboratory allows for rapid and efficient identity testing of key intermediates and proprietary bulk drug substances.

B. Identity Testing

Identity testing of compounds is one role of the pharmaceutical spectroscopy laboratory. Testing can be accomplished with methods utilizing the mid- and near-IR spectral regions, and more recently utilizing FT-Raman spectroscopy. As stated in the USP23/NF18 (41), the infrared absorption spectrum of a substance provides the most conclusive evidence of identity that can be obtained from a single test. For bulk drug substances, the identity test is often listed as ⟨971k⟩, the KBr pellet sampling technique. Identity testing of solid dosage forms of drugs usually involves wet-chemical extraction steps followed by general method ⟨971k⟩. An example is the identity test of Leucovorin calcium tablets, an antidote

Table 1 Listing of Commercially Available IR and Raman Spectral Libraries

Library description	Library type IR	Library type Raman	Number of spectra
Condensed phase	X	X	>15,000
Pharmaceutical excipient	X	X	300 each
Biochemical	X		>13,000
Polymers and rubber compounds	X		>2,000
Polymers and rubber compounds		X	100
Industrial coatings	X		>2,500
Common solvents	X		>200
Vapor phase	X		>5,000
Forensics	X		>3,500
Forensics		X	175
Fibers	X		>350
Food additives	X		>500
Flavors and fragrances	X		>600

to folic acid antagonists and an antianemic (42). The method described in the USP/NF requires that ground tablets be dissolved in water, sonicated, and filtered. The filtrate is mixed with ammonium oxalate, shaken, and centrifuged. Methanol is added to the preparation until it is clear, and then the preparation is placed in a 0°C freezer until a precipitate is formed. The supernatant is decanted, methanol is added to dissolve the precipitate, and the solution is allowed to evaporate. After the remaining precipitate is dried, the potassium bromide pellet is prepared, and the IR spectrum obtained is compared to the IR spectrum obtained from a reference standard prepared following the same procedure.

The near-IR spectral region has been used for chemical identification as well. Along with being used for analysis of polymorphic forms (43) and identification of raw materials and excipients (44,45), near-IR spectroscopy has also been used to test intact tablets (46). Aldridge et al. (47) have also demonstrated that active and placebo tablets can be identified by near-IR spectroscopy nondestructively inside blister packs.

FT-Raman is another technique by which quick and easy identity testing can be performed. A FT-Raman method has been developed to identify the two active components (tegafur and uracil) in formulated capsules (48). Figure 5 displays a spectral region where Raman bands unique to uracil and tegafur can be found in the FT-Raman spectrum of an intact capsule. These bands allow the analyst to confirm that both components are present in the formulated product. Raman spectroscopy has become an essential chemical and physical identification

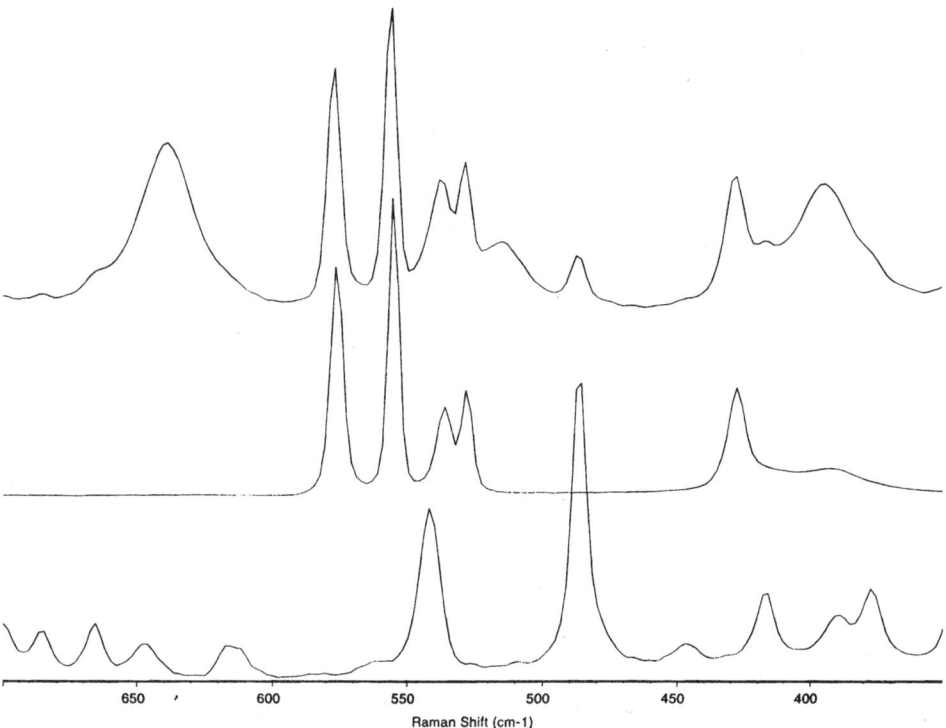

Fig. 5 FT-Raman spectra of an intact capsule (top), bulk uracil (middle), and bulk tegafur (bottom). Raman bands at 576 and 551 cm^{-1} in the spectrum of the capsule indicate the presence of uracil. The Raman band at 487 cm^{-1} indicates the presence of tegafur in the capsule.

tool for the pharmaceutical spectroscopist. Currently, the USP has formed a committee to generate a general chapter on the technique.

C. Polymorphism

Infrared, and more recently Raman spectroscopy, have been widely used for the qualitative and quantitative characterization of polymorphic compounds of pharmaceutical interest (for the sake of brevity, the term polymorphism will encompass polymorphs, pseudo-polymorphs, hydrates, and solvates). Since solid-state vibrational spectroscopy can be used to probe the nature of polymorphism on the molecular level, these methods are particularly useful in instances where full crystallographic characterization of polymorphism was not found to be possible.

Recently, a significant number of publications have appeared where a multidisciplinary, spectroscopic approach to polymorph characterization has taken place. Table 2 presents a list of polymorphic systems studied by vibrational spectroscopy, including a brief summary of the study and appropriate references. A number of these studies are further described within this section.

In the case of DuP 747 (49), X-ray powder diffraction (XRPD), DSC, and thermomicroscopic studies determined the polymorphic system to be monotropic. Distinct diffuse reflectance IR, Raman, and solid-state ^{13}C NMR spectra existed for each physical form. The complementary nature of IR and Raman gave evidence that the polymorphic pair were roughly equivalent in conformation. It was concluded that the polymorphic character of DuP 747 resulted from different modes of packing. Further crystallographic information is required in order to determine the crystal packing and molecular conformation of this polymorphic system.

Analogous to the DuP 747 study, complete crystallographic information was not possible on the fosinopril sodium polymorphic system (1). Two known polymorphs (A and B) were studied via a multidisciplinary approach (XRPD, IR, NMR, and thermal analysis). Complementary spectral data from IR and solid-state ^{13}C NMR revealed that the environment of the acetal side chain of fosinopril sodium differed in the two forms. In addition, possible cis–trans isomerization about the C_6—N peptide bond may exist. These conformational differences are postulated as the origin of the observed polymorphism in fosinopril sodium in the absence of the crystallographic data for form B (single crystals not available).

In each of the aforementioned studies, qualitative IR spectroscopy was used. It is important to realize that IR is also quantitative in nature and a number of quantitative IR assays for polymorphism have appeared in the literature. Sulfamethoxazole (50) exists in at least two polymorphic forms which have been fully characterized. Distinctly different diffuse reflectance, mid-IR spectra exist, permitting quantitation of one form within the other. When working with the diffuse reflectance IR technique, two critical factors must be kept in mind when developing a quantitative assay: (a) the production of homogeneous calibration and validation samples, and (b) consistent particle size for all components, including subsequent samples for analysis. During the assay development for sulfamethoxazole, a number of mixing techniques were investigated in an effort to achieve homogeneous samples. Inhomogeneity of calibration and validation samples can lead to inaccurate IR absorption values and subsequent prediction errors. This is also the case with particle size. Variation in the particle size of the nonabsorbing matrix or sample can influence the diffuse reflectance IR spectrum and again lead to prediction errors. After mixing and particle size factors were optimized, a quantitative diffuse reflectance IR assay was developed in which independent validation samples were predicted within 4% of theoretical values. Other exam-

Table 2 A Selected Listing of Polymorphic Systems Investigated by IR and Raman Spectroscopy

Polymorphic system	Brief description of IR and/or Raman investigation	Ref.
DuP 747	Combined IR and Raman investigation of the monotropic polymorph system. Distinct IR and Raman spectra existed for each form, although the forms were determined to be roughly equivalent in conformation.	49
Fosinopril sodium	Two known polymorphs with distinct IR spectra. X-ray crystallographic data were not obtained on the second crystal form, and spectroscopic evidence led to the conclusion that the polymorphism arises from the acetal side chain.	1
Cefepime·2HCl	Mono- and dihydrated crystal forms of the drug substance existed with distinct IR and XRD patterns. A quantitative diffuse reflectance mid-IR method was developed to determine the presence of the dihydrated form in bulk monohydrate at levels of <1% w/w.	4
Sulfamethoxazole	Distinct IR spectra for the two different crystal forms. Quantitation of one form within the other was performed by the DR technique. Different mixing techniques and particle size issues were investigated in an effort to generate homogeneous calibration samples.	50
SC-41930	Quantitation of one crystal from (low-melting) in another (high-melting) achieved by mid-IR diffuse reflectance analysis. A detection limit of 1% w/w of the low-melting crystal form in the high-melting form was achieved.	16
SC-25469	Near-IR analysis used for the identity of the two different crystal forms which were enantiotropic in nature. Due to the ability of the forms to interconvert, quantitation of the β-form in the solid dosage formulation was necessary.	51
SQ-33600	Variable-temperature and variable humidity, diffuse reflectance experiments in the mid-IR region were used to determine three different crystalline hydrates. Each form had a distinct IR spectrum and definite stability over a range of humidity conditions.	52
Ranitidine HCl	Four crystalline and one noncrystalline form of ranitidine HCl were prepared and characterized by IR spectroscopy.	53
Mefloquine HCl	Different crystal forms of mefloquine HCl were generated by various conditions of recrystallization, leading to distinct IR spectra for each form.	54

Table 2 Continued

Polymorphic system	Brief description of IR and/or Raman investigation	Ref.
Carbovir	Five different crystal forms of the material exist, and four of the forms were studied by IR. Distinct IR spectra existed for each form, and form V was generated at higher levels of humidity.	55
Paroxetine HCl	Distinct IR spectra existed for the two different pseudopolymorphs. The form in the formulated product could be determined if the product contained 10% or more of the drug substance.	56
Griseofulvin	Characterization of the polymorph system by distinct Raman spectra for each form. By observing the lattice vibrations (<500 cm^{-1}), it was observed that upon desolvation, griseofulvin reverted to the lattice structure of the original unsolvated crystal form.	57
Sulfathiazole	The lattice vibrations in the Raman spectra were studied in an effort to characterize the different crystal forms.	?
Cortisone acetate	Raman spectroscopy used to differentiate the different crystal forms of cortisone acetate and quantitate the presence of one form within the other.	58
Ampicillin	Distinct Raman spectra were measured for the two anhydrous and trihydrated forms of ampicillin.	59
Cimetidine	Raman spectroscopy used to differentiate the different crystal forms of cimetidine.	60
Spironolactone	Raman spectroscopy used to differentiate the different crystal forms of spironolactone.	61
Carbamazepine	Distinct Raman spectra exist for the different crystal forms. Significant spectral differences exist within the low-frequency, lattice vibration region of the spectra.	62
Fluconazole	Two different polymorphic forms of fluconazole were identified by Raman spectroscopy. Spectral assignments were made for each form, and the conformational data compared to the crystallographic and thermal analysis data.	63
5-Methyl-2-[(2-nitrophenyl)amino]-3-thiophenecarbonitrile	Five different polymorphs of this compound exist, in which three are described in the reference. The three different crystal forms give rise to different colors of the compound (yellow, orange, and red). Distinct IR spectra exist for each form.	64

ples of quantitative polymorphic analysis by vibrational spectroscopy are listed in Table 2 (16,4,51).

Beside mid-IR, near-IR spectroscopy has been used to quantitate polymorphs at the bulk and dosage product level. For SC-25469 (51), two polymorphic forms were discovered (α and β) and the β-form selected for use in the solid dosage form. Since the β-form can be transformed to the α-form under pressure by its enantiotropic nature, quantitation of the β-form in the solid dosage formulation was necessary. Standard mixtures of both forms in the formulation matrix were prepared and spectra measured in the near-IR via diffuse reflectance. Utilizing a standard, near-IR multiple linear regression, statistical approach, the α- and β-forms could be predicted to within 1% of theoretical. This extension of the diffuse reflectance IR technique shows that quantitation of polymorphic forms at the bulk and/or dosage product level can be performed.

In some pharmaceutical spectroscopy laboratories, TG/IR analysis is routinely used for solvate identification of pseudopolymorphic compounds. In one particular example, a submitted bulk drug substance displayed distinctly different XRPD and DSC data as compared to the research reference standard data. Figure 6 displays the TG weight-loss curve and subsequent mid-IR spectra at various time points/temperatures for this questionable sample. Based on absorption bands at 2972 and 1066 cm^{-1}, an organic species (aliphatic C—H, C—C—O moieties, respectively) is present in the evolved gas. Subsequent analysis of the IR spectra collected at the completion of the weight loss (spectral library matching) reveals that an ethanol solvate was present. Thus, TG/IR provided an unequivocal characterization as to the origin of pseudopolymorphism in this particular compound (7). An obvious extension of the TG/IR technique is the study of pharmaceutical dosage forms subjected to various stress conditions to determine if moisture is released, or more importantly, odor analysis for sulfur-containing active drugs.

D. Particulate Analysis/Contaminant Identification

IR and Raman microspectroscopy are well suited for in-situ analysis of contaminants found in pharmaceutical processes. Due to the nondestructive nature of the analysis, further experiments such as energy dispersive X-ray analysis may be performed on the same sample once IR and Raman investigations are complete. To illustrate the potential of IR microspectroscopy, one application is presented.

A series of foreign particulates was found in several bulk lots and final product lots (tablets) of a developmental drug. Black, red, and brown particles were isolated from the bulk drug material, whereas black particles were observed embedded into the tablets. Only the black particles will be focused upon in this discussion. Since the foreign materials were opaque, IR microspectroscopy data were obtained in the reflectance mode. Figure 7A displays the relatively simple IR spectrum of the isolated black particle. No absorption bands corresponding

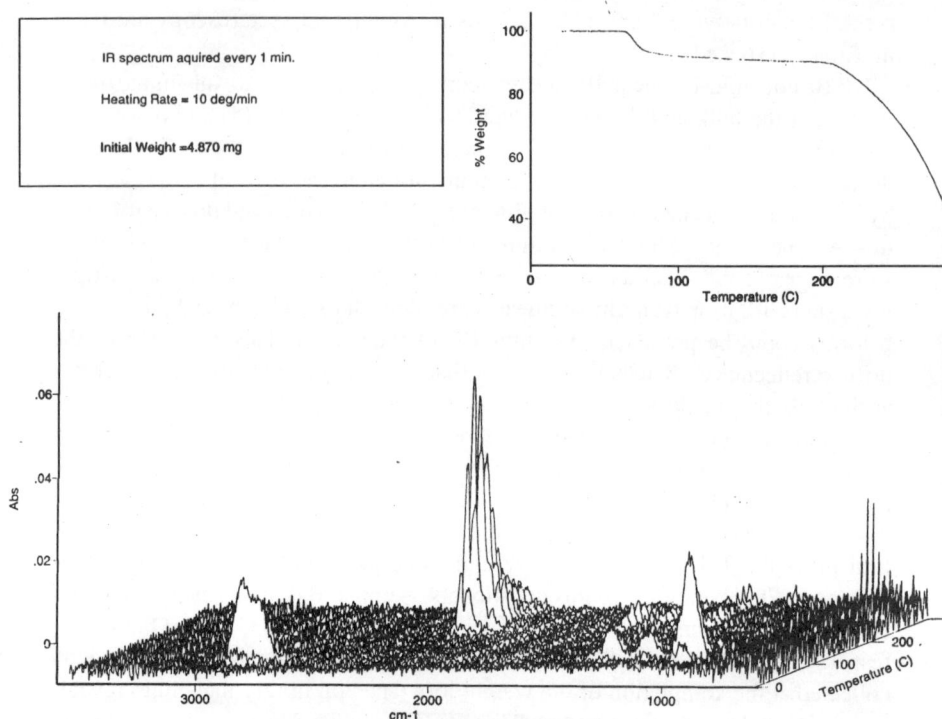

Fig. 6 TG weight-loss curve and subsequent IR spectra measured as a function of temperature. A slight lag time exists between the TG weight loss and IR spectral acquisition, due to the evolved gas being "carried" into the IR gas cell by the He carrier gas. Each IR spectrum is plotted on the same absolute intensity scale (Abs. units).

to aromatic or aliphatic C—H groups (2800–3200 cm^{-1}) were present, whereas a few absorption bands were observed between 1150 and 1250 cm^{-1} corresponding to C—F and C—C functional groups. Computer-aided spectral library search of the spectrum revealed an IR spectral match with polytetrafluroethylene. Subsequent energy dispersive X-ray analysis confirmed that fluorine was present in the black particulate sample. Selectivity of the IR microspectroscopy technique is revealed in Figure 7B. This spectrum represents the reflectance IR measurement of the black particulate which is embedded into the tablet. No sample extraction process was required to obtain the spectrum. A distinct advantage of the IR microspectroscopy technique is the ability to sample only the area of the sample defined by the microscope's aperture. In addition, no destructive sample preparation is required.

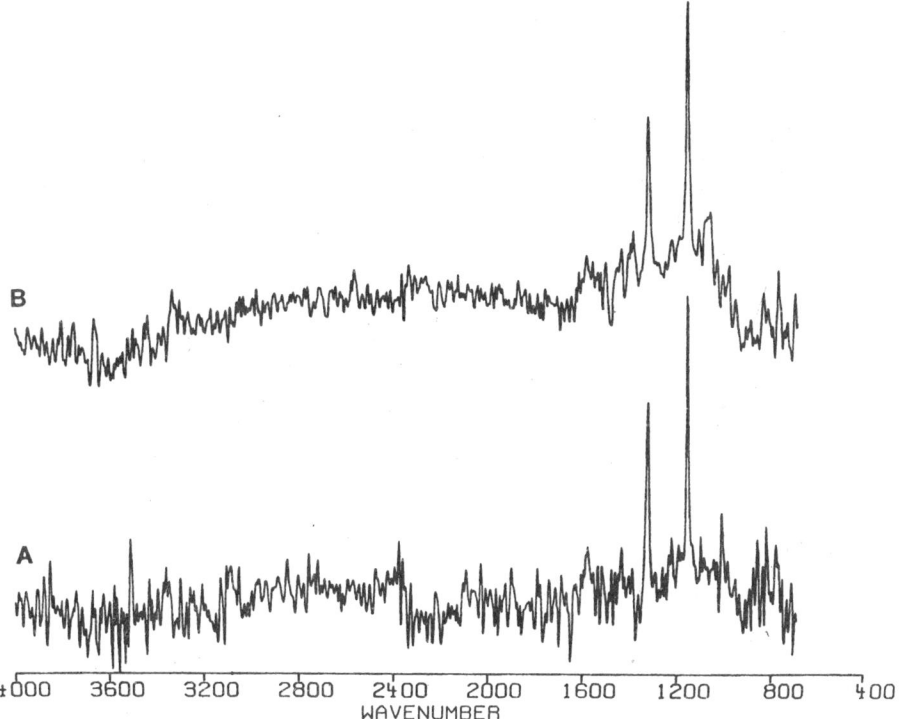

Fig. 7 IR spectra (microspectroscopy) of the isolated black particle (A) and a particle embedded into a pharmaceutical tablet (B).

E. Biopharmaceutical

Pharmaceutical companies are increasingly interested in developing products based on proteins, enzymes, and peptides. With the development of such products comes the need for methods to evaluate the purity and structural nature of these biopharmaceuticals. Proteins, unlike traditional pharmaceutical entities, rely on a specific secondary structure for efficacy. Methods to monitor the secondary structure of pharmaceutically active proteins, thus, is necessary. Infrared spectroscopy provides a way to study these compounds quickly and easily. Byler et al. (65) used second-derivative IR to assess the purity and structural integrity of porcine pancreatic elastase. Seven different lyophilized samples of porcine pancreatic elastase were dissolved in D_2O, placed in demountable cells with CaF_2 windows, and IR spectra obtained. The second derivatives of the spectra were calculated and the spectral features due to residual water vapor and D_2O removed.

Analysis of the spectra showed that one sample was clearly of inferior quality as compared to the other six. The differences noted in the amide I region ($\sim$1650 cm^{-1}) were probably due to structural disorder in the conformation of the protein molecule. That led to the conclusion that the particular protein was no longer in its native state. Analysis of the derivative spectra also revealed evidence of the extent of aggregation of the protein samples. The conclusions drawn from the spectra regarding aggregation were confirmed by gel electrophoresis. Examination of some of the weaker amide I bands in the proteins showed that the samples had varying degrees of α-helix and β-structure present. The relative purity of the various samples was determined by examination of the derivative spectra in the 1620–1300 cm^{-1} region. Coomassie blue dye binding was used to confirm that two of the seven samples demonstrated greater purity than the others. The investigation demonstrated that second-derivative IR is a quick, easy, economical, nondestructive, and reproducible single method that complements traditional methods of analysis such as gel electrophoresis and enzyme activity assays.

Prestrelski et al. (66) used IR to optimize lyophilization conditions for interleukin-2 (IL-2). The authors were able to show that proteins can unfold during lyophilization, and that altering the lyophilization conditions determines whether the proteins refold into the original conformation upon reconstitution. Dong et al. (67) provide a comprehensive review of the use of infrared spectroscopy for the study of the effects of lyophilization on proteins.

F. Other Application Areas

Near-IR spectroscopy is quickly becoming a preferred technique for the quantitative identification of an active component within a formulated tablet. In addition, the same spectroscopic measurement can be used to determine water content, since the combination band of water displays a fairly large absorption band in the near-IR. In one such study (68), the concentration of ceftazidime pentahydrate and water content in physical mixtures has been determined. Due to the ease of sample preparation, near-IR spectra were collected on 20 samples and subsequent calibrations curves constructed for active ingredient and water content, respectively. An interesting aspect of this study was the determination that the calibration samples must be representative of the production process. When calibration curves were constructed from laboratory samples only, significant prediction errors were noted. However, when calibration curves were constructed from laboratory and production samples, realistic prediction values were determined ($\pm$5%).

One major advantage of the near-IR technique is the rapid evaluation of the sample without the need for traditional extraction techniques and subsequent chromatographic or colorimetric analysis. For this reason, near-IR is an ideal form of analysis in a quality control or on-line process analysis environment. In a complementary mid- and near-IR study (69), it was determined that near-IR spectroscopy could not differentiate between simvastatin and lovastatin. These two molecules differ only by the presence and absence of an α-methyl group attached to an ester carbonyl, respectively. Instead, mid-IR diffuse reflectance spectroscopy was used to identify and quantify the active components within a formulation. This study also investigated sample preparation variables, including cup filling, variance in grinding times, and duplicate spectral measurements. In addition to the previously discussed molecules, enalapril maleate and finastride were also studied.

One of the key reasons to implement process analytical spectroscopy in the manufacturing environment is to provide instant data about a product. Spectroscopy, in this capacity, can be used for identity testing and quantitative determinations. González and Pous (70) used near-IR and pattern recognition to evaluate polyalcohol and cellulose content in a pharmaceutical product. They concluded that near-IR is a useful and rapid technique for determining the quality of process intermediates. By evaluating the entire spectrum, they were able to detect most of the process deviations in a short time. Williams and Mason (71) demonstrated the usefulness of FT-Raman for monitoring the reactions of polymers. Using near-IR excitation (1.064 μm) and optimized optical fibers, the authors were able to analyze samples previously incapable of being studied by Raman spectroscopy. The authors recognized the potential of the technique proposing extension to industrial situations such as in-situ reaction monitoring and processes monitoring.

With the ever-increasing need to improve quality and productivity in the analytical pharmaceutical laboratory, automation has become a key component. Automation for vibrational spectroscopy has been fairly limited. Although most software packages for vibrational spectrometers allow for the construction of macro routines for the grouping of repetitive software tasks, there is only a small number of automation routines in which sample introduction and subsequent spectral acquisition/data interpretation are available. For the routine analysis of alkali halide pellets, a number of commercially available "sample wheels" are used in which the wheel contains a selected number of pellets in specific locations. The wheel is then indexed to a sample disk, the IR spectrum obtained and archived, and then the wheel indexed to the next sample. This system requires that the pellets be manually pressed and placed into the wheel before automated spectral acquisition. A similar system is also available for automated liquid analysis in which samples in individual vials are pumped onto an ATR crystal and subsequently analyzed. Between samples, a cleaning solution is passed over the ATR crystal to reduce cross-contamination. Automated diffuse reflectance has also been introduced in which a tray of DR sample cups is indexed into the IR sample beam and subsequently scanned. In each of these cases, manual preparation of the sample is necessary (23). In the field of Raman spectroscopy, automation is being developed in conjunction with fiber-optic probes and accompanying software for identity testing and reaction monitoring. In coming years, advances in hardware and software will permit more widespread use of automation in vibrational analysis.

Addressing the question of amorphous/crystalline content of indomethacin samples, Taylor and Zografi present an excellent utilization of quantitative Raman spectroscopy (72). The paper highlights the quantitative nature of Raman spectroscopy, the need for producing homogeneous calibration/validation samples, and difficulties associated with collecting a Raman spectrum that is truly representative of the concentration level. A linear correlation curve was constructed in which low levels of both amorphous and crystalline material could be detected and predicted in mixtures. The authors felt that the largest source of error in the measurements arose from inhomogeneous mixing of the amorphous and crystalline components in the blends. For solid-state analysis, this conclusion points out the need for a sampling device that collects a Raman spectrum that is truly representative of the sample, in this case, a mixture.

Drug–excipient interactions are typically studied via chromatographic techniques. Unfortunately, this form of analysis requires extraction and/or dissolution techniques that may destroy critical physical and chemical information. Hence, the ability to study drug–excipient interactions noninvasively is crucial. DSC and diffuse reflectance IR have been used to study the yellow or brown color which develops when aminophylline is mixed with lactose (73). The DSC thermogram of the aminophylline/lactose mixture is not a direct superposition of the individual component traces leading to the indication of an incompatibility (74). Although DSC is able to determine a drug–excipient interaction, the exact nature of the interaction is unknown, hence the need for IR spectroscopy. After complete analysis of the IR spectra of individual components, physical mixtures, and various samples subjected to stress conditions (60°C for 3 weeks), it was concluded that ethylenediamine is liberated from the aminophylline complex and reacts with lactose through a Schiff base intermediate. This reaction, in turn, results in brown discoloration of the sample. With currently available diffuse reflectance environmental chambers providing control of temperature, pressure, and relative humidity conditions, IR spectroscopy is now a first-line approach to the determination of physicochemical interactions between drugs and excipients.

Salt selection for a pharmaceutical substance is a critical component of drug development because selection of the wrong salt can lead to chemical and physical stability issues as well as formulation problems. Raman spectroscopy has been used to assess the molecular nature of salts crystallized from salbutamol base (75). Variations in vibrational frequencies due to electron-withdrawing or -donating substituents were clearly evident; the C—C—O stretching vibration shifted from 776 cm^{-1} in the free base to 756 cm^{-1} in the benzoate salt. The C=C stretching frequency also shifted from 1610 cm^{-1} to 1603 cm^{-1} with the benzoate ion but showed an increase to 1616 cm^{-1} with sulfate ion. Clearly, the choice of salt affects the molecular nature of the drug, with obvious implications for its physicochemical properties.

Combinatorial chemistry is rapidly becoming a reliable way for pharmaceutical companies to discover and identify drug candidates. With advances in synthetic procedures and automation, the production of very large libraries of compounds is possible by even the smallest pharmaceutical companies (76). Yan et al. (77,78) have described an infrared spectroscopic method for analyzing solid-phase organic reactions that occur on resin beads. The method is an improvement over the previous IR methods of preparing a KBr pellet from >10 mg of resin beads. In the described method, a drop of resin solution is removed from the reaction vessel, washed in organic solvents, and dried under vacuum. The dried resin beads are placed on a NaCl window, flattened to 10–15µm thickness, and spectra acquired in the transmission mode of the IR microscope. The authors demonstrated that the flattened beads provided superior results to nonflattened beads.

Probably one of the most exciting new applications of IR and Raman spectroscopy is chemical imaging. By utilizing one or the other of the vibrational spectroscopy techniques, one is now able to generate a chemical image of a two-dimensional area of a sample (79). The spectroscopic chemical imaging technique relies on the interface of an optical microscope, equipped with a motorized stage, to an IR or Raman spectrometer. The operator is able to focus visually on a sample of interest, and optical observations about the sample are made (morphology, separation of layers, etc.). Subsequently, the same visual area (two-dimensional) is defined for spectroscopic analysis. Utilizing a raster pattern (step-wise movement of the motorized stage along the x- and y- axes), individual spectra are acquired for each spatial location within the two-dimensional area. Spatial resolution is typically defined by the technique (~1 µm for Raman, ~5–10 µm for IR). Once the individual spectra are obtained for the two-dimensional area, the intensity of a specific spectral feature within each spectrum can be plotted versus the spatial position. In this manner, one is able to obtain a contour plot showing the spatial position of a chemical entity.

In a recent Raman application of chemical imaging, the content uniformity of an active drug substance within a compressed tablet was investigated (80). A pharmaceutical tablet containing cyclobenzaprine HCl was shaved such that a chemical image was obtained for each sampled depth within the tablet. Since the Raman spectrum of cyclobenzaprine HCl had distinct spectral features different from those of the excipients present, a chemical image specific to the drug substance could be obtained, or for that matter, for each excipient. Figure 8 displays the chemical image for one plane (depth) of the tablet. It is clearly observed that cyclobenzaprine HCl is detected at the 0,0 spatial position within the sampled plane of the tablet. The corresponding excipient chemical image also shows the lack of excipient concentration (signal intensity) at the same spatial position.

A

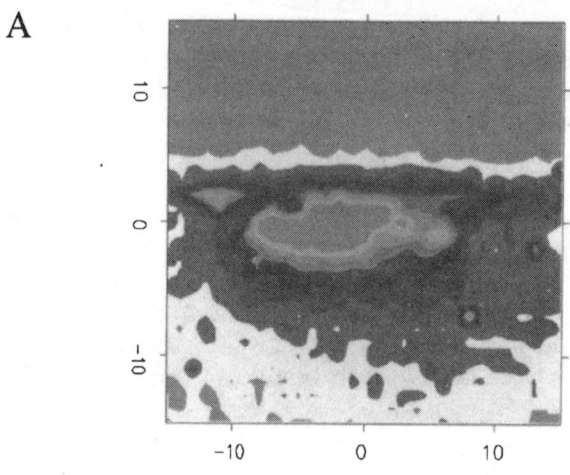

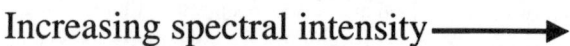

Increasing spectral intensity ⟶

B

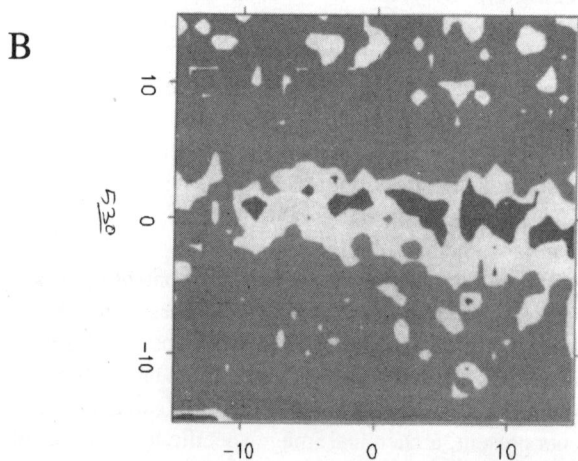

Fig. 8 Chemical images, obtained by dispersive Raman spectroscopy, of a shaved surface of a pharmaceutical tablet containing cyclobenzaprine HCl. Image (A) maps the intensity of a spectral feature specific to the drug substance (cyclobenzaprine HCl), whereas image (B) is an image of the same spatial area, but mapping a spectral feature of the excipients.

These results demonstrate the excellent sensitivity and selectivity (spectrally and spatially) for the spectroscopic, chemical imaging technique. Additional studies utilizing spectroscopic chemical imaging include: (a) the determination of pharmaceutical film thicknesses, (b) the physical migration of active drug substance in a formulation as a function of stress conditions (high temperature/humidity), and (c) the possible physical transformation (polymorphic or crystalline/amorphous) of a drug substance upon formulation (direct compression or wet granulation techniques). It is anticipated that this form of spectroscopic analysis will become an integral technique for pharmaceutical analysis in the very near future.

V. SUMMARY

Clearly, the potential applications for vibrational spectroscopy techniques in the pharmaceutical sciences are broad, particularly with the advent of Fourier transform instrumentation at competitive prices. Numerous sampling accessories are currently available for IR and Raman spectroscopy, enabling the analysis of virtually any type of pharmaceutically based sample. In addition, new sampling devices are rapidly being developed for at-line and on-line applications.

ACKNOWLEDGMENTS

The authors would like to thank Chris Rodriguez for acquiring the TG/IR data set and Melissa A. Houghtaling for the Raman spectroscopic images.

REFERENCES

1. HG Brittain, KR Morris, DE Bugay, AB Thakur, ATM Serajuddin. Solid-state NMR and IR for the analysis of pharmaceutical solids: Polymorphs of fosinopril sodium. J Pharm Biomed Anal 11:1063–1069, 1993.
2. DA Roston, MC Walters, RR Rhinebarger, LJ Ferro. Characterization of polymorphs of a new anti-inflammatory drug. J Pharm Biomed Anal 11:293–300, 1993.
3. BA Bolton, PN Prasad. Laser Raman investigation of drug-polymer conjugates: Sulfathiazole-povidone coprecipitates. J Pharm Sci 73:1849–1851, 1984.
4. DE Bugay, AW Newman, WP Findlay. Quantitation of cefepime · 2HCl dihydrate in cefepime · 2HCl monohydrate by diffuse reflectance IR and powder X-ray diffraction techniques. J Pharm Biomed Anal 15:49–61, 1996.
5. PR Griffiths, AJ Lange. On-line use of the concentric flow nebulizer for direct deposition liquid chromatography-Fourier transform infrared spectrometry. J Chromatogr Sci 30:93–97, 1992.

6. DJ Johnson, DAC Compton. Solvent retention studies for pharmaceutical samples using an integrated TGA/FT-IR system. Spectroscopy 3:47–50, 1988.

7. C Rodriguez, DE Bugay. Characterization of pharmaceutical solvates by combined thermogravimetric and infrared analysis. J Pharm Sci 86:263–266, 1997.

8. Y Kim, CA Rose, Y Liu, Y Ozaki, G Datta, AT Tu. FT-IR and near-infrared FT-Raman studies of the secondary structure of insulinotropin in the solid state: α-helix to β-sheet conversion induced by phenol and/or by high shear force. J Pharm Sci 83:1175–1180, 1994.

9. J Lynch, S Riseman, W Laswell, D Tschaen, R Volante, G Smith, I Shinkai. Mechanism of an acid chloride-imine reaction by low-temperature FT IR: β-lactam formation occurs exclusively through a ketene intermediate. J Org Chem 54:3792–3796, 1989.

10. MA Dempster, BF MacDonald, PF Gemperline, NR Boyer. A near-infrared reflectance analysis method for the noninvasive identification of film-coated and non-filmcoated, blister-packed tablets. Anal Chim Acta 310:43–51, 1995.

11. CV Raman, KS Krishnan. A new type of secondary radiation. Nature 121:501, 1928.

12. G Herzberg. Molecular spectra and molecular structure. II. Infrared and Raman spectra of polyatomic molecules. 6th ed. New York: D Van Nostrand Co, 1954.

13. NB Colthup, LH Daly, SE Wiberley. Introduction to Infrared and Raman Spectroscopy, 3rd ed. New York: Academic Press, 1990.

14. DA Long. Raman Spectroscopy. New York: McGraw-Hill, 1977.

15. P Hendra, C Jones, G Warnes. Fourier Transformation Raman Spectroscopy: Instrumental and Chemical Applications. New York: Ellis Horwood, 1991.

16. EA Cutmore, PW Skett. Application of Fourier transform Raman spectroscopy to a range of compounds of pharmaceutical interest. Spectrochim Acta 48A:809–818, 1993.

17. PR Griffiths, JA de Haseth. Fourier Transform Infrared Spectrometry. New York: Wiley, 1986.

18. JE Stewart. Infrared Spectroscopy: Experimental Methods and Techniques. New York: Marcel Dekker, 1970.

19. VA Bell, VR Citro, GD Hodge. Effect of pellet pressing on the infrared spectrum of kaolinite. Clays and Clay Minerals 39:290–292, 1991.

20. SC Mutha, WB Ludemann. Solid-state anomalies in IR spectra of compounds of pharmaceutical interest. J Pharm Sci 65:1400–1403, 1976.

21. J Hlavay, J Inczédy. Pellet preparation for quantitative determination of inorganic solid substances by infrared spectroscopy. Spectrochim Acta 41A:783–787, 1985.

22. MP Fuller, PR Griffiths. Diffuse reflectance measurements by infrared Fourier transform spectrometry. Anal Chem 50:1906–1910, 1978.

23. Spectra-Tech, Inc. Product Catalog. The Complete Guide to FT-IR. Stamford, CT, 1993.

24. R Barer, ARH Cole, HW Thompson. Infra-red spectroscopy with the reflecting microscope in physics, chemistry and biology. Nature 163:198–201, 1949.

25. JA Reffner, JP Coates, RG Messerschmidt. Chemical microscopy with FTIR microspectrometry. Am Lab 19:5–11, 1987.

26. SV Compton, DAC Compton. Optimization of data by internal reflectance spectros-

copy. In: PB Coleman, ed. Practical Sampling Techniques for Infrared Analysis. Boca Raton; FL. CRC Press, 1993; pp. 55–92.

27. AG Bell. On the production and reproduction of sound by light. Am Assoc Proc 29:115–136, 1881.

28. DW Vidrine. Photoacoustic Fourier transform infrared spectroscopy of solid samples. Appl Spectrosc 34:314–319, 1980.

29. MG Rockley. Fourier-transformed infrared photoacoustic spectroscopy of solids. Appl Spectrosc 34:405–406, 1980.

30. JF McClelland, S Luo, RW Jones, LM Seaverson. A tutorial on the state-of-the-art of FTIR photoacoustic spectroscopy. In: D Bićanić, ed. Photoacoustic and Photothermal Phenomena III. Berlin: Springer-Verlag, 1992, pp. 113–124.

31. P MacLaurin, NC Crabb, I Wells, PJ Worsfold, D Coombs. Quantitative in situ monitoring of an elevated temperature reaction using a water-cooled mid-infrared fiber-optic probe. Anal Chem 68:1116–1123, 1996.

32. JR Ferraro. A history of Raman spectroscopy. Spectroscopy 11:18–25, 1996.

33. DB Chase, JF Rabolt. Fourier Transform Raman Spectroscopy. New York: Academic Press, 1994.

34. T Hirschfeld. Raman microprobe: Vibrational spectroscopy in the femtogram range. J Opt Soc Am 63:476–477, 1973.

35. M Lankers, D Göttges, A Materny, K Schaschek, W Kiefer. A device for surface-scanning micro-Raman spectroscopy. Appl Spectrosc 46:1331–1334, 1992.

36. SD Schwab, RL McCreery. Versatile, efficient Raman sampling with fiber optics. Anal Chem 56:2199–2204, 1984.

37. SD Schwab, RL McCreery. Remote, long-pathlength cell for high-sensitivity Raman spectroscopy. Appl Spectrosc 41:126–130, 1987.

38. DD Archibald, LT Lin, DE Honigs. Raman spectroscopy over optical fibers with the use of a near-IR FT spectrometer. Appl Spectrosc 42:1558–1563, 1988.

39. WA Warr. Computer assisted structure elucidation. Library search and spectral data collections. Anal Chem 65:1045A–1050A, 1993.

40. DE Bugay, WP Findlay. Pharmaceutical Excipients: Characterization by IR, Raman, and NMR Spectroscopy. New York: Marcel Dekker, 1999.

41. USP 23/NF 18. Rockville, MD: United States Pharmacopeial Convention, Inc., 1995.

42. S Budavari. The Merck Index. 11th ed. Rahway, NJ: Merck & Co., 1989.

43. EW Ciurczak. Uses of near-infrared spectroscopy in pharmaceutical analysis. Appl Spectrosc Rev 23:147–163, 1987.

44. RE Schirmer. Modern Methods of Pharmaceutical Analysis. Boca Raton, FL: CRC Press, 1991.

45. PJ Gemperline, LD Webber, FO Cox. Raw materials testing using soft independent modeling of class analogy analysis of near-infrared reflectance spectra. Anal Chem 61:138–144, 1989.

46. RA Lodder, GM Hieftje. Analysis of intact tablets by near-infrared reflectance spectrometry. Appl Spectrosc 42:556–558, 1988.

47. PK Aldridge, RF Mushinsky, MM Andino, CL Evans. Identification of tablet formulations inside blister packages by near-infrared spectroscopy. Appl Spectrosc 48: 1272–1276, 1994.

48. CJ Petty, DE Bugay, WP Findlay, C Rodriguez. Applications of FT-Raman spectroscopy in the pharmaceutical industry. Spectroscopy 11:41–45, 1996.
49. K Raghavan, A Dwivedi, GC Campbell Jr., G Nemeth, MA Hussain. A spectroscopic investigation of DuP 747 polymorphs. J Pharm Biomed Anal 12:777–785, 1994.
50. KJ Hartauer, ES Miller, JK Guillory. Diffuse reflectance infrared Fourier transform spectroscopy for the quantitative analysis of mixtures of polymorphs. Int J Pharm 85:163–174, 1992.
51. R Gimet, AT Luong. Quantitative determination of polymorphic forms in a formulation matrix using the near infra-red reflectance analysis technique. J Pharm Biomed Anal 5:205–211, 1987.
52. KR Morris, AW Newman, DE Bugay, SA Ranadive, AK Singh, M Szyper, SA Varia, HG Brittain, ATM Serajuddin. Characterization of humidity-dependent changes in crystal properties of a new HMG-CoA reductase inhibitor in support of its dosage form development. Int J Pharm 108:195–206, 1994.
53. T Madan, AP Kakkar. Preparation and characterization of ranitidine HCl crystals. Drug Dev Ind Pharm 20:1571–1588, 1994.
54. A Kiss, J Répási, Z Salamon, C Novák, G Pokol, K Tomor. Solid state investigation of mefloquine hydrochloride. J Pharm Biomed Anal 12:889–893, 1994.
55. NAT Nguyen, S Ghosh, LA Gatlin, DJW Grant. Physicochemical characterization of the various solid forms of carbovir, an antiviral nucleoside. J Pharm Sci 83:1116–1123, 1994.
56. IR Lynch, PC Buxton, JM Roe. Infrared spectroscopic studies on the solid state forms of paroxetine hydrochloride. Anal Proc 25:305–306, 1988.
57. BA Bolton, PN Prasad. Laser Raman investigation of pharmaceutical solids: Griseofulvin and its solvates. J Pharm Sci 70:789–792, 1981.
58. CM Deeley, RA Spragg, TL Threlfall. A comparison of Fourier transform infrared and near-infrared Fourier transform Raman spectroscopy for quantitative measurements: an application in polymorphism. Spectrochim Acta 47A:1217–1223, 1991.
59. JC Bellows, FP Chen, PN Prasad. Determination of drug polymorphs by laser Raman spectroscopy. I. Ampicillin and griseofulvin. Drug Dev Ind Pharm 3:451–458, 1977.
60. AM Tudor, MC Davies, CD Melia, DC Lee, RC Mitchell, PJ Hendra, SJ Church. The applications of near-infrared Fourier transform Raman spectroscopy to the analysis of polymorphic forms of cimetidine. Spectrochim Acta 47A:1389–1393, 1991.
61. GA Neville, HD Beckstead, HF Shurvell. Utility of Fourier transform-Raman and Fourier transform-infrared diffuse reflectance spectroscopy for differentiation of polymorphic spironolactone samples. J Pharm Sci 81:1141–1146, 1992.
62. LE McMahon, P Timmins, AC Williams, P York. Characterization of dihydrates prepared from carbamazepine polymorphs. J Pharm Sci 85:1064–1069, 1996.
63. XJ Gu, W Jiang. Characterization of polymorphic forms of fluconazole using Fourier transform Raman spectroscopy. J Pharm Sci 84:1438–1441, 1995.
64. GA Stephenson, TB Borchardt, SR Byrn, J Bowyer, CA Bunnell, SV Snorek, L Yu. Conformational and color polymorphism of 5-methyl-2-[(2-nitrophenyl)amino]-3-thiophenecarbonitrile. J Pharm Sci 84:1385–1386, 1995.
65. DM Byler, RM Wilson, CS Randall, TD Sokoloski. Second derivative infrared spectroscopy as a non-destructive tool to assess the purity and structural integrity of proteins. Pharm Res 12:446–450, 1995.

66. SJ Prestrelski, KA Pikal, T Arakawa. Optimization of lyophilization conditions for recombinant human interleukin-2 by dried-state conformational analysis using Fourier-transform infrared spectroscopy. Pharm Res 12:1250–1259, 1995.

67. A Dong, SJ Prestrelski, SD Allison, JF Carpenter. Infrared spectroscopic studies of lyophilization- and temperature-induced protein aggregation. J Pharm Sci 84:415–424, 1995.

68. S Lonardi, R Viviani, L Mosconi, M Bernuzzi, P Corti, E Dreassi, C Murratzu, G Corbini. Drug analysis by near-infra-red reflectance spectroscopy. Determination of the active ingredient and water content in antibiotic powders. J Pharm Biomed Anal 7:303–308, 1989.

69. JA Ryan, SV Compton, MA Brooks, DAC Compton. Rapid verification of identity and content of drug formulations using mid-infrared spectroscopy. J Pharm Biomed Anal 9:303–310, 1991.

70. F González, R Pous. Quality-control in manufacturing process by near-infrared spectroscopy. J Pharm Biomed Anal 13:419–423, 1995.

71. KPJ Williams, SM Mason. Future directions for Fourier transform Raman spectroscopy in industrial analysis. Spectrochim Acta 46A:187–196, 1990.

72. LS Taylor, G Zografi. The quantitative analysis of crystallinity using FT-Raman spectroscopy. Pharm Res 15:755–761, 1998.

73. KJ Hartauer, JK Guillory. A comparison of diffuse reflectance FT-IR spectroscopy and DSC in the characterization of a drug-excipient interaction. Drug Dev Ind Pharm 17:617–630, 1991.

74. AA Van Dooren, BV Duphar. Design for drug-escipient interaction studies. Drug Dev Ind Pharm 9:43–55, 1983.

75. AB Brown, P York, AC Williams, HGM Edwards, H Worthington. Solid state characterization of salbutamol salts using FT-Raman and SSNMR spectroscopy. J Pharm Pharmacol 45(suppl. 2):1135, 1993.

76. SE Blondelle, E Pérez-Payá, CT Dooley, C Pinilla, RA Houghten. Soluble combinatorial libraries of organic, peptidomimetic and peptide diversities. Trends Anal Chem 14:83–92, 1995.

77. B Yan, G Kumaravel. Probing solid-phase reactions by monitoring the IR bands of compounds on a single "flattened" resin bead. Tetrahedron 52:843–848, 1996.

78. B Yan, G Kumaravel, H Anjaria, A Wu, RC Petter, CF Jewell, JR Wareing. Infrared spectrum of a single resin bead for real-time monitoring of solid-phase reactions. J Org Chem 60:5736–5738, 1995.

79. K Krishnan, JR Powell, SL Hill. Infrared microimaging. In: H Humecki ed. Practical Guide to Infrared Microspectroscopy. New York: Marcel Dekker, pp. 85–110, 1995.

80. MA Houghtaling, DE Bugay. Raman spectroscopic imaging: A potential new technique for content uniformity testing. 14th Annual American Association of Pharmaceutical Scientists Annual Meeting and Exposition, New Orleans, LA, Nov 14–18, 1999.

12
Statistical Considerations in Pharmaceutical Process Development and Validation

Gerald J. Mergen
McNeil Consumer Healthcare Company, Fort Washington, Pennsylvania

I. INTRODUCTION

It has been more than a decade since the publication of the Food and Drug Administration (FDA) *Guidelines on Process Validation* (1). The intervening years have seen the creation of a whole industry devoted to educating pharmaceutical scientists in the areas of preparing validation protocols and validation reports. These scientists include the R&D analytical chemist in methods development and stability services, the formulation scientist, the analytical chemist in the quality control laboratory, and the validation specialist. However, guidance has been lacking on methods to use for establishing the ''predetermined specifications'' crucial to any validation study. We appear to have the format and syntax of the protocols correct, but are still failing to receive passing grades in this all-important area, as measured by the number of FD-483's and ''Warning Letters'' issued in the area of process validation.

Additionally, the issuance of guidelines for pre- and post-New Drug Application approval inspections do provide, as the title suggests, guidance to the field inspector (and the pharmaceutical industry) on what should be addressed in the product development report (2,3). These considerations include the characterization of the final blend of the granulation and establishment of specifications to be used in the validation study. Establishing finished product specifications and in-process requirements are among the most important and critical steps in prod-

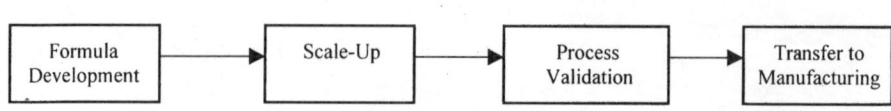

Fig. 1   Linear representation of product development or technical transfer process.

uct development. Let's examine the typical linear representation of the product development process as shown in Fig. 1.

A more accurate portrayal of the process is in keeping with the legacy of Walter Shewhart's ''Plan–Do–Check–Act'' cycle, which reflects the continuous improvement spirit advocated by W. Edwards Deming (4) and others. The product development process is nonlinear and a continuous cycle, see Fig. 2, as products, after being transferred to manufacturing, are constantly changing, whether these changes have been planned or not. The specifications developed during the formula development and scale-up phase, will have long-lasting effects throughout the product life cycle. Not only must they be adequate for the short term (process validation), but for the long term (transfer to manufacturing) as well.

Let's discuss now some possible statistical approaches for evaluating the data obtained from development or scale-up activities, to learn all there is about the process prior to transfer to manufacturing. In addition, let's see how this data can be used to establish practical and reasonable in-process requirements and acceptance criteria for the process and product as we enter the process validation phase. The examples used in the following sections are of solid dosage forms, although the techniques are certainly applicable to other dosage forms and also analytical method development.

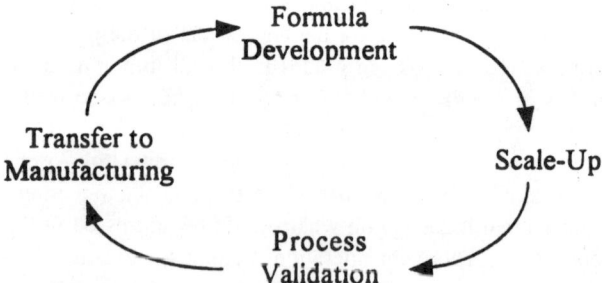

Fig. 2   Nonlinear representation of the product development process using the ''Plan–Do–Check–Act'' Shewhart cycle.

II. GRANULATION FINAL BLEND

A. Blend Uniformity

A strength of the FDA guidelines is their lack of prescriptions for a particular course of action. The agency recognizes full well that a one-size approach does not necessarily fit all situations, so it leaves the methodology and reasoning for determining specifications up to each firm. Unfortunately, without any guidance, some firms have resorted to the only source of specifications available, namely, the U.S. Pharmacopeia (USP) (5). The FDA has commented repeatedly about the unacceptability of using the USP "Dose Uniformity by Content Uniformity," general chapter ⟨905⟩, requirements for blend uniformity. In addition, the agency has also "suggested" limits of 90.0–110.0% of label claim and a relative standard deviation of 4–5% (6,7) as appropriate limits. The number of samples to be taken during a blend uniformity study has not been specified, so it is not clear whether these suggested limits are applicable for every possible sampling scheme.

Although there is philosophical agreement that the sample weight analyzed should equal the weight of the solid dosage unit, there has been no recent discussion of a minimum number of samples to be taken out of the blender or drums of powder/granulation.

The guidelines also present the agrument that the predetermined limits for the final blend uniformity should be tighter than the USP requirements because of the increase in process variability as one moves downstream. Let's assume that the geometry of the blender is such that 15 unit dose samples adequately represents the stratification of the blender: five samples from the top, the middle, and the bottom of the bin. Reasonable acceptance criteria for the 15 unit dose samples is the stage 1 dose uniformity attributes requirements of no unit outside 85.0–115.0% of label claim. But what do we do about setting requirements for the relative standard deviation (RSD)? It is known from statistical theory, refer to Larson (8), that the relationship between the population standard deviation and the sample standard deviation is given by Eq. 1.

$$\chi_\alpha^2 = \frac{(n-1)s^2}{\sigma^2} \tag{1}$$

where

χ_α^2 = critical value of the chi-square distribution with $(n-1)$ degrees of freedom at a stated confidence level α

s^2 = sample variance

σ^2 = population variance

For a given confidence level α, sample size, n, and population variance, σ^2, it is an easy task to rearrange Eq. (1) to derive the appropriate sample variance and consequently the relative standard deviation. The USP Committee of Revision published an article in *Pharmacopeial Forum* (PF), Stimuli to the Revision Process (9), where they presented the statistical analysis and justification of the requirements specified in the current "Uniformity of Dosage Units" general chapter ⟨905⟩. The current general chapter of the USP (5) added the variables acceptance criteria of the relative standard deviation to the attributes plan already in place from the previous edition. In this article the authors assumed the population standard deviation, σ, to be 10% for computing the relative standard deviation (RSD) applicable for a sample size of 10 and 30 dosage units. Table 1 presents the corresponding sample standard deviation (equivalent to the RSD when the sample mean = 100% of label claim) for various sample sizes and population standard deviation.

Thus, if we wish to be 95% confident ($\alpha = 0.05$) that our sample standard deviation does not exceed the population standard deviation of 8% (as compared to the assumed population standard deviation of 10%), then we should choose the maximum RSD for $n = 15$ to be 5.5%. The relationship expressed in Eq. (1) may be used to find any desired RSD for a given confidence level and sample size, as shown in Eq. (2).

$$s = \sigma\sqrt{\frac{\chi_\alpha^2}{(n-1)}} \tag{2}$$

The complete acceptance criteria for blend uniformity then becomes: Obtain 15 unit dose samples and assay for label claim of the active ingredient(s).

Table 1 Relationship Between Sample Size, Population Standard Deviation, and the Relative Standard Deviation

Sample size	Population standard deviation (%)	Relative standard deviation (%)
5	10	4.22
5	9	3.79
10	10	6.08
15	8	5.48
15	9	6.16
15	9	6.16
30	10	7.81

The requirements are met if no unit falls outside the range of 85.0–115.0% of label claim and the relative standard deviation is less than 5.5%. Requirements for the average assay are determined from the stability profile for the particular drug substance, so the potency range may differ depending on product.

B. Particle Size Distribution

The most common method for obtaining the particle size distribution of a powder or granulation is still the sieve analysis determination. Let's assume we have measured the weight percent retained of stratified samples from a blender for multiple-batches during development and/or scale-up. By transforming the individual weight percent retained measurements into a cumulative frequency distribution, we can assess the fit of the data to two commonly used probability models, the long-normal and normal distributions. These two models have wide applicability in this area and are suitable for most pharmaceutical powders and granulations. The cumulative weight percentages are considered the cumulative probabilities or the cumulative area under the normal curve. The value of the standard normal random variable, z value, which corresponds to these areas, can be found in any statistics textbook, e.g., Larson (8).

One then plots the z value for the cumulative weight percent finer against the diameter of the screen opening. In the case of the log-normal distribution, the z value for the cumulative percent finer would be plotted against the natural logarithm of the diameter of the screen opening. The original data are shown in Tables 2, 3, and 4, and the distribution plots of the normal and log-normal are shown in Figs. 3 and 4, respectively.

Table 2 Sieve Analysis Summary: Percent Retained on Screens

	Screen mesh						
Sample	20	40	50	60	80	100	Pan
1	4.4	24.4	24.4	18.8	20.6	3.8	3.6
2	5.3	26.4	23.4	17.9	20.6	3.0	3.5
3	4.4	25.0	25.1	18.9	19.5	3.9	3.1
4	3.6	21.9	24.6	19.6	21.6	4.6	4.1
5	1.0	13.0	17.0	19.0	32.0	11.0	7.0
6	1.0	15.2	21.2	17.2	29.3	10.1	6.1
7	2.8	19.8	20.8	17.4	25.5	8.5	4.7
Average (μm)	3.21	20.81	22.36	18.40	24.16	6.41	4.59
Standard deviation (μm)	1.697	5.105	2.895	0.904	4.887	3.343	1.457

Table 3 Cumulative Fraction Finer Than Stated Particle Diameter

	Micron (µm)					
Sample	149	177	250	297	420	840
1	0.036	0.074	0.280	0.468	0.712	0.956
2	0.035	0.065	0.271	0.450	0.684	0.948
3	0.031	0.070	0.265	0.454	0.705	0.955
4	0.041	0.087	0.303	0.499	0.745	0.964
5	0.070	0.180	0.500	0.690	0.860	0.990
6	0.061	0.162	0.455	0.627	0.839	0.991
7	0.047	0.132	0.387	0.561	0.769	0.967

A similar method has been proposed in the PF (10). The technique just described differs from the PF proposal in that it does not rely on subjectivity to draw the best-fitting straight line through the data points; instead, it uses simple linear regression to find the best fit. Also, the scale for the cumulative percent finer is linear and not logarithmic, which the subsequent discussion will show leads to a more thorough analysis and interpretation.

In this example one can see that the log-normal model, Fig. 4, is a better fit to the data than the normal distribution, Fig. 3. The parameter estimates for the median and standard deviation of each sample can be found from a straightforward application of the simple linear regression model expressed by equation (3):

$$y = \beta_0 + \beta_1 x + \varepsilon \tag{3}$$

where

y = response or value of the standard normal deviate for the cumulative percent finer

Table 4 Standard Normal Deviate for the Corresponding Cumulative Percent Finer

	Micron (µm)					
Sample	149	177	250	297	420	840
1	−1.799	−1.447	−0.583	−0.080	0.559	1.706
2	−1.812	−1.514	−0.610	−0.126	0.479	1.626
3	−1.866	−1.476	−0.628	−0.116	0.539	1.695
4	−1.739	−1.359	−0.516	−0.003	0.659	1.799
5	−1.476	−0.915	0.000	0.496	1.080	2.236
6	−1.546	−0.986	−0.113	0.324	0.990	2.366
7	−1.675	−1.117	−0.287	0.154	0.736	1.838

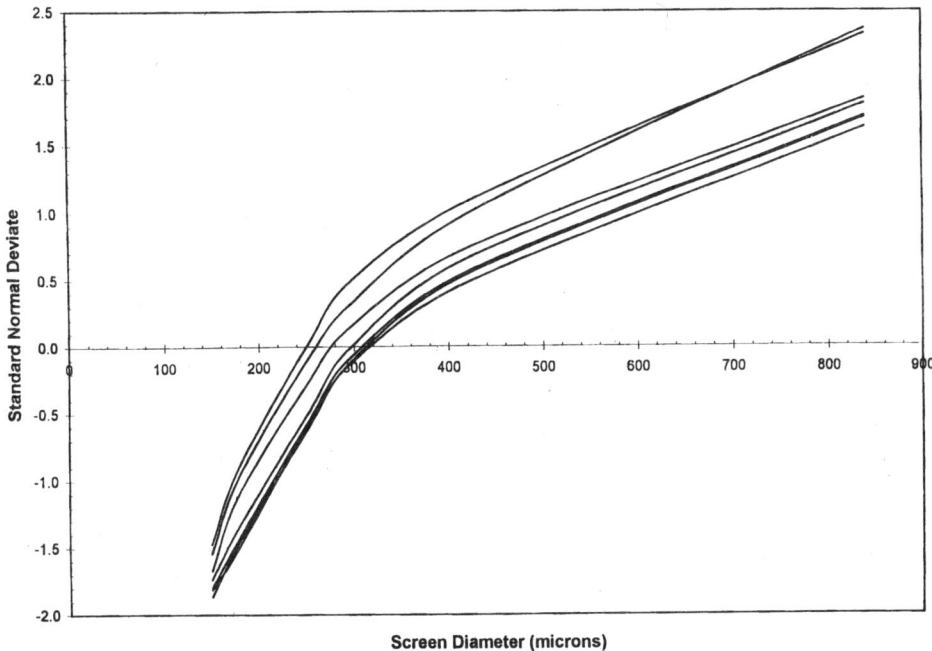

Fig. 3 Normal distribution model for cumulative percent finer (z value) versus the diameter of the sieve screen opening.

x = independent variable or the screen opening diameter (μm)
β_0 = intercept of the regression model when x = 0
β_1 = slope of the regression model, which measures the change in y with a unit change in x

The median of the distribution, or the particle size of the 50th percentile, is found by setting the value of z equal to 0 and solving for x, the screen opening or particle diameter. It can be shown that the standard deviation of the distribution is the reciprocal of the slope of the line, β_1. The calculations necessary to compute the parameter estimates are given in Eq. (4). The parameter estimates for the lognormal and normal distributions for this example are presented in Table 5.

$$\text{median} = -\frac{\beta_0}{\beta_1} \tag{4.a}$$

$$\text{standard deviation} = \frac{1}{\beta_1} \tag{4.b}$$

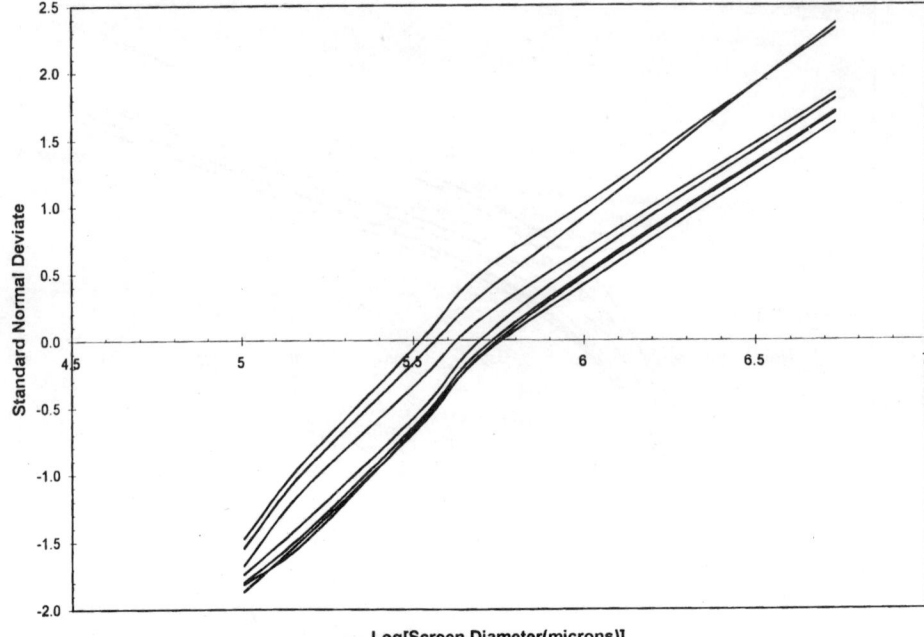

Fig. 4 Log-normal distribution model for cumulative percent finer (z value) versus the natural logarithm of the sieve screen opening.

Table 5 Parameter Estimates for the Particle Size Distribution Models

| | Lognormal | | Normal | |
	Median (μm)	Standard deviation (μm)	Median (μm)	Standard deviation (μm)
Sample				
1	5.828	0.486	413.1	210.4
2	5.856	0.494	425.2	213.7
3	5.843	0.480	419.6	207.8
4	5.788	0.482	395.8	208.8
5	5.579	0.461	305.0	200.5
6	5.617	0.449	322.2	193.1
7	5.724	0.500	368.3	218.0
Average (μm)	5.748	0.479	378.47	207.48
Standard deviation (μm)	0.112	0.018	48.396	8.315

An advantage to using the distribution model approach rather than the percent retained on the sieve screens is its superiority in the visual display of the data. Since the screen openings increase geometrically, the histogram class intervals from screen to screen are not equal. For example, the screen diameter changes from 840 to 420 μm for the change from 20 to 40-mesh screens, whereas the diameter changes from only 177 to 149 μm for the change from 80- to 100-mesh screens. A histogram of percent retained by screen would be a very deceiving representation of the true particle size distribution, and makes for difficult sample-to-sample comparisons. This is evident from a comparison of the log-normal plots, Fig. 4, with the usual weight percent retained-versus-screen mesh bar chart, Fig. 5. The parameter estimates for the particle size distribution from the log-normal or normal plot can be used to construct more accurate histograms. Using the parameter estimates as inputs into a random number generator for the appropriate model, a histogram of the distribution can prepared.

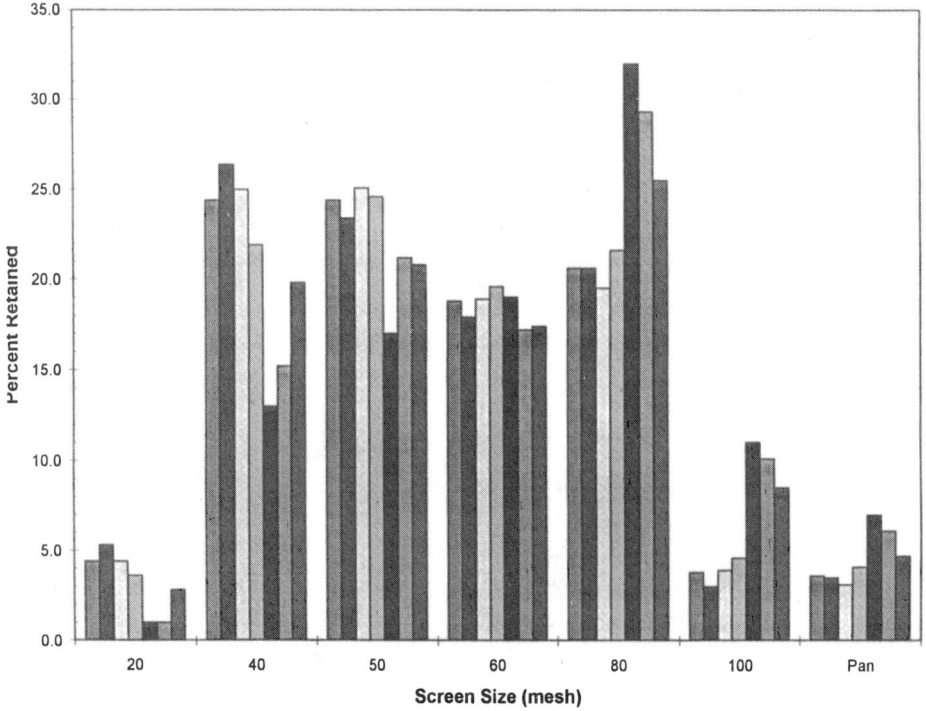

Fig. 5 Bar chart representation of the percent retained versus the screen mesh for the data in Table 2.

Consider the first sample, which has a log-normal distribution with median = 5.83 µm and standard deviation = 0.49 µm. To visualize this distribution, conduct a simulation by generating 10,000 observations from a log-normal distribution using the parameter estimates given above. The histogram from the simulation, Fig. 6, shows the skewness inherent in the log-normal distribution. There is excellent agreement between the actual data and the simulation; see Table 6. The same approach may be used for data that is normally distributed.

One can calculate the acceptance criteria for the particle size distribution median and standard deviation by use of a technique described by Hahn and Meeker (11). Their work describes three types of statistical interval: confidence, prediction, and tolerance. The authors maintain that the choice of the appropriate statistical interval to use depends on the nature of the parameters to be estimated. The confidence interval is used when trying to find bounds on a population parameter—for example, the population mean or standard deviation. The confidence interval is the most commonly appearing of the three intervals and is the interval

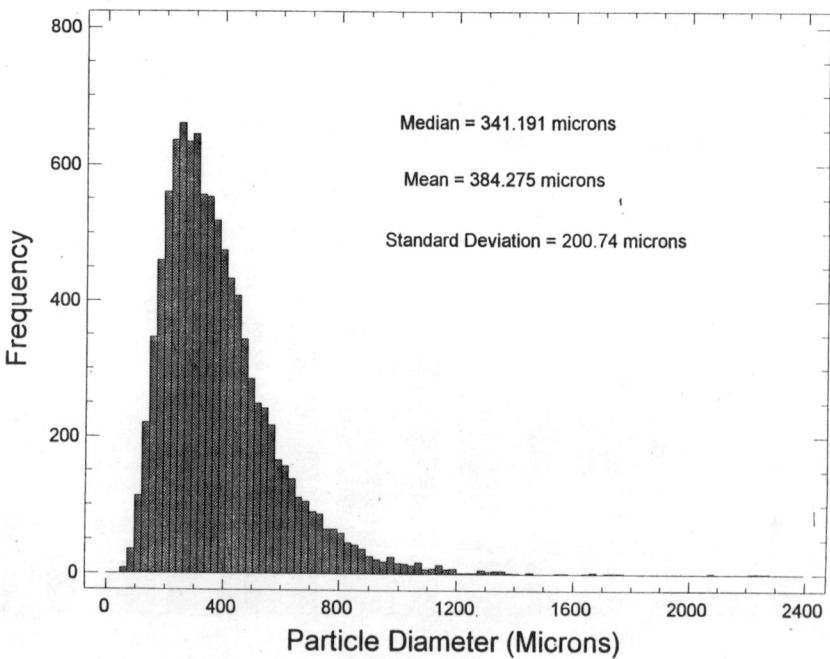

Fig. 6 Results of the random number simulation for the log-normal distribution with parameters: median = 5.83 µm, standard deviation = 0.49 µm.

Table 6 Cumulative Percent Finer:
Observed Versus Estimated

Screen diameter (μm)	Observed (%)	Estimated (%)
149	3.6	4.6
177	7.4	9.4
250	28.0	26.5
297	46.8	39.8
420	71.1	66.4
840	95.6	96.8

usually referenced in the literature; one finds statements such as "the 95% confidence interval around the average." The interpretation of this statement is that if one repeated this experiment an infinite number of times, 95% of these intervals would contain the true or population mean.

A prediction interval is fundamentally different from a confidence statement. It is appropriate when trying to find the bounds for future outcomes, given that some knowledge of past performance exists. A prediction interval is a confidence statement about future individual samples. In both instances, the method for interval estimates assumes that the data are normally distributed. The tolerance interval is used when one is interested in finding the proportion of future samples falling within limits, with a stated confidence level.

In this particular application to particle size distributions, we desire to predict the limits for the future parameter estimates of the particle size distribution based on our prior data that reflects the process being developed. The 95% prediction interval for each of the next samples is found from Eq. (5).

$$[\tilde{T}_L, \tilde{T}_U] = \overline{X} \pm r_{(1-\alpha;m,n)}S \tag{5}$$

where
$\overline{X}$ = sample average
$r_{(1-\alpha;m,n)}$ = factor for calculating the $(1 - \alpha)\%$ prediction interval for m future samples based on a prior sample of n observations
s = the sample standard deviation

and $r_{(1-\alpha;m,n)}$ is found from Table 7, which is an abridgment of Table A.13 of Hahn and Meeker (11). Referring again to Table 5, we find the average particle size median is 5.748 μm and the sample standard deviation of the medians is 0.0.112 μm. Consider sampling three additional batches from the top, middle,

Table 7 Factors $r(1 - \alpha; m, n)$ for Calculating Normal Distribution Two-Sided 95% Prediction Intervals for m Future Observations Using the Results of a Previous Sample of n Observations

	m					
n	8	9	10	12	16	20
7	4.108	4.191	4.265	4.391	4.319	4.457
8	3.374	3.449	3.516	3.631	3.811	3.948
9	3.250	3.321	3.384	3.492	3.660	3.790
10	3.157	3.224	3.284	3.387	3.547	3.670
15	2.904	2.962	3.013	3.101	3.237	3.342
20	2.790	2.843	2.891	2.972	3.098	3.194

Source: Adapted from Table A.13 of J. Hahn and W. Q. Meeker, *Statistical Intervals: A Guide for Practitioners*, John Wiley & Sons, New York (1991). Reprinted permission of John Wiley & Sons, Inc.

and top of a blender. A 95% prediction interval for the sample median for each of the next nine samples would be

$$[\tilde{T}_L, \tilde{T}_U] = 5.748 \pm 4.191 \cdot 0.112$$
$$= 5.279, 6.216$$

and similarly, the 95% prediction interval for the particle size standard deviation would be

$$\tilde{T}_L, \tilde{T}_U = 0.479 \pm 4.191 \cdot 0.018$$
$$= 0.403, 0.555$$

Reasonable and justifiable acceptance criteria for the particle size distribution of samples from a future study, either scale-up or validation, would be the following: each sample median should fall within the range 5.28–6.22 µm, and the particle size standard deviation should fall with the range 0.40–0.56 µm.

This same technique can be applied to the individual weight percent retained on the sieve screens. Using the data from Table 2, the following prediction intervals are obtained as shown in Table 8. However, there is a drawback to this approach. Since we are computing prediction intervals for at least five screen sizes, there is an increase in the chance of incorrectly claiming a statistical difference as compared to the distribution model approach, where two intervals are computed. As long as the distribution model provides a reasonable fit to the data, there is less chance of incorrectly concluding that a significant difference exists among the samples when in fact none really exists. The choice of using the distri-

Table 8 95% Percent Prediction Interval for the Percent Retained on the Sieve Screens for the Next 9 Samples

Sample	Screen mesh					
	20	40	50	60	80	100
1	4.4	24.4	24.4	18.8	20.6	3.8
2	5.3	26.4	23.4	17.9	20.6	3.0
3	4.4	25.0	25.1	18.9	19.5	3.9
4	3.6	21.9	24.6	19.6	21.6	4.6
5	1.0	13.0	17.0	19.0	32.0	11.0
6	1.0	15.2	21.2	17.2	29.3	10.1
7	2.8	19.8	20.8	17.4	25.5	8.5
Average (μm)	3.21	20.81	22.36	18.40	24.16	6.41
Standard deviation (μm)	1.697	5.105	2.895	0.904	4.887	3.343
Lower Limit (μm)	0	0	10.2	14.6	3.7	0
Upper Limit (μm)	10.3	42.2	34.5	22.2	44.6	20.4

bution model or weight percent retained as acceptance criterion is left to the discretion of the user.

C. Unit Dose Sampling

What happens when the blend uniformity assay results do not agree with the tablet uniformity results? Several possibilities can occur:

1. The average potency of the blend versus tablets could be different, but the unit-to-unit variability may be the same.
2. The average potency of the blend versus tablets could be the same, but the unit-to-unit variability may be different.
3. The average potency of the blend versus tablets could be different, and the unit-to-unit variability may be different.

In the discussion which follows, it is assumed that a more representative and "believable" assay is obtained from the compressed tablets than from the final blend. The difficulties in obtaining a representative sample are greater with a powder than with an intact dosage unit. Any time one samples a static bed of powder, there is always the potential for a non representative sample to be obtained. This is especially true with unit dose sampling, where multiple penetrations disturb the original distribution of particles. Later samples may not be truly representative of the blend originally sampled. This situation is reminiscent of

Table 9 Assay Determination of Material Retained on Sieve Screen

Screen diameter (μm)	Sample number		
	1	2	3
840	108.8	108.6	107.7
420	105.3	106.0	105.0
250	100.5	99.5	100.5
177	90.4	92.2	91.4
Pan	80.6	82.8	80.9

Heisenberg's uncertainty principle in quantum mechanics, where the measurement system interacts with the system being measured and it is difficult to distinguish between the subject and the object—much like trying to read the meniscus on a mercury thermometer with a match.

Harwood and Ripley (12) have described experiments conducted to measure the errors associated with the sampling thief and have concluded that there may be significant migration of particles as the sampling device moves through the bed of powder. In addition, there is always the possibility that the particle size distribution in the sample does not mirror the particle size distribution of the bulk state. This phenomenon is difficult to verify directly, since the unit dose

Table 10 Reliability of Selected Sampling Methods of a Binary Sand Mixture

Method	Standard deviation (%)
Cone and quartering	6.81
Scoop sampling	5.14
Table sampling	2.09
Chute sampling	1.01
Spinning riffling	0.125
Random variation	0.076

Source: Adapted from Table 1.1 from T. Allen, *Particle Size Measurement*, 4th ed., Chapman & Hall, New York (1990). Reprinted permission of Chapman & Hall.

samples are relatively small in comparison to the quantity required for sieve analysis. However, if one assayed the material retained on the nest of sieve screens, one could measure the effect of particle size on potency. An example of this situation is given in Table 9.

What other alternatives exist for obtaining unit dose samples? Terence Allen (13) has published the results of a study which compared five common methods of obtaining a representative lab sample from a larger sample. The results of the study are presented in Table 10.

As can be seen from Table 10 and Fig. 7, the spinning riffler is the best of the five methods in minimizing overall sampling variance and therefore provides a more representative sample. A commercially available apparatus is shown in Fig. 8 and consists of a discharge hopper which holds about 25 g of material

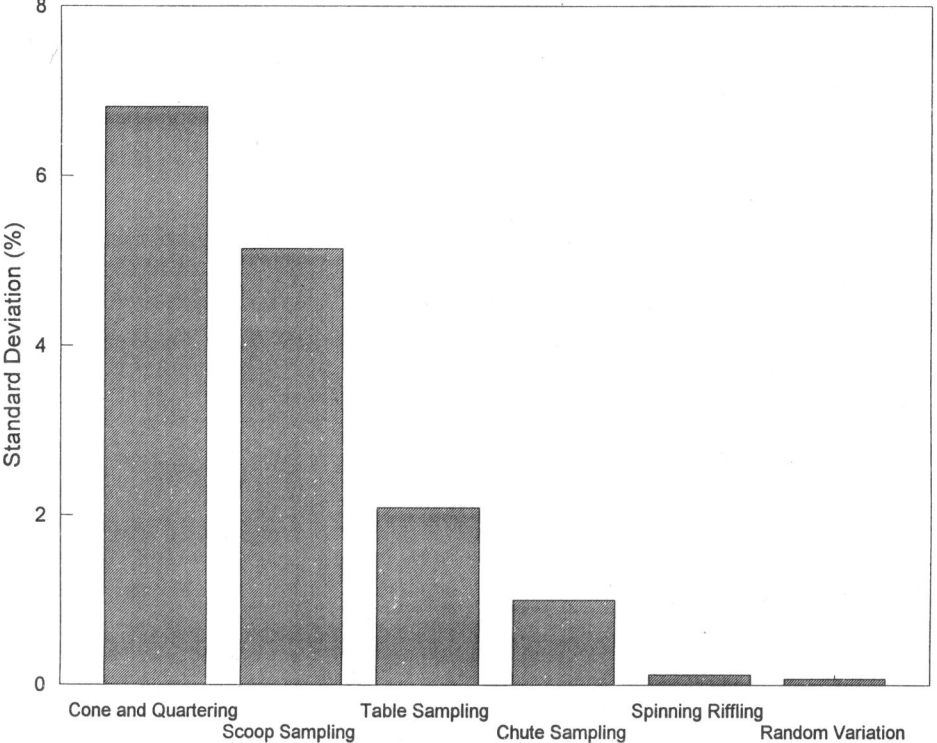

Fig. 7 Reliability, expressed as the percent standard deviation, of five common methods for obtaining representative samples. Adapted with permission from Table 1.1 of T. Allen, Particle Size Measurement, 4th ed., Chapman & Hall, New York, (1990.)

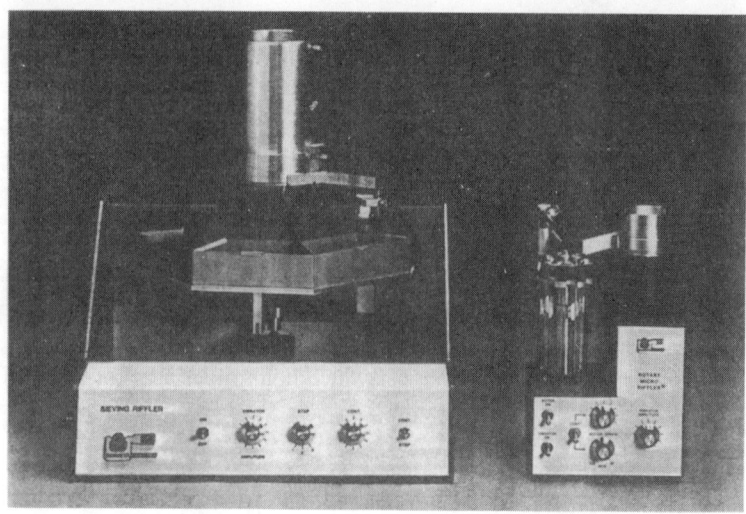

Fig. 8 Permission granted from Quantachrome Corp. to use a photograph of their commercially available spinning rifflers. The instruments are called the Sieving Riffler and the Rotary Micro Riffler respectively.

and a circular rack of test tubes. The powder sample is added to the mass flow hopper of the spinning riffler. The hopper is vibrated and the sample is discharged from the sample cup. The sample is collected into a series of rotating test tubes. The segment-to-segment variability is minimized by the continuous rotation of the test tube holder, assuring equal distribution of powder in each test tube. The subdivision process may be repeated as often as necessary to obtain the desired sample weight.

Lately there has appeared in the pharmaceutical literature examples of problems associated with unit dose sampling of powders (14,15). Let's assume that the blend is uniform at a specific site or location, and that we are looking for evidence of nonhomogeneity at different locations; see Fig. 9. The circles represent locations in the V-blender being sampled, and within each segment, the drug is homogeneously distributed. The source of variability, if it exists at all, is segment-to-segment. If we do not assume homogeneity at the site or location, then what conclusions may be drawn from the blend uniformity study? Every result could be considered untrustworthy. A technique such as the spinning riffler could be used to obtain a representative sample from a location, and a more accurate estimate of site-to-site differences could be measured. It must be remembered that we are no longer measuring variability at the unit dose level, but are examining location-to-location differences. Further work is needed in the area

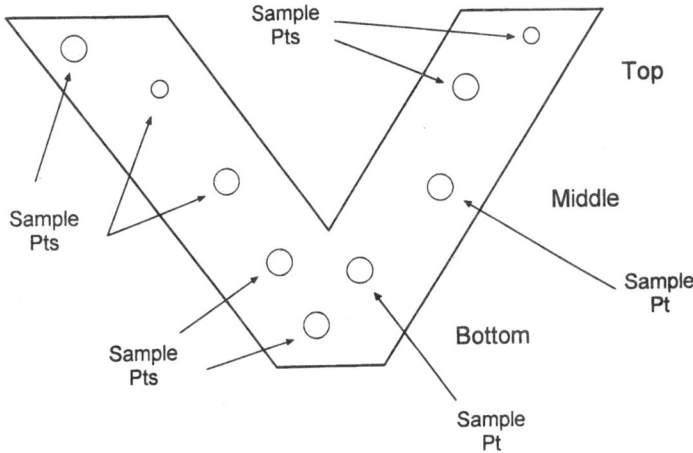

Fig. 9 Sampling locations from a V-shaped blender for a blend uniformity study. The open circles represent sample locations.

of sampling granulation and powder blends using alternative techniques to unit dose sampling, such as the spinning riffler, to measure blend uniformity.

III. COMPRESSED TABLETS

Evaluating the results from development and/or scale-up batches of compressed tablets employs the use of control charting techniques, process capability analysis, and prediction intervals as described earlier. The techniques apply equally well to in-process measurements of hardness, thickness, and tablet weight, and compendial requirements such as assay, dose uniformity by content uniformity, and dissolution.

As with any project, effective planning is essential for success. Examples of planning for the statistical sampling of the process include: (a) determining the sampling frequency, (b) the number of tablets to be sampled, and (c) the isolation of the side of the compression machine if a double-sided press is used. Operational considerations may include: (a) type of compression machine(s) to be evaluated, (b) range of compression speeds included in the study, (c) adjusting tablet hardness over the range of the various compression speeds, and (d) compression forces employed (precompression and main, where applicable).

If one can demonstrate that the batches manufactured during scale-up define a process which is reliable, repeatable, and predictable, then we have tilted the odds in our favor that the requirements for validation will be met. But how do we know when we have a reliable, repeatable, and predictable process? Since we have collected our samples in a rational way, i.e., we have considered all possible sources of process variability and sampled accordingly, we can use control charts to assess the stability of the process.

A. Control Charts

Control charts are a plot of sample subgroup averages and subgroup variability (measured as the subgroup range or standard deviation) against the order of production. The control limits displayed on the charts are set at 3 standard deviations about the sample average and process variation; see Eqs. (6) and (7), respectively, and Evans (16).

$$UCL_x = \hat{\mu} + 3\,\frac{\hat{\sigma}}{\sqrt{n}} \tag{6a}$$

$$CL_x = \hat{\mu} \tag{6b}$$

$$LCL_x = \hat{\mu} - 3\,\frac{\hat{\sigma}}{\sqrt{n}} \tag{6c}$$

and

$$UCL_\sigma = \hat{\sigma}[c_4 + 3\,\sqrt{(1 - c_4^2)}] \tag{7a}$$

$$CL_\sigma = \hat{\sigma} \tag{7b}$$

$$LCL_\sigma = \max\{\hat{\sigma}[c_4 - 3\sqrt{(1 - c_4^2)}], 0\} \tag{7c}$$

where UCL_x, CL_x, and LCL_x are the upper, center, and lower control limits for the process average, respectively; $\hat{\mu}$ is the estimate of the process average; and $\hat{\sigma}$ is the estimate of the process standard deviation. Similarly, UCL_σ, CL_σ, and LCL_σ are the upper, center, and lower control limits for the process standard deviation, respectively; $\hat{\sigma}$ is the estimate of the process standard deviation; and c_4 is a constant which corrects for the bias in the estimation of $\hat{\sigma}$. Values for c_4 may be found in Table 8.3 in Evans (14) and in most texts on statistical process control.

Normal distribution theory states that the chances of falling outside these control limits is 0.27%. If the subgroup ranges or standard deviations all fall within the upper and lower control limits, then there is evidence that the process variability is stable. It is unlikely that any point would fall beyond the control limits purely due to chance causes. Likewise, if the chart for subgroup averages shows all the points within the control limits, then the process is stable with respect to location or target value. Any point(s) outside the control limits for the average or variability should be a rare event if the occurrence were due to chance alone. Therefore this should indicate that the process is not stable, and the cause of the instability should be determined before drawing any conclusions about the reliability of the process. The control limits for the sample average provide the natural process limits for validation. We can be 99.73% confident that all future subgroups of the same size will fall within these limits if the process is operating under the same sources of variability.

A hypothetical example of these techniques is now presented. Let's consider three scale-up batches which have potency averages (μ) of 100.0%, 97.0%, and 104.0% of label claim, and the same process variability or standard deviation, $\sigma = 3\%$. Samples of five tablets are taken throughout the compression operation; control charts for subgroup average and subgroup standard deviation are prepared as shown in Figs. 10 and 11; 20 subgroups of five tablets (100 tablets) are assayed for each batch.

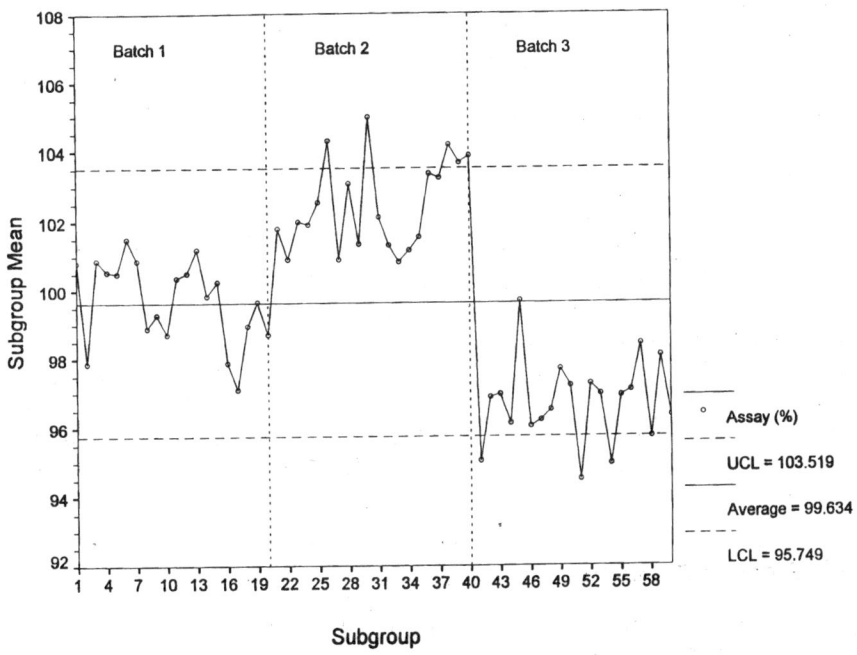

Fig. 10 Control chart for subgroup average label claim for samples taken throughout compressing of three batches. Twenty subgroups of size 5 were taken systematically from each batch and plotted in the order of production.

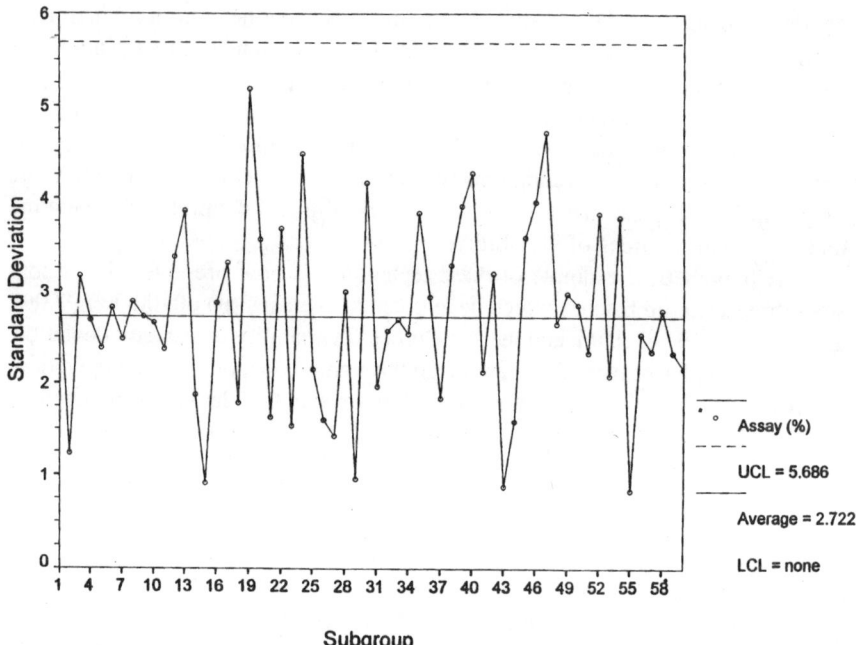

Fig. 11 Control chart for the subgroup standard deviation for the batches described in Fig. 10.

Is there any evidence that the process is unstable with respect to the amount of variability inherent in the process? Can the process be considered reliable to meet potency requirements? From the control chart for subgroup standard deviation, the process appears stable. The control chart for the subgroup average certainly detected the potency differences among the batches. Whether these process shifts in potency are of practical significance to manufacturing and quality assurance may depend on the chemical stability of the product and the ability of the process to remain at the current level of variability, namely, the standard deviation of 3%.

B. Process Capability

What about the capability of the process to meet the USP "Uniformity of Dosage Units" requirement? A technique useful in this area is process capability analysis. The relationship between the spread or variability of a process and the specifications imposed on the process is defined by the process capability. Define process capability, C_p, by Eq. (8):

$$C_p = \frac{USL - LSL}{6\hat{\sigma}} \tag{8}$$

where

USL = upper specification limit
LSL = lower specification limit
$\hat{\sigma}$ = estimate of the population standard deviation

The value of C_p measures how much of the specification width, USL − LSL, is filled by the variability in the process, $6\hat{\sigma}$. If $C_p = 1$, then the entire specification range is consumed by the process variability. Since the data are assumed to be normally distributed, statistical theory will provide an estimate of how much of the population will fall outside specification limits. For this example, the width of the specification limits is $6\hat{\sigma}$, so it can be expected that 0.27% of the units will fall outside specifications. No amount of process adjustment will reduce this figure, unless, of course, action is taken to identify and reduce the amount of variability in the process itself.

Many practitioners in the area of quality engineering recommend that a minimum value for C_p should be 1.33 in order for virtually no defects to occur due purely to the variability in the process. This value of C_p implies that the width of the specification should be eight times that of the estimated process standard deviation. With this value for C_p, only 0.006% defects will be produced. One can see immediately the nonlinear relationship between C_p and the percent defective. A more comprehensive list of C_p values versus percent of the population falling outside specification limits is given in Table 11. The results of a

Table 11 Process Capability (C_p)
and Parts per Million (PPM) Outside
Specification Limits

Process capability (C_p)	PPM outside limits
0.50	133,614
0.75	24,449
1.00	2,700
1.25	177
1.50	6.8
1.75	0.152
2.00	0.002
2.50	0

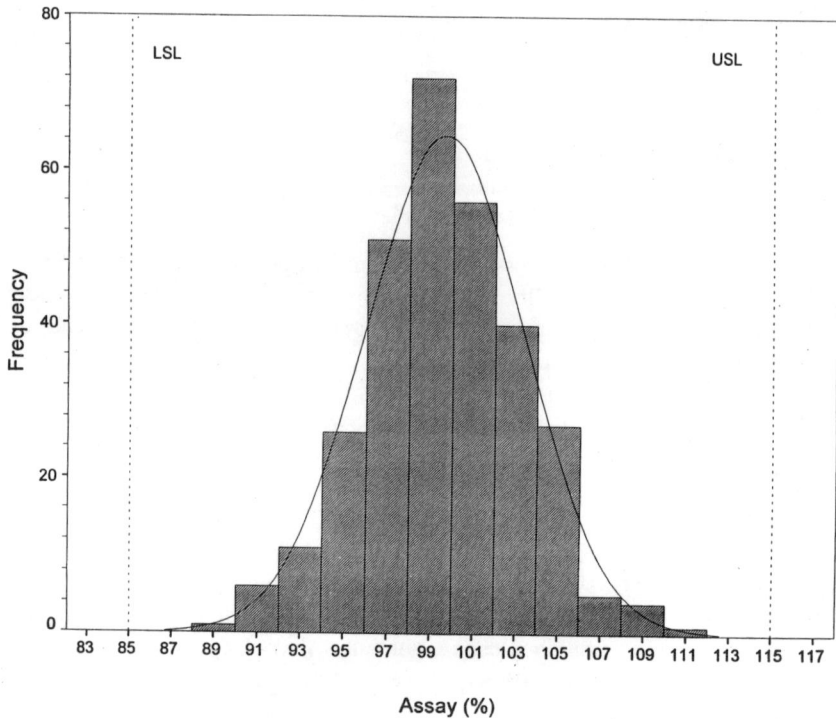

Fig. 12 Process capability analysis for the tablet assays described in Section III.B. The width of the process spread is compared to the width of the product specifications.

process capability analysis for the three batches described above are shown in Fig. 12. The assumption of normality is satisfied from an examination of the normal probability plot in Fig. 13, as well it should since the samples were chosen from a normal distribution. The assumption of normality should always be examined, since non-normality may seriously affect the estimated percentage of units outside specifications. This topic will not be considered here; consult the "Further Reading" for further information. It can be seen that the process is capable of consistently meeting the dose uniformity stage 1 attribute requirements with a $C_p = 1.34$. There are many other considerations in this exciting area, and the interested reader is referred to the "Further Reading" section for sources treating this topic more in depth.

C. Setting Specifications

What if there are no compendial or regulatory requirements for a particular property, such as tablet hardness? How can we use these techniques to set specifica-

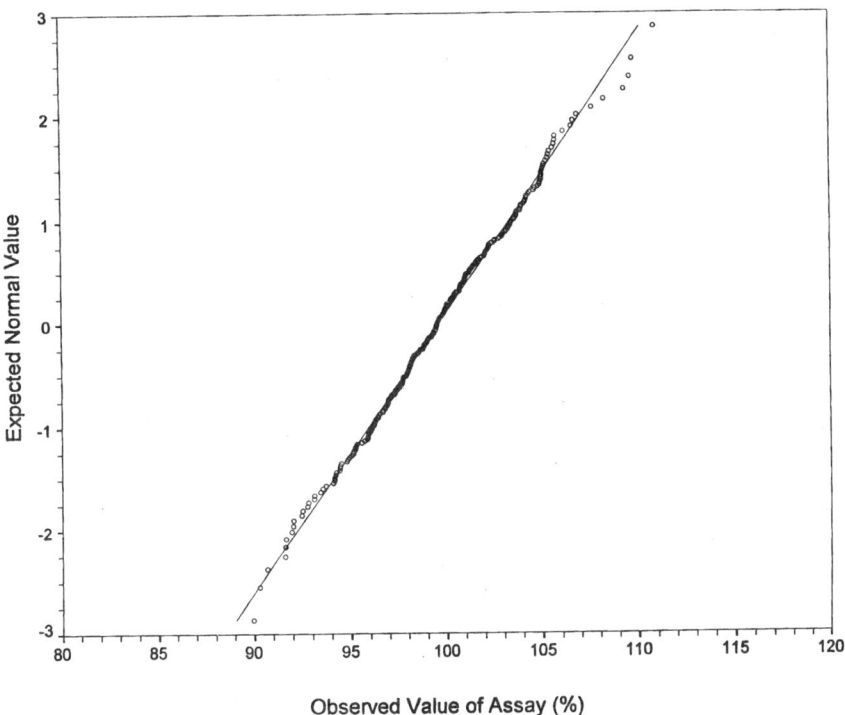

Fig. 13 Normal probability plot of the cumulative percent (z value) less than the ordered tablet assay values.

tions or acceptance limits for individual values and subgroup averages? Consider the situation for a process from which systematic samples are taken and the tablet hardness measured. The tablet hardness was targeted at 8.0 kp. Hourly samples were taken and the tablet hardness for each of five tablets was determined; 30 subgroups were obtained.

The control charts for subgroup average and range from the compression run are shown in Figs. 14 and 15. The overall tablet hardness is 7.9 kp and the average range for all the subgroups is 2.47 kp. The manufacturing process appears stable, as no points fall outside the upper or lower control limits for subgroup average or subgroup range. Therefore we can proceed to assess the capability of the process to meet specifications. But there are no specifications yet! Computing the sample standard deviation, s, for the 150 tablet hardness measurements can provide an estimate of the process variability. In this example, $s = 0.99$ kp. If we choose the capability of the process to be 1.33, or some other desired value, we can compute the specification width by Eq. (9).

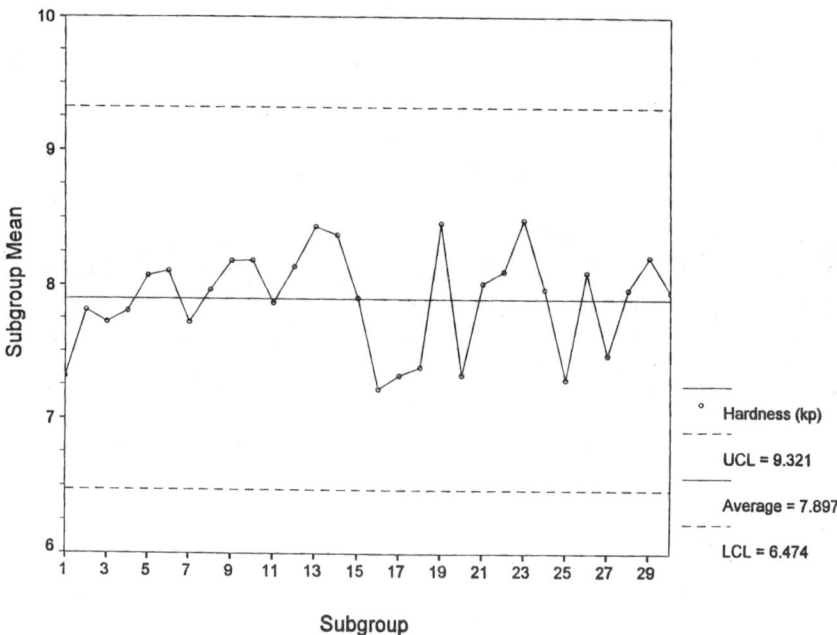

Fig. 14 Control chart for subgroup average hardness (kp) for samples taken throughout compressing of a batch. Thirty subgroups of size 5 were taken systematically from the batch and plotted in the order of production.

$$\text{USL} - \text{LSL} = 8\hat{\sigma} \tag{9a}$$

$$\frac{(\text{USL} - \text{LSL})}{2} = 4\hat{\sigma} \tag{9b}$$

Assuming symmetrical limits are placed around the target hardness of 8 kp, the limits for individual target hardness should be 4–12 kp. We can use the upper and lower control limits from the control chart analysis to establish the average hardness range: 6.5–9.3 kp. The normal probability plot for tablet hardness, Fig. 16, does not show any significant departure from normality, so the proposed limits for individual tablet hardness are consistent with our assumptions.

Another approach to setting specifications involves the use of tolerance intervals mentioned in Section II.B. Again, a tolerance interval provides the bounds or confidence limits to contain a stated proportion of future samples. The equation for the tolerance limits looks very similar to that for the prediction interval; it is given by Eq. (10).

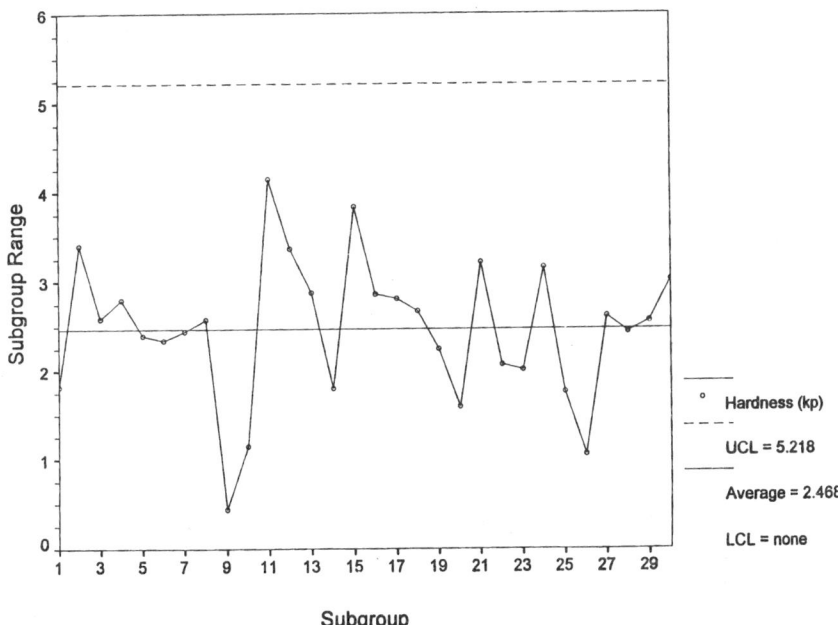

Fig. 15 Control chart for the subgroup range of tablet hardness for the batch described in Fig. 14. The range is the highest minus the lowest value obtained for each subgroup.

$$[\tilde{T}_L, \tilde{T}_U] = \overline{X} \pm g_{(1-\alpha;p,n)} \cdot s \tag{10}$$

where

$\tilde{T}_L, \tilde{T}_U$ = lower and upper limits of the tolerance interval
$\overline{X}$ = sample average
$g_{(1-\alpha,p,n)}$ = factor for calculating the $(1 - \alpha)\%$ tolerance interval for p proportion of future samples, based on a prior observation of n samples
s = sample standard deviation

The value for $g_{(1-\alpha;p,n)}$ can be found in Table 12, which is an abridgment of Table A.10b of Hahn and Meeker (11). We can be 99% confident that 99% of all future tablets produced from this process will fall within the limits

$$[\tilde{T}_L, \tilde{T}_U] = 7.9 \pm 3.043 \cdot 0.99$$
$$= 4.9, 10.9$$

Are the limits consistent with the functionality of the tablets? Will tablets produced from this process remain intact throughout the downstream operations?

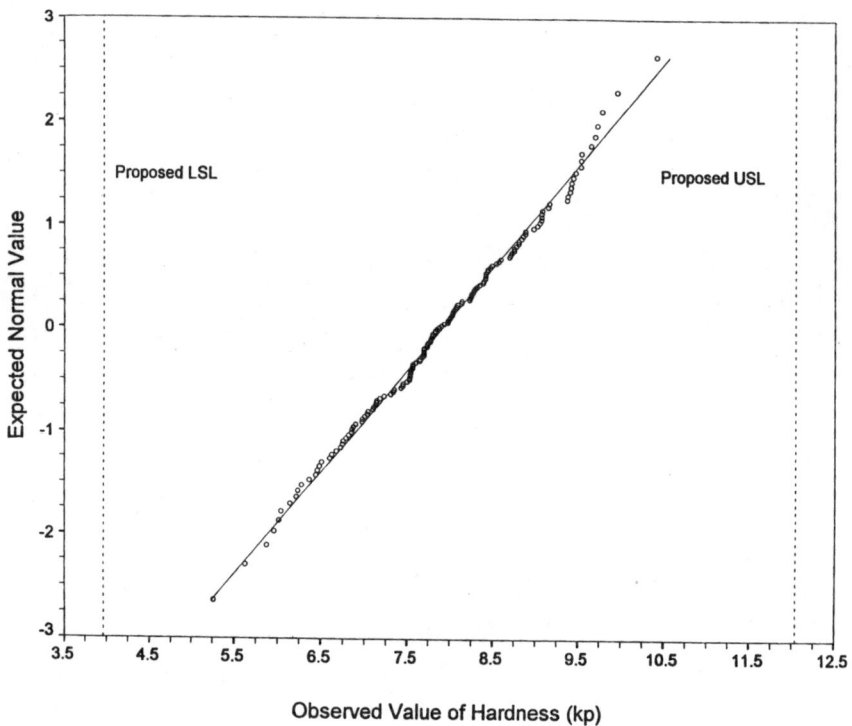

Fig. 16 Normal probability plot of the cumulative percent (z value) less than the ordered tablet hardness. Also shown are the proposed upper and lower tablet hardness limits.

Statistics cannot answer these questions. A statistical analysis of the data can only provide some measure of the impact variability may have on the output of the process. If the output is unsatisfactory, then the process, not the statistics, must be changed.

IV. PROCESS SIMULATION

We can use the power and ease of use of today's personal computer to provide an estimate or prediction of the future performance of a process through simulation. The example will consider compliance to a specification for which there is a known downstream effect. In the case of dose uniformity, it may be the additional testing imposed on the quality control laboratory for stage 2 testing, or it could be tablet hardness where values below the specified limit may result in tablet breakage on the packaging equipment, or if the tablets are too hard, may not have sufficient drug release.

Table 12 Factors ($g(1 - \alpha; p, n)$ for Calculating Normal Distribution Two-Sided $100(1 - \alpha)$% Tolerance Intervals for 99% of Future Observations Using the Results of a Previous Sample of n Observations

	$(1 - \alpha)$				
n	0.50	0.80	0.90	0.95	0.99
7	2.876	3.813	4.508	5.241	7.191
8	2.836	3.670	4.271	4.889	6.479
9	2.806	3.562	4.094	4.633	5.980
10	2.783	3.478	3.958	4.437	5.610
15	2.714	3.229	3.565	3.885	4.621
20	2.680	3.104	3.372	3.621	4.175

Source: Adapted from Table A.10b of J. Hahn and W. Q. Meeker, *Statistical Intervals: A Guide for Practitioners*, John Wiley & Sons, New York, (1991). Reprinted permission of John Wiley & Sons, Inc.

Let us "manufacture" batches by simple random sampling from a process with average assay = 100.0% of label claim and population standard deviation $\sigma = 1\%, 3\%, 6\%,$ and 10%. These tablets should meet all dose uniformity requirements, since the potency is on target and the variability does not exceed the assumed maximum value stated in ref. 9. The simulation consists of generating 500 random numbers from a normal distribution with average potency = 100.0% of label claim and $\sigma = 1\%, 3\%, 6\%,$ and 10%. Box-whisker plots for the potency for each population standard deviation are presented in Fig. 17. It is clear from this plot that the number of dosage units falling outside the stage 1 attribute limits of 85.0–115.0% of label claim increases dramatically with an increase in the standard deviation. Table 13 summarizes the results of conducting a process capability analysis for each simulation. The percentage of dosage units falling outside stage 1 limits is also provided. Only the processes with a standard deviation less than 6% are capable of consistently meeting stage 1 requirements.

What happens if we repeatedly draw samples, with replacement, from each simulation? Can an estimate of the impact of increasing variability be detected in the number of "batch" failures and/or additional stage of testing? The objective of this simulation is the estimation of the effect of the batch-to-batch variability on the dose uniformity testing level, stage 1 versus stage 2, and the batch failure rate for what appear to be relatively small changes in the standard deviation. From each set of random numbers generated from each standard deviation, sample sizes of 10 and 30 units were selected, with replacement. This "resampling" procedure was repeated until a total of 1000 batches had been formed for each of the four standard deviations under consideration. Table 14 contains the

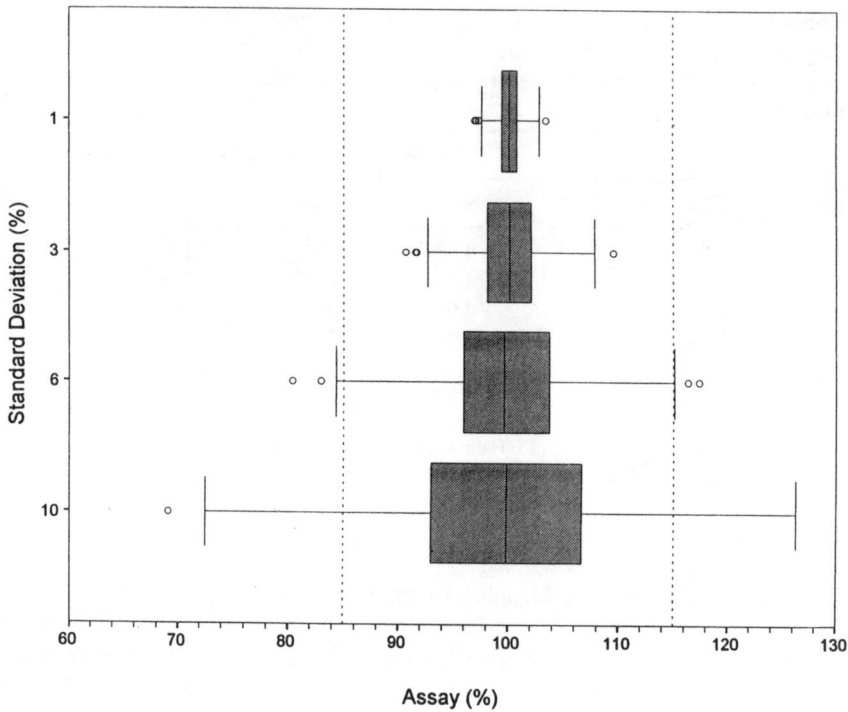

Fig. 17 Box plots for the simulated tablet assays for batches with mean potency of 100% label claim and standard deviations of 1%, 3%, 6%, and 10%.

Table 13 Process Capability Summary for Process Standard Deviation Simulation

Process standard deviation (%)	C_p	Observed beyond specification (%)	Estimated beyond specification (%)
1	4.98	0.0	0.0
3	1.69	0.0	0.0
6	0.84	1.4	1.21
10	0.50	14.2	13.75

Table 14 Potency Range for Resampling Procedure for
Different Sample Sizes and Process Standard Deviation

Process standard deviation (%)	Potency range (n = 10)	Potency range (n = 30)
1	99.02–101.08%	99.53–100.71%
3	96.78–103.07%	98.26–102.00%
6	93.74–105.46%	96.39–103.58%
10	90.34–108.81%	93.86–105.04%

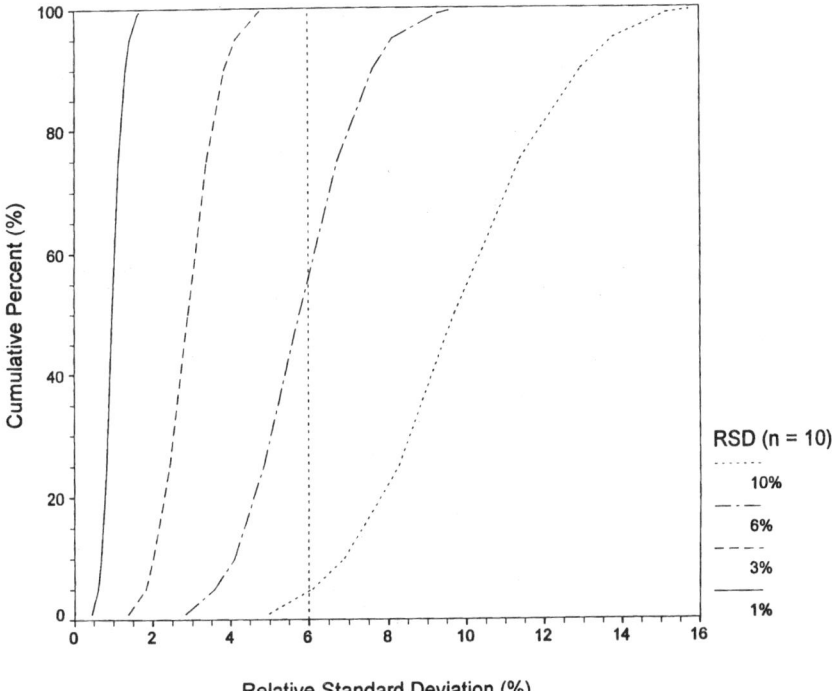

Fig. 18 Cumulative percent of batches having a relative standard deviation (RSD) less than 6%, the stage 1 dose uniformity maximum. The batches were designed with mean potency of 100% label claim and standard deviations of 1%, 3%, 6%, and 10%.

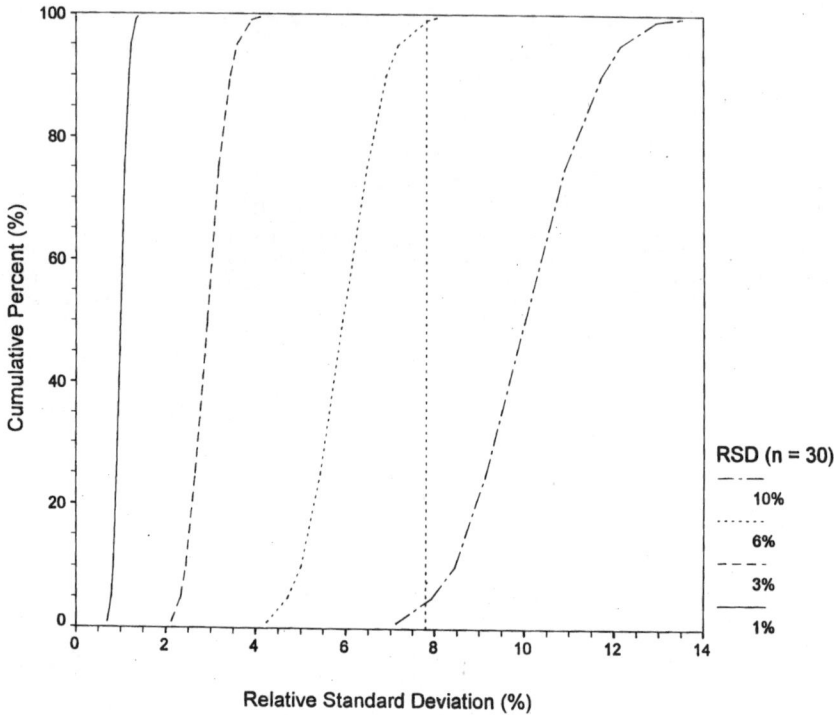

Fig. 19 Cumulative percent of batches having a relative standard deviation (RSD) less than 7.8%, the stage 2 dose uniformity maximum. The batches were designed with mean potency of 100% label claim and standard deviations of 1%, 3%, 6%, and 10%.

range of potency determinations for the two sample sizes. The cumulative percentiles for the relative standard deviation as well as the critical value of the RSD for each stage of testing are presented in Figs. 18 and 19.

Using the RSD requirement alone, for the testing performed at stage 1, over 40% of the batches manufactured from a process standard deviation of 6% would require stage 2 testing; virtually all the batches would meet stage 2 requirements and be released. The cost to the quality control lab will be enormous in time, resources, missed schedules, and an increased burden on the hazardous waste disposal system. Quite an expensive false alarm! Roughly 95% of all batches from a process with a standard deviation of 10% would require stage 2 testing, and approximately 95% of all batches tested at this level would fail to meet requirements. Contrast this performance with those batches from a 3% or less standard deviation: all batches would meet stage 1 requirements, and no additional testing would be required!

A similar analysis using the attributes portion of the acceptance plan is just as disturbing. Table 15 summarizes the number of dosage units outside 85.0–115.0% and 75.0–125.0% for the 1000 batches produced under the four conditions.

As long as the process variability is 3% or less, no units fall outside the stage 1 limits. But as soon as we reach $s = 6\%$, the situation changes: over 10% of the batches will require additional testing, and 0.5% will be rejected outright. At $s = 10\%$, over 35% of the batches will require stage 2 testing and 40% of the batches will be rejected. At stage 2 testing, about 95% of the batches contain 2 or more units outside the 85.0–115.0% limits.

Now consider the effect of the lack of discrimination of the pseudo-operating characteristic curves for σ at 6% and 10%; refer to Figs. 18 and 19. Since we are trying to form a conclusion about an unknown process by sampling it, if the true state is σ = 6%, then it is quite possible by the luck of the draw to obtain an estimate of $s = 4.5\%$ from a sample. Half the time a batch will be released at stage 1, while half the time it will require additional testing—might as well flip a coin! Someone investigating this pattern may concentrate efforts on what is wrong with the batches at stage 2 while ignoring those at stage 1 because they met requirements. In fact, all the batches are coming from the same population or distribution. Every batch is coming from the same process. It is just the luck or unluck of the draw which value of the standard deviation or relative standard deviation we obtain. Again we see the inappropriateness of using compendial requirements to set limits on the process. The use of compendial

Table 15 Number of Units Outside Attribute Range for Resampling Procedure for Different Sample Sizes and Process Standard Deviation

Process standard deviation (%)	No. of units outside range	Potency range ($n = 10$)		Potency range ($n = 30$)	
		85.0–115.0%	75.0–125.0%	85.0–115.0%	75.0–125.0%
1	0	1000	1000	1000	1000
	1	0	0	0	0
3	0	1000	1000	1000	1000
	1	0	0	0	0
6	0	8708	1000	661	1000
	1	116	0	275	0
	>1	6	0	64	0
10	0	223	897	11	717
	1	379	94	50	236
	>1	398	9	939	47

specifications should be restricted to evaluating a given batch with respect to its release into the market and not a measure of process performance.

The impact to manufacturing and the quality control laboratory can be measured or at least estimated with this technique. One could establish acceptance criteria on this basis: minimize the total cost to the system by designing and developing products which do not result in excessive batch rejections or heightened level of testing with no return to the organization.

V. CONCLUSION

Product development is too important an activity to be left to the formulation scientist alone. Today's sophisticated manufacturing technology and more complicated formulas require a truly multidisciplinary approach. This is not the first time this statement has been made, but it nonetheless is as true as ever. We have long passed the era (indeed, if it ever existed at all) when the formulator could develop products from lab to plant floor almost single-handed. The formulator needs the research analytical chemist to develop analytical methods that will accurately and precisely determine the drug content and rate of release properties of the drug product. The quality control laboratory requires analytical methods which are rapid, robust, and rugged. Manufacturing wants a process that will work the first time, every time.

Every step along this journey is not unlike walking through quicksand: the moment we get into trouble and squirm and wiggle to get out, the faster and deeper we sink. The way out of the quicksand is the same as the way out of a minefield: carefully. There is no substitute for good planning, knowing where it is you want to go, a map to get you there (but knowing there may be unforeseen detours along the way—at least anticipated), and knowing when you have arrived at your destination.

REFERENCES

1. FDA. "Guidelines on General Principles of Process Validation." Center for Drugs and Biologics and Center for Devices and Radiological Health, May, 1987.
2. FDA. "Guide to Inspections of Oral Solid Dosage Forms, Pre- and Post Approval Issues for Development and Validation." Office of Regional Operations, January 1994.
3. FDA. Pre-Approval Inspections/Investigations. Center for Drug Evaluation and Research and the Office of Regulatory Affairs, August, 1994.
4. E. Deming. Out of the Crisis. Cambridge, MA: Massachusetts Institute of Technology, 1986.

5. The United States Pharmacopeia. Rockville, MD: United States Pharmacopeial Convention, Inc., 1995.
6. Need to tighten blend uniformity assays debated. Drug GMP Report 15, October 1993.
7. Generic Drug Approval Workshop. Regulatory Affairs Professionals Society, May 1993.
8. J Larson, Introduction to Probability Theory and Statistical Inference. 3rd ed. New York: Wiley & Sons, 1982.
9. S Cowdery, T Michaels. An evaluation of the proposed dose uniformity acceptance sampling plan for tablets and capsules. Pharmacopeial Forum, Stimuli to the Revision Process, p. 614, Nov–Dec 1980.
10. Report and Recommendation of the USP Advisory Panel on Physical Test Methods III. Reporting Particle Size Distribution Data. Pharmacopeial Forum, Stimuli to the Revision Process 19(1):614, 1993.
11. J Hahn, WQ Meeker. Statistical Intervals: A Guide for Practitioners. New York: John Wiley, & Sons 1991.
12. F Harwood, T Ripley. Errors associated with the thief probe for bulk powder sampling. J Powder Bulk Solids Technol 20(1):20, 1977.
13. T Allen. Particle Size Measurement. 4th ed., New York: Chapman & Hall, 1990.
14. T Garcia, B Elsheimer, F Tarczynski. Examination of components of variance for a production scale, low dose powder blend and resulting tablets. Drug Development and Industrial Res. 21(18):2035, 1995.
15. J Berman, JA Planchard. Blend uniformity and unit dose sampling. Drug Dev. Ind. Res. 21(11):1257, 1995.
16. DH Evans. Probability and Its Applications for Engineers. New York: Marcel Dekker, 1992.

FURTHER READING

T Pyzdek, RW Berger, eds., Quality Engineering Handbook. New York: Marcel Dekker, ASQC Quality Press, 1992.

DJ Wheeler, DS Chambers. Understanding Statistical Process Control. 2nd ed., Knoxville, TN: SPC Press, 1992.

DJ Wheeler. Advanced Topics in Statistical Process Control. Knoxville, TN: SPC Press, 1995.

DC Montgomery. Introduction to Statistical Quality Control. 2nd ed., New York: J Wiley, 1991.

BH Kaye. The many measures of fine particles. Chem Eng, p. 78, April 1995.

ZT Chowhan. Segregation of particulate solids. Part I and Part II, Pharma Technol, 19(5): 56, 1995, Part II, 19(6):80, 1995.

FF Pitard. Pierre Gy's Sampling Theory and Sampling Practice. 2nd ed. Boca Ration, FL: CRC Press, 1993.

ZT Chowhan. Sampling of particulate systems. Pharm Technol 18(4):48, 1994.

Drug Index

This index contains a listing of all the drugs, whether marketed or not, and other biologically active molecules, intended for administration to animals or humans, which appear in the text, tables or figures of this book. In some cases salts of certain drugs have been specified within the body of the book; however, the salt name has been eliminated within the index.

Acetaminophen, 77, 334, 489
Acetylstrophanthidin, 270
Actiimmune, 342
Activase, 340
Acyclovir, 271, 272
Adriamycin, 273
β-Agonists, 155
AJ-2615, 158
Albuterol, 26, 264
Alendronate, 330
Alfatronol, 342
Alfentanil, 165, 265
Alprazolam, 269
Alprenolol, 328
Altiopril, 270
Amikacin, 271
Amiloride, 330
2-Amino-1-benzylbenzimidazole, 158
3′-Amino-3′-deoxythymidine, 485

Aminoglycoside antibiotics, 332
Aminophylline, 528
p-Aminosalicylic acid, 26
Amiodarone, 485
Amitriptyline, 265, 330, 337
Amoxicillin, 330
Amphetamine, 78, 330
Ampicillin, 522
Amylobarbitone, 265
Androstanolone, 488
Antacid(s), 419–421
Antibiotics, 331
Antihistamines, 333, 423
Antipsychotics, 333
Antipyrine, 265, 275
Aprophen, 167
Arsenamide, 421
Ascorbic acid, 333
Aspartame, 491

Aspirin, 489
Atenolol, 261, 264, 328
Atosiban, 158
Atracurium, 330
Atrial natriuretic peptide, 261, 270
Atropine, 264
Aurothiomalate, 421
Avonex, 342
Avoparcin, 327
Azithromycin, 159

Bambuterol, 330
Barbiturates, 265, 330–332, 423
Bendroflumethiazide, 330
Benefix, 340
Benidipine , 261, 270
Benoxaprofen, 487
Benzalkonium, 335
Benzobromarone, 168
Benzodiazepines, 251, 261, 265, 330, 332
Benzphetamine, 264
Benztropine, 266
Benzylpenicillin, 423
Berlopentin, 271
Beromun, 346
Beta blockers, 330
Betaferon, 342
Betamethasone, 330, 463
Betaseron, 342
Biantrazole, 169
Bile acids, 330
Bioclate, 340
BioTropin, 341
Bleomycin, 273
BMS-180431, 25, 28
BMS-193884, 524
Bromocriptine, 266
Bromopheniramine, 324
Bucromarone, 167
Bumetanide, 270, 330
Buprenorphine, 266
Bupropion, 266
Butorphanol, 266
BW 1370U87, 160
BW942C, 275

Cadralazine, 271
Caffeine, 266, 330
Calcium antagonist, 330
Calcium channel blockers, 261
Cannabinoids, 251, 266
Captopril, 270, 330, 488
Carbamazepine, 266, 489, 522
Carbenoxolone, 275
Carbovir, 522
S-Carboxymethyl-L-cysteine, 169
Cardiac glycosides, 270
CEA-scan, 344
Cefaclor, 487
Cefazolin, 487
Cefepime, 521
Ceftazidime, 526
Cephalosphorin, 330, 332
Cephalothin, 463
Cerezyme, 346
Ceruletide, 270
Chloramphenicol, 271
Chlordiazepoxide, 266
Chlorhexidine, 14, 15
1-(2-Chloroethyl)-3-cyclohexyl-1-nitrosourea, 159, 169
N,N'-bis(2-Chloroethyl)-N-nitrosourea, 157
Chloroquine, 271, 324
Chlorothiazide, 330
Chlorpheniramine, 167, 269, 324, 334
Chlorpromazine, 168, 266
CI-906, 270
CI-937, 164, 271
CI-941, 169
CI-976, 167
Cilazapril, 270
Cimetidine, 330, 488, 522
Cimetropium, 158
Cinobufagin, 272
Cisplatin, 158
Clenbuterol, 262, 264, 330
Clindamycin, 272
Clodronate, 485
Clofibric acid, 166
Clomiphene, 330

Clomipramine, 264, 337
Clonazepam, 266
Clonidine, 165, 264
Cocaine, 158, 264, 330
Codeine, 266, 330
Colchicine, 269
Colony-stimulating factor-1, 275
Comvax, 342
Corticosteroids, 331, 334
Cortisol, 261, 273, 334
Cortisone, 251, 334, 488, 490, 522
Cotinine, 264
CP-68,722, 167
Crisnatol, 168
Cyanocobalamine, 333
Cyclobenzaprine, 529, 530
Cyclopenthiazide, 488
Cyclophosphamide, 46, 47
Cyclosporin(e), 156, 158, 165, 168, 273, 351
Cytokines, 347

Danazol, 273
Daunomycin, 273
Dehydroepiandrosterone, 34–37
Desiclovir, 272
Desipramine, 264
Desmosine, 275
Detirelix, 273
Detomidine, 264
Dexamethasone, 273, 330, 335
Dexmedetomidine, 165
Dextromethorphan, 334
Dextrorphan, 155
Dialysis solutions, 419, 420
cis-Diaminodichloroplatinum(II), 421
Diazepam, 266
Diazepines, 423
Dibekacin, 272
Didanosine, 468
Dienogest, 273
Diethylstilbestrol, 273
Diflunisal, 168
Digitoxin, 225, 235, 270
Digoxin, 270
Dihydrocodeine, 157

1-(2,3-Dihydro-5-methoxybenzo[b]furan-2-ylmethyl)-4-(o-methoxyphenyl)piperazine, 26
1,2-Dimethyl-3-hydroxy-4-pyridone, 488
5,6-Dimethylxanthenone, 156, 163
Diphenhydramine, 488
Dirithromycin, 41, 42
Disopyramide, 337
Disulfiram, 157
Doxepin, 266
Doxorubicin, 441
Doxycycline, 31
Doxylamine, 324
DuP 747, 520, 521

Ebrotidine, 169
Ecokinase, 340
Enalapril, 330, 489, 527
Enalaprilat, 271
Enbrel, 346
Ephedrine, 328, 330
Epinephrine, 15
Epogen, 341
Ergotamine, 264, 330
Erythromycin, 26, 272, 485
Erythropoietin, 275, 347, 348
Estradiol, 236, 274
Estriol, 274
Estrogens, 261, 274
ET18-OME, 163
Ethimizol, 163
Ethinyl estradiol, 274
Etoposide, 273
Etorphine, 266

Famotidine, 156
Felbamate, 167
Fenoprofen, 27
Fenoterol, 264, 330
Fentanyl, 267
Finasteride, 160, 527
FK506, 164, 276
Flavonoids, 331, 332
Flecainide, 271
Fluconazole, 522

Flufenamic acid, 330
Flumethasone, 330
Flunitrazepam, 267
1-[2-[*bis*(4-Fluorophenyl)methoxy]ethyl]-
  4-(3-phenylpropyl)piperazine (GBR-
  12909), 158
Fluoxymesterone, 274
Flupenthixol, 267
Fluphenazine, 163, 168, 267
Flurazepam, 267
Flurbiprofen, 330
Fluticasone-17-propionate, 274
Follistim, 341
Fosinopril, 271, 475–477, 488, 520, 521
Fursemide, 488

Galanthamine, 267
Gallopamil, 163
Genotropin, 341
Gentamicin, 251, 272
Gliclazide, 485
Glipizide, 275
Glisoxepide, 275
Glucagen, 341
Glyburide, 487
Gonal F, 341
Gramicidin A, 485
Griseofulvin, 485, 522
Guanethidine, 271

Haloperidol, 156, 164, 267, 437
Herceptin, 345
Heroin, 330
Hexobarbital, 337
Humalog, 340
Human growth hormone, 274, 341, 347
Humaspect, 345
Humatrope, 341
Humulin, 340
Hydrochlorothiazide, 72, 330, 506, 509
Hydrocodone, 267
Hydrocortisone, 463, 485
4-Hydroxyandrostenedione, 274
5-Hydroxymethyldeoxyuridine, 272
4-Hydroxy-2-(4-methylphenyl)benzo-
  thiazole, 275

25-Hydroxyvitamin D_3, 274
L-Hyoscyamine, 264

Ibuprofen, 164, 167, 330, 335
ICI-200,800, 275
Idazoxan, 275
Iloperidone, 169
Iloprost, 262, 269
Imidapril, 167
Imipramine, 264, 330
Indacrinone, 275
Indimacis 125, 343
9-[2-(Indol-3-yl)ethyl]-1-oxa-3-oxo-4,9-
  diazaspirol[5,5]undecane, 26
Indomethacin, 269
Infergen, 342
Infusion solutions, 420
Insulin, 225, 274, 421
Insuman, 340
Interleukin-2, 526
Intron A, 342
Irinotecan, 167
ISF-2405, 271
Isoniazid, 272, 423, 463
Isradipine, 271

Kanamycin, 327
Ketoprofen, 26
Kogenate, 340

L-660,711, 488
L-739,010, 164
L-746,530, 164
Labetalol, 163, 264
Lacidipine, 267
β-Lactam antibiotics, 330
Lamictal, 267
Lamivudine, 33–35
Lamotrigine, 163, 267
Lanatoside C, 271
Lantus, 340
Leucinostatins, 330
Leucovorin, 261, 517
Leukine, 341
LeukoScan, 344
Levomethadyl, 275

Levorphanol, 156, 267
Lidocaine, 16, 21, 267, 330
Liprolog, 340
Lisinopril, 271
Lithium, 387, 419, 420
Lomotil, 267
Loperamide, 267
Lorazepam, 27, 269
Lormetazepam, 267
Losartan, 164, 488
Lovastatin, 527
Lymerix, 343
Lysergic acid diethylamide, 266

Mabthera, 346
Mabuterol, 264
Maprotiline, 265
MDL-28,050, 167
Medigoxin, 271
Mefloquine, 521
Meperidine, 267
Mepiradipine, 271
6-Mercaptopurine, 331
Mesoridazine, 267
Mestranol, 274
Methadone, 267, 331
Methamphetamine, 265, 330, 423
Methaqualone, 267
Methotrexate, 273
4′-Methyl-2′-nitroacetanilide, 488
5-Methyl-2-[(2-nitrophenyl)amino]-3-
    thiophenecarbonitrile, 32–34, 491, 522
Methylprednisolone, 274
N-Methyl pyridinium-2-aldoxime, 20,
    21, 27
Methylthiouracyl, 331
Metoclopramide, 267
Metoprolol, 337
Metyrapone, 167
Mexiletine, 167
Mifentidine, 158, 164
Minoxidil, 271
Mithramycin, 272
Mitoxantrone, 157, 168, 169
MK-0499, 157
MK-571, 488

MK-852, 275
Mofebutazone, 488
Mometasone, 274
Monensin, 272
Morphine, 268, 331, 333
Multivitamin, 419
MyoScint, 344

N-0923, 155, 163
Nabilone, 487
Naftopidil, 271
Nalmefene, 268
Naloxone, 268
Naphazoline, 335
Navelbine, 272
Nedocromil, 28, 29
Neorecormon, 341
Netilmicin, 272
Neumega, 342
Neupogen, 341
Nicardipine, 271
Nicergoline, 265
Nicotinamide, 333
Nicotine, 265
Nifedipine, 271
Niridazole, 165
Nitrazepam, 268
Nitrofurantoin, 31–33
Nocloprost, 262
Nomifensine, 268
Nordiazepam, 268, 269
Norditropin, 341
Norethindrone, 274
Norgestimate, 165
D-Norgestrel, 273
Normegesterol, 274
Normetanephrine, 265
Norphenylephrine, 328
Nortestosterone, 274
Nortriptyline, 268, 337
Noscapine, 423
Novolin, 340
NovoRapid, 340
NovoSeven, 340
Nutropin, 341
Nuvenzepine, 268

Omeprazole, 156
OncoScint CR/OV, 343
Ontak, 346
Opiates, 251
Opioid peptides, 261, 268
Orthoclone OKT3, 343
Ouabain, 271
Oxazepam, 27
Oxitropium, 261, 265
Oxprenolol, 337
Oxyphenbutazone, 488

Pantothenic acid, 333
Paracetamol, 331
Paroxetine, 522
Patellin, 485
Penciclovir, 331
Penicillin G, 158, 463
Penicillin V, 463
Penicillin(s), 272, 422
Pentazocine, 268
Pentobarbital, 268, 337
Perindopril, 271
Phencyclidine, 268
Pheniramine, 324
Phenobarbital, 72
Phenobarbitone, 169, 268
Phenothiazine(s), 268, 331, 423
Phenylbutazone, 488
Phenytoin, 268
Pilocarpine, 331
Pimozide, 269
α-(2-Piperidyl)-3,6-*bis*(trifluoromethyl)-
    9-phenanthrenemethanol, 13, 14, 16,
    27
Pirenzepine, 265
Platelet-derived growth factor (PDGF),
    341, 347, 348
Polymyxins, 331
Porcine pancreatic elastase, 525
Pravastatin, 155, 169, 262
Praziquantel, 168
Prednisolone, 274, 487, 489
Prednisone, 251, 274
Primavax, 343

Primidone, 169
Procaine, 331
Procaterol, 265
Procomvax, 343
Procrit, 341
Progabide, 269
Progesterone, 274
Proleukin, 342
Promethazine, 463
Propranolol, 265, 337
Propylthiouracil, 274, 331
Prostacyclins, 262
ProstaScint, 344
Protropin, 341
Pseudoephedrine, 265, 334
Pulmozyme, 346
Puregon, 341
Pyridoxine, 333
Pyrilamine, 166

Ranitidine, 331, 521
Rapilysin, 340
Rebetron, 342
Rebif, 342
Recombinate, 340
Recombivax, 342
ReFacto, 340
Refludan, 340
Regranex, 341
Remicade, 345
ReoPro, 343
Reserpine, 265
Retavase, 340
Retrovir, 272
REV 5901, 13
Revasc, 340
Ribavirin, 272
Riboflavin, 333
Rifamycin B, 328
Rifamycin(s), 327–329
Rilmazafone, 269
Rimantadine, 169
Ristocetin A, 327–329
Ritodrine, 265
Ritonavir, 39, 40

Rituxan, 345
Roferon, A, 341
Rogletimide, 158
Rolipram, 269
RP-42068, 265
RS-26306, 167
RS-82856, 13, 26

Saizen, 341
Salbutamol, 328, 331, 528
Saralasin, 271
SC-25469, 521, 523
SC-41930, 521
SC-42867, 156, 164
SC-51089, 156, 164
SCH40120, 269
Scopolamine, 265
Selegiline, 331
Semduramicin, 156
Serostim, 341
Sevofluorane, 157, 166, 168
Simulect, 345
Simvastatin, 490, 527
Sisomicin, 251, 272
Somatomedin-C, 274
Somatostatin, 275
Somatotropin, 347
Spironolactone, 522
SQ27,519, 271
SQ29,852, 271
SQ-33600, 521
Stanozolol, 169
Stiripentol, 166
Streptomycin, 327, 422
Sufentanil, 269
Sulfadiazine, 331
Sulfamethoxazole, 331, 520, 521
Sulfanilamide, 488
Sulfathiazole, 522
Sulfonamides, 155, 157, 331
Sulforidazine, 269
Sulindac, 40, 41, 269
Sulpiride, 269
Suramin, 331
Synagis, 345

Tacrine, 334
Tacrolimus, 164, 273
Tamoxifen, 157, 161, 165, 166
Taxol, 168, 273, 331
Tecnemab KI, 344
Tegafur, 518, 519
Teicoplanin, 272, 327–329
Terbutaline, 328, 331
Testesterone, 251, 275, 487, 488
Tetracycline(s), 272, 331, 332
Thalidomide, 321
Theophylline, 269, 331
Thiamine, 333
Thioridazine, 269
Thromboxane B_2 analogs, 262
Thyrogen, 341
Thyroid stimulating hormone (TSH), 275
Thyroliberin, 275
Thyroxine, 465
Tiaprofenic, 331
Timolol, 159
Tiospirone, 164, 169
Tissue plasminogen activator, 275
Tobramycin, 272
Triamcinolone, 275
Triamterene, 21, 331
Triazolam, 269
Tricyclic antidepressants, 269
Trifluoperazine, 269
Trimipramine, 166
Tripelennamine, 166
Tritanrix, 342
Trospectomycin, 485
Tubocurarine, 265
Twinrix, 343

UK47880, 14, 15
Uracil, 518, 519
Ursodeoxycholic acid, 36, 38, 46, 485

Valproic acid, 157, 165, 169, 269
Vancomycin, 327–329
Vasopressin, 270
Venlafaxine, 331
Verapamil, 163

Verlukast (MK-679), 165, 488
Verluma, 344
Vinblastine, 273
Vincristine, 273
Vinpocetine, 271
Virtron, 342
Vitamin A, 27
Vitamin B_{12}, 463
Vitamin C, 334
Vitamins, 331, 333
Vitravene, 346

WR 238605, 164

Xilobam, 15, 16, 27

Zabicipril, 271
Zalcitabine, 272
Zenapax, 345
Zidovudine (AZT), 165, 272
Zinc, 421
Zonisamide, 269

Subject Index

Absorbance, 194
Accuracy, 229, 230
Active pharmaceutical ingredient (API), 1
ADME, 316
Affinity constant, 242
Amorphous form, 30, 51
Amorphous solid, 30
Amorphous state, 29, 36
Antibodies, chimeric, 241, 242
Antibody production, 240–242
Antibody(ies), 225–232, 234–237, 239, 241–253, 260–262, 350, 351, 455
  monoclonal, 230, 235, 240–242, 250
  polyclonal, 240–242, 250
Antigen, 225–228, 231, 237, 239, 243–249, 260, 261
Antiserum, 230, 240, 242
Assay
  enzyme-linked immunosorbent (ELISA), 244, 347
  excess-reagent, 227–229, 241, 245
  gravimetric, 61
  immunoradiometric, 228
  limited-reagent, 227–229, 245

Assays
  disequilibrium, 247
  heterogeneous, 260
  homogeneous, 248, 259
Auxochromes, 195

Batch variability, 24
Beer-Lambert law, 219, 220, 222–224
Binding capacity, 253
Binding, nonspecific, 249
Bioavailability, 12, 13, 22, 26, 51, 53, 60, 251, 277, 315, 485
Bioequivalence, 152, 277
Biopharmaceuticals, 525, 526
Biotransformation(s), 155, 165
Boltzmann constant, 415
Bond(s)
  hydrogen, 15, 30, 32, 202
  π, 188
  σ, 188, 198
  chemical, 188
  covalent, 189
  dative, 189
  electrostatic, 243
  hydrophilic, 243
  hydrophobic, 243

Bragg equation, 6
Broadening
  collisional, 401
  Doppler, 401, 402
  Holtzmark, 401, 402
  Lorentz, 401
  Stark, 401, 402
  Zeeman, 401, 402
Brønsted base, 188
Burger-Ramburger rules, 49, 51

Calibration
  external standard, 80–82
  internal standard, 80, 82, 83
  standard addition, 80, 81, 83
Calorimetry, differential scanning, 9; 11,
    22, 41–43, 45, 49, 513, 520, 523
Capacity factor, 90, 102
Capillary ion analysis, 335
Cell volume, 118
Charts, control, 554–556
Chemiluminescence, 457–466
Chemometrics, 84
Chiral drugs, 152
Chirality, 490
Chromatography
  adsorption, 127, 128
  affinity, 350
  bonded-phase, 131–134
  capillary electro-, 318, 336–338
  gas, 9, 62, 64, 75, 78, 79, 87, 88,
    264, 265, 267–269, 317, 318,
    321
  ion exchange, 73, 128–130, 235
  micellar electrokinetic, 318, 322,
    330–334
  normal-phase, 73, 127, 134–138
  reversed-phase, 135, 136, 138–141
  size-exclusion, 77, 87, 130, 131, 235
  supercritical fluid, 87
  thin-layer, 22, 87
Chromophore, 195–197, 217, 219, 226,
    349
Collimator, 211, 212
Collision-induced dissociation, 152,
    156–158

Color, 2
Colorimetry, 421
Column switching, 67, 74
Combinatorial chemistry, 313, 315, 529
Common ion effect, 17
Compactibility, 2
Conjugate acid, 17, 439
Conjugate base, 438
Conjugate preparation, 232–235
Conjugates, drug, 153, 154, 157, 163
Conjugating reagents, 233
Coordination, metal ions, 209, 210
Counterion(s), 17, 20, 142
Coupling, j-j, 394, 395, 403
Coupling, Russell-Saunders, 394, 395
Couplings, spin-spin, 474
Cross-polarization, 474, 477–481, 483–
    485, 487–491
Cross-reactivity, 230, 231, 242
Crystal, 5, 6
Crystalline organic solids, 5
Crystallinity, 12, 23, 25
Crystallization, 40, 41, 51, 52
Cyclodextrins, 322–327, 452

Decoupling, dipolar, 474, 475
Density, 2
Derivatization, 67, 78
  postcolumn, 349
Detection limit, 116, 451
Detectors
  conductance, 125
  electrochemical, 123–125, 349
  enzyme chemiluminescence, 250
  fluorescence, 122, 123, 349
  imaging, 445
  photo-, 445, 446
  refractive index, 120
  ultraviolet/visible, 119–121, 349
  photodiode array, 214, 215
  photoemissive tubes, 214
  photomultiplier tubes, 214
  photovoltaic cells, 214
Dialysis, 67, 76
Diastereomer(s), 321
Dipole moment(s), 190, 191, 400, 507

Direct injection, 67, 77, 79
Disposition, drug, 151, 155
Dissolution
   data, 15
   rate, 2, 12, 13, 22, 25, 48, 52, 53
   testing, 39
Distribution coefficient, 89, 90
Drug
   loading, 17
   product, 1, 17, 54, 339, 489
   substance, 1, 4, 16, 17, 20, 25, 51,
      52, 54, 70, 339, 485–487, 489,
      501, 529, 531, 541

Electrokinetic loading, 80
Electron(s)
   n-, 189, 195
   nonbonded, 189, 428, 434, 435, 437
   unpaired, 191, 393
   -π, 188, 189, 192, 428
   -σ, 188, 189
Electronic transition, 190
Electron-pair, 189
Electrons, lone pair, 197
Electrophoresis, 243
   capillary affinity, 318, 330, 348
   capillary array, 353
   capillary gel, 318, 348
   capillary zone 324, 325, 330–335,
      348
   capillary, 61, 77, 79, 313–386
   free-zone, 318
   immunoaffinity capillary, 350, 351
   microchip capillary, 355
   polyacrylamide gel, 339
Elemental analysis, 22
Enantiomer(s), 321, 322, 324, 325, 327,
   329
Enantiotropic system, 49–51
Environmental Protection Agency
   (EPA), 79
Equilibrium constant, 242
Excimer, 438, 439
Excipient(s), 2, 17, 24, 25, 218, 489,
   529
Exciplex, 439

Extraction
   liquid-liquid, 65–68, 71–73, 77, 250
   liquid-solid, 65, 66
   solid-phase micro-, 67, 68, 75, 77, 79
   solid-phase, 61, 66–68, 73, 74, 77,
      80, 250
   Soxhlet, 65, 71

Fab fragments, 241, 350
Fc fragment, 241
Fick's law, 96
Field-amplified sample injection (FASI),
   80
Flowability, 48
Fluorimetry, 437, 438, 441, 443, 446, 447
   phase, 447
   spectro-, 453
   time-resolved, 446, 447
Fluorophore, 446, 447, 454, 460
Food and Drug Administration (FDA),
   3, 70, 321, 335, 492, 537, 539
Formulation(s), 1, 12, 17, 51

Gastrointestinal tract, 17
Genomics, 313
Gradient elution, 142–144
Gravimetry, 421

Hapten, 232, 234, 235, 242, 243, 248
Headspace methods, 75, 79
Heavy atom effect, 436
High-performance liquid chromatogra-
   phy (HPLC), 22, 52, 62–64, 68,
   72–74, 76–78, 83, 84, 87–149,
   154, 166, 168, 231, 241, 250, 262,
   263, 265–270, 272–276, 317, 318,
   320–322, 329, 333, 334, 336, 337,
   339, 347, 360
High-performance liquid chromatogra-
   phy-mass spectrometry, 83
High-throughput screening, 313, 315
Hook effect, high-dose, 241, 258
Hooke's law, 503, 507
Hot-stage microscopy, 9–11, 22, 45
Hund's rules, 392, 396
Hydrodynamic loading, 80

Hydrogen bonding, 5, 31, 40, 66, 68,
    207, 208, 321, 487
Hydrolysis, drug conjugates, 67, 77
Hygroscopicity, 2, 11–13, 16, 17, 22,
    24, 25

Immunization, 230
Immunoassay, 225–312
    enzyme multiplied, 248, 265
    enzyme, 231, 237, 238, 251, 264–276
    enzyme-channeling, 248
    fluorescence polarization, 237, 457
    fluorescence, 455–457
    radio-, 225, 227, 231, 235, 237, 243,
        251, 252, 254, 262–276, 347
    chemiluminescent, 465, 466
Immunocomplexation, 235
Immunogen, 235
Interaction(s)
    chemical shift, 472, 473
    dipole-dipole, 66, 207, 208, 321, 436,
        470–472, 474
    hydrophobic, 66, 68, 321
    ionic, 66
    quadrupolar, 474
    Van der Waals, 5, 30, 207
    Zeeman, 468, 470, 474, 477
Internal standard(s), 152, 159
International Committee on Harmoniza-
    tion (ICH), 3, 70, 492
Ion pair, 72
Ion pairing, 141, 142
Ionization
    electron, 153, 166, 167–169
    electron-capture negative chemical, 153
    electrospray, 161–163
    positive chemical, 153, 167, 169
Isobestic point, 223
Isoelectric focusing, capillary, 348
Isotachophoresis, capillary, 318
Isotope cluster, 152
Isotope(s)
    radio-152, 225, 237, 455
    stable, 152, 159, 160

Jablonski energy-level diagram, 504

Karl Fisher titration, 9, 22

Lambert-Beer law, 428, 448
Lamps
    deuterium, 210, 211
    gas discharge, 442
    high pressure mercury vapor, 442
    hydrogen, 210, 211
    incandescent filament, 210, 211, 443
Lasers, tunable dye, 443
Law of mass action, 253
Lewis acid, 209
Lewis base, 209
Linearity, 117
Liquid chromatography (LC), 87, 88,
    154, 155, 241
Liquid chromatography/mass spectrome-
    try (LC/MS), 151, 155, 156, 162,
    164, 169–171
Liquid chromatography/nuclear mag-
    netic resonance spectroscopy (LC/
    NMR), 169, 170

Maclaurin series, 448
Mass spectrometry, 151–185, 347, 351
    atmospheric pressure chemical ioniza-
        tion, 151, 158–160
    atmospheric-pressure ionization, 151,
        152, 154–163, 165–170
    atmospheric-pressure ionization, ion-
        spray, 151
    atmospheric-pressure ionization, turbo
        ionspray, 151
    capillary electrophoresis, 154, 160–
        162, 170, 330, 331, 351–353, 357,
        358
    chemical reaction interface, 152, 160,
        161, 170
    electrospray, 151
    fast atom bombardment, 153, 154,
        158, 161, 163–165, 167, 168
    Fourier-transform ion cyclotron res-
        onance, 162
    gas chromatography, 157, 164, 169,
        170, 262–264, 266–269, 274,
        276

[Mass spectrometry]
  ion trap, 162, 170
  liquid secondary-ion, 153, 154, 165, 167–169
  tandem, 152
  thermospray, 151, 152, 154, 155, 158
  time-of-flight, 154, 162, 163, 170
Mass spectroscopy, 9, 45, 59
Matrix effect(s), 231, 248–250, 252, 258
Melting point, 2, 12–15, 22, 49
Metabolism, 60, 163–170
Metabolite(s), 60, 151, 154–161, 163–166, 230, 242, 250, 261, 262, 266, 267, 269–273, 335, 440, 453
Metastable form, 51, 54
Microcentrifuge, 61
Microtiter plate(s), 231, 244
Mobility, molecular, 490
Moisture
  desorption, 11, 24, 45, 46
  sorption, 11, 24, 45, 46
Molecular ions, 152
Monochromator, 211, 442–444
Monotropic system, 49–51, 520
Mull, mineral oil, 510, 511
Multiple reaction monitoring, 157, 159

New Drug Application, 3, 537
Noise, 116

Orbital(s), 390–396, 398, 403
  antibonding, 189
  atomic, 188
  molecular, 188, 197, 435
  n, 192
  p, 188
  unoccupied, 190
  π-, 191, 195, 196, 199
  σ-, 189, 191
Ostwald's rule, 38

Particle morphology, 48
Particle size, 2, 48, 53, 486, 520, 541–548
Partition coefficient, 69

Pauli exclusion principle, 188, 392
Pellet, alkali halide, 510
Permeability, 315
pH, 12, 21, 24, 62, 64, 68, 72, 74, 134, 219, 223, 231, 234, 244, 325, 328, 329, 332, 333, 352, 437–439, 453, 464
Pharmacodynamic(s), 251, 253, 261
Pharmacokinetic(s), 152, 158, 251, 253, 261
Phase equilibrium, 68–70
Phosphorimetry, 437, 444
Photometry
  flame, 388
  spectro-, 453
  ultraviolet-visible spectro-, 187–224
Photon, 191
pK_a, 17, 68, 219, 438, 460
Planck's constant, 194, 389, 394, 399, 468, 504
Plate count, 93
Plate height, 94
Plate number, 91, 94, 99
Plate theory, 91
Polarography, 421
Polymorph(s), 3, 4, 25, 30, 31, 37–48, 54, 485, 487–489, 510, 511, 519, 520
Polymorphic forms, 24, 31–33, 52, 518
Polymorphism, 23, 36, 485–487, 501, 508, 512, 519, 520
Postcolumn reaction detection, 125
Precision, 228, 229, 231
Process capability, 556–558, 564
Process control, 23
Process simulation, 562–568
Protein binding, 76, 77
Protein precipitation, 61, 67, 76
Proteomics, 313, 314, 339
Purity, 12

Quantitation, 118
Quantitation, low limit of, 228, 229
Quenching, 237, 440, 441, 447, 452, 455
Q-values, 217

Racemate, 321
Reflectance
    attenuated total, 513, 514
    diffuse, 511, 512
Relaxation, 479–481, 489, 490
Response time, 117

Salt
    exchange, 20
    formation, 20
    preparation, 20–22
    selection, 12, 22–25, 528
Samples, quality-control, 251, 252, 258, 259
Sampling, unit dose, 549–553
Scatchard plots, 242
Scatter, Raman, 447
Scatter, Rayleigh, 447, 454
Sensitivity, 116, 228, 230, 278
Signal amplification, 239
Solid form selection, 29–54
Solubility, 2, 12–14, 16, 17, 20–25, 27, 48, 49, 51–54, 66, 68, 315, 485
Solvent effects, 435–437
Specificity, 230, 231, 248, 278
Spectra
    absorption, 194, 195, 387, 397, 399, 433
    constant neutral loss, 152
    electronic absorption, 188
    emission, 388, 397–399
    luminescence, 433, 437
    precursor ion, 152
    product ion, 152
Spectrofluorimeters, 445
Spectrophotometer, 210–217, 444
    double-beam, 216
    single-beam, 216
Spectroscopy
    arc emission, 404–408
    atomic absorption, 388, 389, 398, 404, 415–418, 420, 421, 423
    atomic emission, 388
    atomic fluorescence, 389, 404, 405, 418, 419
    atomic, 387–426
    fiber-optic fluorescence, 447, 448

[Spectroscopy]
    flame absorption, 405
    flame emission, 388, 404, 405, 413–415, 420
    fluorescence, 427, 448
    furnace atomic absorption, 404, 405, 422
    inductively coupled plasma emission, 404, 405, 408–413
    infrared, 10, 11, 22, 41, 44, 45, 59, 84, 488, 501–529, 531
    luminescence, 427–466
    micro-, 512, 513, 516, 523
    nuclear magnetic resonance (NMR), 10, 11, 22, 41, 45, 59, 155, 164, 168, 170, 358, 359, 467–499
    phosphorescence, 427
    photoacoustic, 514
    Raman, 10, 41, 501, 502, 508, 509, 515–519, 521, 522, 527–531
    spark emission, 405–408
    thermogravimetric infrared, 513, 523
    ultraviolet/visible, 52, 62
    vibrational, 501–535
Spin angular momentum, 189–191, 392, 403, 432
Spinning, magic angle, 474–477, 481–485, 487–492
Spline functions, 255
Stability
    chemical, 2, 12, 13, 17, 22, 24, 486, 528
    oxidative, 2
    physical, 2, 12, 22–24, 486, 528
    solid-state, 25, 27
    thermodynamic, 2, 48, 53
Stark effect, 401
State(s)
    electronic, 191
    excited, 389, 390, 398, 427, 428
    ground, 191, 389, 390, 429
    metastable, 398
    triplet, 191, 429, 432
Steric hindrance, 241

Taste, 12
Theoretical plates, 91

Thermal methods, 9, 10, 486
Thermogravimetry, 9–11, 22, 41–43, 45
Titrimetry, 421
Toxicity, 2, 12, 315, 351
Toxicology, 1

U.S. Pharmacopeia, 70, 217, 332, 420, 519, 539, 556
Ultrafiltration, 61, 67, 76
Uniformity
  blend, 539–541, 549, 553
  content, 529, 539, 553
  dose, 539, 553, 562, 563

Valency, 189

Validation
  analytical methods, 70
  process, 537, 538
Van Deemter plot, 96, 97

Wavelength(s), 193–213, 215–223, 389, 427, 429, 432, 441–444, 448, 449, 451–454
Wavenumber, 503
Wettability, 22

X-ray diffraction, 5–9, 11, 22, 41–46, 49, 485, 486, 520, 523

Yield, 12

Zeeman effect, 399, 418, 470